Nonperturbative Quantum Field Theory and the Structure of Matter

Fundamental Theories of Physics

An International Book Series on The Fundamental Theories of Physics:
Their Clarification, Development and Application

Editor:
ALWYN VAN DER MERWE, *University of Denver, U.S.A.*

Volume 114

Nonperturbative Quantum Field Theory and the Structure of Matter

by

Thomas Borne

Institute of Theoretical Physics,
University of Tübingen,
Tübingen, Germany

Georges Lochak

Fondation Louis de Broglie,
Paris, France

and

Harald Stumpf

Institute of Theoretical Physics,
University of Tübingen,
Tübingen, Germany

KLUWER ACADEMIC PUBLISHERS
DORDRECHT / BOSTON / LONDON

A C.I.P. Catalogue record for this book is available from the Library of Congress.

ISBN 0-7923-6803-7

Published by Kluwer Academic Publishers,
P.O. Box 17, 3300 AA Dordrecht, The Netherlands.

Sold and distributed in North, Central and South America
by Kluwer Academic Publishers,
101 Philip Drive, Norwell, MA 02061, U.S.A.

In all other countries, sold and distributed
by Kluwer Academic Publishers,
P.O. Box 322, 3300 AH Dordrecht, The Netherlands.

Printed on acid-free paper

Table of Contents

Preface

Nonperturbative quantum field theory means the evaluation and investigation of quantum field theory beyond perturbation theory, and includes a wide range of problems. It has been one of the most fascinating topics of theoretical physical research in the past few decades and, in addition, it is closely linked with the investigation of the structure of matter. That is why an enormous number of papers has been devoted to the treatment of these problems, and it would go beyond the scope of this preface—and even of the whole book—to give a review of these developments.

Thus we confine ourselves to referring to some modern books on quantum field theory where these topics are treated; for instance, the books by Weinberg [Wein 95], [Wein 96], Dobado, Gomez Nicola, Maroto and Pelaez [Dob 97] and Latal and Schweiger [Lat 97], and where further references can be found. But summarizing the results obtained so far in this regard one can see that inspite of great efforts there is still need of improvement and further progress.

In this book a new method of nonperturbative quantum field theory is developed which offers an alternative version of the common treatment of these problems, and which, in addition, is adapted to the investigation of composite particle dynamics and reactions, or, synonymously, to the investigation of the structure of matter. This new method is based on the Algebraic Schrödinger Representation of canonically quantized (relativistic) fields, and the derivation, formulation and application of this representation will be extensively discussed in Chapters 3, 4 and 5. In this method use is made of the Hamiltonian formalism in combination with elements of the algebraic representation theory of quantum fields and Heisenberg's equation of motion.

The field dynamics is formulated by functional equations in functional spaces which are isomorphic to the original state spaces of the quantum fields under consideration. This formalism can be applied to conventional gauge theories as well as to other types of fields. Composite particle dynamics is introduced by Weak Mappings which are defined by a reordering of the generating functional states and corresponding equations according to certain bound state structures. From these transformations effective field theories (in functional space) for composite particles can be derived if the problem under consideration involves very disparate mass scales and if the physics which is to be described happens at much lower energies than the scale set by some heavy particles (fields) in the theory.

In order to make use of this decoupling effect the formalism of the Algebraic Schrödinger Representation and the corresponding transformations are deduced for the case of a NJL model with very heavy constituent fermions (subfermions); to keep the theory finite a new nonperturbative Pauli–Villars regularization is applied, see Chapter 2.

It is beyond the scope of the book to treat applications of the Algebraic Schrödinger Representation to gauge theories. In this respect we can only refer to the literature:

the treatment of QED by Stumpf, Fauser and Pfister [Stu 93], Stumpf and Borne [Stu 94], Stumpf and Pfister [Stu 96], Fauser and Stumpf [Fau 97], Fauser [Fau 98]; the treatment of QCD by Stumpf and Pfister [Stu 97].

The remaining chapters of the book are about some selected topics of the structure of matter, in particular with respect to the unification problem of matter and forces.

At present the number of so called elementary particles in the phenomenology and the number of corresponding elementary processes is rather large. Thus one cannot expect to cover this subject down to the last detail. It is, rather, our aim to demonstrate the applicability of our method by means of the treatment of these selected topics, and this is the only way by which any new theoretical scheme can be introduced.

The standard model of matter contains at least 61 'elementary' particles and more than 20 parameters. This strongly supports the assumption that this model describes composite particles and resulting forces instead of elementary particles. But the crucial question is: are the gauge bosons composites? Deep inelastic lepton–photon scattering experiments, photon–proton interactions, etc., see Erdmann [Erd 97], Drees and Godbole [Dre 95], Sjöstrand, Storrow and Vogt [Sjö 96], seem to indicate that this is the case. Indeed, lots of models were designed with gauge bosons as composites, see Lyons [Lyo 83]. But the most ingenious idea came from Louis de Broglie.

Louis de Broglie was the first theoretical physicist to derive dynamical equations for relativistic composite particles. In particular he showed that classical electrodynamics results from the fusion of two spin $\frac{1}{2}$ fermions, i.e., from composite photons [Brog 34]. In Chapter 1 de Broglie's original fusion theory is described and some extensions of this theory are reviewed.

In the language of quantum mechanics de Broglie's theory deals with wave functions for single composite particles. A generalization of this approach has to include many-particle systems and interactions. A suitable framework for such an extension is relativistic quantum field theory, and in accordance with de Broglie's fusion idea this leads to the investigation of self-interacting spinor fields. In Chapter 2 a review of various approaches in this field is given.

If by generalization of de Broglie's assumption the bosons and fermions of the standard model are considered as composite, a suitable ansatz for a quantitative formulation of this assumption is a NJL-like spinor field model with additional isospin and nonperturbative Pauli–Villars regularization.

This choice calls for a comment. Usually nonrenormalizable NJL models are applied to the description of one-boson exchange processes in a low energy approximation, and are thus regarded as not fundamental. But in this book we show that this is a misunderstanding of the role of these models caused by crude approximation. If one applies the abovementioned formalism to a regularized NJL model, then with high precision effective gauge theories for composite particles or fields can be derived. This means that in the 'low' energy sector local symmetries are generated by the effective composite particle dynamics, which justifies the use of such models as a starting point for an investigation of the structure of matter.

In the treatment of such a model, as a first step composite particle states are defined by means of solutions of generalized de Broglie–Bargmann–Wigner equations which result from the diagonal part of the corresponding functional equation, see Chapter 6. Afterwards, by Weak Mapping the spinor field can be transformed into a left–right symmetric $SU(2) \times U(1)$ unbroken effective local gauge theory describing the dynamics of composite bosons and fermions, see Chapter 7.

Symmetry breaking comes about by a change of the algebraic representation, *i.e.*, by a change of the physical vacuum according to Heisenberg, [Hei 66]. If for simplicity only $SU(2)$ is broken, apart from left–right asymmetry, this yields the electro–weak sector of the standard model without the use of the Higgs mechanism, see Chapter 8. The calculation is performed for one lepton–quark generation but the inclusion of the other generations and of strong interactions is outlined. In this way the basis for a quantitative explanation of the standard model as an effective theory is provided.

In addition, in Chapter 9 linearized gravity theory is quantitatively interpreted as an effective field theory resulting from subfermion composites in accordance with de Broglie's fusion theory.

Summarizing the investigations with respect to the structure of matter, it is legitimate to ask whether this formalism leads to new insights into the structure of matter apart from reproducing already known results. To answer this question one should first of all acknowledge that the derivation of phenomenological theories as effective theories of composite particles is a nontrivial matter. But with these results the theory is not exhausted. Apart from the task to reveal the phenomenological theory as an effective theory in all ramifications, the functional formalism leads to form factors, residual forces and exchange forces for composite particles which offer new aspects of their dynamics.

In the present case the corresponding effects are negligible, or at least very small, but may be observable under suitable circumstances. In particular, such forces become important if, for instance, instead of high energy models, models of superconductivity, *etc.*, are treated. Then these forces may have a considerable influence on the physical behavior of the system under consideration. In Chapters 5 and 8 these forces are given in terms of functional operators and some estimates are performed, but more detailed calculations of the composite particle spectrum and effects of residual forces are tasks of future research.

Finally, one might ask: is such a model a candidate for a theory of everything? Clearly not, and that for philosophical reasons. Nobody knows the exact theory up to infinitely high energies. Thus any realistic calculation is carried out using an effective theory. In view of this fact the nonperturbative Pauli–Villars regularization is a parametrization of our ignorance which replaces Heisenberg's fundamental smallest length. Nevertheless, within this limit much remains to be detected.

With respect to the presentation of the material we emphasize that we have based the theoretical discussion on propositions and proofs, as far as doing so is possible for the very difficult mathematical structure of quantum fields. Nevertheless, by such propositions we want to express our intention of providing a rigorous basis for composite particle theory which can claim greater generality than, for instance, approximation methods. In particular, the central idea of Weak Mapping is based on a sound mathematical basis by Weak Mapping theorems.

As far as applications are concerned, explicit calculations are required. They can be divided up into algebraic calculations and integrations over spacial or space–time functions. Apart from the basic problem of regularization in such cases, we use strongly simplified expressions for the orbital parts of the wave functions in order to avoid difficult and laborious evaluations of integrals. Such calculations can be improved if necessary, or if technically possible; in general these approximations have no structural consequences for the theory as long as one does not intend to derive mass spectra, form factors, coupling constants, etc..

On the other hand, the algebraic calculations are carried out exactly throughout, as they are crucial for the structure of the resulting effective theories, and any deviation from this rule would damage the verification of the corresponding effective theory. So the material presented in this book can still be improved, but already in this form the progress in the understanding of the unification problem can be clearly demonstrated.

Notation

General conventions:

In the first chapters, which provide the mathematical and physical framework of the theory, we use c.g.s.-units, whilst the applications discussed in the subsequent chapters are formulated in natural units, *i.e.*, with $\hbar = c = 1$.

As a rule we use the summation convention and sum over double discrete indices and integrate over two continuous variables. Exceptions from this rule will be explicitly quoted.

A contravariant Minkowski space–time vector is given by its coordinates

$$x = (x^\mu) = (x_0, x_1, x_2, x_3) = (ct, \mathbf{r}) .$$

The Minkowski metric is described by the diagonal tensor

$$\eta_{\mu\nu} = \mathrm{diag}(1, -1, -1, -1) ,$$

whilst the metric tensor of a Riemann–Cartan space is denoted by $g_{\mu\nu}(x)$.

Partial derivatives with respect to space–time coordinates are written as

$$\partial_\mu \equiv \partial_\mu(x) = \frac{\partial}{\partial x^\mu} .$$

In general, small greek indices $\alpha, \beta \ldots$ take the values $0, 1, 2, 3$, and small Latin indices $i, j, k \ldots$ the values $1, 2, 3$; exceptions from this usage will be explicitly announced.

For Fourier transformations we use the convention

$$\tilde{f}(p) = \frac{1}{(2\pi)^4} \int d^4x \, f(x) e^{ipx}$$

with

$$f(x) = \int d^4p \, \tilde{f}(p) e^{-ipx} .$$

The complex conjugate of a complex number z is given by z^*; for an operator $\mathbf{A}$ the Hermitian conjugate is denoted by $\mathbf{A}^+$ and the transposition of a matrix A by ^TA.

Furthermore, for spinors $\psi \equiv (\psi_\alpha)$ we introduce $\bar{\psi} := \psi^+ \gamma^0$ and the charge conjugated spinor $\psi^C := C\bar{\psi}$ with the charge conjugation matrix C, see below.

Commutators and anticommutators are given by

$$[A, B]_- = AB - BA, \qquad [A, B]_+ = AB + BA.$$

Finally, symmetrization and antisymmetrization of a quantity M_{kl} in its indices is denoted by $M_{(kl)}$ and $M_{[kl]}$, respectively. Alternatively, we use the symmetrization operator S or the antisymmetrization operator A.

For the total antisymmetric tensor we use the convention

$$\varepsilon^{0123} = -\varepsilon_{0123} = -1 \ .$$

Pauli and Dirac matrices:

We use the representation

$$\sigma^1 = \begin{pmatrix} 0 & 1 \\ 1 & 0 \end{pmatrix}, \quad \sigma^2 = \begin{pmatrix} 0 & -i \\ i & 0 \end{pmatrix}, \quad \sigma^3 = \begin{pmatrix} 1 & 0 \\ 0 & -1 \end{pmatrix}$$

of the Pauli matrices. The Dirac matrices satisfy the relations

$$[\gamma^\mu, \gamma^\nu]_+ = 2\eta_{\mu\nu} \ ,$$

and we choose the Dirac representation

$$\gamma^0 = \begin{pmatrix} 1 & 0 \\ 0 & -1 \end{pmatrix}, \quad \gamma^k = \begin{pmatrix} 0 & \sigma_k \\ -\sigma_k & 0 \end{pmatrix} \ .$$

Furthermore, we use the matrices $\gamma^5 = i\gamma^0\gamma^1\gamma^2\gamma^3$ and the charge conjugation matrix $C = i\gamma^2\gamma^0$. In the Dirac representation they read

$$\gamma^5 = \begin{pmatrix} 0 & 1 \\ 1 & 0 \end{pmatrix}, \quad C = -i \begin{pmatrix} 0 & \sigma_2 \\ \sigma_2 & 0 \end{pmatrix} \ .$$

In the following we list further important symbols used in the text. This list is not complete; for the remaining symbols, indices, and abbreviations we refer to the text.

General symbols:

$\mathbb{N}$	set of natural numbers
$\mathbb{R}$	set of real numbers
$\mathbb{C}$	set of complex numbers
$\mathbb{M}^4$	Minkowski space
$\otimes$	direct product
$\oplus$	direct sum
S_n	n-dimensional permutation group
p_i	permutation of i
T	time ordering operator
$\overset{!}{=}$	postulated equivalence

State space symbols:

$\mathbb{H}$	(indefinite) inner product space
$|a\rangle, \dots$	vectors from $\mathbb{H}$
$|0\rangle$	vacuum, or ground state, of $\mathbb{H}$
$S, \dots$	linear operators on $\mathbb{H}$
$|a), \dots$	vectors from $\mathbb{L}$

Function symbols:

$\tau^{(n)}(I_1 \ldots I_n\|a)$	τ functions, time ordered transition matrix elements
$\varphi^{(n)}(I_1 \ldots I_n\|a)$	φ functions, normal ordered transition matrix elements
$\tau_t^{(n)}, \ldots$	corresponding single–time functions
$\rho^{(n)}, \ldots$	correlation functions of effective theories
$\Delta(x, m), F_{I_1 I_2}$	propagators
$K_i(r)$	modified Bessel functions
$G_{I_1 I_2}$	Green's functions
$C_J^I, C_k^{I_1 I_2}, \ldots$	composite particle (field) wave functions
$R_I^J, R_{I_1 I_2}^k, \ldots$	corresponding duals

Functional symbols:

$\mathbb{K}_F$	fermion functional space
$\mathbb{K}_B$	boson functional space
$\|0\rangle_F, \ldots$	functional ground states
$\|\mathcal{T}(j; a)\rangle$	functional state for $\tau^{(n)}$ functions
$\|\mathcal{F}(j; a)\rangle$	functional state for $\varphi^{(n)}$ functions
$\|\mathcal{A}(j; a)\rangle$	functional state for one-time $\tau^{(n)}$ functions
$\|\mathcal{B}(b; a)\rangle$	effective boson functional state
$\|\mathcal{G}(b, f; a)\rangle$	effective boson–fermion functional state
$j_I, \partial_I \equiv \delta/\delta j_I$	generating functional fermion operators
$b_k, \partial_k^b \equiv \delta/\delta b_k$	generating functional boson operators
$f_l, \partial_l^f \equiv \delta/\delta f_l$	functional dressed fermion operators
$\mathcal{H}, \mathcal{P}, \ldots$	linear operators on functional space

For the definition and explanation of all other technical or mathematical terms used in this book we refer to the index.

1 The Theory of Light and Wave Mechanics

1.1 Theory of Light and Wave Mechanics

1.1.1 Historical Background

In this century two theories were developed which are crucial for the theoretical understanding of the structure of matter: relativity theory and quantum theory. The first attempts to combine both theories in order to investigate the structure of matter beyond perturbation theory were made by de Broglie in his theory of fusion, starting with his theory of light.

De Broglie's line of approach is a precursor of a quantum field theoretical treatment of these problems, and it already contains the derivation of effective theories, which is at the center of interest in modern theoretical investigations.

In the first chapter we thus give a review of de Broglie's theory and its subsequent developments. Owing to historical reasons, in this review a slightly different notation with respect to the metric and the definition of γ matrices is used, the definitions of which are given in Section 1.3. For the other chapters the notation holds which is given at the beginning.

To understand de Broglie's theory of light it is interesting to remember that wave mechanics is an extrapolation of a dynamical theory of Einstein's light quanta [Brog 22], [Brog 24], [Brog 25]. De Broglie tried, at the beginning, to go as far as possible with radiation theory on a purely *corpuscular* basis, making use only of relativistic mechanics, kinetic theory, and thermodynamics, without electromagnetism: *i.e.*, *without waves*.

He considered photons to be true particles (he called them 'atoms of light') with a *small proper mass*[1], obeying the laws of relativistic mechanics. Starting from this corpuscular point of view he proved some results previously considered consequences of electromagnetism.

For instance:

— The kinetic energy of an atom of light is given by the relativistic formula $E_{\mathrm{kin}} = m_0 c^2 [(1 - \beta^2)^{-1/2} - 1]$, where m_0 is its proper mass. For $m_0 \ll 1$ and $\beta = v/c \approx 1$ one obtains $E_{\mathrm{kin}} \approx m_0 c^2 (1 - \beta^2)^{-1/2}$ and since $m = m_0 (1 - \beta^2)^{-1/2}$ we have $E_{\mathrm{kin}} \approx mc^2 = E$, *i.e.*, the total energy. The momentum G may thus be written: $G = mc = E/c \approx E_{\mathrm{kin}}/c$, from which de Broglie obtained the correct relation: $p = \rho/3$ between pressure and energy density of black body radiation [Brog 22], derived by Boltzmann from Maxwell's theory.

Using classical mechanics Planck found double this result, because: $E_{\mathrm{kin}} = mv^2/2 \Rightarrow$

---

[1]The objections related to a massive photon will be examined in Section 1.7 in more detail.

$G = mv = 2E_{kin}/v \Rightarrow p = 2\rho/3$. This factor 2, owing to the neglect of relativity (quite astonishing for Max Planck) was considered to be an argument against the photon theory of light [Pla 59].

— Applying relativity, de Broglie obtained the correct mean energy $3kT$ of a photon, instead of the value $3/2kT$ of the classical theory of gases. This $3kT$ is usually considered to be the sum of electric and magnetic energies, but it is a simple consequence of relativistic kinematics.

— Eventually, de Broglie obtained the formulae of the Doppler effect from the relativistic *addition of velocities and Planck's law*.

Only after having derived these results, de Broglie had the idea of defining a frequency for every material particle by the equality: $mc^2 = h\nu$. *This relation brought him back to waves,* because this equality is not relativistically invariant if ν is the 'internal' frequency of a particle: in this case the transformation law would be $\nu' = \nu(1 - \beta^2)^{1/2}$, i.e., the frequency would be coupled to the slowing down of clocks, whilst m increases with the velocity.

On the contrary, the frequency of a wave has the same transformation property as m. The equality becomes invariant and uniquely defines ν from m.

This was the starting point of wave mechanics, and we can see that for de Broglie, from the very beginning, the photon had a mass and was included in a uniform description of particles in the universe.

Unfortunately such a theory, including light amongst other particles, cannot be developed by means of the Schrödinger or Klein–Gordon equations, because the first equation is non-relativistic and in the second there is no polarization.

The situation changed with the discovery of the Dirac equation. De Broglie was very enthusiastic (see *The Magnetic Electron*, [Brog 34a]) and there he saw a possible beginning for a theory of light: The Dirac equation has a four-component wave function (a polarization), a spin (an axial vector) and a second rank tensor $M_{\mu\nu} = \bar{\psi}\gamma_\mu\gamma_\nu\psi$, antisymmetric as is the electromagnetic tensor.

In his 1922 paper, de Broglie wrote:

"A more complete theory of quanta of light must introduce polarization in such a way that to each atom of light should be linked an internal state of right or left polarization represented by an axial vector with the same direction as the propagation velocity." [Brog 22]

It was *the* correct idea because it was shown later that when the velocity of a particle approaches the velocity of light the space components of the spin vector become parallel to the velocity.

The elements of the Dirac equation were not directly useful for a photon: the wave has neither the transformation property of a vector nor of an antisymmetric tensor; such a tensor ($\bar{\psi}\gamma_\mu\gamma_\nu\psi$) is present, but it is not a wave; the spin has only half the rotation frequency of the photon's spin and the Dirac particle is a fermion. Nevertheless, the way was not obstructed as before: though disordered, the different elements did exist.

1.1.2   The New Theory of Light

After some first attempts, [Brog 32a,b], [Brog 33], [Brog 34b], de Broglie realized that
a photon is not an elementary particle, but the fusion of a pair of a spin $\frac{1}{2}$ corpuscle
and of its anticorpuscle[2], both obeying a Dirac equation [Brog 34c].

The creation and annihilation of pairs suggested that a photon could be the 'fusion' of an electron–positron pair linked by an electrostatic force. The smallness of
the photon mass would be a consequence of the mass defect. Without analytic formulation a similar proposal was made by Eddington and Heisenberg in 1928 in order to
explain the Bose–Einstein statistics of photons by the fusion of positive and negative
electronic charges, see Jordan [Jor 28]. But the introduction of an electrostatic force
is a boot strap, because a theory of the photon is a theory of electromagnetism, and
the electrostatic force must be a *consequence* of the theory.

Thus de Broglie assumed a photon to be a neutrino–antineutrino pair; but the
choice of conjugated particles is not useful, and, finally, *he considered a photon as
the center of mass of a pair of massive neutrinos (Dirac particles).* He published the
equation in 1934 [Brog 34d] and developed the theory during later years [Brog 36],
[Brog 40], [Brog 43], [Brog 49].

1.2   The Photon Equations

1.2.1   The Method of Fusion

Let us, at first, take a couple of identical, free particles of mass m, obeying the
Schrödinger equation, with coordinates x_1, y_1, z_1 and x_2, y_2, z_2. Consider their center
of mass:

$$x = \frac{x_1 + x_2}{2}; \qquad y = \frac{y_1 + y_2}{2}; \qquad z = \frac{z_1 + z_2}{2}. \tag{1.1}$$

The Schrödinger equation of this center of mass reads:

$$-i\hbar \frac{\partial \Phi}{\partial t} = \frac{1}{2M} \Delta \Phi \quad (M = 2m). \tag{1.2}$$

But this procedure could not be extended to a pair of neutrinos because a relativistic
quantum theory for systems of particles analogous to the classical theory did not
exist. So de Broglie suggested a formal procedure which could easily be generalized
to an arbitrary number of constituents.

He associated to the particles two different waves ψ and φ, without making a
distinction between their coordinates. So we have the following equations with the
same x_k:

$$-i\hbar \frac{\partial \psi}{\partial t} = \frac{1}{2M} \Delta \psi; \qquad -i\hbar \frac{\partial \varphi}{\partial t} = \frac{1}{2M} \Delta \varphi. \tag{1.3}$$

---

[2]The term 'antiparticle' appeared there for the first time.

Now, the *fusion* conditions, expressing (in the case of plane waves) the equality of momentum and energy, are:

$$\frac{\partial \psi}{\partial t}\varphi = \psi\frac{\partial \varphi}{\partial t} = \frac{1}{2}\frac{\partial(\psi\varphi)}{\partial t}, \tag{1.4}$$

$$\frac{\partial^2 \psi}{\partial x_k^2}\varphi = \frac{\partial \psi}{\partial x_k}\frac{\partial \varphi}{\partial x_k} = \psi\frac{\partial^2 \varphi}{\partial x_k^2} = \frac{1}{4}\frac{\partial^2(\psi\varphi)}{\partial x_k^2}.$$

Finally, multiplying the first equation (1.3) by φ and the second by ψ, we find equation (1.2). This is the method we shall apply to the relativistic case.

1.2.2 De Broglie's Equations of the Photon

Consider the Dirac equations of two particles of mass $\mu_0/2$:

$$\frac{1}{c}\frac{\partial \psi}{\partial t} = \alpha_k\frac{\partial \psi}{\partial x_k} + i\frac{\mu_0 c}{2\hbar}\alpha_4\psi, \tag{1.5}$$

$$\frac{1}{c}\frac{\partial \varphi}{\partial t} = \alpha_k\frac{\partial \varphi}{\partial x_k} + i\frac{\mu_0 c}{2\hbar}\alpha_4\varphi,$$

where $\{\alpha_k, \alpha_4\}$ are the Dirac matrices:

$$\alpha_k = \begin{pmatrix} 0 & \sigma_k \\ \sigma_k & 0 \end{pmatrix}; \quad \alpha_4 = \begin{pmatrix} I & 0 \\ 0 & I \end{pmatrix}; \quad (\sigma_k = \text{Pauli matrices}). \tag{1.6}$$

In analogy to (1.4), we impose the *fusion* conditions on two solutions ψ_n and φ_m of (1.5) with spinor indices n and m:

$$\frac{\partial \psi_n}{\partial t}\varphi_m = \psi_n\frac{\partial \varphi_m}{\partial t} = \frac{1}{2}\frac{\partial(\psi_n\varphi_m)}{\partial t}, \tag{1.7}$$

$$\frac{\partial \psi_n}{\partial x_k}\varphi_m = \psi_n\frac{\partial \varphi_m}{\partial x_k} = \frac{1}{2}\frac{\partial(\psi_n\varphi_m)}{\partial x_k},$$

and we find for $\Phi = \{\Phi_{nm} = \psi_n\varphi_m\}$ a new equation, that will be extended to all the Φ functions by postulate, even if the components are not of the form $\Phi_{nm} = \psi_n\varphi_m$:

$$\frac{1}{c}\frac{\partial \Phi}{\partial t} = a_k\frac{\partial \Phi}{\partial x_k} + i\frac{\mu_0 c}{\hbar}a_4\Phi, \tag{1.8}$$

$$\frac{1}{c}\frac{\partial \Phi}{\partial t} = b_k\frac{\partial \Phi}{\partial x_k} + i\frac{\mu_0 c}{\hbar}b_4\Phi.$$

The matrices a and b are defined as:

$$a_r = \alpha_r \times I, \quad (a_r)_{ik,lm} = (\alpha_r)_{il}\delta_{km} \ (r = 1, 2, 3, 4), \tag{1.9}$$

$$b_r = I \times \alpha_r, \quad (b_r)_{ik,lm} = (-1)^{r+1}(\alpha_r)_{km}\delta_{il} \ (r = 1, 2, 3, 4).$$

They satisfy the same relations as Dirac matrices, and a and b commute:

$$a_r a_s + a_s a_r = 2\delta_{rs}; \quad b_r b_s + b_s b_r = 2\delta_{rs}; \quad a_r b_s - b_s a_r = 0. \tag{1.10}$$

Furthermore, it is easy to prove that the components of Φ obey the Klein–Gordon equation.

The equations (1.8) are the *de Broglie photon equations*. We shall see that they include the Maxwell equations, but first we must examine some preliminary properties.

1.3 Other Representations of the Photon Equations

1.3.1 A First 'Quasi-Maxwellian' Form

First of all it has to be noted that there are too many equations: 32 equations for only 16 components of the wave function Φ. There is a problem of compatibility. To solve this problem we can add and substract the two systems in (1.8):

$$\text{(A)} \qquad \frac{1}{c}\frac{\partial \Phi}{\partial t} = \frac{a_k + b_k}{2}\frac{\partial \Phi}{\partial x_k} + i\frac{\mu_0 c}{\hbar}\frac{a_4 + b_4}{2}\Phi, \tag{1.11}$$

$$\text{(B)} \qquad 0 = \frac{a_k - b_k}{2}\frac{\partial \Phi}{\partial x_k} + i\frac{\mu_0 c}{\hbar}\frac{a_4 - b_4}{2}\Phi. \tag{1.12}$$

In the following it will be shown that (1.8) contains exactly the Maxwell equations (up to mass terms), but (1.11) is already an outline of these equations, because this system is divided into a group (A) of 'evolution equations' that looks like the equations for $\partial \mathbf{E}/\partial t$ and $\partial \mathbf{H}/\partial t$, and a group (B) of 'constraint equations', of the same kind as $\nabla.\mathbf{E} = 0$ and $\nabla.\mathbf{H} = 0$.

In his first paper of 1934 [Brog 34c] de Broglie discussed only the group (A) of evolution equations, but it is easy to prove that, in analogy with the Maxwell equations:

1) The time derivative of the second term of (B) is equal to zero by virtue of (A), as a consequence of (1.10).

2) (B) is thus satisfied by all the solutions of (A), the Fourier expansion of which does not contain a zero frequency. But such a frequency is evidently absent from all the solutions of (A) if $\mu_0 \neq 0$.

Therefore if $\mu_0 \neq 0$ the condition (B) is a consequence of the evolution equations (A).

1.3.2 The Canonical Form

The equations (1.8) may also be transformed in the following way:

$$\text{(C)} \qquad \frac{1}{c}\frac{a_4 + b_4}{2}\frac{\partial \Phi}{\partial t} = \frac{b_4 a_k + a_4 b_k}{2}\frac{\partial \Phi}{\partial x_k} + i\frac{\mu_0 c}{\hbar}a_4 b_4 \Phi, \tag{1.13}$$

$$\text{(D)} \qquad \frac{1}{c}\frac{a_4 - b_4}{2}\frac{\partial \Phi}{\partial t} = \frac{b_4 a_k - a_4 b_k}{2}\frac{\partial \Phi}{\partial x_k}.$$

This new system will be the basis of the Lagrangian formulation of the theory and of its tensorial version. It was used by de Broglie to *quantize* the photon field and to describe the *photon–electron interaction* [Brog 49].

Just as in (1.11), (D) is a consequence of (C): it may be proved by applying to (C) the operator $(2c)^{-1}(a_4 - b_4)\partial/\partial t$, taking account of (1.10), but it requires $\mu_0 \neq 0$.

It is noteworthy that the arguments given here in favor of a massive photon are imposed by the fusion theory. They are in accordance with the preceding arguments, requiring relativity and a uniform description of particles, but they have nothing to do with them: it is a simple convergence of two sets of different arguments.

1.3.3 Introduction of a Square Matrix Wave Function

Let us go back to the initial system (1.5) and write it in terms of relativistic coordinates $x_k = (x, y, z)$, $x_4 = ict$ and γ matrices with:

$$\gamma_\mu \gamma_\nu + \gamma_\nu \gamma_\mu = 2\delta_{\mu\nu} \quad \text{for} \quad \mu, \nu = 1, ..., 4; \tag{1.14}$$

$$\gamma_k = i\alpha_4 \alpha_k; \qquad \gamma_4 = \alpha_4; \qquad \gamma_5 = \gamma_1 \gamma_2 \gamma_3 \gamma_4.$$

Then (1.5) can be equivalently written as:

$$\partial_\mu \gamma_\mu \psi - \frac{\mu_0 c}{2\hbar} \psi = 0, \tag{1.15}$$

$$\partial_\mu \gamma_\mu \phi - \frac{\mu_0 c}{2\hbar} \phi = 0.$$

If again we multiply the first equation by φ from the left hand side and the second by ψ, we obtain (with $\Psi = \psi \otimes \varphi$, if the fusion conditions (1.7) are applied), the equations

$$\partial_\mu \gamma_\mu \Psi - \frac{\mu_0 c}{\hbar} \Psi = 0, \tag{1.16}$$

$$\partial_\mu \Psi \tilde{\gamma}_\mu - \frac{\mu_0 c}{\hbar} \Psi = 0,$$

with $\mu, \nu = 1, ..., 4$; $\tilde{\gamma}_\mu = \gamma_\mu$ transposed.

The transposed matrices $\tilde{\gamma}$ are easily eliminated because, if two sets of Dirac matrices γ_μ and $\tilde{\gamma}_\mu$ satisfy the relations (1.15), there are two non-singular matrices Λ and Γ, such that:

$$\tilde{\gamma}_\mu = \Lambda \gamma_\mu \Lambda^{-1}, \tag{1.17}$$

$$\tilde{\gamma}_\mu = -\Gamma \gamma_\mu \Gamma^{-1},$$

where $\mu, \nu = 1, 2, 3, 4$, $\Gamma = \Lambda \gamma_5$, and where γ_5 is given in (1.15). Of course (1.17) is true if $\tilde{\gamma}_\mu$ are the transposed γ_μ matrices. When the γ_μ are defined by (1.15), with the classical Dirac matrices α_k, a solution for Γ and Λ is:

$$\Gamma = -i\gamma_2 \gamma_4, \tag{1.18}$$

$$\Lambda = \Gamma \gamma_5 = -i\gamma_3 \gamma_1.$$

The Λ case in (1.17) was given by Pauli [Pau 36] and the Γ case was used by de Broglie to eliminate $\tilde{\gamma}_\mu$ in (1.16). Indeed, introducing Γ into (1.16), we find the system given by de Broglie, Tonnelat and Petiau ([Brog 43], Ch. VII):

$$\partial_\mu \gamma_\mu \left(\Psi\Gamma\right) - \frac{\mu_0 c}{\hbar}\left(\Psi\Gamma\right) = 0, \tag{1.19}$$

$$\partial_\mu \left(\Psi\Gamma\right) \gamma_\mu + \frac{\mu_0 c}{\hbar}\left(\Psi\Gamma\right) = 0.$$

The system obtained by the use of Λ was given only recently [Loch 95a]:

$$\partial_\mu \gamma_\mu \left(\Psi\Lambda\right) - \frac{\mu_0 c}{\hbar}\left(\Psi\Lambda\right) = 0, \tag{1.20}$$

$$\partial_\mu \left(\Psi\Lambda\right) \gamma_\mu - \frac{\mu_0 c}{\hbar}\left(\Psi\Lambda\right) = 0.$$

The only formal difference is a minus sign in the second group of equations. But this entails a great physical difference because the wave functions, solutions of (1.19) and (1.20), transform themselves between each other by a mutiplication by γ_5: *They are dual in space–time*, so that, as will be proved, they exchange electricity and magnetism.

The representation (1.8), where Φ is a *second order spinor*, was extended in the *General Theory of Spin Particles* [Brog 43]. For evident reasons, the representation (1.19)–(1.20) can be generalized only for the fusion of an *even* number of spin $\frac{1}{2}$ particles.

1.4   Two Kinds of Photons

1.4.1   The Electromagnetic Form of the Photon Equations

The electromagnetic form was given by de Broglie in his first papers, starting from (1.8) [Brog 34d], [Brog 36], [Brog 40]. For the sake of simplicity we shall start from (1.19) and (1.20), and apply a procedure suggested by Tonnelat for (1.19) which afterwards was used by de Broglie [Brog 43].

So let us expand a 4×4 matrix Θ in a Clifford algebra:

$$\Theta = I\varphi_0 + \gamma_\mu \varphi_\mu + \gamma_{[\mu\nu]}\varphi_{[\mu\nu]} + \gamma_\mu \gamma_5 \varphi_{\mu 5} + \gamma_5 \varphi_5. \tag{1.21}$$

φ_0 is a scalar, φ_μ a polar vector, $\varphi_{[\mu\nu]}$ a tensor of rank two, $\varphi_{\mu 5}$ an axial vector and φ_5 a pseudo-scalar. From this we can define electromagnetic quantities in $\mathbb{R}^3$:

$$\begin{aligned}
\mathbf{H} &= K k_0 \left(\varphi_{23}, \varphi_{31}, \varphi_{12}\right), \tag{1.22}\\
\mathbf{E} &= K k_0 \left(\varphi_{14}, \varphi_{24}, \varphi_{34}\right),\\
\mathbf{A} &= K \left(\varphi_1, \varphi_2, \varphi_3\right),\\
iV &= K \varphi_4,
\end{aligned}$$

$$-i\mathbf{B} = K\left(\varphi_{15}, \varphi_{25}, \varphi_{35}\right),$$
$$W = K\varphi_{45},$$
$$I_1 = \varphi_0,$$
$$iI_2 = \varphi_5,$$

where $k_0 = \mu_0 c\hbar^{-1}$; $K = \hbar\mu_0^{-1/2}/2$. (**B** is *not an induction*, but an axial potential).

If we now substitute (1.21) and (1.22) into (1.19) and (1.20) by identifying $\Theta \equiv \Psi$, and if we afterwards project these equations onto the set of Clifford elements, we find two sets of equations in terms of field equations for the quantities (1.22).

1.4.2   The Equations of the 'Electric Photon'

The expansion of the matrix wave function Ψ of (1.19) according to (1.21) splits the equation (1.19) into two systems [Brog 36], [Brog 40], [Brog 43] that we call the *electric photon* for reasons to be given below:

$$\frac{1}{c}\frac{\partial \mathbf{H}}{\partial t} = \mathrm{rot}\,\mathbf{E}, \qquad (1.23)$$

$$\frac{1}{c}\frac{\partial \mathbf{E}}{\partial t} = \mathrm{rot}\,\mathbf{H} + k_0^2\mathbf{A},$$

$$\mathrm{div}\,\mathbf{H} = 0,$$

(M) $$\mathrm{div}\,\mathbf{E} = -k_0^2 V,$$

$$\mathbf{H} = \mathrm{rot}\,\mathbf{A},$$

$$\mathbf{E} = -\mathrm{grad}\,V - \frac{1}{c}\frac{\partial \mathbf{A}}{\partial t},$$

$$\frac{1}{c}\frac{\partial V}{\partial t} + \mathrm{div}\,\mathbf{A} = 0,$$

and

$$\frac{1}{c}\frac{\partial I_1}{\partial t} = 0, \qquad (1.24)$$

$$\mathrm{grad}\,I_1 = 0,$$

$$k_0 I_1 = 0 \quad (k_0 \neq 0 \Rightarrow I_1 = 0),$$

(NM) $$-\frac{1}{c}\frac{\partial I_2}{\partial t} = k_0 W,$$

$$\mathrm{grad}\,I_2 = k_0\mathbf{B},$$

$$\frac{1}{c}\frac{\partial W}{\partial t} + \mathrm{div}\,\mathbf{B} = k_0 I_2,$$

$$\mathrm{rot}\,\mathbf{B} = 0,$$

$$\mathrm{grad}\,W + \frac{1}{c}\frac{\partial \mathbf{B}}{\partial t} = 0.$$

These are two different systems of equations because the fundamental equations (1.8) are not the equations of a spin 1 particle, but of *a particle of maximum spin* 1.

In a sense that will be specified in Section 1.6, the equations denoted by (M) ('Maxwellian') correspond to the spin 1 and they are strictly speaking the *equations of the photon*. The equations denoted by (NM) ('Non–Maxwellian') correspond to the spin 0.

a) The (M) equations are Maxwell's equations, with two differences:

1) The presence of *mass terms*, which introduce a link between fields and potentials. The latter become physical quantities and lose their gauge invariance.

2) The definition of fields by Lorentz potentials and the Lorentz gauge condition:

$$\mathbf{H} = \operatorname{rot}\mathbf{A}, \tag{1.25}$$

$$\mathbf{E} = -\operatorname{grad}V - \frac{1}{c}\frac{\partial \mathbf{A}}{\partial t},$$

$$\frac{1}{c}\frac{\partial V}{\partial t} + \operatorname{div}\mathbf{A} = 0.$$

These relations are not arbitrarily added to the field equations, as they were in the classical theory, *they appear automatically as field equations*. Of course, they were already present in a hidden form in (1.8), (1.11), (1.13) and (1.19).

As a consequence of (1.23) the fields and the potentials do not obey the d'Alembert wave equation but the Klein–Gordon equation:

$$\Box F + k_0^2 F = 0 \quad (F = \mathbf{B},\ \mathbf{H},\ \mathbf{A},\ V). \tag{1.26}$$

The electrostatic potential is now the Yukawa potential $V = e^{-r/k_0}/r$, but it is nevertheless a long range potential because of the smallness of the Compton wave number $k_0 = \mu_0 c/\hbar$ (see Section 1.7).

b) The (NM) equations were considered, at first, by de Broglie, as describing a spin 0 meson (μ_0 is the photon mass). The particle is chiral because:

I_1 is an invariant, but $I_1 = 0$; $I_2 \neq 0$ is a pseudo-invariant and $\{\mathbf{B}, W\}$ is a pseudo-quadrivector, dual of a tensor of rank 3.

But de Broglie remarked [Brog 40] that the situation may be interpreted in another way, defining a second electromagnetic field (he said an 'anti-field'):

$$\mathbf{H}' = \frac{1}{c}\frac{\partial \mathbf{B}}{\partial t} + \operatorname{grad}W, \tag{1.27}$$

$$\mathbf{E}' = \operatorname{rot}\mathbf{B},$$

which *is equal to zero* by virtue of (1.24).

We shall follow this second interpretation, on the basis of a symmetry between electricity and magnetism developed in our papers concerning the photon [Loch 95a] and the magnetic monopole [Loch 85], [Loch 95b]. The de Broglie definition (1.27) of $\mathbf{H}'$ and $\mathbf{E}'$ in terms of a *pseudo* quadripotential $\{W, \mathbf{B}\}$, rediscovered later by Cabibbo and Ferrari [Cab 62], plays a central role in theories of monopoles.

We call the system (1.23)–(1.24) *the electric state of the photon* for the following reasons:

1) We have an electromagnetic field $\{\mathbf{E}, \mathbf{H}\}$ and a *polar* 4-potential related to $\{\mathbf{E}, \mathbf{H}\}$ by the Lorentz formulae (1.25). These fields and potentials enter in the dynamics of an *electric* charge. Because $k_0 \neq 0$ we have, in general, $\mathrm{div}\mathbf{E} \neq 0$, so that the *electric* field $\mathbf{E}$ is not transversal, contrary to the magnetic field: $\mathbf{E}$ has a small longitudinal component of the order of k_0.

2) In the (NM) equations we have a *pseudo* invariant I_2 and an *axial* 4-potential $\{W, \mathbf{B}\}$, to which the invariant I_1 and the 'anti-field' $\{\mathbf{E}', \mathbf{H}'\}$ may be added. The latter are defined in (1.27) and will be related to magnetism. But $I_1 = \mathbf{E}' = \mathbf{H}' = 0$, which confirms the electric character of this photon.

Equations (1.23) and (1.24) are historically the first example of the derivation of an effective theory for relativistic composite particles. With respect to the physical interpretation it is somewhat surprising that in addition to the conventional physically relevant equations (1.23), the set of equations (1.24) appears which is to be connected with a magnetic monopole interpretation.

Clearly this is a consequence of using the complete basis (1.21) for the representation of the spin tensor Ψ. In subsequent applications of de Broglie's equations (1.8) by Bargmann and Wigner and other authors, compare Lurié [Luri 68], instead of the complete set (1.21) only the symmetric elements of the Dirac algebra were used for the representation of the bi-spinor Ψ and thus led to experimentally confirmed effective equations only.

But of course this is an arbitrary act which does not follow from an inherent logic of the system itself. In particular for the local fusion of two fermions one would expect the Pauli principle to work leading to antisymmetric spin matrices and not to symmetric ones. But within the present scheme this would lead to no photon at all.

A satisfactory explanation and removal of these difficulties can only be given in the quantum field theoretical formulation of fusion theory and by embedding the photon into the electro–weak theory. At the same time this leads to a strong confirmation of the idea of a composite photon as its existence can be only justified with regard to the modern electro–weak unification.

The discussion of these aspects will be the topic of the subsequent chapters. In this chapter we continue the review of de Broglie's theory and subsequent developments, because it offers insights which may be of importance in actual and future duality research *etc.*.

1.4.3   The Equations of the 'Magnetic Photon'

This second photon corresponds to (1.20), with $\Lambda = \Gamma\gamma_5$, instead of Γ in (1.16) [Loch 95a]. The new (primed) field components are *dual* to the preceding fields. They are de Broglie's anti-fields and they exchange electricity and magnetism:

$$-\frac{1}{c}\frac{\partial \mathbf{H}'}{\partial t} \;=\; \mathrm{rot}\mathbf{E}' + k_0^2 \mathbf{B}', \tag{1.28}$$

$$\frac{1}{c}\frac{\partial \mathbf{E'}}{\partial t} = \mathrm{rot}\mathbf{H'},$$

$$\mathrm{div}\mathbf{H'} = k_0^2 W',$$

(M) $$\mathrm{div}\mathbf{E'} = 0,$$

$$\mathbf{H'} = \mathrm{grad}W' + \frac{1}{c}\frac{\partial \mathbf{B'}}{\partial t},$$

$$\mathbf{E'} = \mathrm{rot}\mathbf{B'},$$

$$0 = \frac{1}{c}\frac{\partial W'}{\partial t} + \mathrm{div}\mathbf{B'},$$

and

$$\frac{1}{c}\frac{\partial I_2}{\partial t} = 0, \qquad\qquad (1.29)$$

$$\mathrm{grad}I_2 = 0,$$

$$k_0 I_2 = 0 \quad (k_0 \neq 0 \Rightarrow I_2 = 0),$$

(NM) $$-\frac{1}{c}\frac{\partial I_1}{\partial t} = k_0 V',$$

$$\mathrm{grad}I_1 = k_0 \mathbf{A'},$$

$$k_0 I_1 = \frac{1}{c}\frac{\partial V'}{\partial t} + \mathrm{div}\mathbf{A'},$$

$$\mathrm{rot}\mathbf{A'} = 0,$$

$$0 = \frac{1}{c}\frac{\partial \mathbf{A'}}{\partial t} + \mathrm{grad}V'.$$

As before, the new photon is associated with two fields. But the situation is inverted:

1) The 'anti-field' $\{\mathbf{E'}, \mathbf{H'}\}$ and the *axial* 4-potential $\{W', \mathbf{B'}\}$ now obey the (M) system (1.28). The *a priori* definition (1.26) of the anti-fields now appears in (1.28) automatically, as one of the field equations. Of course, $\{\mathbf{E'}, \mathbf{H'}\}$ are not equal to zero.

The fields $\{\mathbf{E'}, \mathbf{H'}\}$, defined by $\{W', \mathbf{B'}\}$, are exactly those that enter in the dynamics of a magnetic charge [Loch 85], [Loch 95b]: a *monopole*.

Moreover, 'symmetrically' with respect to the electric case, we have now $\mathrm{div}\mathbf{H'} \neq 0$, so that, in a plane wave, the *magnetic* field $\mathbf{H'}$ (instead of the electric field) will now have a small longitudinal component, of the order of k_0, whilst $\mathbf{E'}$ is transversal. Thus we have a *magnetic photon*.

2) The *polar* potentials $\{V', \mathbf{A'}\}$, dual to $\{W, \mathbf{B}\}$, now appear in the (NM) system, the *spin* 0 *state*. The invariant I_1 and the pseudo-invariant I_2 invert their roles: We have now $I_2 = 0$ and $I_1 \neq 0$. And the electromagnetic field $\{\mathbf{E}, \mathbf{H}\}$ defined by the classical Lorentz formulae (1.25) gives now: $\mathbf{E} = \mathbf{H} = 0$, just as we had $\mathbf{E'} = \mathbf{H'} = 0$ in the electric case.

It is a remarkable fact that in the sense of de Broglie, the *fusion* of two Dirac equations does not only give the classical Maxwell equations, but also defines *two classes* of photons, corresponding to the interaction with an electric or a magnetic charge respectively. The algebraic procedure excludes any other possibility.

The symmetry between these two electromagnetic fields is all the more interesting, as such a symmetry already appears in the Dirac equation itself, in the form of two minimal interactions corresponding to an electric and a magnetic charge, to which are associated the two kinds of fields [Loch 85], [Loch 95b]. The symmetries of Dirac's and de Broglie's equations are thus linked together. We take up these problems again in Section 1.6.

1.5 Hamiltonian, Lagrangian, Current, Energy, Spin

1.5.1 The Lagrangian

Now, let us go back to the 16–line column wave function and the canonical form (1.13), keeping only (C), because (D) can be deduced from it:

$$\frac{1}{c}\frac{a_4 + b_4}{2}\frac{\partial \Phi}{\partial t} = \frac{b_4 a_k + a_4 b_k}{2}\frac{\partial \Phi}{\partial x_k} + i\frac{\mu_0 c}{\hbar}a_4 b_4 \Phi. \tag{1.30}$$

Note the anomaly of the presence of the operator $(a_4 + b_4)$, as a factor of the time derivative $\partial/\partial t$. This factor is necessary in order to obtain coherent definitions for tensor densities and as we will see it expresses the fact that in linear theories the indefiniteness of energy can be shifted to the state norms. The Hamiltonian operator is thus:

$$H = i\hbar \left[\frac{b_4 a_k + a_4 b_k}{2}\frac{\partial}{\partial x_k} + ik_0\, a_4 b_4\right] \qquad \left(k_0 = \frac{\mu_0 c}{\hbar}\right) \tag{1.31}$$

and the Lagrangian density is:

$$L = -i\hbar c \left[\Phi^+ \left(\frac{1}{c}\frac{a_4 + b_4}{2}\frac{\partial \Phi}{\partial t} - \frac{b_4 a_k + a_4 b_k}{2}\frac{\partial \Phi}{\partial x_k} - ik_0\, a_4 b_4 \Phi\right) + \text{c.c.}\right], \tag{1.32}$$

where $\Phi^+ = \Phi$ h.c..

1.5.2 The Current Density Vector

The general formula

$$J_\mu = \frac{i}{\hbar}\left[\frac{\partial L}{\partial \Phi_{,\mu}}\Phi - \frac{\partial L}{\partial \Phi^+_{,\mu}}\Phi^+\right] \tag{1.33}$$

gives with (1.32):

$$J_k = -c\Phi^+ \frac{b_4 a_k + a_4 b_k}{2}\Phi, \tag{1.34}$$

$$J_4 = ic\rho,$$

$$\rho = \Phi^+ \frac{a_4 + b_4}{2}\Phi.$$

Therefore $\int \rho dv$ *is not positive definite*, but on the other hand we shall find a *positive definite energy* $\int \rho E dv \geq 0$, contrary to what happens in the Dirac electron. This result will be generalized in Section 1.8 in the general theory of particles with spin $s = n/2$.

In terms of electromagnetic quantities (1.33) is given by the *Gehéniau formulae* [Brog 40], with two kinds of terms corresponding to spin 1 and spin 0:

$$\mathbf{J} = \frac{i}{\hbar c} [\mathbf{A}^* \times \mathbf{H} + \mathbf{H}^* \times \mathbf{A} + V^* \mathbf{E} - \mathbf{E}^* V] + \frac{c}{4} (I_2^* \mathbf{B} + \mathbf{B}^* I_2), \qquad (1.35)$$

$$\rho = \frac{i}{\hbar c} [(\mathbf{A}^* \cdot \mathbf{E}) - (\mathbf{E}^* \cdot \mathbf{A})] + \frac{1}{4} (I_2^* W + W^* I_2).$$

These formulae correspond to the case of an *electric photon*. The *magnetic case* has not been calculated.

Again this result demonstrates that de Broglie's theory contains more than the Maxwell theory. In Maxwell's theory bilinear expressions are to be performed by real field quantities only, whereas the above formalism basically deals with field quantities and their conjugates, see (1.32).

This of course induces nonvanishing currents and 'charges' as a result of a global $U(1)$ invariance of (1.32). If the (N) field quantities are restricted to be real and the (NM) set is assumed to be zero, then $\mathbf{J}$ and ρ vanish as is the case in Maxwell's theory.

Actually the equations like (1.8), (1.19), (1.20), (1.23), (1.24) or (1.29) are not at all the Maxwell equations (even not, if the latters are completed by mass terms). This means, that these equations do not describe the electromagnetic Maxwell field, but a de Broglie wave associated with *a single photon*.

In order to obtain the Maxwell field one has to collect a great number of such single photon waves. This was performed by de Broglie in his book *Une nouvelle théorie de la lumière* [Brog 40]. De Broglie showed that the imaginary part of the complex photon wave tends to zero for an increasing number of photons, so one ends up with the real Maxwell field.

Remark: Speaking about *measuring an electromagnetic field* never means measuring the de Broglie wave of one photon, but measuring the sum of the effects of a great number of photons associated or carried by one wave. *This* is the Maxwell field, which is always a real quantity.

1.5.3 The Energy Tensor

Consider the expression of the energy tensor:

$$T_{\mu\nu} = -\frac{\partial L}{\partial \Phi_{,\mu}} \Phi_{,\nu} - \frac{\partial L}{\partial \Phi_{,\mu}^+} \Phi_{,\nu}^+ + L \delta_{\mu\nu}. \qquad (1.36)$$

For the Lagrangian (1.32) it explicitly reads:

$$T_{ik} = -\frac{i\hbar c}{2}\left[\Phi^+\frac{b_4 a_i + a_4 b_i}{2}\frac{\partial\Phi}{\partial x_k} + \frac{\partial\Phi^+}{\partial x_k}\frac{b_4 a_i + a_4 b_i}{2}\Phi\right], \tag{1.37}$$

$$T_{i4} = \frac{\hbar}{2}\left[\Phi^+\frac{b_4 a_i + a_4 b_i}{2}\frac{\partial\Phi}{\partial t} + \text{c.c.}\right],$$

$$T_{4i} = -\frac{\hbar c}{2}\left[\Phi^+\frac{a_4 + b_4}{2}\frac{\partial\Phi}{\partial x_i} + \text{c.c.}\right],$$

$$T_{44} = -w = i\hbar\Phi^+\frac{a_4 + b_4}{2}\frac{\partial\Phi}{\partial t} = -\Phi^+ H\Phi.$$

In the electromagnetic form (once more for the *electric photon*), we have:

$$T_{\mu\nu} = \frac{1}{2}\left(F_{\mu\lambda}\frac{\partial A_\lambda}{\partial x_\nu} - A_\lambda\frac{\partial F_{\lambda\mu}}{\partial x_\nu}\right) - \frac{i\hbar}{8}\left(I_2^*\frac{\partial B_\mu}{\partial x_\nu} - B_\mu^*\frac{\partial I_2}{\partial x_\nu}\right) + \text{c.c.}, \tag{1.38}$$

where $F_{\mu\nu} = \partial_\nu A_\mu - \partial_\mu A_\nu$.

In particular, the energy density ρW takes the form:

$$T_{44} = \frac{1}{2c}\left[\mathbf{A}^*\cdot\frac{\partial\mathbf{E}}{\partial t} - \mathbf{E}^*\cdot\frac{\partial\mathbf{A}}{\partial t} + \mathbf{A}\cdot\frac{\partial\mathbf{E}^*}{\partial t} - \mathbf{E}\cdot\frac{\partial\mathbf{A}^*}{\partial t}\right] \tag{1.39}$$

$$+ \frac{i\hbar c}{2}\left[I_2^*\frac{\partial W}{\partial t} + W^*\frac{\partial I_2}{\partial t} - I_2\frac{\partial W^*}{\partial t} - W\frac{\partial I_2^*}{\partial t}\right].$$

There are arguments in favor of this nonsymmetric tensor, developed by Costa de Beauregard and de Broglie [Brog 43], [Cos 43], but we can also symmetrize the tensor, putting: $T'_{\mu\nu} = \frac{1}{2}(T_{\mu\nu} + T_{\nu\mu})$.

And we can also find other tensors, the integrals of which are equal to the integrals of the above expressions (they differ by a divergence). One of these tensors is[3]:

$$M_{ik} = M_{ki} = \mu_0 c^2\Phi^+\frac{a_i b_k + a_k b_i}{2}\Phi \quad (i,k = 1,2,3), \tag{1.40}$$

$$M_{i4} = M_{4i} = \mu_0 c^2\Phi^+\frac{a_i + b_i}{2}\Phi \quad (i = 1,2,3),$$

$$M_{44} = \mu_0 c^2\Phi^+\Phi.$$

This tensor is of Maxwell type because in electromagnetical terms, we have (for the electric photon):

$$M_{i4} = \left[(\mathbf{E}^*\times\mathbf{H})_i - (\mathbf{E}\times\mathbf{H}^*)_i - k_0^2(V^*\mathbf{A}_i + V\mathbf{A}_i^*)\right], \tag{1.41}$$

$$M_{44} = |\mathbf{E}|^2 + |\mathbf{H}|^2 - k_0^2\left(|\mathbf{A}|^2 + |V|^2\right).$$

Except for the mass terms, we recognize the Maxwellian form. And we have:

$$\int T_{\mu\nu}d\tau = \int M_{\mu\nu}d\tau. \tag{1.42}$$

---

[3]The factor μ_0 seems surprising, but according to (1.22) it will disappear from the fields and potentials.

1.5.4   The Photon Spin

Let us calculate the angular momentum with the nonsymmetric tensor $T_{\mu\nu}$:

$$m_{ik} = -\frac{i}{c} \int \left[x_i T_{4k} - x_k T_{4i} \right] d\tau \quad (i, k = 1, 2, 3).$$ (1.43)

m_{ik} is *not* a constant of motion. But, as in Dirac's theory, we can find a constant of motion m'_{ik} if we add an appropriate term:

$$m'_{ik} = m_{ik} + S_{ik}$$ (1.44)

with

$$S_{ik} = i\hbar \int \Phi^+ \frac{b_4 a_i a_k + a_4 a_i b_k}{2} \Phi \quad (i, k = 1, 2, 3).$$ (1.45)

The dual $s_j = \varepsilon_{jik} S_{ik}$ of this tensor in $\mathbb{R}^3$ is a pseudo-vector. By analogy with the Dirac spin, we find a *space–time pseudo-vector*, by adding a time component:

$$s_4 = c\hbar \int \Phi^+ \frac{b_4 a_1 a_2 a_3 + a_4 b_1 b_2 b_3}{2} \Phi.$$ (1.46)

If we replace in (1.43) the tensor $T_{\mu\nu}$ by the symmetric tensor $T'_{\mu\nu} = \frac{1}{2}(T_{\mu\nu} + T_{\nu\mu})$, the new momentum

$$m'_{ik} = -\frac{i}{c} \int \left[x_i T'_{4k} - x_k T'_{4i} \right] d\tau \quad (i, k = 1, 2, 3)$$ (1.47)

is nothing but (1.44). Of course, it is conservative. As opposed to the theory of the electron, the eigenvalues of the matrices in the integrals (1.45) are: -1, 0, $+1$, instead of $\pm\frac{1}{2}$. We have a particle of maximum spin 1.

The space–time pseudovector $s_\mu = \{ \mathbf{s}, s_4 \}$ has the following form in terms of electromagnetic quantities (for the *electric photon*):

$$\mathbf{s} = \frac{1}{c} \left[\mathbf{E}^* \times \mathbf{A} - \mathbf{A}^* \times \mathbf{E} + V^* \mathbf{H} + \mathbf{H}^* V \right],$$ (1.48)

$$s_4 = \frac{1}{c} \left[\mathbf{A}^* \cdot \mathbf{H} + \mathbf{H}^* \cdot \mathbf{A} \right].$$

Only terms corresponding to spin 1 appear in this formula: The terms corresponding to spin 0 vanish because $I_1 = 0$. It is not astonishing that the value 0 automatically disappears from the spin formulae, but it must be emphasized that it is owed to the presence of mass terms $\mu_0 \neq 0$: see (1.24).

If, instead of starting from the canonical equations (1.13), we start from (1.11A), we can consider the orbital momentum operator:

$$\mathbf{M}_{\mathrm{op}} = \mathbf{r} \times \mathbf{p}$$ (1.49)

which does not commute with the second term of (1.11A) and is thus not an integral of motion.

But we can find a conserved quantity if we add to $\mathbf{M}_{\mathrm{op}}$ the new spin operator:

$$S = \left\{ -\frac{i\hbar}{2} \left(a_2 a_3 + b_2 b_3 \right), \ -\frac{i\hbar}{2} \left(a_3 a_1 + b_3 b_1 \right), \ -\frac{i\hbar}{2} \left(a_1 a_2 + b_1 b_2 \right) \right\} \tag{1.50}$$

that may be completed by:

$$S_4 = -\frac{i\hbar}{2} \left(a_1 a_2 a_3 + b_1 b_2 b_3 \right) \tag{1.51}$$

in order to form with S a relativistic quadri-vector.

Of course, the components of S satisfy the spin commutation relations and this is finally the definition that will be used in the generalized theory of fusion.

1.6 Relativistic Noninvariance of the Decomposition Spin 1– Spin 0

The spin operators $\{ s \left(s_j = \varepsilon_{jik} S_{ik} \right), s_4 \}$ in (1.45), (1.46) obey the commutation rules of an angular momentum and they have the eigenvalues $\{ -1, 0, 1 \}$. The total spin s^2 has the eigenvalues $l(l+1) = \{ 2, 0 \}$, corresponding to $l = 1, 0$.

In the case of a plane wave in (1.23), (1.24) and (1.28), (1.29), one can show that the group of equations (M) is associated with $l = 1$, with the projections $s = -1, 0, +1$ on the direction of propagation of the wave: $s = -1 \Rightarrow$ right-circular wave, $s = +1 \Rightarrow$ left-circular wave, $s = 0 \Rightarrow$ small longitudinal *electric* wave or *magnetic* wave for the second photon respectively. The group (NM) is associated to $l = 0$.

In this sense, we can speak of (M) as a 'spin 1 particle' and of (NM) as a 'spin 0 particle'.

However, de Broglie made an important remark: *Although the equations (M) and (NM) are relativistically invariant, their separation into 'spin 1' and 'spin 0' is not covariant* [Brog 43], [Brog 49]. The reason is the following: the distinction between the values 2 and 0 of the total spin is based on the operator $s^2 = s_1^2 + s_2^2 + s_3^2$ which is not a relativistic invariant.

If we now examine field quantities and eigenvalues of s^2 we find the following correspondence :

— for the electric photon, see (1.22), (1.27):

A	V	E	H	I_1	B	W	I_2	E'	H'
2	0	2	2	0	2	0	0	2	2

$$\tag{1.52}$$

— for the magnetic photon, see (1.28), (1.29):

B'	W'	H'	E'	I_2	A'	V'	I_1	H	E
2	0	2	2	0	2	0	0	2	2

$$\tag{1.53}$$

In both cases, the first group corresponds to (M) equations and the second group to (NM) and when passing from (1.52) to (1.53), the following exchanges take place:

— between potentials $\mathbf{A}$, V and pseudo-potentials $\mathbf{B}'$, W' and conversely;

— between fields $\mathbf{E}$, $\mathbf{H}$ and anti-fields $\mathbf{E}'$, $\mathbf{H}'$ (we know that $\mathbf{E}'$, $\mathbf{H}' = 0$ in (1.52) and $\mathbf{E}$, $\mathbf{H} = 0$ in (1.53));

— between the invariant I_1 and the pseudo-invariant I_2, inside the group (NM) (with $I_1 = 0$ in (1.52) and $I_2 = 0$ in (1.53)).

But the most important fact is that in both groups (M) and (NM) there are field quantities with $\mathbf{s}^2 = 2$ and $\mathbf{s}^2 = 0$. This means that *in both groups (M) and (NM), there are spin 1 and spin 0 components*: there is no true separation between the two values of spin.

Following de Broglie one can show (for both photons) that this separation occurs only in the rest system, because:

a) For the electric photon, the potential $(\mathbf{A}, V)$ is *space-like*, and the pseudo-potential $(\mathbf{B}, W)$ *time-like*, so that V and B disappear from (1.52) and only $\mathbf{s}^2 = 2$ remains in (M) and in (NM) because in (NM), we know that $\mathbf{E}' = \mathbf{H}' = 0$.

b) For the magnetic photon, the same happens, because this case follows from the preceding by multiplying an electric solution by γ_5, exchanging polar and axial quantities:

$$(\mathbf{E}, \mathbf{H}) \longleftrightarrow (\mathbf{H}', \mathbf{E}') ; \quad (V, \mathbf{A}) \longleftrightarrow (W, \mathbf{B}) ; \quad (I_1, I_2) \longleftrightarrow (I_2, I_1) \tag{1.54}$$

so that the potential $(\mathbf{A}, V)$ becomes *time-like* and the pseudo-potential $(\mathbf{B}, W)$ *space-like*. And we have once more, in the rest frame, $\mathbf{s}^2 = 2$ in (M) and $\mathbf{s}^2 = 0$ in (NM), taking into account that we have $\mathbf{E} = \mathbf{H} = 0$ instead of $\mathbf{E}' = \mathbf{H}' = 0$.

Thus it follows that, apart from the rest frame, the (M) and (NM) groups of equations cannot be rigorously separated by their spin properties and therefore they must probably be considered as forming one block for two reasons:

1) The difficulty of separating spin 1 and spin 0 finally shows that the composite photon cannot be considered as a spin 1 particle, but as a particle with *maximum spin* 1, just as a two-electron atom or a two-atom molecule. It is noteworthy that the proper state, in which the components 1 and 0 are separated, is obviously the same for both components.

2) On the other hand, the presence of two photons (electric and magnetic) is contained in the very structure of the theory. Although their separation is covariant and it seems more radical than the separation of spin states, the simultaneous presence, in (M) and (NM) equations, of potentials and pseudo-potentials, of fields and anti-fields (even if half of them equal zero) and the 'migration' of these quantities from one group of equations to the other, according to the type of photon, all constitutes another link.

Of course, a question remains unsolved: What is, physically speaking, this spin 0 photon component?

It must be stressed that all these questions — one can say 'these difficulties' — are raised by the hypothesis $\mu_0 \neq 0$: *If, instead of basing the theory on the hypothesis of a composite photon, we started from the theory of group representations, admitting as a postulate that $\mu_0 = 0$, we would certainly avoid such questions.* But it would be a bad idea to shield a theory from a physical problem by a formal condition at the expense of the more synthetic structure as we have already emphasized before.

1.7 The Problem of a Massive Photon

We have seen that many features of de Broglie's theory of the photon, including its logical coherence, are owed to the hypothesis $\mu_0 \neq 0$.

But, even if μ_0 is small, this implies many differences with ordinary electromagnetism. These differences were examined by de Broglie [Brog 40], [Brog 43], [Brog 49] and coworkers, principally Costa de Beauregard [Cos 95], [Cos 97a,b].

1.7.1 Gauge Invariance

Obviously this invariance disappears, but it requires some comments:

1) First of all, why do we find the Lorentz gauge? Simply because it is the only relativistically invariant, linear, first order differential law. It was the only possible law.

2) Some practical problems: The relations between potentials and fields show that they are of the same order of magnitude. The mass terms are thus very small. Therefore, in general, the gauge symmetry remains, up to a negligible error, and we can still choose the convenient gauge for each practical problem, provided that physics does not impose some particular choice.

3) In this theory, potentials are deducible from fields, thus from observable phenomena: They are no longer mathematical fictions, but physical quantities.

This must be already conjectured to hold for zero field phenomena, which arise only as a result of the presence of a potential, such as the Aharonov–Bohm effect. The fact that this effect is gauge invariant is not an objection, because we know of other physical quantities that are only partially defined by some effects and exactly defined by others: for instance, energy is defined by spectral laws up to an additive constant, but exactly fixed by relativistic effects.

De Broglie gave another example of a physically defined potential: the electron gun. The potential V between the electrodes is exactly defined for several reasons: a) the *measurable* velocity of the emerging electron is given by the increase of energy, which is equal to eV; b) the phase of the wave associated to the electron is relativistically invariant *only if* the frequency and the phase velocity obey the classical de Broglie formulae, which imposes the gauge of V (the same as above); c) the fundamental reason for all of this is that the inertia of energy does not allow an arbitrary choice of the origin of electrostatic potentials, which actually are not gauge invariant. They are physical quantities, related to mesurable effects.

More recently, Costa de Beauregard published many other impressive experimental examples in favor of the physical importance of electromagnetic potentials [Cos 95], [Cos 97a,b].

4) A remark on the neutrino: from the very beginning, de Broglie supposed that the Dirac particles that by fusion constitute photons and gravitons, were *neutrinos*. For a long time the neutrino was considered as a massless particle, with arguments based on gauge invariance, separation of chiral components, *etc.*. But the ideas have changed: new theoretical arguments based on the supposed oscillations between different kinds of neutrinos, the subsequent need of coupling constants and some experimental evidences, tend to a common belief in the neutrino mass.

If this idea is confirmed by facts, de Broglie's fusion theory will have as a consequence the prediction of a photon and a graviton mass, which will become in turn a credible conception.

1.7.2 Vacuum Dispersion

If $\mu_0 \neq 0$ the dispersion relations for plane waves can be exactly calculated. One obtains for the phase and group velocities v_{ph}, v_g:

$$v_{ph} = \frac{\omega}{k} = c \left(1 - \frac{\mu_0^2 c^4}{h^2 \nu^2}\right)^{-1/2} \geq c, \tag{1.55}$$

$$v_g = \frac{d\omega}{dk} = c \left(1 - \frac{\mu_0^2 c^4}{h^2 \nu^2}\right)^{1/2} \leq c. \tag{1.56}$$

Their product is (in contrast to the massless case $v_g = v_{ph} = c$):

$$v_g v_{ph} = c^2. \tag{1.57}$$

The vacuum is thus dispersive, which has not yet been observed.

It seems that if $\mu_0 < 10^{-45}$g one can explain that there is no evidence of such a fact, but it must be stressed that this conclusion of de Broglie is at least fifty years old. It might be revived on the basis of present knowledge. In any case, with this limit for mass, the Compton wavelength is $\lambda_{ph} > 10^8$cm $= 10^3$km, which means that the replacement of the Coulomb potential r^{-1} by the Yukawa potential $e^{-k_0 r} r^{-1}$ has no effect on electrostatics.

Another question is that one could, in principle, observe a photon with a velocity smaller than c in the vacuum. In de Broglie's time his estimates proved that this should be impossible if $\mu_0 < 10^{-45}$g [Brog 49], but with the progress of experimental physics such a possibility must be re-examined and perhaps may be considered rather as a question than as an objection.

1.7.3 Relativity

In practice, the velocity predicted for the photon is so close to c that the difference will have not any consequence. But the problem is: how shall we build the theory of relativity? De Broglie's answer is summarized by one of his favorite jokes: "Light is not obliged to go at the velocity of light". In other words, we need, in relativity, a maximum invariant velocity, but we do not need it to be the speed of light. It only happens that, in vacuum, the velocity of light is very close to this limit.

1.7.4 Black Body Radiation

In a unit volume there are $dn_\nu = 4\pi c^{-3}\nu^2 d\nu$ stationary waves. This number must be multiplied by 2 because of the transversality of light waves, which gives, in Planck's law, a factor 8. But if $\mu_0 \neq 0$ we must multiply by 3 because there is a longitudinal electric component. And it gives 12 in Planck's law, which is certainly false.

The answer is the following: If we apply the formula (1.41) for energy, it can be shown that the longitudinal part of the field (likewise the one corresponding to potentials) is of the order of k_0, i.e., negligible [Brog 49], so that it takes no part in the observed equilibrium and the factor 8 remains the correct one. De Broglie's argument was later independently confirmed by Schrödinger [Bass 55].

1.7.5 A Remark on Structural Stability

A physical theory has three criteria of truth: *experiment, logical consistency and structural stability.*

The first two points are evident, the third is less. It means that a theory needs a minimum of adaptability in order to resist slight experimental deviations without destroying its mathematical framework.

Actually, most physical theories are too rigid and have structural instabilities: for instance, Hamiltonian dynamics is structurally unstable because its formalism does not allow the slightest dissipation. But, at least, one must eliminate *algebraic conditions*, or *exact symmetries*, which can never be verified experimentally.

An example is the mass of the photon: It may be proved experimentally that the mass is *small*, but it cannot be proved that the mass is *exactly zero*, which would be an *algebraic condition*. In other words, electromagnetic gauge invariance, as a law of symmetry, may be proved approximately, not exactly.

It would be extremely worrisome if electromagnetism required exactly zero mass and gauge invariance[4]. But it is not so, as is proved by de Broglie's theory of the photon and arguments proving that, if μ_0 is small, the deviations with respect to the experimental facts are negligible.

A recent review of the theory and experiment concerning duality, magnetic monopoles and photon mass is contained in the book by Lehner [Leh 90]. By improved experimental techniques, at present an upper limit for the photon mass is $\mu_0 \leq 8.10^{-49}$g. But experimental precision cannot be boundlessly increased. Due to the fact that the photon rest mass is associated with the limit frequency $\nu = h^{-1}c^2\mu_0$, the observation of such a limit frequency would require a measuring time over one period, i.e., $\tau = \frac{1}{\nu}$. For such a measurement the upper limit of τ can at the utmost be the age of the

---

[4]An old example is a theory of Eddington, based on 16 degrees of freedom and the *exact* formula $1/\alpha = \frac{1}{2}16(16+1) + 1 = 137$ ($\alpha =$ the fine structure constant). Unfortunately $1/\alpha = 137,036\ldots$

universe. This leads to a lower bound of the photon mass which cannot be less than 4.10^{-66}g. Therefore ultimately the experiment seems to be of no help with respect to the decision wether the photon mass vanishes or not.

On the other hand in the mathematical treatment idealizations are preferred. So conventional electrodynamics and quantum electrodynamics (QED) are based on the massless photon. In particular the latter depends crucially on $\mu_0 \equiv 0$, as an appropriate renormalization is only possible for massless vector bosons and related exact gauge invariance. So the question is: should one consider QED as the final theory of electromagnetism, which, of course, includes $\mu_0 \equiv 0$?

There are good reasons to be doubtful about this. Apart from other problems, QED suffers from the infrared catastrophe. This catastrophe is cured by the introduction of a small photon mass which provides for a convenient regularization of the infrared divergence; see, for instance, Itzykson and Zuber [Itz 80]. But this assumption immediately causes the existence of a third state of polarization and a corresponding modification of the black body radiation spectrum, and this difficulty is bypassed by the argument that this effect is unobservable (see above).

From this it follows that the infrared and ultraviolet treatments of QED are not compatible, which is a strong argument against the idea that QED is the final theory of electromagnetism. For a more detailed discussion of the problems connected with QED, see Prugovecki [Pru 94]. In this paper the criticism of Dirac, Schwinger, Rohrlich, Pauli, Feynman and others are reviewed and discussed. In any case, in view of this criticism there are no convincing arguments against de Broglie's assumption of a very small photon mass μ_0.

1.8   General Theory of Particles with Maximum Spin n

The general theory is the subject of the second part of de Broglie's book on fusion [Brog 43]. Here we can give only a short survey, even shorter than for the case of spin 1.

1.8.1   Generalized Method of Fusion

Extending (1.7), the fusion of n Dirac equations gives a generalization of the equations (1.8):

$$\frac{1}{c}\frac{\partial \Phi_{ikl...}}{\partial t} = a_k^{(p)}\frac{\partial \Phi}{\partial x_k} + i\frac{\mu_0 c}{\hbar}a_4^{(p)}\Phi_{ikl...}, \qquad (1.58)$$

where $p = 1, 2, \ldots n$. Thus we have n matrix equations and a 4^n component wave function (a spinor of nth rank) instead of 16 components for the photon. And there are $4n$ matrices $\left(a_r^{(p)}\right)$, with 4^{2n} elements:

$$\left(a_r^{(p)}\right)_{ik...opq...,i'k'...o'p'q'...} = \delta_{ii'}\delta_{kk'}\ldots\delta_{oo'}\left(\alpha_r\right)_{pp'}\delta_{qq'}\ldots \qquad (1.59)$$

They obey the relations (1.10):

$$a_r^{(p)} a_s^{(p)} + a_s^{(p)} a_r^{(p)} = 2\delta_{rs}; \qquad a_r^{(p)} a_s^{(q)} - a_s^{(q)} a_r^{(p)} = 0 \quad \text{(if } p \neq q\text{).} \qquad (1.60)$$

The same problem as in equations (1.8) occurs: We have n times too many equations (for the photon, we had twice as many). We have, indeed, $n4^n$ equations for 4^n components of the wave function. The answer will be almost the same.

1.8.2 'Quasi-Maxwellian' Form

We shall proceed as in 3.1, but let us first posit:

$$F^{(p)} = a_k^{(p)} \frac{\partial}{\partial x_k} + i \frac{\mu_0 c}{\hbar} a_4^{(p)}. \qquad (1.61)$$

We have the relations:

$$F^{(p)} F^{(q)} = F^{(q)} F^{(p)}, \; \forall p, q; \qquad \left(F^{(p)} \right)^2 = \Delta - k_0^2, \qquad (1.62)$$

which means, by the way, that the wave components obey the Klein–Gordon equation. Now, (1.58) takes the form:

$$\frac{1}{c} \frac{\partial \Phi}{\partial t} = F^{(p)} \Phi; \; p = 1, 2, \dots, n. \qquad (1.63)$$

By adding these equations, we find a new evolution equation generalizing equation (A) in (1.11):

$$\text{(A)} \qquad \frac{1}{c} \frac{\partial \Phi}{\partial t} = F\Phi, \qquad (1.64)$$

where $F = \dfrac{1}{n} \sum\limits_{p=1}^{n} F^{(p)}$.

Now, subtracting equations (1.63) one from another in a suitable way, we can eliminate the time derivatives and find $n - 1$ 'constraint conditions'. It may be done in several ways. For instance, we can choose the following system, similar to equation (B) in (1.11):

$$\text{(B)} \qquad G^{(p)} \Phi = \frac{F^{(1)} - F^{(p)}}{2} \Phi = 0 \quad (p = 2, 3, \dots, n), \qquad (1.65)$$

and it is easy to prove that the new system (A), (B) is equivalent to (1.58) or (1.63).

Owing to (1.60) one can see that F and $G^{(p)}$ commute, but their product does not equal zero, contrary to what happened with the operators of the right hand sides of (1.11) in the particular case $n = 2$:

$$G^{(p)} F = F G^{(p)} \neq 0. \qquad (1.66)$$

This means that, contrary to (1.11), we cannot prove, using (1.64) and (1.65), that the conditions (B) are deducible from the evolution equation (A). However, as a consequence of (1.66), the right hand sides of (1.65) are solutions of (1.64), so that if the conditions (B) are satisfied at an initial time $t = 0$ they are satisfied for all time.

On the other hand, one can prove the compatibility of the $(n-1)$ equations (B), so that the compatibility of the system (1.58) — or, equivalently of (1.64), (1.65) — is proved.

1.8.3   The Density of Quadri-current

Generalizing (1.34), de Broglie introduces a new set of matrices:

$$B_4^{(p)} = a_4^{(1)} a_4^{(2)} \ldots a_4^{(p-1)} a_4^{(p+1)} \ldots a_4^{(n)}, \tag{1.67}$$

and the quadri-current density is:

$$J_k = -c\Phi^* \frac{1}{n} \sum_{p=1}^{n} a_k^{(p)} B_4^{(p)} \Phi, \tag{1.68}$$

$$\rho = \Phi^* \frac{1}{n} \sum_{p=1}^{n} B_4^{(p)} \Phi.$$

It is easy to verify that this current is conserved:

$$\frac{\partial \rho}{\partial t} + \partial_k J_k = 0. \tag{1.69}$$

Generalizing a remark made in Section 1.5.2, it is interesting to examine the *density* ρ. Following de Broglie, we shall do it in the case of a plane wave. Let us note, by the way, that it is not difficult to calculate a plane wave for a particle of maximum spin $n/2$: The phase is evident and the amplitudes are given by the n products of 4 amplitudes of n Dirac plane waves, which gives 2^n constants restricted by the fusion conditions. Nevertheless, the calculation is rather long ([Brog 43] Ch. IX), but the result is simple:

$$\rho = \left(\frac{\mu_0 c^2}{W} \right)^{n-1} |\Phi|^2, \tag{1.70}$$

where μ_0 = mass of the particle, W = energy, n = number of spin $\frac{1}{2}$ particles composing the considered particle. We see that:

— If n is odd, the sign of ρ is positive definite, as in the case $n = 1$ of a Dirac electron.

— If n is even, ρ has the same sign as energy, it is indefinite: this was the case for a photon (spin 1) and it will be the case for a graviton (spin 2).

It is interesting to note, with de Broglie, the curious presence in (1.70), of the $(n-1)$th power of the Lorentz contraction, which means that the density ρ, integrated over a volume ($\int \rho dv$), will be contracted exactly n times (the number of elementary spin $\frac{1}{2}$ particles). The exception is the Dirac particle, for which $n - 1 = 0$, so that the factor disappears and the integral is only contracted by the integration volume itself. De Broglie conjectured that this factor is perhaps an echo of a complicated spatial structure of the composite particle which we cannot describe except as a point in the present state of linear quantum mechanics.

1.8.4 The Energy Density

We shall begin with an elementary calculation of the *energy density*, using the preceding density ρ for a plane wave. The definition of the ρ density shows that all the mean values will be obtained by integration of a physical quantity multiplied by ρ.

The energy density is thus obtained (in the case of a plane wave) using the formula (1.70):

$$\rho E = \left(\frac{\mu_0 c^2}{E} \right)^{n-1} E \, |\Phi|^2 . \tag{1.71}$$

Here, the power of E is not $(n - 1)$ but $(n - 2)$, so that we find a result opposite to the result for ρ:

— If n is odd, ρE has the same indefinite sign as energy: this was the case for $n = 1$, for a Dirac electron.

— If n is even, the sign of ρE is positive definite as for the photon and as will be the case for the graviton.

This is confirmed by more sophisticated calculations using the energy tensor density.

We shall introduce two classes of tensors. The first was named 'corpuscular' by de Broglie: This is a tensor like (1.36), given by recipes of quantum mechanics. The second class, called 'type M' by de Broglie (M for Maxwell), is wider and looks like (1.40); it is inspired by electromagnetism.

1.8.5 The 'Corpuscular' Tensor

We shall make use of real space-time coordinates, as in Section 1.5.3, of B matrices defined in (1.67) and of the following notations:

$$U_i^{(p)} = a_i^{(p)} \quad (i = 1, 2, 3), \quad U_i^{(p)} = 1 \quad (p = 1, 2, \ldots, n). \tag{1.72}$$

The tensor is then [Brog 43]:

$$T_{\mu\nu} = T_{\nu\mu} = \frac{\hbar c}{4in} \sum_{p=1}^{n} \left[\Phi^* U_\mu^{(p)} B_4^{(p)} \frac{\partial \Phi}{\partial x_\nu} - \frac{\partial \Phi^*}{\partial x_\nu} U_\mu^{(p)} B_4^{(p)} \Phi \right. \tag{1.73}$$

$$\left. + \Phi^* U_\nu^{(p)} B_4^{(p)} \frac{\partial \Phi}{\partial x_\mu} - \frac{\partial \Phi^*}{\partial x_\mu} U_\nu^{(p)} B_4^{(p)} \Phi \right]$$

and we can verify its conservation by virtue of the wave equations:

$$\partial_\nu T_{\mu\nu} = \partial_\mu T_{\mu\nu} = 0. \tag{1.74}$$

It is interesting to verify that the tensor takes, for a plane wave, the form that is to be expected. It is so and we find indeed the following matrix for the components of T ($\mathbf{p}$ = momentum, $\mathbf{v}$ = group velocity):

$$\left\{ \begin{array}{cccc} \rho p_1 v_1 & \rho p_1 v_2 & \rho p_1 v_3 & \rho p_1 c \\ \rho p_2 v_1 & \rho p_2 v_2 & \rho p_2 v_3 & \rho p_2 c \\ \rho p_3 v_1 & \rho p_3 v_2 & \rho p_3 v_3 & \rho p_3 c \\ \rho p_1 c & \rho p_2 c & \rho p_3 c & \rho W \end{array} \right\} \tag{1.75}$$

In particular, we see that T_{44} is the quantity (1.71).

1.8.6 The Tensors 'of type M'

First, we shall generalize formula (1.67) by defining a set of operators of rank m:

$$B_4^{(pq\ldots)} = a_4^{(1)} a_4^{(2)} \ldots a_4^{(p-1)} a_4^{(p+1)} \ldots a_4^{(q-1)} a_4^{(q+1)} \ldots a_4^{(n)}. \tag{1.76}$$

It is the product of all the $a_4^{(r)}$ ($r = 1, 2, \ldots, n$), with the exception of those for which $4r$ is equal to one of the m indices p, q of B.

Now making use of these operators and of (1.72) again, we define a set of tensors of rank m [Brog 43]:

$$M_{ij\ldots} = \mu_0 c^2 \Phi^* \frac{\sum\limits_{p,q\ldots} U_i^{(p)} U_j^{(q)} \ldots B_4^{(pq\ldots)}}{a_n^m} \Phi, \quad \text{where} \quad a_n^m = \frac{n!}{(n-m)!}. \tag{1.77}$$

These tensors are obviously symmetric, but we keep only those for which the rank $m = 2r$ is even. Thus we have defined (for a particle of maximum spin n) $n/2$ tensors if n is even and $(n-1)/2$ tensors if n is odd.

Finally, we contract each of these tensors of rank $2r$, over $2r - 2$ of its indices, which gives the number defined just above (half of the greatest even number contained in n) of tensors of rank two, according to the formula:

$$M_{ij}^{(r)} = \sum_{k,l\ldots=1}^{4} M_{ijkl\ldots}^{kl\ldots}. \tag{1.78}$$

These tensors were defined by de Broglie as tensors 'of type M'. By virtue of the general equations (1.58), we have, just as for the tensor T:

$$\partial_\nu M_{\mu\nu}^{(r)} = \partial_\mu M_{\mu\nu}^{(r)} = 0. \tag{1.79}$$

A priori, each of these conserved tensors may be considered as an energy momentum tensor. It can be shown that, *for a plane wave*, every tensor $M_{\mu\nu}^{(r)}$, $\forall r$, gives exactly the table of components (1.75). This is not true for other solutions, but it remains true that

$$\int T_{\mu\nu} d\tau = \int M_{\mu\nu}^{(r)} d\tau, \ \forall r. \tag{1.80}$$

1.8.7 Spin

Starting from equation (1.64A) — a generalization of (1.11A) — we have the same orbital operator (1.49), and the spin operators are now:

$$S_i = \hbar \sum_{p=1}^{n} s_i^{(p)} \ (i = 1, 2, 3); \qquad S_4 = \hbar \sum_{p=1}^{n} s_4^{(p)} \qquad (1.81)$$

with

$$s_1^{(p)} = -\frac{i}{2} a_2^{(p)} a_3^{(p)}, \qquad (1.82)$$

$$s_2^{(p)} = -\frac{i}{2} a_3^{(p)} a_1^{(p)},$$

$$s_3^{(p)} = -\frac{i}{2} a_1^{(p)} a_2^{(p)},$$

$$s_4^{(p)} = -\frac{i}{2} a_1^{(p)} a_2^{(p)} a_3^{(p)}.$$

It would be difficult to reproduce the general nomenclature of spin states and (for an even number of spin $\frac{1}{2}$ particles) the decomposition of wave functions in terms of tensor components. This nomenclature is based on the Clebsch–Gordon theorem on the product of irreducible representations, but it is completed in de Broglie's book [Brog 43] by a study of the number of independent constants on which a plane wave depends and of the symmetry of two tensors defined in the case of an even number of particles.

These problems are treated in a different form, by Fierz in [Fie 39a], whose work is based not on the fusion theory, but on the conditions that must be added to a field obeying the Klein–Gordon equation, to describe a spin $n/2$ particle. This point of view was developed by Pauli and Fierz [Fie 39b,c] on the basis of a previous work of Dirac on the generalization of the equation of the electron for higher spin values [Dir 36].

1.9 The Particles with Maximum Spin 2. Graviton

Pauli and Fierz [Fie 39a] were the first to discover the analogy between the equation of a particle of spin 2 and the linear approximation of the Einstein equation of a gravitational field. This approximation was given by Einstein himself [Ein 16], [Ein 18] and it may be found, for instance, in [Laue 22], [Møll 72]. The paper [Ein 16] was the first in which Einstein formulated the idea of gravitational waves. He even took into consideration a possible modification of gravitation theory by quantum effects, in analogy with the modification of Maxwell's electromagnetism.

It should be emphasized that the quantum theory of gravitation developed by de Broglie and Tonnelat [Ton 42] on the basis of the fusion method is quite different at many points from that of Pauli–Fierz. It is not based on a particle of spin 2 but on a particle of *maximum* spin 2, which has a twofold importance:

1) The fusion theory raises the question: is the graviton a *composite particle*, just as the photon and all particles of spin higher than $\frac{1}{2}$?

2) In this theory gravitons do not appear alone. They are linked to photons. Thus we have a *unified theory* of gravitation and electromagnetism (at least in the linear approximation). The fields are not combined by some extended geometry, but by the fusion of spins.

1.9.1   Why Gravitation and Electromagnetism are linked

Formally, one could say that 'fields are linked by the Clebsch–Gordon theorem' because

$$D_{\frac{1}{2}} \otimes D_{\frac{1}{2}} \otimes D_{\frac{1}{2}} \otimes D_{\frac{1}{2}} = D_2 + 3D_1 + 2D_0 \qquad (1.83)$$

so that, in the fusion of four spin $\frac{1}{2}$ particles, we must find one particle of spin 2, three particles of spin 1 and two particles of spin 0: in particular, we have gravitons and photons.

Nevertheless, it is interesting to give an intuitive argument of de Broglie. He defines a particle of maximum spin 2 by the fusion of two particles of spin 1, described by quadri-potentials $A_\mu^{(1)} = \left\{ \mathbf{A}^{(1)}, V \right\}$, $A_\mu^{(2)} = \left\{ \mathbf{A}^{(2)}, V \right\}$ and invariants $I_2^{(1)}, I_2^{(2)}$ ($I_1^{(1)}, I_1^{(2)} = 0$ because $\mu_0 \neq 0$)[5]. The fusion gives:

$$A_\mu^{(1)} \times A_\mu^{(2)}; \quad A_\mu^{(1)} \times I_2^{(2)}; \quad I_2^{(1)} \times A_\mu^{(2)}; \quad I_2^{(1)} \times I_2^{(2)}. \qquad (1.84)$$

The first product is a tensor of rank 2 that defines a symmetric and an antisymmetric tensor:

$$A_{(\mu\nu)} = \frac{A_{\mu\nu} + A_{\nu\mu}}{2}; \qquad A_{[\mu\nu]} = \frac{A_{\mu\nu} - A_{\nu\mu}}{2}. \qquad (1.85)$$

The products $A_\mu^{(1)} \times I_2^{(2)}$ and $I_2^{(1)} \times A_\mu^{(2)}$ are vector-like quantities $P_\mu^{(1)}$, $P_\mu^{(2)}$, and it may be guessed that they will be photon potentials. The antisymmetric tensor $A_{[\mu\nu]}$ suggests the presence of an electromagnetic field.

The symmetric tensor $A_{(\mu\nu)}$ cannot be interpreted at this level of exposition, but actually we know in advance that it will be related to gravitation.

De Broglie showed, using a study of plane waves, that $P_\mu^{(1)}$, $P_\mu^{(2)}$ and the antisymmetric tensor $A_{[\mu\nu]}$ are related to spin 1; $A_{(\mu\nu)}$ is linked to spin 2, only if it is reduced to a tensor with vanishing trace because $\operatorname{tr} A_{(\mu\nu)} = A_{(\mu\mu)}$ is an invariant and it will actually be related to spin 0, just like the invariant $I_2^{(1)} \times I_2^{(2)}$.

---

[5] Here we consider only the electric case.

Now it must be remembered that, as was shown in the case of the photon, *the splitting between different spin states is not relativistically covariant*, because it is based on the total spin operator which is not a relativistic invariant. Therefore in the fusion theory gravitation cannot appear without electromagnetism. Furthermore, it will be shown that if $\mu_0 \neq 0$ the splitting between spin 2 and spin 0 is impossible, and the interpretation of this fact is highly interesting.

1.9.2   The Tensorial Equations

We do not write down the wave equations for a particle of maximum spin 2. They are of the type (1.58) with $n = 4$. The wave function has, in principle $4^4 = 256$ components, but, finally, there are only 168 independent quantities[6].

We give only the tensorial form, generalizing the procedure of Section 1.4.1. The equations are reduced to three relatively simple systems (A), (B), (C):

$$(A) \qquad \partial_\mu \Phi_{(\nu\rho)} - \partial_\nu \Phi_{(\mu\rho)} = k_0 \Phi_{[\mu\nu]\rho}, \qquad (1.86)$$

$$\partial_\rho \Phi_{[\rho\mu]\nu} = k_0 \Phi_{(\mu\nu)},$$

$$\partial_\mu \Phi_{[\rho\sigma]\nu} - \partial_\nu \Phi_{[\rho\sigma]\mu} = k_0 \Phi_{([\mu\nu][\rho\sigma])},$$

$$\partial_\varepsilon \Phi_{([\varepsilon\rho][\mu\nu])} = k_0 \Phi_{[\mu\nu]\rho}.$$

$\Phi_{(\mu\nu)}$ is a symmetric tensor of rank 2, $\Phi_{[\mu\nu]\rho}$ a tensor of rank 3 antisymmetric with respect to the two first indices, $\Phi_{([\mu\nu][\rho\sigma])}$ a tensor of rank 4 antisymmetric with respect to $\mu\nu$ and $\rho\sigma$, but symmetric with respect to these pairs. A consequence of (1.86) is:

$$\partial_\nu \Phi_{(\mu\nu)} = \partial_\rho \partial_\nu \Phi_{[\rho\mu]\nu} = 0, \qquad (1.87)$$

$$\Phi_{(\rho\rho)} = \frac{1}{2} \Phi_{([\mu\rho][\mu\rho])},$$

$$\partial_\nu \Phi_{(\rho\rho)} = k_0 \Phi_{[\nu\rho]\rho}.$$

The group (B) is divided in three subgroups where new tensors of rank 2, 3 and 4 appear:

$$(B1) \qquad \partial_\mu \Phi^{(1)}_{[\nu\rho]} - \partial_\nu \Phi^{(1)}_{[\mu\rho]} = k_0 \Phi^{(1)}_{[\mu\nu]\rho}, \qquad (1.88)$$

$$\partial_\rho \Phi^{(1)}_{[\rho\mu]\nu} = k_0 \Phi^{(1)}_{[\mu\nu]} \left(= \frac{1}{2} \left(\partial_\rho \Phi^{(1)}_{[\rho\mu]\nu} - \partial_\rho \Phi^{(1)}_{[\rho\nu]\mu} \right) \right),$$

$$\partial_\mu \Phi^{(1)}_{[\rho\sigma]\nu} - \partial_\nu \Phi^{(1)}_{[\rho\sigma]\mu} = k_0 \Phi^{(1)}_{([\mu\nu][\rho\sigma])},$$

$$\partial_\varepsilon \Phi^{(1)}_{([\mu\nu][\rho\sigma])} = k_0 \Phi^{(1)}_{[\mu\nu]\rho}.$$

Note the antisymmetries (square brackets). From (1.88) we deduce the identities:

$$\Phi^{(1)}_{[\nu\mu]\nu} = \Phi^{(1)}_{([\mu\nu][\rho\nu])} = 0. \qquad (1.89)$$

---

[6]In this section the essential bibliographic reference is [Brog 43].

The equations (B2) and (B3) are identical and we have:

$$(\text{B2}, \text{B3}) \qquad \partial_\mu \chi_\nu^{(1)} - \partial_\nu \chi_\mu^{(1)} = k_0 \chi_{[\mu\nu]}^{(1)}, \tag{1.90}$$

$$\partial_\rho \chi_{[\rho\nu]}^{(1)} = k_0 \chi_\nu^{(1)},$$

$$\partial_\mu \chi_\nu^{(1)} = k_0 \chi_{\rho\nu}^{(1)},$$

$$\partial_\rho \chi_{[\mu\nu]}^{(1)} = k_0 \chi_{[\mu\nu]\rho}^{(1)}.$$

In the third equation $\chi_{\rho\nu}^{(1)}$ is neither symmetric nor antisymmetric. (1.90) entails:

$$\chi_{\rho\rho}^{(1)} = 0, \tag{1.91}$$

$$\chi_{\mu\nu}^{(1)} - \chi_{\nu\mu}^{(1)} = \chi_{[\mu\nu]}^{(1)},$$

$$\chi_{[\mu\nu]\rho}^{(1)} = -\chi_\nu^{(1)},$$

$$\chi_{[\mu\nu]\rho}^{(1)} + \chi_{[\nu\rho]\mu}^{(1)} + \chi_{[\rho\mu]\nu}^{(1)} = 0.$$

Finally, we find a last group of equations:

$$(\text{C}) \qquad \partial_\mu \varphi_\nu^{(0)} = \partial_\nu \varphi_\mu^{(0)} = k_0 \varphi_{(\mu\nu)}^{(0)}, \tag{1.92}$$

$$\partial_\mu \varphi_\mu^{(0)} = k_0 \partial_\mu \varphi^{(0)},$$

$$\partial_\mu \varphi^{(0)} = k_0 \varphi_\mu^{(0)}.$$

The equations (B1), (B2), (B3) are three realizations of total spin 1. It is evident for (B2), (B3) because putting:

$$F_\mu = K \chi_\mu^{(1)}; \qquad F_{[\mu\nu]} = k_0 K \chi_{[\mu\nu]}^{(1)} \tag{1.93}$$

and defining potentials and fields as we did in (1.22), we find the Maxwell equations with mass (we shall see that this requires some comments).

The correspondence is a little less evident for (1.88). Instead of (1.93) we must write:

$$F_\mu = \frac{K}{6} \varepsilon_{\mu\lambda\nu\rho} \Phi_{[\lambda\nu]\rho}^{(1)}; \qquad F_{[\mu\nu]} = k_0 K \Phi_{[\mu\nu]}^{(1)} \tag{1.94}$$

($\varepsilon_{\mu\lambda\nu\rho}$ is the Levi–Civita symbol). Applying (1.22) again, we find the Maxwell equations (same remark as above).

Now, (C) is a realization of spin 0, as may be seen by comparison of (1.92) with (1.24) or (1.29). But here we find a difficulty (which justifies the preceding remarks):

De Broglie (who did not know the magnetic case), considered only the electric photon (1.23) and he identified (1.92) with the non-Maxwellian equations (1.24). But this implies the identity $\varphi^{(0)} = I_2$, where $\varphi^{(0)}$ is a *scalar* whilst I_2 is a *pseudo-scalar*.

At the time when de Broglie's book was published [Brog 43] (1942) people were less careful than now with parity and de Broglie wrote that (1.92) and (1.24) *"are entirely equivalent (at least when vectors and pseudo-vectors are assimilated)"*. Nowadays we pay more attention to parity and we cannot neglect such a discrepancy: an equality like $\varphi^{(0)} = I_2$ is unacceptable.

It seems that two solutions may be suggested:

1) The photon is an electric photon, but the spin 0 component must vanish.

We can admit that $\varphi^{(0)} = I_2$, if $\varphi^{(0)} = I_2 = 0$. Thus the spin 0 component (C) vanishes. But there is a second spin 0 component, hidden in the equations (A) in the form of an invariant $\Phi^{(0)}$, a vector $\Phi_\mu^{(0)}$, and a symmetric tensor $\Phi_{(\mu\nu)}^{(0)}$, which may be defined as:

$$\Phi^{(0)} = \Phi_{(\rho\rho)}^{(0)}, \tag{1.95}$$

$$\Phi_\mu^{(0)} = \Phi_{[\mu\rho]\rho},$$

$$\Phi_{(\mu\nu)}^{(0)} = \Phi_{([\mu\nu][\nu\rho])} - \Phi_{(\mu\nu)}.$$

One can show, using (1.86), that these tensors obey the group C of equations (1.92). But, once more, if $\Phi^{(0)}$ is a true scalar, we can write $\Phi^{(0)} = I_2$ only if $\Phi^{(0)} = I_2 = 0$.

This implies that (1.86) is subject to the condition $\mathrm{tr}\Phi_{\rho\rho}^{(0)} = 0$. This was assumed *a priori* by Fierz and Pauli who based their theory on a spin 2 (and not maximum spin 2) particle. De Broglie criticized this postulate as artificial. The above suggestion, based on parity, could be considered as a justification. However, it may be objected that, as was shown, the splitting of spin components is not covariant. This is the case for the condition $\varphi^{(0)} = I_2 = 0$, even if, on the other hand, the equality $\mathrm{tr}\Phi_{\rho\rho}^{(0)} = 0$ is covariant: the problem thus remains unsolved. But we have another proposal.

2) The photon is a magnetic photon.

We can ask the question: Is $\varphi^{(0)} = I_2$ a good equality? Perhaps it is rather $\varphi^{(0)} = I_1$, which is covariant because I_1 is a true invariant. In such a case (1.92) need not be identified with (1.24), but with (1.29). Is that possible? It seems that it is.

Let us go back to (1.84). The products $A_\mu^{(1)} \times I_2^{(2)}$ and $I_2^{(1)} \times A_\mu^{(2)}$, denoted $P_\mu^{(1)}$, $P_\mu^{(2)}$, were considered by de Broglie to be vectors, but we said, more prudently, 'vector-like': actually, they are *pseudo-vectors, because they are products of a polar vector by a pseudo-scalar.* Therefore $P_\mu^{(1)}$, $P_\mu^{(2)}$ are not polar potentials but pseudo-potentials of magnetic type as are those that appear in (1.28). On the contrary, the product $I_2^{(1)} \times I_2^{(2)}$ of two pseudo-scalars is a *true* scalar, of the same type as I_1, which appears in (1.29), and they may be identified.

The answer to the difficulty seems to be that the photon associated to the graviton is not of electric but of magnetic type.

Now assume that, instead of electric photons, we introduce magnetic photons in the symbolic formulae (1.84): pseudo-potentials $B_\mu^{(1)}$, $B_\mu^{(2)}$, and scalars $I_1^{(1)}$, $I_1^{(2)}$. The fusion gives:

$$B_\mu^{(1)} \times B_\mu^{(2)}; \quad B_\mu^{(1)} \times I_1^{(2)}; \quad I_1^{(1)} \times B_\mu^{(2)}; \quad I_1^{(1)} \times I_1^{(2)} \tag{1.96}$$

and we see that:

— the spin 2 product $B_\mu^{(1)} \times B_\mu^{(2)}$ has the same symmetry as $A_\mu^{(1)} \times A_\mu^{(2)}$, because the axial character of $B_\mu^{(1)}$, $B_\mu^{(2)}$ is annihilated by the product;

— for the same reason the spin 0 product $I_1^{(1)} \times I_1^{(2)}$ is a scalar, as was $I_2^{(1)} \times I_2^{(2)}$;

— the spin 1 products $B_\mu^{(1)} \times I_1^{(2)}$; $I_1^{(1)} \times B_\mu^{(2)}$ are pseudo-vectors, as $A_\mu^{(1)} \times I_2^{(2)}$; $I_2^{(1)} \times A_\mu^{(2)}$: they are products of a pseudo-vector by a scalar, whilst the latter were products of a polar vector by a pseudo-scalar.

Thus we find a magnetic photon, whether we start from electric or from magnetic photons: *the photon associated to the graviton is not electric but magnetic, which is a new orientation for a unitary field.*

1.9.3 Gravitation

We shall now follow de Broglie and Tonnelat and consider the general equations (A), when $\mathrm{tr}\Phi_{\rho\rho}^{(0)} \neq 0$. But we shall not be able to separate the spin 2 component from its spin 0 part!

We start from (1.86), (1.87), and the Klein–Gordon equation which is satisfied by all the field quantities:

$$\Box\Phi = -k_0^2\Phi \quad (\Box = -\partial_\rho\partial_\rho). \tag{1.97}$$

The metric tensor $g_{(\mu\nu)}$ will be taken in the linear approximation:

$$g_{(\mu\nu)} = \delta_{\mu\nu} + h_{(\mu\nu)} \quad \left(\left|h_{(\mu\nu)}\right| \ll 1\right). \tag{1.98}$$

At this limit the propagation of gravitational waves is given (in 'isothermal' or 'harmonic' coordinates [Brog 43], [Laue 22], [Ton 42]) by:

$$\Box g_{(\mu\nu)} = -2R_{(\mu\nu)} \quad \left(R_{(\mu\nu)} = g^{\rho\sigma} R_{([\mu\rho][\nu\sigma])}\right), \tag{1.99}$$

where $R_{([\mu\rho][\nu\sigma])}$ is the Riemann–Christoffel tensor; for vanishing curvature we have the d'Alembert equation $\Box g_{(\mu\nu)} = 0$ without the inhomogeneous term.

Now, it might seem that metrics may be defined by:

$$g_{(\mu\nu)} = \Phi_{(\mu\nu)}, \tag{1.100}$$

but Tonnelat remarked that according to (1.87) this implies: $\partial_\mu g_{(\mu\nu)} = 0$, which is false because 'isothermal' coordinates obey the relation:

$$\partial_\mu g_{(\mu\nu)} = \frac{1}{2}\partial_\nu g_{(\rho\rho)} \quad \left(g_{(\rho\rho)} = g_{(\mu\nu)}\gamma^{(\mu\nu)}\right), \tag{1.101}$$

and the right hand side is not equal to zero. (1.101) thus contradicts (1.100). This is why Tonnelat suggested the following metrics (which is possible because $k_0 \neq 0$!):

$$g_{(\mu\nu)} = \Phi_{([\mu\rho][\nu\rho])} = \Phi_{(\mu\nu)} + \frac{1}{k_0^2}\partial_\mu\partial_\nu\Phi_{(\rho\rho)}. \tag{1.102}$$

It follows immediately:

$$\partial_\mu g_{(\mu\nu)} = \partial_\mu\Phi_{([\mu\rho][\nu\rho])} = \partial_\nu\Phi_{(\rho\rho)}, \tag{1.103}$$

and we obtain from (1.87), (1.102) and (1.103):

$$g_{(\rho\rho)} = 2\Phi_{(\rho\rho)} \rightarrow \partial_\mu g_{(\mu\nu)} = \frac{1}{2}\partial_\nu g_{(\rho\rho)} \tag{1.104}$$

in accordance with (1.101).

Now, from (1.102) we deduce that $g_{(\mu\nu)}$ obeys the Klein–Gordon equation:

$$\Box g_{(\mu\nu)} = -k_0^2 g_{(\mu\nu)}. \tag{1.105}$$

Identifying (1.105) with (1.99) we obtain:

$$R_{(\mu\nu)} = \frac{k_0^2}{2} g_{(\mu\nu)}. \tag{1.106}$$

Furthermore, the Riemann–Christoffel tensor may be deduced in the linear approximation, from (1.102), (1.86) and (1.87):

$$\Phi_{([\mu\rho][\nu\rho])} \cong \frac{2}{k_0^2} R_{([\mu\rho][\nu\rho])}. \tag{1.107}$$

This formula is possible only if $\mu_0 \neq 0$, which imposes a curvature of the universe. Indeed, $k_0^2/2$ is nothing but the cosmological constant, defined by:

$$R_{(\mu\nu)} = \lambda g_{(\mu\nu)}. \tag{1.108}$$

λ is related to a 'natural curvature' of space–time. In Euclidian space we have $\lambda = 0$; in a de Sitter space of radius R we have $\lambda = 3/R^2$. Therefore:

$$\lambda = \frac{k_0^2}{2} = \frac{\mu_0^2 c^2}{2\hbar^2}, \tag{1.109}$$

and the mass of the graviton is related to a natural curvature of radius R:

$$\mu_0 = \frac{\hbar\sqrt{6}}{Rc}. \tag{1.110}$$

If $R \cong 10^{26}$cm, the graviton mass is:

$$\mu_0 = 10^{-66}\text{g}, \tag{1.111}$$

in accordance with the approximation given in 1.7.5. (Do not forget that the photon's mass and the graviton's mass must be of the same order).

Now let us go back to the definitions (1.95) that gave the spin 0 part of (1.86) (A). In order to separate a 'pure' spin 2 component, we could write:

$$\Phi_{(\mu\nu)} = \Phi_{(\mu\nu)}^{(2)} + \Phi'_{(\mu\nu)}, \tag{1.112}$$

$$\Phi_{(\mu\nu)}^{(2)} = \Phi_{(\mu\nu)} - \delta_{\mu\nu}\Phi_0,$$

$$\Phi'_{(\mu\nu)} = \delta_{\mu\nu}\Phi_0.$$

Thus we have:

$$\Phi^{(2)}_{(\mu\mu)} = 0, \tag{1.113}$$

$$\Phi'_{(\mu\mu)} = \Phi_0.$$

It looks quite well, as a separation between spin 2 and spin 0, and it is easy to find the same decomposition for all the other tensors [Brog 43]. But, unfortunately, *in the general case*, the spin 2 components of type $\Phi^{(2)}$ do not obey equations (1.86): there are additional terms of type $\Phi^{(0)}$, corresponding to zero spin.

The spin 0 may be eliminated from the equations of spin 2 only in two cases:

— either by the *a priori* supposition that $\Phi^{(0)} = 0$ (Fierz equations);

— or *in the limit case* $\mu_0 = 0$, when the radius of the universe is infinite: the Euclidian case[7].

In conclusion, the quantum theory of gravitation based on de Broglie's fusion theory raises the important question of a composite nature of photons and gravitons, and, above all, this theory furnishes the beginning of a unified quantum field theory of electromagnetism and gravitation. Only the beginning, because it is linear.

Two remarks may be made about all this:

— It could be asked if the obstinate efforts of Einstein and other great physicists and mathematicians towards a unified field theory were justified, given that we know hundreds of elementary particles and it seems that there is no reason to pay particular attention to two of them: the photon and the graviton. De Broglie's theory gives a reason: these particles are those which appear, linked by spin properties, in the fusion procedure. The strength of this argument is that it is absolutely independent of and beyond (at least so it seems!) the geometrical path followed by Einstein.

— The second remark concerns symmetry: the fact that the photon associated with the graviton could be magnetic instead of electric, as was suggested above, signifies the introduction of duality, chirality, magnetic monopoles instead of electric charges, and so on, into the theory. It is certainly of interest that the photon is perhaps not the one that was expected.

1.10   Comparison with Other Theories

First of all we want to emphasize the priority of Louis de Broglie in the quantum theory of the photon considered as a *composite particle*. His first paper appeared in 1934 (*The Wave Equation of the Photon* [Brog 34d]) and the idea of a *fusion* of Dirac particles is the starting point of the theory of particles of higher spin.

A second point is that, unlike other scientists, de Broglie's initial aim was not a generalization of Dirac's equation but a theory of light. This is why he did not introduce any electromagnetic interaction.

---

[7]These problems are carefully examined in de Broglie's book [Brog 43].

For reasons given above, he was the only one to assume a massive photon, contrary to other authors who considered a massless photon as obvious. As for him, he never tried to extend his theory to massless particles and even hardly ever alluded to this possibility.

1.10.1 The 'Proca Equation'

The equations (1.23) and the very idea of a massive photon are often ascribed to Proca. Actually, this is the result of a misunderstanding, if not a 'misreading'.

1) The 'Proca equations' [Pro 36] appeared in 1936, two years after de Broglie's equations [Brog 34d]. Moreover, the paper of Proca was entitled: *On the ondulatory theory of positive and negative electrons*. It was not a theory of photons but of electrons(!): an attempt to avoid the negative energies, as was done frequently at that time[8].

2) Rejecting spinorial wave functions, Proca suggested a *vectorial equation* derived from the Lagrangian:

$$L = \frac{\hbar^2 c^2}{2} G_{rs}^* G_{rs} + m_0^2 c^4 \psi_r^* \psi_r, \qquad (1.114)$$

$$G_{rs} = (\partial_r - iA_r)\psi_s - (\partial_s - iA_s)\psi_r \quad (r, s = 1, 2, 3, 4),$$

where the complex vectorial function ψ_r of the *electron* takes the place of de Broglie's *photon* potential $(\mathbf{A}, V)$; A_r is the real potential of an *external* electromagnetic field acting on the electron ψ_r.

From (1.114) Proca derived the equations:

$$(\partial_r - iA_r)G_{rs} = k^2 \psi_s, \qquad (1.115)$$

$$(\partial_r + iA_r)G_{rs}^* = k^2 \psi_s^*, \quad \left(k = \frac{m_0}{\hbar c}\right),$$

and he remarked that "they have the form of Maxwell's equations [...] completed by an external potential (A_r)". But in no way did he consider (1.115) to be the equations of a massive photon.

He then gave a spin operator, but without calculating its eigenvalues and thus ignoring that *his electron had a spin 1!* Very astonishing because de Broglie worked on one floor above Proca and had found, two years before, this value 1 for an equivalent equation describing a photon [Brog 34e].

---

[8]Heisenberg and de Broglie were amongst the few who immediately adopted Dirac's equation, whatever difficulties there might be with negative energies.

1.10.2   The Bargmann–Wigner Equations

The Bargmann–Wigner equations for higher values of spin published in 1948 [Barg 48], are exactly de Broglie's equations published in 1943 [Brog 43], but without the idea of fusion and restricted by an *a priori* condition of symmetry on the indices of the wave function $\Phi_{ikl...}$. Thus there are only half of de Broglie solutions.

As may be verified in Lurié's *Particles and Fields* [Luri 68], the equations 1(97) p. 27 are exactly the equations (1.58), (1.59), taken from de Broglie's *Théorie générale des particules à spin*, [Brog 43] p. 138.

When they applied the general theory to the spin 1 case, Bargmann and Wigner found the equations 1(108a), 1(108b) [Luri 68] which are, of course, identical to equation (1.16) quoted from de Broglie's book [Brog 43] p. 106, with a difference:

By virtue of the condition of symmetry, Bargmann and Wigner did not expand $\Phi_{ikl...}$ for the 16 Clifford matrices, as we did in (1.21), but only for 10 of them: $\gamma_\mu C$, $\gamma_{[\mu\nu]} C$. The other 6 were rejected. Thus, only the Maxwell type of equations (1.23) were obtained (see 1(122) [Luri 68] p. 34), but not the non-Maxwellian (1.24), corresponding to spin 0.

One could be satified that spin 0 components vanish if one feels uncomfortable because we do not know what may be done with them. But we cannot forget that:

1) In a formal equation such as Bargmann–Wigner's, any new axiom may be added, provided it is not contradictory. But with a model, like de Broglie's fusion, we are obliged to accept all its consequences, even those that we dislike, and we must try to understand them. Especially when we know that the splitting of spin components is not covariant.

2) There is a path to an explanation: the magnetic photon[9]. We know that it leaves a mark on the electric spin 0 (the 'anti-fields') and, *vice versa*, the electric photon prints its mark on the magnetic spin 0. It would be a dubious decision to prefer a formal symmetry at the expense of the symmetry between electricity and magnetism.

In past few decades de Broglie's fusion equations (often falsely called Bargmann–Wigner equations) and similar equations have been studied by many authors. It would exceed the scope of this book to repeat the numerous references. For a review see the introduction of the paper by Pfister, Rosa, and Stumpf [Pfi 89]. With respect to generalizations the papers of Bopp and von Weizsäcker should be mentioned. In the sense of de Broglie, Bopp [Bopp 58] investigated the fusion of neutrinos by assuming δ forces between them, and von Weizsäcker [Weiz 85] discussed the fusion of 'uralternatives' which can be considered as an abstraction of de Broglie's neutrino fusion. Von Weizsäcker succeeded in deriving Dirac and Maxwell equations from his principle.

---

[9]Bargmann–Wigner only find the electric photon: their C matrix is the Γ matrix of de Broglie (1.18).

2 The Spinor Field Model

2.1 Fusion Concepts in Relativistic Quantum Field Theory

The wave mechanical (or quantum mechanical) formulation and evaluation of de Broglie's fusion theory produces impressive results, as was demonstrated in the preceding sections. In particular de Broglie was the first to derive an effective theory, namely electrodynamics, by means of his fusion principle.

If, however, de Broglie's theory is considered from the point of view of present elementary particle physics, it is obvious that a generalization of de Broglie's fusion method is necessary: In modern language de Broglie's fusion theory and corresponding theories of his theoretical successors are concerned with the construction of single local one-particle states for arbitrary spin. But in order to obtain a theory which can be compared with microphysical experiments the inclusion of interactions between these single particle states is needed.

If this idea is consequently pursued it necessarily leads to a quantum field-theoretic formulation of de Broglie's fusion principle. Indeed the fusion idea was applied in relativistic quantum field theory under various points of view and led to an enormous production of papers in this field. In some decades one might even speak of a fusion mania; for a review see Lyons [Lyo 83].

But in the course of this development its origin was forgotten: the original fusion principle of de Broglie was completely ignored and, much worse, the mathematical difficulties of conventional quantum field theories, as for instance those of quantum electrodynamics, were not solved, but simply shifted to the level of subquarks, preons, *etc.*. In addition, new, even more hazardous, methods were introduced.

In this situation it seems not to be appropriate to present a review of all these different approaches. Rather we will concentrate on those lines of development which *are in agreement with de Broglie's original fusion idea* and which at least can be considered as generalizations of de Broglie's idea, even if in the course of investigations the corresponding mathematical tools were incomplete, inadequate, or wrong.

We remember that de Broglie's fusion principle is exclusively a spin $\frac{1}{2}$ fusion theory which makes use of linear Dirac equations. If de Broglie's theory is consequently extended to describe interactions, any basic interaction can therefore take place only between spin $\frac{1}{2}$ particles (or fields).

As the example of Fermi's β-decay model shows, this leads to nonlinear spinor equations with Dirac operators in their kinetic parts. These equations are in general nonrenormalizable in perturbation theory. In former times this property was considered to be an absolute and strict argument against the use of such equations.

However, in the meantime the attitude towards such equations has changed. At present the strict refusal of nonrenormalizable theories has been given up in favour of the impression that nonrenormalizable interactions and their consequences are unavoidable, see for instance Weinberg [Wein 96]. Hence spinor field models are in accordance with the modern theoretical trends. So the only question is how to treat them successfully. In the following we will shortly review the attempts in this direction and point out their drawbacks.

We begin with a review of Heisenberg's nonlinear spinor theory and discuss it a little bit more in detail, because this discussion sheds some light on the situation in theoretical physics at the beginning and in the middle of the century and on problems which also at present are not satisfactorily solved.

The story starts with the discovery that in classical electrodynamics the electromagnetic self-energy of point charges diverges. This property led to attempts to construct unified models of fields and matter where this drawback is avoided. The earliest attempt was the theory of Mie [Mie 12a,b,13]. It was followed by the nonlinear electrodynamics of Born [Born 34a] and Born and Infeld [Born 34b] and by the higher order derivative electrodynamics of Bopp and of Podolski, see [Bopp 40], [Pod 41]. At the same time Einstein tried to find a unified theory of electromagnetism, gravity, and matter in terms of differential geometric quantities.

All these attempts were made at the classical level and had to compete with the quantum revolution. Dirac, Heisenberg, and Pauli constructed the scheme of quantum electrodynamics, and it was soon discovered that in the quantum version the divergence of classical electrodynamics was softened, but not removed, and that, on the other hand, new divergencies occurred.

It was that situation which Heisenberg hoped to cure by the proposal of a unified spinor field quantum theory.

i) Heisenberg's spinor theory

Heisenberg was guided to assume a nonlinear spinor field theory by a critical analysis of the concept of 'elementary' particles. The (relatively) great number of such 'elementary' particles as well as the relativistic equivalence of mass and energy and their mutual transmutation in various interactions suggest considering them as secondary quantities derived from an unobservable (?) matter field and its corresponding dynamical law.

In the terminology of de Broglie: in Heisenberg's theory any 'elementary' particle must be created by fusion of the basic matter fields. This condition and the existence of relativistic conservation laws, which were experimentally well confirmed in high energy processes, led to the assumption that the basic matter field had to be a relativistic spin $\frac{1}{2}$ field.

Its field equation was proposed to be of the form, see [Heis 54], [Dürr 59]:

$$\gamma^\mu \partial_\mu \psi \pm l^2 \gamma_\mu \gamma^5 \psi \left[\bar{\psi} \gamma^\mu \gamma^5 \psi \right] = 0 \tag{2.1}$$

with a four-component spinor field $\psi_\alpha(x)$, $\alpha = 1 \ldots 4$ and with the coupling constant l, which was considered as a fundamental length.

This equation is invariant under two chiral transformations: the Touschek transformation and the Pauli–Gürsey transformation. The chiral invariance allows us to formulate (2.1) in terms of Weyl spinors, see [Dürr 61]. In this form equation (2.1) can be written equivalently as

$$-i\sigma^\mu_{\alpha\beta}\partial_\mu\chi_{A\beta}(x) \pm l^2(\sigma_\mu)_{\alpha\beta}\chi_{A\beta}(x)\left[\chi^+_{B\gamma}(x)\sigma^\mu_{\gamma\delta}\chi_{B\delta}(x)\right] = 0 \qquad (2.2)$$

with the four-component Weyl spinor–isospinor $\chi_{A\alpha}(x)$, $A, \alpha = 1, 2$, and the Pauli matrices σ^μ. This version shows that equation (2.1) contains an elementary algebraic isospin degree of freedom.

In (2.1) and (2.2) the interaction term is strictly local in order not to violate relativistic microcausality. Canonically quantized, such local equations and their perturbation series are nonrenormalizable. To avoid divergences Heisenberg decided to give up canonical quantization. After some preliminary studies he introduced regularizing dipole ghosts, which in consequence led to the indefinite metric of the theory.

After the destruction of the canonical structure of the theory the only means of obtaining concrete numerical information was the New Tamm–Dancoff Method, *i.e.*, an approximate evaluation of the many-time Schwinger–Dyson–Freese equations of the theory. In the lowest orders of Tamm–Dancoff approximations some successful calculations of states, coupling constants and scattering processes in the pre-standard model physics were performed, see the book by Heisenberg [Heis 66], but the theory has serious drawbacks which did not admit a further continuation along these lines.

The striking drawbacks of this aproach are:

α) The dipole ghost regularization destroys the canonical structure of the theory. As a consequence no explicit Hamiltonian operator and no Lagrangian function can be given. No well defined equal time anticommutator exists. This fact prevents an algebraic treatment and an evaluation of the theory beyond the Tamm–Dancoff calculations.

β) The dipole ghosts are no mass eigenstates of the free Dirac operator. They appear in every scattering process and no possibility exists of applying a decoupling theorem in order to avoid their permanent presence. Although Heisenberg discussed an indefinite metric for various model equations he did not succeed in giving a final answer with respect to this problem of the spinor field.

γ) The elementary constituents of the theory are Weyl spinors which belong to representations of massless states. Thus the formation of bound states is prevented because bound states arise from mass defects of the constituents by the binding force. This mechanism cannot be applied in Heisenberg's theory. In addition the Weyl constituents can escape the bound states at any energy, thus hypothetical bound states cannot be stable.

δ) Apart from some lowest order Tamm–Dancoff calculations, no mathematical method was developed for mastering the problems of a systematic derivation of effective dynamics for composite particles. A result comparable to de Broglie's derivation of electrodynamics by means of fusion theory has not been attained in the Heisenberg theory.

ii) Fusion by operator products

The objects of de Broglie's fusion theory are quantum mechanical wave functions. In quantum field theory transition matrix elements correspond to the quantum mechanical wave functions, and these transition matrix elements are the objects of the new Tamm–Dancoff calculations. Although the new Tamm–Dancoff Method is not sufficient for a successful treatment of spinor field models, one can at least in lowest approximations calculate some of the transition matrix elements.

By such calculations Heisenberg and coworkers derived for instance the 'wave function' of the photon [Heis 66] as follows:

$$\varphi^\mu(x|y) = \text{const.}\, e^{-iyJ} \int d^4p\, e^{-ip(x-y)}(u+T_3)\frac{\text{tr}[\sigma^\mu p_\rho \bar{\sigma}^\rho \sigma_\nu B^\nu(J-p)_\tau \bar{\sigma}^\tau]}{(p^2)^2(p^2-\kappa^2)(J-p)^2}. \tag{2.3}$$

Irrespective of the question whether such states are stable or immediately decay into ghost states, it is remarkable that de Broglie, as well as Heisenberg, describes composite particles by local or nonlocal wave functions (i.e., by states), and this view is shared by other authors who calculated composite photon functions by the Bethe–Salpeter equations etc..

Perhaps the most cited example is the pion wave function of Nambu and Jona–Lasinio [Nam 61], which is nonlocal, too. In addition, Huang and Weldon [Hua 75] defined a relativistic S matrix for such extended structures.

On the other hand, in the first decades of quantum field theory it was assumed that composite particles can be described by local or nonlocal field operator products, which is a completely different description compared to that by means of wave functions. The first attempts in this direction were made by the interpretation of de Broglie's bi-spinors as products of neutrino field operators. Later on, by inclusion of interactions equivalences between nonlinear spinor fields and coupling theories were considered.

Let us assume that the Lagrangian density of a self-coupled spinor field is of the type

$$\mathcal{L}[\psi] = \bar{\psi}(i\gamma^\mu\partial_\mu - m)\psi - \frac{g}{4}(\bar{\psi}\gamma^5\psi)^2, \tag{2.4}$$

and that the self-interaction is strong enough to produce a stable pseudo-scalar bound state from a fermion–antifermion pair. Can the four-fermion interaction then be replaced by an equivalent Yukawa coupling leading to a Lagrangian density

$$\mathcal{L}'[\psi, \Phi] = \bar{\psi}(i\gamma^\mu\partial_\mu - m)\psi - \frac{1}{2}(\partial^\mu\Phi\partial_\mu\Phi + \mu^2\Phi) + iG(\bar{\psi}\gamma^5\psi)\Phi, \tag{2.5}$$

in which the bound state is represented by an elementary field Φ?

In various forms this problem has been considered by numerous authors. It is strongly related to the replacement of spinor field operators by bound state operators, and in a preliminary way this procedure can be denoted as a strong mapping of the spinor field theory onto the Yukawa theory. In particular, Dürr and Saller, as members of the Heisenberg group, deviated from the original strategy of Heisenberg to work with states and used strong mappings for the investigation of effective theories, symmetry breaking, *etc.*. For a review of such attempts see the introductions of [Stu 85], [Stu 86].

But from the early beginning of neutrino operator fusion, strong mappings were seriously criticized by various authors, and it was also acknowledged by Dürr and Saller that 'the inflation of dynamical degrees of freedom referring to a dynamical deduction of the phenomenologically established local fields and their particular properties as effective local compounds of the fundamental fields, of course, constitutes an extremely complicated problem' [Dürr 80].

Without treating the critical arguments of various authors in more detail we only mention some obvious drawbacks of strong mappings:

α) The matrix elements of operator products (composite operators), as for instance $\bar{\psi}\psi$, describe different physical situations in different algebraic operator representations, and one cannot understand how the replacement of such operator products by one boson operator $\Phi \approx \bar{\psi}\psi$ is now referred only to one special class, namely to bound state matrix elements.

β) If bosons or (and) fermions are genuine bound states of elementary fermion fields then only the latter fields can be assumed to obey strict canonical commutation relations and to exhibit strict relativistic microcausality. Any effective field must violate these conditions because under suitable conditions the behaviour of bound states will differ from that of elementary pointlike particles. The variety of such phenomena cannot be appropriately described by local operator products.

γ) Maintaining the assumption of elementary four-fermion interactions no way is known of treating more complex situations by strong mapping, as for instance three-fermion bound states, *etc.*, dressed fermion bound states, or even several kinds of bound states, simultaneously.

δ) Calculations with simplified models show that no true equivalences between the fundamental theory and the effective theory can be obtained by strong mappings. Rather, even in very simple models the situation is more complicated and needs a careful examination of the spectrum, *etc.*.

ϵ) The inverse problem, *i.e.*, a strong mapping of coupling theories onto spinor theories, leads to nonlocal four-fermion interactions.

iii) Fusion and path integrals

The discussion of strong mappings in ii) leads up to path integrals. In the previous decades path integrals have been increasingly used to treat fusion problems. This approach originated from the papers of Zimmermann [Zim 66a,b,67] and was further developed by Coleman, Jackiw and Politzer [Cole 74] and by Gross and Neveu [Gro 74], and at present the problem of deriving effective interactions of composite particles from a given dynamics of subcomponents is practically exclusively treated by means of path integrals.

The derivation of effective actions from path integrals rests on a simple trick which we demonstrate for a global $O(N)$-invariant scalar field theory. Its Lagrangian density is given by

$$\mathcal{L}[\Phi] = \frac{1}{2}\partial_\mu\Phi^a\partial^\mu\Phi^a - \frac{1}{2}m^2\Phi^a\Phi^a - \frac{1}{8}\lambda(\Phi^a\Phi^a)^2, \tag{2.6}$$

and the composite field is defined by $\sigma = (g/2N)\Phi^a\Phi^a$.

If the auxiliary field Lagrangian density

$$\mathcal{L}'[\Phi,\sigma] = \mathcal{L}[\Phi] + \frac{N}{2g}\left(\sigma - \frac{g}{2N}\Phi^a\Phi^a\right)^2 \tag{2.7}$$

is introduced, the corresponding generating functionals (*i.e.*, the path integrals)

$$W[J] = \int D[\Phi^a] \exp\left\{i\int d^4x\,[\mathcal{L}[\Phi] + J^a\Phi^a]\right\} \tag{2.8}$$

and

$$W'[J,h] = \int D[\Phi^a]D[\sigma] \exp\left\{i\int d^4x\,[\mathcal{L}'[\Phi,\sigma] + J^a\Phi^a + h\sigma]\right\} \tag{2.9}$$

are equivalent for $h \equiv 0$, *i.e.*, $W'[J,0] = W[J]$. This can be verified by the translation $\sigma \to \sigma + (g/2N)\Phi^a\Phi^a$, provided the measure D is invariant against translations.

Under the integral and for $g = \lambda N$ the Lagrangian density (2.7) can be rearranged in the form

$$\mathcal{L}'[\Phi,\sigma] = -\frac{1}{2}\Phi^a(\partial_\mu\partial^\mu + m^2 + \sigma)\Phi^a + \frac{g}{2N}\sigma^2, \tag{2.10}$$

and in this way (2.9) becomes a Gaussian integral in Φ^a. This Gaussian integral can be directly integrated and yields

$$W'[J,0] = \int D[\sigma] \exp[iNI(\sigma)] \tag{2.11}$$

with

$$I(\sigma) := \frac{1}{2}\left[i\operatorname{tr}\ln(\partial_\mu\partial^\mu + m^2 - \sigma) + \frac{1}{g}\int d^4x\,\sigma^2(x)\right. \tag{2.12}$$

$$\left. -\frac{1}{N}\int d^4x\,d^4y\,J^a(x)(\partial_\mu\partial^\mu + m^2 - \sigma)^{-1}J^a(y)\right],$$

where the formula (2.12) is derived by the use of

$$\det(\partial_\mu\partial^\mu + m^2 - \sigma) = \exp\left[\operatorname{tr}\ln(\partial_\mu\partial^\mu + m^2 - \sigma)\right]. \tag{2.13}$$

The expression (2.12) is the effective action for the composite field σ and the corresponding trick is applied in numerous versions to other systems in order to obtain effective actions for corresponding composite fields.

Although this mathematical technique dominates the discussion of effective actions for composite particles or fields at present, in particular in application to spinor field models of quark interactions, severe objections to it have been raised.

α) From the definition of the path integral itself several mathematical problems arise. These are mainly concerned with the analytic continuation from Minkowski space into Euclidean space, the definition of the measure, the nonexistence of classical paths for field theories, *etc.*. Rivers has devoted a chapter in his book [Riv 88] to the discussion of these difficulties, see also Cheng [Che 90] and the critical remarks of Roepstorff [Roep 91].

β) The determinantal formula (2.13) does not exist for unbounded operators and the evaluation of $\mathrm{tr}\ln(\dots)$ in (2.12), which is usually carried out by series expansion of the logarithm, exceeds its radius of convergence, compare Prop. 9.3.1 of Glimm and Jaffe [Gli 87].

Closely related to the evaluation of the determinantal expression is the generation of the kinetic energies of the composite fields which are derived from these expressions by means of doubtful methods.

γ) No method is known to derive effective actions for hard core particles or (with respect to strong mappings) for composite fields with more than two elementary fields. There also exists no concept for the treatment of dressed particles.

Whilst dressed particles are completely unattainable, one can try to decompose hard core states (field operator products) of more than two constituents into parts. However, this decomposition violates the inherent antisymmetry of the composite states or fields, respectively.

δ) No way is known of deriving interactions which depend on the quantum effects of exchange forces between composite particles. Furthermore, the inclusion of inequivalent representations of the field operator algebra, which is an important feature of relativistic quantum field theory, cannot be treated along the lines which at present govern the evaluation of path integrals.

iv) Fusion and chiral symmetry breaking

In i) we outlined the reformulation of the original Heisenberg equation (2.1) as the Weyl spinor–isospinor equation (2.2) of Dürr. This reformulation was fully acknowledged by Heisenberg, see [Hei 66]. Although such an equation seems to be very fundamental, it is the source of serious difficulties; in particular, it prevents the formation of bound states as we already remarked.

In 1961 Nambu and Jona–Lasinio made an attempt to remove these difficulties and proposed a nonlinear spinor field equation for massive Dirac spinors which was afterwards in different versions denoted as the NJL model (equation). We briefly describe their motivation and techniques to arrive at this result, and for simplicity we omit the isospinor part because it plays no role with respect to the arguments of Nambu and Jona–Lasinio [Nam 61a,b].

These authors start from the Lagrangian density

$$\mathcal{L}[\psi] = -\bar{\psi}\gamma^\mu\partial_\mu\psi + g\left[(\bar{\psi}\psi)^2 - (\bar{\psi}\gamma^5\psi)^2\right], \tag{2.14}$$

which is invariant under the symmetry transformations

$$\psi' = e^{i\alpha\gamma^5}\psi, \qquad\qquad \bar{\psi}' = \bar{\psi}e^{-i\alpha\gamma^5}, \tag{2.15}$$
$$\psi' = e^{i\alpha}\psi, \qquad\qquad \bar{\psi}' = \bar{\psi}e^{-i\alpha}.$$

If one introduces left and right handed fields

$$\psi_L = \frac{1}{2}(1-\gamma^5)\psi, \tag{2.16}$$
$$\psi_R = \frac{1}{2}(1+\gamma^5)\psi,$$

the symmetry transformations (2.15) go over into the chiral symmetry transformations

$$\psi'_L = e^{i\alpha_L}\psi_L, \tag{2.17}$$
$$\psi'_R = e^{i\alpha_R}\psi_R,$$

i.e., (2.14) is a chiral invariant Lagrangian. The introduction of a mass term $\mathcal{L}_m = -m_0\bar{\psi}\psi$ in (2.14) would destroy the axial γ^5 symmetry.

By considering a mean field approximation of (2.14) which is defined by

$$\left[i\gamma^\mu\partial_\mu + 2G\langle 0|\bar{\psi}\psi|0\rangle\right]\psi = 0. \tag{2.18}$$

Nambu and Jona–Lasinio showed that an axial symmetry breaking solution of (2.18) exists and they concluded that an 'effective' Lagrangian should contain a chiral symmetry breaking mass term. Therefore nonlinear spinor field equations of the Heisenberg type (2.1) with an additional mass term are denoted as NJL equations.

For the evaluation of their theory Nambu and Jona–Lasinio took the theory of superconductivity as an example, and one can argue about whether their assumptions and calculations are justified or not. So we do not give more details but emphasize only that the NJL model works with a broken chiral symmetry.

In the terminology of this book the effective Lagrangians or Hamiltonians are created by transforming the original 'sub-fermions' into dressed ones. These transformations are the topic of this book, and in the introductory sections we do not want to burden the discussion with the details of a dressed particle transformation and the justification of a chiral symmetry breaking.

Rather, by pragmatic reasoning, we assume that such a discussion is possible and that it justifies the desired symmetry breaking. Thus in the following we take the NJL equation for granted and modify it only by the introduction of Heisenberg's isospin degrees of freedom. But in addition we solve the regularization problem of such an equation in another way as Nambu and Jona–Lasinio did. This then leads to the equation of Section 2.3 which will be extensively discussed.

v) Fusion versus GUTs, strings, and membranes

Although in some periods wild speculations about fusion models and theories were made, owing to the lack of an adequate mathematical tool for the description of fusion in the realm of quantum field theory, the majority of high energy physicists tried to derive the manifold of elementary particles from ever increasing symmetry groups.

This led to so called grand unification theories (GUTs) and in connection with regularization to supersymmetric unification theories. Both approaches have in common that they generalize the schemes of quantum electrodynamics, *etc.*, to higher gauge symmetry groups without essentially improving the mathematical formalism; on the other hand, they lead to the necessity of breaking down these higher symmetry groups by cascades of symmetry breaking steps.

Regardless of whether the symmetries are broken or not the elementary particles of such theories are described by the field representations of these groups with high numbers of elements. Apart from the common difficulties of quantum field theories, from the atomistic point of view the inflation of elementarity being induced by these group representations is no recommendation for the acceptance of such theories.

A comparable attempt, for instance, at describing the atoms of the periodic table as the members of a representation of a higher symmetry group and their reactions by a corresponding dynamical law for this group, may illustrate the dubiousness of this strategy. Not enough with that, the efforts in this direction were even continued by the design of new mathematical 'machines', the strings and membranes, in order to streamline and automate the production of elementary fields.

The elementary fields are now produced by the vibrations of strings and membranes, an idea as dubious as the inflation of symmetry groups. Although the introduction of these kinds of theories was accompanied by a singular degree of advertising, desire and reality are somewhat contradictory; *cf.* also the comments of Hehl, McCrea, Mielke and Ne'eman on superstrings [Hehl 95]. At the moment when experiments reveal the internal structure of one of the so called elementary particles such artificial theoretical constructs will break down under the weight of evidence and the era of fusion theory will be renewed.

2.2 Review of Regularization

How to deal with nonrenormalizable interactions is one of the central questions of fusion theory if it is to be extended to relativistic quantum field theory. In any case, an appropriate regularization of such interactions is needed. Before we review regularization procedures, we summarize the results of the preceding section, as they are the basis of further investigations.

In the preceding section we arrived at the conclusion that a nonlinear spinor field forms a suitable candidate for an extension of the quantum mechanical fusion conception to relativistic quantum field theory. This choice is enforced by group theory:

the essential symmetry group of relativistic quantum fields is the Poincaré group. All representations of this group can be generated by direct products of elementary spin $\frac{1}{2}$ representations, *cf.* Wigner [Wign 39], Fonda and Ghirardi [Fon 70], Bogoliubov, Logunov and Todorov [Bogo 75], Haag [Haag 92]. Therefore, as this group is fundamental one expects that nature reflects this composition principle, *i.e.*, that spin $\frac{1}{2}$ fermions are the basic quantities for any attempt to generate the hierarchy of observable and unobservable (quarks, gluons!) physical particles.

This idea motivated de Broglie to formulate his fusion principle in relativistic quantum mechanics, and Heisenberg's spinor theory was the first attempt at realizing this principle in quantum field theory. Heisenberg's philosophical ideas were originally distinct from de Broglie's, but from a modern theoretical point of view the philosophical differences are irrelevant. However, owing to the shortcomings of Heisenberg's theory in spite of a general agreement we are compelled to develop a new theoretical scheme for the treatment of the field-theoretic fusion principle as well as to choose another spinor field equation as the starting point of our investigations.

This new theoretical scheme is the subject of this book and the field equations are those of a nonperturbatively regularized Nambu–Jona–Lasinio (NJL) model. However, the NJL model was inspired by Heisenberg's line of approach, and, in particular, we take over Heisenberg's isospin concept. But instead of massless spinor–isospinor fields we consider massive fields. Furthermore, it will turn out that Heisenberg's non-canonical quantization can be replaced by a systematical regularization which respects the canonical quantization.

After these preliminaries we turn to the discussion of regularization. Both Heisenberg's equation and the NJL equation are formulated by means of Dirac operators with first order derivatives and local interaction terms in order to guarantee relativistic invariance and causality. Canonically quantized, such equations are nonrenormalizable in (perturbative) quantum field theory.

As we already mentioned, in spite of this drawback such models are frequently used in high and medium energy physics, and to circumvent nonrenormalizability and divergences, in the great majority of applications the spinor fields are regularized by various cut-off prescriptions and (or) nonlocal interactions. An exception is the regularization prescription by non-canonical quantization of Heisenberg (and coworkers). Whilst the former prescriptions lead to the breakdown either of relativistic invariance or of relativistic causality, or of both of them, Heisenberg's prescription destroys the canonical structure of the theory, which produces considerable difficulties.

In order to avoid these difficulties we apply a new nonperturbative Pauli–Villars regularization which respects relativistic invariance and does not lead to nonlocal interactions. It can be introduced in a systematic way by a decomposition of a higher order nonlinear spinor field equation; this decomposition preserves the canonical structure of the theory and at the same time allows a well defined interpretation of the Heisenberg regularization.

The decomposition of higher order nonlinear spinor field equations enables one to transform such equations into a set of first order equations. The corresponding decomposition theorem enforces the introduction of auxiliary fields which is similar to the introduction of auxiliary fields in the perturbative Pauli–Villars regularization of renormalizable quantum field theories, *cf.* for instance [Gup 77]. But if these procedures are similar or even identical for renormalizable and nonrenormalizable theories this raises the question about the difference between these classes of theories.

The answer is that in the case of a renormalizable theory one can take the auxiliary field masses to infinity and obtain finite results that are insensitive to the cut-off. This statement is false for a nonrenormalizable theory: in this case the limit at infinity is sensitive to the cut-off and can in general not be performed, see [Coll 87]. Hence one has to expect that in nonlinear spinor theories with nonperturbative Pauli–Villars regularization the auxiliary field masses can be made very large but are not allowed to go to infinity.

This fact has serious consequences. Irrespective of whether one considers renormalizable or nonrenormalizable theories, finite masses of the auxiliary fields bring in its train an indefinite metric into the theory. If the limit at infinity can be performed, one can get rid of this indefinite metric, in the opposite case one has to deal with this question.

Thus for nonlinear spinor fields the prize of this kind of regularization is the introduction of an indefinite metric into the theory. An indefinite metric also appears in the relativistic invariant formulation of gauge theories (Lorentz gauge, *etc.*), where it is removed by constraints. Constraints may be interpreted as measuring limitations formulated theoretically and imposed on the state space. In the spinor field model such constraints cannot be analogously derived at present. Nevertheless, the interpretation that an indefinite metric imposes limitations of measurement remains possible, *cf.*, also Feynman [Feyn 87] and Section 6.8.

In the natural units $\hbar = c = 1$ length has the dimension m^{-1}. This reminds us of the smallest fundamental length of Heisenberg and its possible connections with the auxiliary field masses m_i. Owing to nonrenormalizability the theory depends on these masses and thus on length parameters $[\ell_i] = [m_i^{-1}]$. As, on the other hand, this dependence is connected with limitations of measurement one can detect the influence of these parameters.

By a proper interpretation of regularization we treat these problems in Section 6.8. But we avoid the detailed investigation of the state space, which is rather troublesome. For brevity we take a more pragmatic point of view and investigate effective composite particle theories resulting from the spinor field model. If these effective theories are identical with phenomenological quantum field theories which allow a statistical interpretation, then in an indirect way in this sector of the spinor field state space the compatibility of the spinor field metric with a statistical interpretation is secured.

Finally, one can ask the question about the nature of the 'constituents' of the spinor field. We simply call them subfermions, because we expect them to be unobservable, *i.e.*, break up reactions of bound states must be avoided as they lead to the dangerous area of indefiniteness. To achieve this we assume the subfermions to be very heavy. With such a sub-structure mass scale the subfermions need no confining forces and it is merely their large mass which prevents them from being detected. The consequences of this assumption will be studied in the following in detail.

As we take over Heisenberg's isospin this corresponds, figuratively speaking, to the introduction of subfermion 'protons' and subfermion 'neutrons'. The term 'subfermion' associates a certain relationship with the 'rishon' concept of Harari and Shupe [Har 79], [Shu 79]. But the original rishon concept was only of a combinatorial nature, and as we will see later on there exists a correspondence only to some extent. In other cases the dynamical calculations require quite different configurations of subfermions compared with those of the rishon model.

And a last point: owing to their assumed unobservability, the subfermions have no direct physical meaning. Thus a renormalization is not needed for the (auxiliary) spinor field itself, and the term 'subfermion' is used illustratively for the set of auxiliary spinor fields.

2.3   The Decomposition Theorem

The basic quantities of the model are two four-component Dirac spinor–isospinor fields $\Psi_{A\alpha}(x)$, where $\alpha = 1, \ldots, 4$ is the spinor index and $A = 1, 2$ the isospinor index. We postulate the following nonlinear field equation for these fields:

$$(i\hbar\gamma^\mu\partial_\mu - cm)^{\text{reg}}_{\alpha\beta}\delta_{AB}\Psi_{B\beta}(x) \tag{2.19}$$
$$= g\left[\Psi_{A\alpha}(x)\overline{\Psi}_{C\gamma}(x)\Psi_{C\gamma}(x) - \gamma^5_{\alpha\beta}\Psi_{A\beta}(x)\overline{\Psi}_{C\gamma}(x)\gamma^5_{\gamma\delta}\Psi_{C\delta}(x)\right],$$

which exhibits a scalar and a pseudo-scalar coupling of the Dirac spinors and a scalar coupling of the isospin components. The superscript 'reg' denotes the special kind of regularization already announced and will be defined below.

Equation (2.19) may be abbreviated as

$$(i\hbar\gamma^\mu\partial_\mu - cm)^{\text{reg}}_{\alpha\beta}\delta_{AB}\Psi_{B\beta}(x) = gV_{\substack{ABCD \\ \alpha\beta\gamma\delta}}\Psi_{B\beta}(x)\overline{\Psi}_{C\gamma}(x)\Psi_{D\delta}(x), \tag{2.20}$$

where the vertex is defined by

$$V_{\substack{ABCD \\ \alpha\beta\gamma\delta}} := \sum_{h=1}^{2}\delta_{AB}\delta_{CD}v^h_{\alpha\beta}v^h_{\gamma\delta}, \tag{2.21}$$

$$v^1_{\alpha\beta} := \delta_{\alpha\beta},$$

$$v^2_{\alpha\beta} := i\gamma^5_{\alpha\beta}.$$

The adjoint spinor–isospinor field

$$\overline{\Psi}_{A\alpha}(x) := \Psi^+_{B\beta}(x)\gamma^0_{\beta\alpha}\delta_{AB} \tag{2.22}$$

satisfies the adjoint equation

$$(-i\hbar\,{}^{\text{T}}\gamma^\mu\partial_\mu - cm)^{\text{reg}}_{\alpha\beta}\delta_{AB}\overline{\Psi}_{B\beta}(x) = gV_{\substack{ABCD \\ \alpha\beta\gamma\delta}}\overline{\Psi}_{D\delta}(x)\Psi_{C\gamma}(x)\overline{\Psi}_{B\beta}(x). \tag{2.23}$$

The coupling constant g is real and will be fixed later. With respect to the choice of the vertex we refer to Grimm [Gri 94a], who performed a systematic discussion of vertex expressions. As we will perform a quantization of the $\Psi_{A\alpha}$ as fermion fields we have to consider them on the 'classical' level as anticommuting Grassmann variables; this leads to the antisymmetrization $V_{ABCD} = -V_{ADCB}$ of the vertex.

The invariance groups of the field equations (2.20) and (2.23) are the Poincaré group, a global $SU(2)$ isospin group and a global $U(1)$; the mass terms, however, break the chiral symmetries. The invariance of (2.20) and (2.23) with respect to the discrete C-, P- and T-transformations is established on the quantum level, *i.e.*, if (2.20) and (2.23) are considered as equations for anticommuting operators. A detailed account of discrete transformations is given by Grimm [Gri 94a].

Finally, we explain the super-index 'reg' of the Dirac operator in (2.20) and (2.23). It indicates that the two equations have to be regularized if applied in their quantized version. We define the regularization by

$$(i\hbar\gamma^\mu\partial_\mu - cm)^{\text{reg}}_{\alpha\beta} := \left[(i\hbar\gamma^\mu\partial_\mu - cm_1)(i\hbar\gamma^\nu\partial_\nu - cm_2)(i\hbar\gamma^\rho\partial_\rho - cm_3)\right]_{\alpha\beta} \qquad (2.24)$$

with

$$(m_i)_{\alpha\beta} := m_i\delta_{\alpha\beta}\,, \quad i = 1,2,3, \qquad (2.25)$$

where m_i, $i = 1,2,3$, are real and $m_1 \neq m_2 \neq m_3$, *i.e.*, we regularize (2.20) by a third order differential operator. For later applications we set $m_1 > m_2 > m_3$.

Higher order differential equations were first introduced in electrodynamics by Bopp [Bopp 40] and Podolski [Pod 41]. The canonical formalism for such higher order theories was developed by various authors, *e.g.*, de Wet [Wet 48], Chang [Cha 48], Borneas [Bor 69], Riewe and Green [Riew 72a,b], Bigi, Dürr and Winter [Bigi 74], Rodrigues [Rod 77], Aldaya and Azcarraga [Ald 80]. It is based on the introduction of derivatives of the original fields as independent fields which are 'canonically' quantized. This leads to commutation relations between the original fields and their derivatives and to rather complicated expressions for the conserved quantities like energy–momentum tensor, *etc.*. Thus we have developed a new method for canonical quantization which is based on a decomposition theorem and which has the advantage of generating a nonperturbative Pauli–Villars regularization.

A theorem which decomposes a linear Dirac equation with higher order derivatives, *i.e.*, (2.20) with $V \equiv 0$, into a set of linear Dirac equations with first order derivatives was first proved by Wildermuth [Wild 50]. The extension to the decomposition of nonlinear field equations is nontrivial. As a preliminary step we prove the following proposition:

- **Proposition 2.1:** Let $\{m_i, i = 1, \ldots, N, \ N \geq 2\}$ be a set of real parameters with $m_i \neq m_k$, $\forall\, i \neq k$ and D a complex variable. For the decomposition of the following fraction into partial fractions

$$\prod_{i=1}^{N}(D - m_i)^{-1} = \sum_{i=1}^{N}\lambda_i(D - m_i)^{-1}, \qquad (2.26)$$

the relation

$$\sum_{i=1}^{N} \lambda_i \prod_{\substack{k=1 \\ k \neq i}}^{N} (D - m_k) = 1 \tag{2.27}$$

holds with

$$\lambda_i = \prod_{\substack{k=1 \\ k \neq i}}^{N} (m_i - m_k)^{-1}. \tag{2.28}$$

Proof: Let $F(D)$ be defined by $F(D) := \prod_{i=1}^{N}(D - m_i)$. If (2.26) is multiplied with $F(D)(D - m_k)$, one obtains

$$D - m_k = \lambda_k F(D) + \sum_{\substack{i=1 \\ i \neq k}}^{N} \frac{\lambda_i}{D - m_i}(D - m_k)F(D). \tag{2.29}$$

Differentiation with respect to D yields for $D = m_k$

$$1 = \lambda_k F'(m_k) \tag{2.30}$$

with

$$F'(m_k) = \prod_{\substack{i=1 \\ i \neq k}}^{N} (m_k - m_i). \tag{2.31}$$

Thus λ_i is given by (2.28). Multiplication of (2.26) with $F(D)$ gives the relation (2.27).
◇

We mention that Proposition 2.1 holds for operator-valued D, too.

The relation between a higher order differential (field) equation and first order equations can now be established by the following 'decomposition theorem' which was proved for $N = 3$ by Stumpf [Stu 82] and for general N by Grosser [Gro 83].

- **Proposition 2.2:** Under the conditions of Proposition 2.1 the following statements hold:

i) Let $\Psi \equiv \Psi(x)$ be a solution of a N-th order equation

$$\prod_{i=1}^{N} (D - m_i)\Psi = V[\Psi], \tag{2.32}$$

where $V[\Psi]$ is a (nonlinear) interaction term, and let $\psi_i \equiv \psi_i(x)$, $i = 1, \ldots N$, be auxiliary fields which are defined by

$$\psi_i := \lambda_i \prod_{\substack{k=1 \\ k \neq i}}^{N} (D - m_k)\Psi, \quad i = 1, \ldots N, \tag{2.33}$$

with

$$\lambda_i := \prod_{\substack{k=1 \\ k \neq i}}^{N} (m_i - m_k)^{-1}, \quad i = 1, \ldots N. \tag{2.34}$$

Then there holds

$$\Psi = \sum_{i=1}^{N} \psi_i \tag{2.35}$$

and the auxiliary fields are solutions of the first order equations

$$(D - m_i)\psi_i = \lambda_i V \left[\sum_{j=1}^{N} \psi_j \right], \quad i = 1, \ldots N, \tag{2.36}$$

with no summation over i on the left hand side.

ii) If the fields $\psi_i \equiv \psi_i(x)$, $i = 1, \ldots N$, are solutions of equations (2.36) with λ_i of (2.34), and if $\Psi \equiv \Psi(x)$ is defined by (2.35), then the relations (2.33) hold and Ψ is a solution of equation (2.32).

Proof: i) With (2.27) and (2.33) we have

$$\Psi = \sum_{i=1}^{N} \lambda_i \prod_{\substack{k=1 \\ k \neq i}}^{N} (D - m_k)\Psi = \sum_{i=1}^{N} \psi_i, \tag{2.37}$$

and owing to (2.32) and (2.33) there holds

$$(D - m_i)\psi_i = \lambda_i \prod_{k=1}^{N} (D - m_k)\Psi = \lambda_i V \left[\sum_{j=1}^{N} \psi_j \right] \tag{2.38}$$

with no summation over i on the left hand side. Thus (2.35) and (2.36) hold.

ii) With (2.27), (2.35) and (2.36) we have

$$\begin{aligned}
\prod_{i=1}^{N} (D - m_i)\Psi &= \prod_{i=1}^{N} (D - m_i) \sum_{j=1}^{N} \psi_j \\
&= \sum_{j=1}^{N} \prod_{\substack{i=1 \\ i \neq j}}^{N} (D - m_i)\lambda_j V \left[\sum_{k=1}^{N} \psi_k \right] \\
&= V \left[\sum_{k=1}^{N} \psi_k \right] \\
&= V[\Psi],
\end{aligned} \tag{2.39}$$

i.e., equation (2.32) holds.

For $i = 1$ and $j = 2, \ldots N$ it follows from (2.36)

$$\lambda_1^{-1}(D - m_1)\psi_1 = \lambda_j^{-1}(D - m_j)\psi_j, \quad j = 2, \ldots N, \tag{2.40}$$

with no summation over j on the right hand side. With (2.40) and (2.27) there holds

$$\lambda_1 \prod_{k=2}^{N}(D - m_k)\Psi = \lambda_1 \prod_{k=2}^{N}(D - m_k)\sum_{j=1}^{N}\psi_j \tag{2.41}$$

$$= \lambda_1 \prod_{k=2}^{N}(D - m_k)\psi_1 + \sum_{j=2}^{N}\lambda_1 \prod_{\substack{k=2 \\ k\neq j}}^{N}(D - m_k)\frac{\lambda_j}{\lambda_1}(D - m_1)\psi_1$$

$$= \sum_{j=1}^{N}\lambda_j \prod_{\substack{k=1 \\ k\neq j}}^{N}(D - m_k)\psi_1$$

$$= \psi_1.$$

The same conclusions can be drawn for $i = 2, \ldots N$. Thus (2.33) holds. $\Diamond$

Proposition 2.2 can be reformulated as follows: Equation (2.33) defines a bijective map between the sets of solutions of equations (2.32) and (2.36). The inverse map of (2.33) is given by (2.35).

For the proof of this statement let M_1 be the set of solutions of (2.32), M_2^i be the set of solutions of (2.36) for $i = 1, \ldots N$, and f_1^i and f_2 the maps which are defined by (2.33) and (2.35). Then according to Proposition 2.2 we have for $M_2 := \oplus_{i=1}^{N} M_2^i$ and $f_1 = (f_1^1, \ldots f_1^N)$

$$f_1 : M_1 \to M_2, \qquad f_2 : M_2 \to M_1, \tag{2.42}$$

and owing to (2.33) and (2.35) we have

$$f_2 \circ f_1 = \mathbf{1}_{M_1}, \qquad f_1 \circ f_2 = \mathbf{1}_{M_2}. \tag{2.43}$$

Thus f_1 defines a bijective map from M_1 onto M_2 with $f_1^{-1} = f_2$.

Conversely we can regain the statements of Proposition 2.2 if the relations (2.42) and (2.43) are valid.

The decomposition theorem is also valid for operator-valued D and can be easily extended to multi-component equations. Thus we can apply it to the field equations (2.20) and (2.23) by defining the auxiliary fields

$$\psi_{A\alpha i}(x) := \lambda_i \prod_{\substack{k=1 \\ k\neq i}}^{3}(i\hbar\gamma^\mu\partial_\mu - cm_k)_{\alpha\beta}\delta_{AB}\Psi_{B\beta}(x), \tag{2.44}$$

$$\bar{\psi}_{A\alpha i}(x) := \lambda_i \prod_{\substack{k=1 \\ k\neq i}}^{3}(-i\hbar\,{}^{\mathrm{T}}\gamma^\mu\partial_\mu - cm_k)_{\alpha\beta}\delta_{AB}\overline{\Psi}_{B\beta}(x) \tag{2.45}$$

with

$$\lambda_i := \prod_{\substack{k=1 \\ k\neq i}}^{3}(cm_i - cm_k)^{-1}. \tag{2.46}$$

Then according to Proposition 2.2 we have the inverse relations

$$\Psi_{A\alpha}(x) = \sum_{i=1}^{3} \psi_{A\alpha i}(x), \qquad \overline{\Psi}_{A\alpha}(x) = \sum_{i=1}^{3} \bar{\psi}_{A\alpha i}(x), \tag{2.47}$$

and the auxiliary fields satisfy the first order differential equations

$$(i\hbar\gamma^{\mu}\partial_{\mu} - cm_i)_{\alpha\beta}\delta_{AB}\psi_{B\beta i}(x) \tag{2.48}$$

$$= g\lambda_i V_{\substack{ABCD \\ \alpha\beta\gamma\delta}} \sum_{j,k,l=1}^{3} \psi_{B\beta j}(x)\bar{\psi}_{C\gamma k}(x)\psi_{D\delta l}(x)$$

and

$$(-i\hbar\ ^{T}\gamma^{\mu}\partial_{\mu} - cm_i)_{\alpha\beta}\delta_{AB}\bar{\psi}_{B\beta i}(x) \tag{2.49}$$

$$= g\lambda_i V_{\substack{ABCD \\ \alpha\beta\gamma\delta}} \sum_{j,k,l=1}^{3} \bar{\psi}_{D\delta l}(x)\psi_{C\gamma k}(x)\bar{\psi}_{B\beta j}(x)$$

with no summation over i on the left hand sides. Owing to (2.22) from (2.47) the relation

$$\bar{\psi}_{A\alpha i}(x) = \psi^{+}_{A\beta i}(x)\gamma^{0}_{\beta\alpha} \tag{2.50}$$

follows, i.e., $\bar{\psi}$ has to be the adjoint auxiliary field spinor.

Finally, we mention that the λ_i of (2.46) satisfy the Pauli–Villars regularization conditions, see Pauli and Villars [Pau 49], Rayski [Ray 48], Bogoliubov and Shirkov [Bogo 59]:

$$\sum_{i=1}^{3} \lambda_i = 0, \qquad \sum_{i=1}^{3} \lambda_i m_i = 0. \tag{2.51}$$

Thus in combination with the Lagrange formulation for the auxiliary fields our regularization (2.24) turns out to be a nonperturbative Pauli–Villars regularization. For a detailed discussion of nonperturbative regularization, probability interpretation and observable quantities we refer to Sections 6.8 and 8.5.

2.4   Lagrange Formalism

In order to derive conservation laws and to carry out canonical quantization, we need the Lagrangian of the fields. This program can be successfully performed only if the field dynamics is based on first order differential equations. Thus the decomposition theorem serves as a means of simultaneously applying spinor field regularization and the canonical formalism. So in the following all derivations and conclusions will be based on the auxiliary spinor field dynamics given by equations (2.48) and (2.49), and in order to perform the regularization we sum over the auxiliary fields and specify the masses m_1, m_2 and m_3.

One can easily verify that (2.48) and (2.49) follow by variation with respect to $\psi_{A\alpha i}(x)$ or $\bar{\psi}_{A\alpha i}(x)$ from an action with the Lagrangian density

$$
\begin{aligned}
\mathcal{L}[\psi, \bar{\psi}] &= \mathcal{L}_{\text{kin}}[\psi, \bar{\psi}] + \mathcal{L}_{\text{int}}[\psi, \bar{\psi}] \\
&= \sum_{i=1}^{3} \lambda_i^{-1} \bar{\psi}_{A\alpha i}(x) (i\hbar\gamma^\mu \partial_\mu - cm_i)_{\alpha\beta} \delta_{AB} \psi_{B\beta i}(x) \\
&\quad - \frac{g}{2} V_{ABCD}_{\alpha\beta\gamma\delta} \sum_{i,j,k,l=1}^{3} \bar{\psi}_{A\alpha i}(x) \psi_{B\beta j}(x) \bar{\psi}_{C\gamma k}(x) \psi_{D\delta l}(x),
\end{aligned}
\tag{2.52}
$$

where $\mathcal{L}_{\text{kin}}$ has the structure of a Pauli–Villars regularized Lagrangian. But as far as $\mathcal{L}_{\text{int}}$ is concerned in conventional quantum field theory $\mathcal{L}_{\text{int}}$ is regularized by hand and not automatically by the formalism itself, see, *e.g.*, Itzykson and Zuber [Itz 80]. In contrast to this common usage the Lagrangian (2.52) automatically also provides a regularization of the interaction term, and this regularization is independent of the kind of calculation. In particular, it does not depend on the application of perturbation theory, as will be demonstrated in the following. Thus by the decomposition theorem we have achieved a nonperturbative Pauli–Villars regularization.

In order to examine the conserved quantities we define

$$
\pi_{A\alpha i}^\mu(x) := \frac{\delta \mathcal{L}}{\delta \partial_\mu \psi_{A\alpha i}(x)}.
\tag{2.53}
$$

Owing to the translational invariance of the Lagrangian and its action, it follows that according to Noether's theorem the energy–momentum vector

$$
P^\mu := \int \mathcal{T}^{0\mu}(\mathbf{r}, t) d^3 r
\tag{2.54}
$$

with the canonical energy–momentum tensor density

$$
\mathcal{T}^{\mu\nu}(x) := \pi_{A\alpha i}^\mu(x) \partial^\nu \psi_{A\alpha i}(x) - \eta^{\mu\nu} \mathcal{L}(x)
\tag{2.55}
$$

is a conserved quantity.

The canonically conjugated momentum $\pi_{A\alpha i}(x) := \pi_{A\alpha i}^0(x)$ to $\psi_{A\alpha i}(x)$ can be calculated by means of (2.52) and (2.53) and reads

$$
\pi_{A\alpha i}(x) = \frac{i\hbar}{\lambda_i} \bar{\psi}_{A\beta i}(x) \gamma_{\beta\alpha}^0 = \frac{i\hbar}{\lambda_i} \psi_{A\alpha i}^+(x)
\tag{2.56}
$$

with no summation over i on the right hand side. The Hamiltonian density follows from $\mathcal{H}(x) := c\mathcal{T}^{00}(x)$ and is given by

$$
\begin{aligned}
\mathcal{H}(x) &= c\sum_{i=1}^{3} \pi_{A\alpha i}(x) \partial_0 \psi_{A\alpha i}(x) - c\mathcal{L}(x) \\
&= c\sum_{i=1}^{3} \lambda_i^{-1} \bar{\psi}_{A\alpha i}(x) (-i\hbar\gamma^k \partial_k + cm_i)_{\alpha\beta} \delta_{AB} \psi_{B\beta i}(x) - c\mathcal{L}_{\text{int}}(x) \\
&= \frac{c}{\hbar} \sum_{i=1}^{3} \pi_{C\gamma i}(x) \gamma_{\gamma\alpha}^0 \delta_{CA} (-\hbar\gamma^k \partial_k - icm_i)_{\alpha\beta} \delta_{AB} \psi_{B\beta i}(x) - c\mathcal{L}_{\text{int}}(x).
\end{aligned}
\tag{2.57}
$$

If by means of Noether's theorem one evaluates the consequence of the invariance of the action under homogeneous Lorentz transformations one obtains a conservation law for the angular momentum tensor

$$M_{\mu\nu} := \int \mathcal{M}^0_{\mu\nu}(\mathbf{r}, t) d^3r \qquad (2.58)$$

with the canonical angular momentum tensor density

$$\mathcal{M}^\rho_{\mu\nu}(x) := \pi^\rho_{A\alpha i}(x)\Sigma^{\alpha\beta}_{\mu\nu}\psi_{A\beta i}(x) + \pi^\rho_{A\alpha i}(x)(x_\mu\partial_\nu - x_\nu\partial_\mu)\psi_{A\alpha i}(x) \qquad (2.59)$$
$$+ (\delta^\rho_\mu x_\nu - \delta^\rho_\nu x_\mu)\mathcal{L}(x),$$

where $\Sigma_{\mu\nu} := -\frac{1}{4}[\gamma_\mu, \gamma_\nu]_-$ is the representation of generators of the homogeneous Lorentz group in the space of Dirac spinors.

So far we have derived some important properties and consequences of the quasiclassical spinor field theory given by equation (2.20), but of course our aim is the quantization of (2.20).

2.5   Canonical Spinor Field Quantization

With the decomposition theorem we are prepared to canonically quantize the regularized spinor field which is dynamically characterized by the Lagrangian density (2.52). Canonical quantization is performed by considering the spinor fields $\psi_{A\alpha i}(x)$ and their conjugate momenta $\pi_{A\alpha i}(x)$ as the constituting elements of a CAR algebra with corresponding anticommutation relations.

Because according to (2.56) we have $\pi_{A\alpha i}(x) = i\hbar\lambda_i^{-1}\psi^+_{A\alpha i}(x)$ (with no summation over i), the adjoint spinor fields $\bar{\psi}_{A\alpha i}(x) = \psi^+_{A\beta i}(x)\gamma^0_{\beta\alpha}$ depend linearly on the conjugate momenta $\pi_{A\alpha i}(x)$ and thus need not be additionally included in the algebra of fields. From now on we assume the $\psi_{A\alpha i}$, etc., to be field operators on a state space and no longer classical quantities.

The canonical anticommutation relations read

$$[\pi_{A\alpha i}(\mathbf{r}, t), \psi_{B\beta j}(\mathbf{r}', t)]_+ = i\hbar\delta_{ij}\delta_{AB}\delta_{\alpha\beta}\delta(\mathbf{r} - \mathbf{r}'), \qquad (2.60)$$
$$[\psi_{A\alpha i}(\mathbf{r}, t), \psi_{B\beta j}(\mathbf{r}', t)]_+ = 0,$$
$$[\pi_{A\alpha i}(\mathbf{r}, t), \pi_{B\beta j}(\mathbf{r}', t)]_+ = 0$$

for equal times. These equal time relations indicate that the complete CAR algebra is already defined at an arbitrary, but fixed, moment of time. The dynamical evolution of a quantum system is then described by a continuous group of automorphism of this algebra. This property of canonical quantization, in connection with a corresponding time translation generator (Hamiltonian $\mathbf{H}$), will turn out to be a central point of our treatment of composite particle dynamics.

If one substitutes (2.56) into (2.60) these relations go over into the spinor field relations

$$\left[\psi^+_{A\alpha i}(\mathbf{r},t),\psi_{B\beta j}(\mathbf{r}',t)\right]_+ = \lambda_i\delta_{ij}\delta_{AB}\delta_{\alpha\beta}\delta(\mathbf{r}-\mathbf{r}'), \tag{2.61}$$

$$\left[\psi_{A\alpha i}(\mathbf{r},t),\psi_{B\beta j}(\mathbf{r}',t)\right]_+ = 0,$$

$$\left[\psi^+_{A\alpha i}(\mathbf{r},t),\psi^+_{B\beta j}(\mathbf{r}',t)\right]_+ = 0.$$

Owing to (2.51) not all λ_i can be positive quantities. Thus (2.61) shows that regularization enforces indefinite metric by producing alternating signs in the anticommutation relations of the fields. This feature of spinor field theory will be discussed in detail in Chapter 4.

The regularization effect can be drastically seen if one derives the anticommutation relations for the original field $\Psi_{A\alpha}(x)$ from (2.61). With (2.35) we obtain from (2.61) by means of (2.51)

$$\left[\Psi^+_{A\alpha}(\mathbf{r},t),\Psi_{B\beta}(\mathbf{r}',t)\right]_+ = \sum_{i,j=1}^{3}\left[\psi^+_{A\alpha i}(\mathbf{r},t),\psi_{B\beta j}(\mathbf{r}',t)\right]_+ \tag{2.62}$$

$$= \sum_{i,j=1}^{3}\lambda_i\delta_{ij}\delta_{AB}\delta(\mathbf{r}-\mathbf{r}')$$

$$= 0,$$

$$\left[\Psi^+_{A\alpha}(\mathbf{r},t),\Psi^+_{B\beta}(\mathbf{r}',t)\right]_+ = \left[\Psi_{A\alpha}(\mathbf{r},t),\Psi_{B\beta}(\mathbf{r}',t)\right]_+ = 0. \tag{2.63}$$

Thus the canonical quantization of the auxiliary fields is equivalent to a non-canonical quantization of the original field. The latter relations (2.62) were already assumed by Heisenberg [Heis 66], but without the background of a decomposition theorem and thus without relation to a strictly canonically quantized system. It is just the combination of our regularization with CAR algebra and Hamiltonian formalism which allows a successful treatment of spinor field dynamics.

Finally, we observe that the canonical quantization has an effect on the vertex of equations (2.20) and (2.23), or (2.48) and (2.49), respectively. From (2.22) and (2.62) it follows that

$$\left[\overline{\Psi}_{A\alpha}(\mathbf{r},t),\Psi_{B\beta}(\mathbf{r}',t)\right]_+ = 0 \tag{2.64}$$

holds, too. Together with (2.63) this means that the local operator product of the regularized spinor fields $\Psi_{B\beta}(x)\overline{\Psi}_{C\gamma}(x)\Psi_{D\delta}(x)$ in (2.20) and (2.23), or (2.48) and (2.49), respectively, is antisymmetric under commutation of the pairs $(B\beta)$ and $(D\delta)$. Therefore, if this product is contracted with the vertex $V_{\substack{ABCD \\ \alpha\beta\gamma\delta}}$ an antisymmetric part is projected out. In order to secure this antisymmetry in any step of calculation we assumed $V_{\substack{ABCD \\ \alpha\beta\gamma\delta}}$ to be antisymmetrized in $(B\beta)$ and $(D\delta)$ from the beginning.

2.6 Super-indexing

Formally, super-indexing is a very compact notation for the spinor fields and their equations. Later on, however, it will turn out that super-indexing is indispensable for the derivation of composite particle dynamics with an appropriate phenomenological interpretation, *i.e.*, for the physical understanding of effective theories.

We introduce the charge conjugated spinor fields

$$\psi^c_{A\alpha i}(x) := C_{\alpha\beta}\bar{\psi}_{A\beta i}(x) \tag{2.65}$$

with the charge conjugation matrix C, and rewrite equations (2.48) and (2.49) in terms of ψ and ψ^c. This gives

$$(i\hbar\gamma^\mu\partial_\mu - cm_i)_{\alpha\beta}\delta_{AB}\psi_{B\beta i}(x) \tag{2.66}$$

$$= g\lambda_i V_{\underset{\alpha\beta\gamma\delta}{ABCD}}C^{-1}_{\gamma\epsilon}\sum_{j,k,l=1}^{3}\psi_{B\beta j}(x)\psi^c_{C\epsilon k}(x)\psi_{D\delta l}(x)$$

and

$$(i\hbar\gamma^\mu\partial_\mu - cm_i)_{\alpha\beta}\delta_{AB}\psi^c_{B\beta i}(x) \tag{2.67}$$

$$= -g\lambda_i V_{\underset{\alpha'\beta'\gamma'\delta'}{ABCD}}C_{\alpha\alpha'}C^{-1}_{\beta'\beta}C^{-1}_{\delta'\delta}\sum_{j,k,l=1}^{3}\psi^c_{B\beta j}(x)\psi_{C\gamma k}(x)\psi^c_{D\delta l}(x)$$

with no summation over i on the left hand sides. From (2.62) it follows that also $\Psi^c \equiv \sum_i \psi^c_i$ and Ψ anticommute for equal times. Hence the vertices in (2.66) and (2.67) have the same antisymmetry properties as the old ones.

Equations (2.66) and (2.67) can now be combined into one superspinor equation if we introduce superspinors by the definition

$$\psi_{A\alpha i\Lambda}(x) := \begin{cases} \psi_{A\alpha i}(x), & \Lambda = 1 \\ \psi^c_{A\alpha i}(x), & \Lambda = 2. \end{cases} \tag{2.68}$$

Together with the super-index $Z := (A, \alpha, i, \Lambda)$, where

$$
\begin{aligned}
A &= 1,2 & \text{(isospinor index)} \\
\alpha &= 1,2,3,4 & \text{(spinor index)} \\
i &= 1,2,3 & \text{(auxiliary field index)} \\
\Lambda &= 1,2 & \text{(superspinor index)},
\end{aligned}
$$

equations (2.66) and (2.67) can be rewritten as

$$\left(D^\mu_{Z_1 Z_2}\partial_\mu - m_{Z_1 Z_2}\right)\psi_{Z_2}(x) = \hat{U}_{Z_1 Z_2 Z_3 Z_4}\psi_{Z_2}(x)\psi_{Z_3}(x)\psi_{Z_4}(x) \tag{2.69}$$

with

$$
\begin{aligned}
D^\mu_{Z_1 Z_2} &:= i\hbar\gamma^\mu_{\alpha_1\alpha_2}\delta_{A_1 A_2}\delta_{i_1 i_2}\delta_{\Lambda_1\Lambda_2}\,, \tag{2.70} \\
m_{Z_1 Z_2} &:= cm_{i_1}\delta_{A_1 A_2}\delta_{\alpha_1\alpha_2}\delta_{i_1 i_2}\delta_{\Lambda_1\Lambda_2}\,, \\
\hat{U}_{Z_1 Z_2 Z_3 Z_4} &:= \sum_{h=1}^{2}g\lambda_{i_1}B_{i_2 i_3 i_4}v^h_{\alpha_1\alpha_2}\delta_{A_1 A_2}\delta_{\Lambda_1\Lambda_2}(v^h C)_{\alpha_3\alpha_4}\delta_{A_3 A_4}\delta_{\Lambda_3 1}\delta_{\Lambda_4 2}\,, \\
B_{i_2 i_3 i_4} &= 1 \quad \text{for } i_2, i_3, i_4 = 1,2,3,
\end{aligned}
$$

where $B_{i_2 i_3 i_4}$ is a formal quantity expressing the auxiliary field summation in the vertex, and with

$$\hat{U}_{Z_1 Z_2 Z_3 Z_4} \equiv \hat{U}_{Z_1 [Z_2 Z_3 Z_4]} \tag{2.71}$$

owing to the total anticommutativity of the sum of all auxiliary fields in the vertex.

In contrast to the sum of all auxiliary fields the single auxiliary fields satisfy non-trivial anticommutation relations (2.61) which read in super-index notation

$$[\psi_Z(\mathbf{r}, t), \psi_{Z'}(\mathbf{r}', t)]_+ = A_{ZZ'} \delta(\mathbf{r} - \mathbf{r}') \tag{2.72}$$

with

$$A_{ZZ'} := \lambda_i \delta_{ii'} \delta_{AA'} \sigma^1_{\Lambda\Lambda'} (C\gamma^0)_{\alpha\alpha'}, \tag{2.73}$$

no summation over i on the left hand side and $\sigma^1 = \begin{pmatrix} 0 & 1 \\ 1 & 0 \end{pmatrix}$.

Moreover, as often as possible we still use a more compact notation by combining (Z, x) into only one general index $I := (Z, x)$ and denote the underlying index set with $\mathcal{I}$. In this case (2.72) can be rewritten as

$$[\psi_I, \psi_{I'}]_+^{t=t'} = A_{II'} \tag{2.74}$$

with

$$A_{II'} := A_{ZZ'} \delta(\mathbf{r} - \mathbf{r}'). \tag{2.75}$$

In a similar way equation (2.69) can be compactified and reads

$$K_{I_1 I_2} \psi_{I_2} = W_{I_1 I_2 I_3 I_4} \psi_{I_2} \psi_{I_3} \psi_{I_4}, \tag{2.76}$$

where the following abbreviations are used:

$$K_{I_1 I_2} := (D^\mu \partial_\mu - m)_{I_1 I_2} \tag{2.77}$$

with

$$D^\mu_{I_1 I_2} := D^\mu_{Z_1 Z_2} \delta(x_1 - x_2), \tag{2.78}$$

$$m_{I_1 I_2} := {}^\cdot m_{Z_1 Z_2} \delta(x_1 - x_2), \tag{2.79}$$

and

$$W_{I_1 I_2 I_3 I_4} := \hat{U}_{Z_1 Z_2 Z_3 Z_4} \delta(x_1 - x_2) \delta(x_1 - x_3) \delta(x_1 - x_4), \tag{2.80}$$

and where the summation convention is assumed to range over the whole set of indices in $I = (Z, x)$.

In the course of our investigations we will also need a version of equation (2.76), where (2.76) is resolved with respect to $\dot{\psi}_I$, and t is considered as an evolution parameter. In this case we transform (2.76) into

$$i \frac{\hbar^2}{c} \frac{\partial}{\partial t} \psi_{I_1} = \widehat{K}_{I_1 I_2} \psi_{I_2} + \widehat{W}_{I_1 I_2 I_3 I_4} \psi_{I_2} \psi_{I_3} \psi_{I_4}, \tag{2.81}$$

where all field operators have to be taken at the same time t, and exclude t from the summation convention. The corresponding definitions are

$$\widehat{K}_{I_1 I_2} := iD^0_{Z_1 Z_3}\left(D^k_{Z_3 Z_2}\partial_k - m_{Z_3 Z_2}\right)\delta(\mathbf{r}_1 - \mathbf{r}_2), \qquad (2.82)$$

$$\widehat{W}_{I_1 I_2 I_3 I_4} := -iD^0_{Z_1 Z_5}\widehat{U}_{Z_5 Z_2 Z_3 Z_4}\delta(\mathbf{r}_1 - \mathbf{r}_2)\delta(\mathbf{r}_1 - \mathbf{r}_3)\delta(\mathbf{r}_1 - \mathbf{r}_4), \qquad (2.83)$$

and the summation convention is assumed to range over the indices $(Z, \mathbf{r})$ in $I = (Z, \mathbf{r}, t)$. The transition to equal times t at the quantum level will be discussed in Chapter 4.

2.7 Symmetry Conditions

The state space of a quantum theory is usually assumed to be spanned by a complete set of eigenstates of a maximal set of commuting observables. In general such a maximal set of commuting observables is constituted by a commuting set of generators of the corresponding symmetry group of transformations. In the case of a relativistic quantum field theory the most important symmetry group is the Poincaré group, and in the following we will discuss the resulting symmetry conditions connected with this group. If not otherwise stated we will restrict ourselves to the proper orthochronous part of the Lorentz group. Moreover, we do not discuss the discrete symmetries but refer to Grimm [Gri 94a].

We assume that with respect to the spinor field operators $\psi_{A\alpha i\Lambda}(x)$ there exists a state space $\mathbb{H}$ which is an inner product space. In order to realize symmetries for quantum systems we additionally have to assume that the state space is a representation space of the corresponding symmetry group, i.e., in our case the Poincaré group. Then for any Poincaré transformation

$$x'^\mu = \Lambda^\mu_\nu x^\nu + b^\mu \qquad (2.84)$$

in coordinate space there exists an operator $\mathbf{U}(\Lambda, b)$ which is unitary on a subspace of $\mathbb{H}$ (see Chapter 4) and induces a corresponding transformation in $\mathbb{H}$, i.e.,

$$|a'\rangle = \mathbf{U}(\Lambda, b)|a\rangle, \qquad |a\rangle, |a'\rangle \in \mathbb{H}. \qquad (2.85)$$

Furthermore, we assume that there exists a unique state $|0\rangle \in \mathbb{H}$, the 'vacuum state', with positive norm $\langle 0|0\rangle = 1$ being invariant under Poincaré transformations, i.e.,

$$\mathbf{U}(\Lambda, b)|0\rangle = |0\rangle. \qquad (2.86)$$

According to the correspondence principle one postulates that matrix elements of spinor field operators transform as classical spinorial functions. Therefore if one observes that the abstract operator is independent of representations, in the transformed system the matrix elements are given by $\langle a'|\psi_{A\alpha i}(x')|b'\rangle$, whilst in the original system they are defined to have the form $\langle a|\psi_{A\alpha i}(x)|b\rangle$. The transformation law can then be expressed by

$$\langle a'|\psi_{A\alpha i}(x')|b'\rangle \stackrel{!}{=} S_{\alpha\beta}\langle a|\psi_{A\beta i}(\Lambda^{-1}x')|b\rangle, \tag{2.87}$$

where $S_{\alpha\beta}$ is the classical transformation matrix for Dirac spinors. If (2.85) is substituted into (2.87) one obtains

$$\langle a|U^{-1}(\Lambda, b)\psi_{A\alpha i}(x')U(\Lambda, b)|b\rangle = S_{\alpha\beta}(\Lambda)\langle a|\psi_{A\beta i}(\Lambda^{-1}x')|b\rangle \tag{2.88}$$

$\forall|a\rangle, |b\rangle$, $i.e.$, this equation holds for all states of a given representation. The covariance condition (2.87) for matrix elements can then be formulated in a shorthand form as follows:

$$U(\Lambda, b)\psi_{A\alpha i}(\Lambda^{-1}x')U^{-1}(\Lambda, b) = S_{\alpha\beta}^{-1}(\Lambda)\psi_{A\beta i}(x'). \tag{2.89}$$

This equation is often interpreted as the transformation law of field operators. But as $U(\Lambda, b)$ depends on the respective representation this law is obviously referred to matrix elements of any element of the field algebra and their classical transformation properties.

The invariance of the quantum Lagrangian has to be demonstrated by the invariance of all matrix elements, $i.e.$, $\langle a|\mathcal{L}|b\rangle \stackrel{!}{=} \langle a'|\mathcal{L}|b'\rangle$. But with the help of (2.89) this invariance can be formulated as $\mathcal{L}' := U(\Lambda, b)\mathcal{L}U^{-1}(\Lambda, b) \stackrel{!}{=} \mathcal{L}$ and owing to (2.89) the quantum transformation can be transmuted in the classical transformation which leaves $\mathcal{L}$ invariant.

In super-index formulation equation (2.89) reads

$$U(\Lambda, b)\psi_Z(\Lambda^{-1}x')U^{-1}(\Lambda, b) = L_{ZZ'}(\Lambda^{-1})\psi_{Z'}(x') \tag{2.90}$$

with

$$L_{ZZ'}(\Lambda) := \delta_{AA'}\delta_{ii'}\delta_{\Lambda\Lambda'}S_{\alpha\alpha'}(\Lambda). \tag{2.91}$$

Generators are obtained from infinitesimal transformations of continuous groups. For infinitesimal Poincaré transformations

$$x'^\mu = x^\mu + \omega_\nu^\mu x^\nu + \epsilon^\mu, \quad \epsilon^\mu, \omega_\nu^\mu \in \mathbb{R}, \quad \omega_\nu^\mu = -\omega_\mu^\nu, \tag{2.92}$$

we write for the group element of the corresponding representation in $\mathbb{H}$

$$U(\omega, \epsilon) = 1 + \frac{i}{\hbar}P^\mu \epsilon_\mu + \frac{i}{2\hbar}M^{\mu\nu}\omega_{\mu\nu}, \tag{2.93}$$

where P^μ and $M^{\mu\nu}$ are the generators of the Poincaré group in this representation. In this case the general transformation law (2.89) reads

i) for infinitesimal translations, $i.e.$, $\omega_\nu^\mu = 0$, $\epsilon^\mu \neq 0$:

$$[P^\mu, \psi_{A\alpha i}(x)]_- = P^\mu(x)\psi_{A\alpha i}(x) \tag{2.94}$$

with

$$P^\mu(x)\psi_{A\alpha i}(x) := -i\hbar\partial^\mu\psi_{A\alpha i}(x); \tag{2.95}$$

ii) for infinitesimal rotations, $i.e.$, $\omega_\nu^\mu \neq 0$; $\epsilon^\mu = 0$:

$$[\mathbf{M}^{\mu\nu}, \psi_{A\alpha i}(x)]_{-} = M^{\mu\nu}(x)\psi_{A\alpha i}(x) \tag{2.96}$$

with

$$M^{\mu\nu}(x)\psi_{A\alpha i}(x) := -i\hbar(x^{\mu}\partial^{\nu} - x^{\nu}\partial^{\mu})\psi_{A\alpha i}(x) - i\hbar\Sigma^{\mu\nu}_{\alpha\beta}\psi_{A\beta i}(x). \tag{2.97}$$

Equations (2.94) and (2.96) are conditions which have to be satisfied by the generators $\mathbf{P}^{\mu}$ and $\mathbf{M}^{\mu\nu}$. We show that these conditions are satisfied just when the classical expressions for four-momentum and angular momentum of the spinor fields (2.54), (2.55) and (2.58), (2.59) are transferred into quantum expressions by the correspondence principle and identified with $\mathbf{P}^{\mu}$ and $\mathbf{M}^{\mu\nu}$, see Lauxmann [Lau 86], Borne [Bor 91].

- **Proposition 2.3:** Let $\psi_{A'\alpha'i'}(x')$ be a field operator satisfying equations (2.48). Then the Heisenberg equation

$$\left[\mathbf{P}^{\mu}, \psi_{A'\alpha'i'}(x')\right]_{-} = -i\hbar\partial^{\mu}\psi_{A'\alpha'i'}(x') \tag{2.98}$$

holds, with

$$\mathbf{P}^{\mu} := \int d^3r\left[\pi_{A\alpha i}(\mathbf{r}, t)\partial^{\mu}\psi_{A\alpha i}(\mathbf{r}, t) - \eta^{0\mu}\mathcal{L}(\mathbf{r}, t)\right]. \tag{2.99}$$

Proof: With (2.54), (2.55) and (2.56) we have

$$\left[\mathbf{P}^{\mu}, \psi_{A'\alpha'i'}(x')\right]_{-} = \sum_{i=1}^{3}\frac{i\hbar}{\lambda_i}\int d^3r\left[\psi_{\alpha i}^{+}(x)\partial^{\mu}\psi_{A\alpha i}(x), \psi_{A'\alpha'i'}(x')\right]_{-} \tag{2.100}$$

$$-\eta^{0\mu}\int d^3r\left[\mathcal{L}(x), \psi_{A'\alpha'i'}(x')\right]_{-}.$$

Owing to the conservation law for $\mathbf{P}^{\mu}$ we can choose an arbitrary moment t in $x = (\mathbf{r}, t)$ without loss of generality. We thus choose $t = t'$ in (2.100) and can directly evaluate the commutators in (2.100). One obtains

$$\left[\mathbf{P}^{\mu}, \psi_{A'\alpha'i'}(x')\right]_{-} = \sum_{i=1}^{3}\frac{i\hbar}{\lambda_i}\int d^3r\left\{\psi_{\alpha i}^{+}(x)\left[\partial^{\mu}\psi_{A\alpha i}(x), \psi_{A'\alpha'i'}(x')\right]_{+}^{t=t'} \tag{2.101}\right.$$

$$-\left[\psi_{A\alpha i}^{+}(x), \psi_{A'\alpha'i'}(x')\right]_{+}^{t=t'}\partial^{\mu}\psi_{A\alpha i}(x)\right\}$$

$$-\eta^{0\mu}\int d^3r\left[\mathcal{L}(x), \psi_{A'\alpha'i'}(x')\right]_{-}^{t=t'}.$$

To evaluate the anticommutators we observe that with (2.61) we have for spatial derivatives

$$\partial^k\left[\psi_{A\alpha i}(x), \psi_{A'\alpha'i'}(x')\right]_{+}^{t=t'} = \left[\partial^k\psi_{A\alpha i}(x), \psi_{A'\alpha'i'}(x')\right]_{+}^{t=t'} = 0. \tag{2.102}$$

Together with the $\psi^{+}-\psi$ anticommutator this yields

$$\left[\mathbf{P}^{\mu}, \psi_{A'\alpha'i'}(x')\right]_{-} = -i\hbar\partial^{\mu}\psi_{A'\alpha'i'}(x') \qquad (2.103)$$

$$+\eta^{0\mu}\sum_{i=1}^{3}\frac{i\hbar}{\lambda_i}\int d^3r\,\psi^{+}_{A\alpha i}(x)\left[\partial^0\psi_{A\alpha i}(x), \psi_{A'\alpha'i'}(x')\right]_{+}^{t=t'}$$

$$-\eta^{0\mu}\int d^3r\left[\mathcal{L}(x), \psi_{A'\alpha'i'}(x')\right]_{-}^{t=t'}.$$

Hence we have to show that the last two terms cancel each other. We decompose $\mathcal{L}$ into $\mathcal{L}_{\text{kin}} + \mathcal{L}_{\text{int}}$. With (2.52) there holds

$$\left[\mathcal{L}_{\text{kin}}(x), \psi_{A'\alpha'i'}(x')\right]_{-}^{t=t'} \qquad (2.104)$$

$$= \sum_{i=1}^{3}\lambda_i^{-1}\left[\bar{\psi}_{A\alpha i}(x)(i\hbar\gamma^{\mu}\partial_{\mu} - cm_i)_{\alpha\beta}\psi_{A\beta i}(x), \psi_{A'\alpha'i'}(x')\right]_{-}^{t=t'}$$

$$= \sum_{i=1}^{3}\lambda_i^{-1}\bar{\psi}_{A\alpha i}(x)\left[(i\hbar\gamma^{\mu}\partial_{\mu} - cm_i)_{\alpha\beta}\psi_{A\beta i}(x), \psi_{A'\alpha'i'}(x')\right]_{+}^{t=t'}$$

$$\div \sum_{i=1}^{3}\lambda_i^{-1}\left[\bar{\psi}_{A\alpha i}(x), \psi_{A'\alpha'i'}(x')\right]_{+}^{t=t'}(i\hbar\gamma^{\mu}\partial_{\mu} - cm_i)_{\alpha\beta}\psi_{A\beta i}(x).$$

Using the field equations (2.48) and the properties of the anticommutators we obtain

$$\left[\mathcal{L}_{\text{kin}}(x), \psi_{A'\alpha'i'}(x')\right]_{-}^{t=t'} \qquad (2.105)$$

$$= \sum_{i=1}^{3}\frac{i\hbar}{\lambda_i}\psi^{+}_{A\beta i}(x)\left[\partial_0\psi_{A\beta i}(x), \psi_{A'\alpha'i'}(x')\right]_{+}^{t=t'}$$

$$-g\sum_{i=1}^{3}\left[\bar{\psi}_{A\alpha i}(x), \psi_{A'\alpha'i'}(x')\right]_{+}^{t=t'}V_{ABCD}\sum_{\substack{\alpha\beta\gamma\delta\\ j,k,l=1}}^{3}\psi_{B\beta j}(x)\bar{\psi}_{C\gamma k}(x)\psi_{D\delta l}(x).$$

Finally we evaluate the commutator for $\mathcal{L}_{\text{int}}$. According to (2.52) we have

$$\left[\mathcal{L}_{\text{int}}(x), \psi_{A'\alpha'i'}(x')\right]_{-}^{t=t'} \qquad (2.106)$$

$$= -\frac{g}{2}V_{ABCD}\sum_{\substack{\alpha\beta\gamma\delta\\ i,j,k,l=1}}^{3}\left[\bar{\psi}_{A\alpha i}(x)\psi_{B\beta j}(x)\bar{\psi}_{C\gamma k}(x)\psi_{D\delta l}(x), \psi_{A'\alpha'i'}(x')\right]_{-}^{t=t'}$$

$$= \frac{g}{2}V_{ABCD}\sum_{\substack{\alpha\beta\gamma\delta\\ i,j,k,l=1}}^{3}\left\{\bar{\psi}_{A\alpha i}(x)\psi_{B\beta j}(x)\left[\bar{\psi}_{C\gamma k}(x), \psi_{A'\alpha'i'}(x')\right]_{+}^{t=t'}\psi_{D\delta l}(x)\right.$$

$$\left.+\left[\bar{\psi}_{A\alpha i}(x), \psi_{A'\alpha'i'}(x')\right]_{+}^{t=t'}\psi_{B\beta j}(x)\bar{\psi}_{C\gamma k}(x)\psi_{D\delta l}(x)\right\}.$$

If one substitutes (2.105) and (2.106) into (2.103), the first term on the right hand side of (2.105) and the second term on the right hand side of (2.103) cancel each other. Hence the resulting expression reads

$$\left[\mathbf{P}^\mu, \psi_{A'\alpha'i'}(x')\right]_- \tag{2.107}$$

$$= -i\hbar\partial^\mu\psi_{A\alpha i'}(x')$$

$$+\eta^{0\mu}g\int d^3r V_{\substack{ABCD\\ \alpha\beta\gamma\delta}}\sum_{i,j,k,l=1}^{3}\left[\bar{\psi}_{A\alpha i}(x),\psi_{A'\alpha'i'}(x')\right]_+^{t=t'}\psi_{B\beta j}(x)\bar{\psi}_{C\gamma k}(x)\psi_{D\delta l}(x)$$

$$-\eta^{0\mu}\frac{g}{2}\int d^3r V_{\substack{ABCD\\ \alpha\beta\gamma\delta}}\sum_{i,j,k,l=1}^{3}\left\{\bar{\psi}_{A\alpha i}(x)\psi_{B\beta j}(x)\left[\bar{\psi}_{C\gamma k}(x),\psi_{A'\alpha'i'}(x')\right]_+^{t=t'}\psi_{D\delta l}(x)\right.$$

$$\left.+\left[\bar{\psi}_{A\alpha i}(x),\psi_{A'\alpha'i'}(x')\right]_+^{t=t'}\psi_{B\beta j}(x)\bar{\psi}_{C\gamma k}(x)\psi_{D\delta l}(x)\right\}.$$

According to (2.62) the anticommutators for the sums of auxiliary fields, *i.e.*, for the Ψ fields, vanish. We use this property for the rearrangement

$$V_{\substack{ABCD\\ \alpha\beta\gamma\delta}}\sum_{i,j,k,l=1}^{3}\left[\bar{\psi}_{C\gamma k}(x),\psi_{A'\alpha'i'}(x')\right]_+^{t=t'}\bar{\psi}_{A\alpha i}(x)\psi_{B\beta j}(x)\psi_{D\delta l}(x) \tag{2.108}$$

$$= V_{\substack{ABCD\\ \alpha\beta\gamma\delta}}\sum_{i,j,k,l=1}^{3}\left[\bar{\psi}_{C\gamma k}(x),\psi_{A'\alpha'i'}(x')\right]_+^{t=t'}\psi_{D\delta l}(x)\bar{\psi}_{A\alpha i}(x)\psi_{B\beta j}(x)$$

$$= V_{\substack{ABCD\\ \alpha\beta\gamma\delta}}\sum_{i,j,k,l=1}^{3}\left[\bar{\psi}_{A\alpha i}(x),\psi_{A'\alpha'i'}(x')\right]_+^{t=t'}\psi_{B\beta j}(x)\bar{\psi}_{C\gamma k}(x)\psi_{D\delta l}(x),$$

where in the last step we performed a re-indexing $(A\alpha i),(B\beta j)\rightleftharpoons(C\gamma k),(D\delta l)$ and used $V_{\substack{ABCD\\ \alpha\beta\gamma\delta}}=V_{\substack{CDAB\\ \gamma\delta\alpha\beta}}$. By substitution of (2.108) into (2.107), apart from the first term on the right hand side all other terms cancel each other. $\diamond$

In an analogous way one can show that equations (2.96) can be satisfied with

$$\mathbf{M}^{\mu\nu} = \int d^3r\left[\pi_{A\alpha i}(x)\Sigma_{\alpha\beta}^{\mu\nu}\psi_{A\beta i}(x)+\pi_{A\alpha i}(x)(x^\mu\partial^\nu-x^\nu\partial^\mu)\psi_{A\alpha i}(x)\right.$$

$$\left.+(x^\nu\eta^{0\mu}-x^\mu\eta^{0\nu})\mathcal{L}(x)\right]. \tag{2.109}$$

The operators $\mathbf{P}^\mu$ and $\mathbf{M}^{\mu\nu}$ themselves satisfy the Lie algebra relations

$$[\mathbf{P}^\mu,\mathbf{P}^\nu]_- = 0, \tag{2.110}$$

$$[\mathbf{P}^\rho,\mathbf{M}^{\mu\nu}]_- = i\hbar\left(\eta^{\rho\nu}\mathbf{P}^\mu-\eta^{\rho\mu}\mathbf{P}^\nu\right),$$

$$[\mathbf{M}^{\mu\nu},\mathbf{M}^{\rho\sigma}]_- = i\hbar\left(\eta^{\mu\rho}\mathbf{M}^{\nu\sigma}+\eta^{\nu\sigma}\mathbf{M}^{\mu\rho}-\eta^{\mu\sigma}\mathbf{M}^{\nu\rho}-\eta^{\nu\rho}\mathbf{M}^{\mu\sigma}\right),$$

i.e., $\mathbf{P}^\mu$ and $\mathbf{M}^{\mu\nu}$ given by (2.99) and (2.109) are the generators of a representation of the Poincaré group in the state space $\mathbb{H}$. Furthermore, $\mathbf{P}^\mu$ and $\mathbf{M}^{\mu\nu}$ are assumed to be self-adjoint operators, so they are global observables in our theory.

The proof of the Lie algebra relations (2.110) can be directly performed in analogy to the proof of Proposition 2.3 leading to elementary but extensive calculations. For simplicity we thus confine ourselves to a weaker version of (2.110) which has the advantage of elementary arguments.

- **Proposition 2.4:** Let $A_{Z_1...}(x)$ be any local polynomial of field operators $\psi_Z(x)$ in the field operator algebra $\mathcal{A}$. Then with definitions (2.99) and (2.109) the Lie algebra relations (2.110) hold as 'weak' relations, *i.e.*, in application to $A_{Z_1...}(x)$.

Proof: Without loss of generality we consider the special element $A_{Z_1...}(x) = \psi_Z(x)$. The application of $\mathbf{P}^\mu$ and $\mathbf{M}^{\mu\nu}$ to $\psi_Z(x)$ is given by (2.94)–(2.97). Then we have for instance

$$
\begin{aligned}
[\mathbf{P}^\rho, [\mathbf{M}^{\mu\nu}, \psi_{A\alpha i}(x)]_-]_- &= (-i\hbar)^2 \partial^\rho \left[(x^\mu \partial^\nu - x^\nu \partial^\mu)\delta_{\alpha\beta} + \Sigma_{\alpha\beta}^{\mu\nu} \right] \psi_{A\beta i}(x) \\
&= (-i\hbar)^2 \left[(\eta^{\mu\rho}\partial^\nu - \eta^{\nu\rho}\partial^\mu) \right] \psi_{A\alpha i}(x) \\
&\quad + (-i\hbar)^2 \left[(x^\mu \partial^\nu - x^\nu \partial^\mu)\delta_{\alpha\beta} + \Sigma_{\alpha\beta}^{\mu\nu} \right] \partial^\rho \psi_{A\beta i}(x) \\
&= -i\hbar \left[(\eta^{\mu\rho}\mathbf{P}^\nu - \eta^{\nu\rho}\mathbf{P}^\mu), \psi_{A\alpha i}(x) \right]_- \\
&\quad + [\mathbf{M}^{\mu\nu}, [\mathbf{P}^\rho, \psi_{A\alpha i}(x)]_-]_- \, .
\end{aligned}
\tag{2.111}
$$

Furthermore, there holds

$$
[\mathbf{P}^\rho, [\mathbf{M}^{\mu\nu}, \psi_{A\alpha i}(x)]_-]_- - [\mathbf{M}^{\mu\nu}, [\mathbf{P}^\rho, \psi_{A\alpha i}(x)]_-]_- = [[\mathbf{P}^\rho, \mathbf{M}^{\mu\nu}]_-, \psi_{A\alpha i}(x)]_- , \tag{2.112}
$$

and hence combination with (2.111) yields

$$
[[\mathbf{P}^\rho, \mathbf{M}^{\mu\nu}]_-, \psi_{A\alpha i}(x)]_- = -i\hbar \left[(\eta^{\mu\rho}\mathbf{P}^\nu - \eta^{\nu\rho}\mathbf{P}^\mu), \psi_{A\alpha i}(x) \right]_- , \tag{2.113}
$$

i.e., the 'weak' version of the algebra relation

$$
[\mathbf{P}^\rho, \mathbf{M}^{\mu\nu}]_- = i\hbar \left(\eta^{\rho\nu}\mathbf{P}^\mu - \eta^{\rho\mu}\mathbf{P}^\nu \right) . \tag{2.114}
$$

The other relations of (2.110) can be treated in the same way. $\Diamond$

For the Poincaré group the maximal set of commuting observables is given by the set of generators $\{\mathbf{P}^2, \mathbf{P}_\mu, \mathbf{W}^2, \mathbf{W}_3\}$, where the Casimir operators of the Poincaré group are defined by

$$
\mathbf{P}^2 = \mathbf{P}_\mu \mathbf{P}^\mu , \quad \mathbf{W}^2 = \mathbf{W}_\mu \mathbf{W}^\mu \tag{2.115}
$$

with

$$
\mathbf{W}_\mu := \frac{1}{2} \left(\mathbf{P}^2 \right)^{-\frac{1}{2}} \epsilon_{\mu\nu\rho\sigma} \mathbf{M}^{\nu\rho} \mathbf{P}^\sigma . \tag{2.116}
$$

As commuting observables can have common eigenstates, we consider those states $|p\rangle$ which simultaneously diagonalize $\mathbf{P}^2, \mathbf{P}_\mu, \mathbf{W}^2$ and $\mathbf{W}_3$, *i.e.*,

$$
\begin{aligned}
\mathbf{P}^2 |p\rangle &= (mc)^2 |p\rangle, \\
\mathbf{P}_\mu |p\rangle &= p_\mu |p\rangle, \\
\mathbf{W}^2 |p\rangle &= \hbar^2 s(s+1) |p\rangle, \\
\mathbf{W}_3 |p\rangle &= \hbar s_3 |p\rangle,
\end{aligned}
\tag{2.117}
$$

where the $\mathbf{W}_3$ equation is referred to the rest system, *i.e.*, for $\mathbf{p} = 0$.

Because of (2.98) and (2.99) all operators of the set of equations (2.117) are referred to a fixed, but arbitrary, instant of time. Hence if we consider the common eigenstates of these operators, they depend on this instant of time, $i.e.$, $|p\rangle \equiv |p, t\rangle$. The time-translation generator $\mathbf{P}^0$ is a conserved quantity, so we are allowed to set $t = 0$. As the eigenstates $|p\rangle$ of (2.117) are assumed to form a basis of the full state space IH (for a detailed discussion see Chapter 4), this means that we work with time independent states, $i.e.$, with Heisenberg states, whilst the field operators carry the full time dependence. On the other hand, in Chapter 4 we will give a description of quantum fields which replaces the Schrödinger equation of quantum mechanics by a functional energy equation for quantum fields. So our formulation may be regarded as a kind of fusion of Heisenberg and Schrödinger picture.

Finally, in order to satisfy (2.85) the eigenvalues of $\mathbf{P}_\mu$ have to be renormalized with respect to the vacuum, $i.e.$, $\mathbf{P}_\mu|0\rangle = 0$ must hold. We will explicitly perform this renormalization in Chapter 4.

In addition to the space–time coordinates x and the spinor indices α the field operators depend upon purely algebraic degrees of freedom A, which are assumed to represent isospin. The corresponding $SU(2)$ transformations for the classical fields are given by

$$\psi'_{A\alpha i}(x) = S_{AA'}(a_k)\psi_{A'\alpha i}(x), \tag{2.118}$$
$$\bar{\psi}'_{A\alpha i}(x) = S_{AA'}(-a_k)\bar{\psi}_{A'\alpha i}(x)$$

with

$$S(a_k) := \exp(ia_k\sigma^k), \tag{2.119}$$

where σ^k, $k = 1, 2, 3$, are the Pauli matrices and $a_k \in \mathrm{IR}$.

Furthermore, for $\sigma^0 \equiv \mathbb{1}$ we have the $U(1)$ transformation

$$\psi'_{A\alpha i}(x) = S_{AA'}(a_0)\psi_{A'\alpha i}(x), \tag{2.120}$$
$$\bar{\psi}'_{A\alpha i}(x) = S_{AA'}(-a_0)\bar{\psi}_{A'\alpha i}(x)$$

with

$$S(a_0) := \exp(ia_0\sigma^0). \tag{2.121}$$

In analogy with (2.89) the algebraic transformations induce unitary transformations $\mathbf{U}(a_k)$ and $\mathbf{U}(a_0)$ in the corresponding Hilbert space. The corresponding transformation laws for the spinor field operators read

$$\psi'_{A\alpha i}(x) := \mathbf{U}(a_k)\psi_{A\alpha i}(x)\mathbf{U}^{-1}(a_k) = S^{-1}_{AA'}(a_k)\psi_{A'\alpha i}(x) \tag{2.122}$$

and

$$\bar{\psi}'_{A\alpha i}(x) := \mathbf{U}(-a_k)\bar{\psi}_{A\alpha i}(x)\mathbf{U}^{-1}(-a_k) = S^{-1}_{AA'}(-a_k)\bar{\psi}_{A'\alpha i}(x). \tag{2.123}$$

The matrix elements of spinor field operators transform by analogy with (2.87).

The generators of the transformations $\mathbf{U}$ are given by

$$I_k = \frac{1}{2} \int d\mathbf{r} \, \psi^+_{A\alpha i}(\mathbf{r}, t) \sigma^k_{AA'} \psi_{A'\alpha i}(\mathbf{r}, t) \qquad (2.124)$$

with commutation relations

$$[I_k, I_j]_- = \varepsilon_{kjl} I_l, \qquad (2.125)$$

whilst I_0 with

$$I_0 = \frac{1}{2} \int d\mathbf{r} \, \psi^+_{A\alpha i}(\mathbf{r}, t) \sigma^0_{AA'} \psi_{A'\alpha i}(\mathbf{r}, t) \qquad (2.126)$$

commutes with all I_k.

The regularized spinor field equations (2.48) and (2.49) are invariant under the transformations (2.122) and (2.123), hence the spinor field equations possess a global $SU(2) \times U(1)$ invariance group.

Apart from these continuous groups there also exist discrete symmetry operations for the spinor field equations (2.48) and (2.49). For a discussion of these symmetries we refer to Grimm [Gri 94a].

3 Covariant Quantum Field Dynamics

3.1 Introduction

Following the rules of quantum mechanics the theoretical description of quantum systems is based on the explicit knowledge of the corresponding state space. For quantum field theories these state spaces are assumed to be spanned by the sets of eigenstates of the commuting Poincaré generators with the corresponding eigenvalue equations (2.117). However, in contrast to quantum mechanics, in quantum field theory these eigenvalue equations cannot be simply integrated; rather the infinite number of degrees of freedom of quantum fields enforces a more sophisticated kind of treatment.

This is because, in contrast to quantum mechanics, systems with an infinite number of degrees of freedom possess infinitely many inequivalent representations, which means infinitely many different state representations. In particular, the only explicitly known state representation, the Fock representation, is a definitely wrong representation for quantum fields with mutual interaction. This result is known as Haag's theorem, cf. Haag [Haag 92].

In conventional quantum field theory this fundamental difficulty has promoted several attempts at extracting information without explicit state construction, and therefore without specifying the representation. Apart from many very specialized considerations there are two mainstreams for dealing with this problem; we briefly describe these approaches in order to contrast them with the conception pursued in this book.

The first one originates from perturbation theory which was covariantly reformulated by Dyson [Dys 49a,b], [Dys 51a–d]. Its basic elements are vacuum expectation values (VEV) of time ordered products of field operators for which corresponding equations can be derived, see Dyson [Dys 49a,b], Schwinger [Schwi 51a,b] and Freese [Free 51]. Therefore these equations may be called Schwinger–Dyson–Freese equations.

The explicit state space construction for the interacting quantum fields is circumvented by considering only asymptotic states of scattering processes. They are assumed to describe free ingoing or outgoing particles, where the state representations are referred to the corresponding free fields, and thus are Fock representations. The connection of the VEV with these asymptotic states, i.e., the S matrix, was formulated in an axiomatic way by Lehmann, Symanzik and Zimmermann [Leh 55] and is referred to as the LSZ formalism. In the subsequent development of the theory the Schwinger–Dyson–Freese equations were replaced by path integrals which, however, did not basically change this line of approach, cf., e.g., Faddeev and Slavnov [Fadd

80].

The second mainstream is the Wightman formalism. In this case the basic entities are the Wightman functions, which are defined as vacuum expectation values of ordinary products of field operators, see Wightman [Wigh 56]. But so far the axiomatic approach to relativistic quantum field theory "has not enabled us yet to construct or even characterize a specific theory" (Haag [Haag 92]) apart from models in lower dimensions, see, *e.g.*, Glimm and Jaffe [Gli 87]. The formalism yields an abstract reconstruction theorem, but no explicit state space construction method.

The common feature of these two approaches is the emphasis laid on a covariant formulation and that they use vacuum expectation values, *i.e.*, classical generalized functions for the characterization of a physical system. This appearance of VEV as the basic quantities of a quantum field indicates a change of the notion of a 'state' which is consequently elaborated in the formalism of Algebraic Quantum Theory. In this formalism a 'state' is a positive linear functional on a C^* algebra, *cf*. Bratteli and Robinson, [Brat 79].

The link between algebraic theory and conventional quantum theory is the GNS construction, see Gelfand and Naimark [Gel 43] and Segal [Seg 47], and also Bogoliubov, Logunov and Todorov [Bogo 75], Bratteli and Robinson [Brat 79], Haag [Haag 92], which depends on an algebraic state and allows the representation of the algebra of observables on a suitable cyclic Hilbert space and of an algebraic 'state' as a set of vacuum expectation values in this Hilbert space. Thus for a definite algebraic state the existence of a suitable corresponding state space is secured. With respect to our problem of inequivalent representations this result makes algebraic theory an attractive candidate for a framework for formulating our theory.

However, the problem of an explicit construction of the GNS Hilbert space is not solved in a general and constructive way, and if one looks more closely there are some additional handicaps in the algebraic theory owing to the presuppositions which have to be fulfilled in order to apply it.

As a consequence algebraic theory has so far been applied only to simple models of statistical mechanics. For a review of this line of research see Emch [Emch 72], Bratteli and Robinson [Brat 79], Sewell [Sew 86], Thirring [Thir 80], Primas [Prim 83], Rieckers [Rie 91], Haag [Haag 92], whilst there were no convincing applications in relativistic quantum field theory. Thus, in order to treat composite particles theoretically and practically in relativistic quantum field theory, we are forced to develop a new formalism which holds an intermediate position between algebraic and conventional theory and which may be called Functional Quantum Theory. Once this theory is established it can, of course, be applied to nonrelativistic models, too.

Functional Quantum Theory is based on the language of conventional quantum field theory. To see in which respect it is referred to the algebraic theory we briefly review the statement of the GNS construction.

The GNS theorem points out that for each algebraic state, *i.e.*, for each positive, linear and normalized functional ω on the algebra $\mathcal{A}$ of observables, there exists a cyclic Hilbert space $\mathbb{H}_\omega$ with a cyclic vector $|\Omega\rangle$, so that $\omega(A) = \langle\Omega|\pi_\omega(A)|\Omega\rangle$, where $\pi_\omega(A)$ is the representation of the algebra element $A \in \mathcal{A}$ in $\mathbb{H}_\omega$. Associated with this $|\Omega\rangle$ there is a set of (algebraic) states ω_a, with $\omega_a(A) = \langle a|\pi_\omega(A)|a\rangle$ for each $|a\rangle = B|\Omega\rangle$, $B \in \Pi_\omega(\mathcal{A})$. So, by means of these relations we may consider the algebraic states ω and ω_a as being represented by vectors $|\Omega\rangle \in \mathbb{H}_\omega$ or $|a\rangle \in \mathbb{H}_\omega$, respectively, and if we put $|\Omega\rangle = |0\rangle$ we obtain the usual vacuum expectation values as states $\omega(A)$.

We now deviate from this concept and assume that each state $|a\rangle$ within a generalized state space $\mathbb{H}$ is represented by the set of projections onto a cyclic basis of $\mathbb{H}$. In the algebraic language these projections are given by the functionals $\tilde{\omega}(A) = \langle 0|\pi_\omega(A)|a\rangle$, where we assume the cyclic vector $|\Omega\rangle$ to be identical with the physical vacuum $|0\rangle$. These functionals may be negative, so they are no longer proper algebraic states, and with respect to the formulation of a covariant theory it is necessary to extend such functionals to time ordered expressions.

For these transition matrix elements we will succeed in deriving corresponding Dyson–Schwinger–Freese equations and we will assume the representation of the theory to be fixed by these equations, including additional boundary conditions.

Within the covariant theory Nishijima [Nish 53] tried to apply covariant state representations based on a formal GNS construction for cyclic basis vectors with time ordered monomials of field operators. But owing to the difficulties of the covariant formalism Nishijima was not able to define and to evaluate the corresponding state representations unambiguously and he therefore returned to Green's functions techniques.

A more successful use of covariant transition matrix elements was made by the calculation of bound state Bethe–Salpeter amplitudes. As these amplitudes are very special matrix elements we will discuss this approach in more detail in the introduction of Chapter 6. Furthermore, as already mentioned above, Schwinger–Dyson–Freese equations can be applied to the calculation of transition matrix elements, too. To solve these equations suitable truncation procedures were proposed which were frequently referred to as New (covariant) Tamm–Dancoff Method, *cf.* Zimmermann [Zim 54], Heisenberg [Heis 66]. But this approach did not lead to widespread applications.

In analogy with statistical mechanics the theory of quantum fields is frequently formulated by means of generating functionals; see, for instance, Schwinger [Schwi 51a,b,c], [Schwi 53a,b,c], [Schwi 54a,b], Symanzik [Sym 54], Lurie [Luri 68], Rzewuski [Rze 69], Fried [Frie 72]. We follow this line and discuss our problem in terms of such functionals. Owing to our new concept we are, however, forced to deviate from the usual practice for functionals which considers functionals only as collections of an infinite number of matrix elements. In contrast to common usage we shall enlarge the concept of the generating functionals by embedding the latter into corresponding functional spaces. In this way we define functional states and use these states to establish a map between the original state space $\mathbb{H}$ and its functional image $\mathbb{K}_F$. As we want each 'state' from this functional space $\mathbb{K}_F$ to represent one state in $\mathbb{H}$, this map has to conserve the quantum numbers of a state in $\mathbb{H}$ and it has to be an isometric map.

This program is performed in this and the next chapter and leads to the Functional Quantum Theory. In this formalism we can synonymously speak of a state in $\mathbb{H}$ or of the corresponding state in the functional space $\mathbb{K}_F$ or of a suitable set of transition matrix elements. As we will derive equations for the functional images of the eigen-states of the Poincaré generators in $\mathbb{H}$, in principle the state space can be explicitly constructed.

In this chapter we will perform the abovementioned program under the additional condition of covariance.

3.2 Construction of Functional States

One of the aims of modern quantum field theory is to provide a manifestly relativistic covariant scheme of calculation. For such a formalism matrix elements of Heisenberg operators are an appropriate starting point. In the usual treatment time ordered vacuum expectation values are considered; see, for instance, Lurié [Luri 68], Itzykson and Zuber [Itz 80]. However, with regard to our concept the suitable quantities are transition matrix elements of time ordered products of field operators.

We define these matrix elements by

$$\tau^{(n)}(I_1 \ldots I_n|a) := \langle 0|T(\psi_{I_1} \ldots \psi_{I_n})|a\rangle, \quad \forall\, |a\rangle \in \mathbb{H}, \tag{3.1}$$

where $|0\rangle$ is a cyclic state of $\mathbb{H}$, the 'vacuum state', and the time ordering operator T is given by

$$T(\psi_{I_1} \ldots \psi_{I_n}) := \sum_{p \in S_n} \mathrm{sgn}(p)\Theta\left(t_{p_1}, \ldots, t_{p_n}\right) \psi_{I_{p_1}} \ldots \psi_{I_{p_n}} \tag{3.2}$$

with

$$\Theta(t_1, \ldots, t_n) := \Theta(t_1 - t_2)\Theta(t_2 - t_3)\ldots\Theta(t_{n-1} - t_n). \tag{3.3}$$

For later applications we introduce the equivalent notation

$$\tau^{(n)}(I_1 \ldots I_n|a) \equiv \tau^{(n)}_{Z_1 \ldots Z_n}(x_1 \ldots x_n|a). \tag{3.4}$$

We add that the time ordered products (3.2) are well defined only for unequal times $t_i \neq t_j$ for $i \neq j$.

The matrix elements (3.1) have remarkable properties. For their transformation properties under Poincaré transformations the following proposition holds:

- **Proposition 3.1:** The transformed matrix elements

$$\tau^{(n)}\left(I'_1 \ldots I'_n|a'\right) := \langle 0|T\left[\psi_{Z'_1}(x'_1) \ldots \psi_{Z'_n}(x'_n)\right]|a'\rangle \tag{3.5}$$

  with

$$x' := \Lambda x + b, \quad |a'\rangle := \mathbf{U}(\Lambda, b)|a\rangle,$$

  where $\mathbf{U}$ is defined in (2.85) and Λ orthochronous, obey the relation

$$\tau^{(n)}\left(I'_1 \ldots I'_n|a'\right) = L_{Z'_1 Z_1} \ldots L_{Z'_n Z_n}\tau^{(n)}(I_1 \ldots I_n|a), \tag{3.6}$$

  with $I = (Z, x)$ and $L_{Z'Z} \equiv L_{Z'Z}(\Lambda)$ given by (2.91).

Proof: Equation (2.90) can be resolved to give

$$\psi_{Z'}(x') = L_{Z'Z}(\Lambda)\mathbf{U}(\Lambda, b)\psi_Z(x)\mathbf{U}^{-1}(\Lambda, b). \tag{3.7}$$

Substitution of (3.7) into (3.5) yields with $I' = (Z', x')$

$$\tau^{(n)}\left(I'_1 \ldots I'_n | a'\right) \tag{3.8}$$

$$= \sum_{p \in S_n} \operatorname{sgn}(p)\Theta\left(t'_{p_1} \ldots t'_{p_n}\right) L_{Z'_1 Z_1} \ldots L_{Z'_n Z_n}$$

$$\times \langle 0 | \mathbf{U}(\Lambda, b)\psi_{Z_{p_1}}(x_{p_1})\mathbf{U}^{-1}(\Lambda, b) \ldots \mathbf{U}(\Lambda, b)\psi_{Z_{p_n}}(x_{p_n})\mathbf{U}^{-1}(\Lambda, b) | a' \rangle$$

$$= \sum_{p \in S_n} \operatorname{sgn}(p)\Theta\left(t'_{p_1} \ldots t'_{p_n}\right) L_{Z'_1 Z_1} \ldots L_{Z'_n Z_n} \langle 0 | \psi_{Z_{p_1}}(x_{p_1}) \ldots \psi_{Z_{p_n}}(x_{p_n}) | a \rangle.$$

For orthochronous transformations we have $\Theta(t'_{p_1} \ldots t'_{p_n}) \equiv \Theta(t_{p_1} \ldots t_{p_n})$, and substitution of this identity into (3.8) leads to (3.6).   ◇

Furthermore, we observe that owing to definition (3.2) for $t_1 \neq t_2 \ldots \neq t_n$ the matrix elements (3.1) are antisymmetric in their arguments:

$$\tau^{(n)}(I_1 \ldots I_i \ldots I_j \ldots I_n | a) = -\tau^{(n)}(I_1 \ldots I_j \ldots I_i \ldots I_n | a) \tag{3.9}$$

for all $i, j = 1, \ldots n$.

Thus these functions not only transform correctly under Poincaré transformations, *i.e.*, as corresponding spin tensors, but also provide a representation of the permutation group. The only drawback of this definition is the difficulty to continue (3.1) to equal times. This can only be achieved by a definite limiting process.

As one sees from definition (3.1) the number n of field operators can take an arbitrary value. Roughly speaking, by (3.1) a state $|a\rangle \in \mathbb{H}$ is characterized by its projections onto the set of basis vectors $\langle 0 | T(\psi_{I_1} \ldots \psi_{I_n})$, $n = 1 \ldots \infty$, where we assume that the state space $\mathbb{H}$ is isomorphic to its dual space $\mathbb{H}^d$. This concept will be discussed more precisely in Chapter 4. But before entering into the discussion of these projections we introduce the generating functionals as a means to formulate the dynamics of quantum fields and of composite particles in a compact form.

In accordance with our intention we endow the commonly used functionals with appropriate transformation properties and take them as states in a functional space with an inner product structure. The direct construction of representations of such spaces was given by Stumpf and Scheerer [Stu 75]. Here we only quote the results of this investigation.

Let $j_Z(x)$ and $\partial_Z(x)$, $x \in \mathbb{M}^4$, be the generators of a CAR algebra with the anticommutation relations

$$[j_Z(x), j_{Z'}(x')]_+ = [\partial_Z(x), \partial_{Z'}(x')]_+ = 0, \tag{3.10}$$

$$[j_Z(x), \partial_{Z'}(x')]_+ = \delta_{ZZ'}\delta(x - x'). \tag{3.11}$$

In addition we assume that these generators satisfy the following transformation rules under Poincaré transformations:

$$\mathcal{V}^{-1}(\Lambda, b) j_Z(x') \mathcal{V}(\Lambda, b) = j_{Z'}(x) L_{Z'Z}^{-1}(\Lambda), \tag{3.12}$$

$$\mathcal{V}^{-1}(\Lambda, b) \partial_Z(x') \mathcal{V}(\Lambda, b) = \partial_{Z'}(x) L_{Z'Z}(\Lambda), \tag{3.13}$$

where $\mathcal{V}(\Lambda, b)$ is a representation of the Poincaré group in functional space and $L_{ZZ'}$ is the transformation matrix (2.91). The anticommutators (3.10) and (3.11) remain invariant under the transformations (3.12) and (3.13).

According to the above quoted paper the compatibility of the postulates (3.10)–(3.13) is guaranteed and the explicit construction of a suitable representation shows that this Poincaré CAR algebra can be represented in a cyclic inner product space space $\mathbb{K}_F$, *i.e.*, there exists a cyclic 'functional ground state' $|0\rangle_F$ with $_F\langle 0|0\rangle_F = 1$ and

$$_F\langle 0| j_Z(x) = \partial_Z(x)|0\rangle_F = 0 \tag{3.14}$$

and with

$$\mathcal{V}(\Lambda, b)|0\rangle_F = |0\rangle_F . \tag{3.15}$$

For brevity we again denote the elements of the abstract algebra and of their representations by the same symbols; this will not give rise to misunderstandings.

By means of the cyclic vector $|0\rangle_F$ we obtain a set of basis vectors of the functional space $\mathbb{K}_F$ by

$$\mathbb{B}_F := \left\{ \frac{i^n}{n!} j_{I_1} \dots j_{I_n} |0\rangle_F , \quad I_1 \dots I_n \in \mathcal{I}, \quad n \in \mathbb{N} \right\} \tag{3.16}$$

with $j_I := j_Z(x)$ and where the factors $i^n/n!$ are introduced for convenience.

Using this construction we can now introduce generating functional states by the definition

$$|\mathcal{T}(j; a)\rangle := \sum_{n=0}^{\infty} \frac{i^n}{n!} \tau^{(n)}(I_1 \dots I_n|a) j_{I_1} \dots j_{I_n} |0\rangle_F \tag{3.17}$$

for all $|a\rangle \in \mathbb{H}$. With this definition we have collected all possible projections of a state $|a\rangle$ onto the dual states $\langle 0|T(\psi_{I_1} \dots \psi_{I_n})$, $n = 1 \dots \infty$, into one functional state which from now on we consider as a representation of $|a\rangle \in \mathbb{H}$ by its map into functional space $\mathbb{K}_F$.

In order to work with such functional states and to elaborate their properties we need some calculation rules in functional space:

- **Proposition 3.2:** For functional states

$$|\mathcal{T}(j; a)\rangle := \sum_{n=0}^{\infty} \frac{i^n}{n!} \tau^{(n)}(I_1 \dots I_n|a) j_{I_1} \dots j_{I_n} |0\rangle_F$$

the following relations hold:

i)
$$i^{-1}\partial_K|\mathcal{T}(j;a)\rangle = \sum_{n=0}^{\infty}\frac{i^n}{n!}\tau^{(n+1)}(K,I_1\ldots I_n|a)j_{I_1}\ldots j_{I_n}|0\rangle_F, \quad (3.18)$$

ii)
$$ij_K|\mathcal{T}(j;a)\rangle \qquad\qquad\qquad\qquad\qquad\qquad (3.19)$$
$$= \sum_{n=0}^{\infty}\frac{i^n}{n!}\sum_{l=1}^{n}(-1)^{l-1}\delta_{KI_l}\tau^{(n-1)}(I_1\ldots \overset{o}{I_l}\ldots I_n|a)j_{I_1}\ldots j_{I_n}|0\rangle_F,$$

iii)
$$i^{-m}{}_F\langle 0|\partial_{K_m}\ldots\partial_{K_1}|\mathcal{T}(j;a)\rangle = \tau^{(m)}(K_1\ldots K_m|a) \qquad (3.20)$$

with
$$\tau^{(n-1)}(I_1\ldots \overset{o}{I_l}\ldots I_n|a) := \tau^{(n-1)}(I_1\ldots I_{l-1},I_{l+1}\ldots I_n|a)$$
and $\delta_{I_1I_2}:=\delta_{Z_1Z_2}\delta(x_1-x_2)$.

Proof: For illustration we give a quite detailed treatment of these functional calculation rules.

i) With (3.11) and (3.14) there holds

$$\partial_K j_{I_1}\ldots j_{I_n}|0\rangle_F = \sum_{l=1}^{n}(-1)^{l-1}\delta_{KI_l}j_{I_1}\ldots \overset{o}{j_{I_l}}\ldots j_{I_n}|0\rangle_F. \qquad (3.21)$$

With that and by use of the antisymmetry (3.9) and re-indexing it follows

$$i^{-1}\partial_K|\mathcal{T}(j;a)\rangle = \sum_{n=1}^{\infty}\frac{i^{n-1}}{n!}\sum_{l=1}^{n}(-1)^{l-1}\tau^{(n)}(I_1\ldots I_{l-1},K,I_{l+1}\ldots I_n|a)$$
$$\times j_{I_1}\ldots \overset{o}{j_{I_l}}\ldots j_{I_n}|0\rangle_F$$
$$= \sum_{n=1}^{\infty}\frac{i^{n-1}}{n!}\sum_{l=1}^{n}\tau^{(n)}(K,I_1\ldots \overset{o}{I_l}\ldots I_n|a)j_{I_1}\ldots \overset{o}{j_{I_l}}\ldots j_{I_n}|0\rangle_F$$
$$= \sum_{n=1}^{\infty}\frac{i^{n-1}}{(n-1)!}\tau^{(n)}(K,I_1\ldots I_{n-1}|a)j_{I_1}\ldots j_{I_{n-1}}|0\rangle_F$$
$$= \sum_{n=0}^{\infty}\frac{i^n}{n!}\tau^{(n+1)}(K,I_1\ldots I_n|a)j_{I_1}\ldots j_{I_n}|0\rangle_F. \qquad (3.22)$$

ii) Owing to (3.10) we have

$$ij_K|\mathcal{T}(j;a)\rangle = \sum_{n=0}^{\infty}\frac{i^{n+1}}{n!}\tau^{(n)}(I_1\ldots I_n|a)j_K j_{I_1}\ldots j_{I_n}|0\rangle_F \qquad (3.23)$$
$$= \sum_{n=0}^{\infty}\frac{i^{n+1}}{n!}\tau^{(n)}(I_1\ldots I_n|a)\delta_{KJ}(-1)^{l-1}j_{I_1}\ldots j_{I_{l-1}}j j j_{I_l}\ldots j_{I_n}|0\rangle_F$$
$$= \sum_{n=0}^{\infty}\frac{i^{n+1}}{n!}\tau^{(n)}(I_1\ldots \overset{o}{I_l}\ldots I_{n+1}|a)\delta_{KI_l}(-1)^{l-1}j_{I_1}\ldots j_{I_{n+1}}|0\rangle_F$$
$$= \sum_{n=1}^{\infty}\frac{i^n}{(n-1)!}\tau^{(n-1)}(I_1\ldots \overset{o}{I_l}\ldots I_n|a)\delta_{KI_l}(-1)^{l-1}j_{I_1}\ldots j_{I_n}|0\rangle_F$$
$$= \sum_{n=1}^{\infty}\frac{i^n}{n!}\sum_{l=1}^{n}\tau^{(n-1)}(I_1\ldots \overset{o}{I_l}\ldots I_n|a)\delta_{KI_l}(-1)^{l-1}j_{I_1}\ldots j_{I_n}|0\rangle_F.$$

iii) We define the dual set of functional basis vectors by

$$\mathbb{B}_F^d := \left\{ {}_F\langle 0|(\partial_{K_1}\ldots\partial_{K_m}), \; K_1\ldots K_m \in \mathcal{I}, \; m \in \mathbb{N} \right\}. \tag{3.24}$$

From (3.10), (3.11) and (3.14) and from ${}_F\langle 0|0\rangle_F = 1$ it follows that

$$ {}_F\langle 0|\partial_{K_m}\ldots\partial_{K_1} j_{I_1}\ldots j_{I_n}|0\rangle_F = \delta_{mn} \sum_{p\in S_n} \mathrm{sgn}(p)\delta_{I_1 K_{p_1}}\ldots\delta_{I_n K_{p_n}}. \tag{3.25}$$

Thus we obtain for ${}_F\langle 0|0\rangle_F = 1$ the customary Fock space scalar product between the basis states from (3.16) and their duals (3.24).

With (3.25) and the antisymmetry (3.9) we obtain

$$i^{-m} {}_F\langle 0|\partial_{K_m}\ldots\partial_{K_1}|\mathcal{T}(j;a)\rangle$$

$$= \sum_{n=0}^{\infty} i^{-m}\frac{i^n}{n!}\tau^{(n)}(I_1\ldots I_n|a) {}_F\langle 0|\partial_{K_m}\ldots\partial_{K_1} j_{I_1}\ldots j_{I_n}|0\rangle_F$$

$$= \frac{1}{m!}\tau^{(m)}(I_1\ldots I_m|a) \sum_{p\in S_m} \mathrm{sgn}(p)\delta_{I_1 K_{p_1}}\ldots\delta_{I_m K_{p_m}}$$

$$= \tau^{(m)}(K_1\ldots K_m|a). \qquad\qquad\qquad \diamond$$

By means of (3.20) the individual $\tau^{(n)}$ functions can be regained from the functional state $|\mathcal{T}(j;a)\rangle$; we call this procedure the projection of functional states into coordinate space.

Finally, one should mention: as in ordinary Minkowski space the scalar products between the original basis states (3.16) and the dual basis states (3.24) give no hint about the metrical structure of $\mathbb{K}_F$. Rather, for this purpose we have to consider the scalar products between the original basis states themselves, i.e., we have to calculate the scalar products ${}_F\langle 0|(j_{K_1}\ldots j_{K_m})^+(j_{I_1}\ldots j_{I_n})|0\rangle_F$. It was shown by Stumpf and Scheerer [Stu 75] that $\partial_Z(x) \neq j_Z(x)^+$ and the metrical tensor for these basis states was derived. Thus with $\langle\mathcal{T}(j,a)| := |\mathcal{T}(j,a)\rangle^+$ one can evaluate scalar products $\langle\mathcal{T}(j,a)|\mathcal{T}(j,a)\rangle$.

However, these scalar products in functional space are completely independent from those of the state space $\mathbb{H}$ and cannot be used for the construction of an isometric map between $\mathbb{H}$ and $\mathbb{K}_F$. The functional space is an artificial construction and one cannot expect to guess the physical scalar products by a priori constructions in functional space, as these physical scalar products depend on the system dynamics itself. The theory of isometric maps will be developed in Chapter 4.

3.3 Symmetries in Functional Space

By means of definition (3.17) a map between the physical state space $\mathbb{H}$ and the functional state space $\mathbb{K}_F$ is established. According to our assumption that the basis states of $\mathbb{H}$ are eigenstates of equations (2.117), $\mathbb{H}$ is a representation space of the

Poincaré group. Because we want to describe a quantum system in the functional space $I\!K_F$ the image state $|\mathcal{T}(j;p)\rangle \in I\!K_F$ of a basis state $|p\rangle \in I\!H$ should be characterized by the same quantum numbers as $|p\rangle$. This implies that $I\!K_F$ has to be a representation space of the Poincaré group, too.

By the postulates (3.12), (3.13), (3.15) $I\!K_F$ is endowed with transformation properties. In this section we explore the consequences of these transformation properties in more detail, based on theorems which were given by Stumpf, Scheerer and Märtl in [Stu 70b] and [Stu 72]. It will turn out that $I\!K_F$ indeed carries a representation of the Poincaré group.

- **Proposition 3.3:** The Poincaré transformed functional state

$$|\mathcal{T}(j;a')\rangle := \sum_{n=0}^{\infty} \frac{i^n}{n!} \tau^{(n)}(I_1' \dots I_n'|a') j_{I_1'} \dots j_{I_n'} |0\rangle_F \qquad (3.26)$$

with $I' = (Z', x')$, $x' = \Lambda x + b$ and Λ orthochronous is related to the original state $|\mathcal{T}(j;a)\rangle$ by the transformation rule

$$|\mathcal{T}(j;a')\rangle = \mathcal{V}(\Lambda, b)|\mathcal{T}(j;a)\rangle. \qquad (3.27)$$

Proof: With (3.6), (3.12) and (3.15) we have

$$\begin{aligned}
|\mathcal{T}(j;a')\rangle &= \sum_{n=0}^{\infty} \frac{i^n}{n!} \tau_{Z_1 \dots Z_n}^{(n)}(x_1 \dots x_n|a) L_{Z_1' Z_1}(\Lambda) \dots L_{Z_n' Z_n}(\Lambda) \qquad (3.28) \\
&\quad \times j_{Z_1'}(x_1') \dots j_{Z_n'}(x_n')|0\rangle_F \\
&= \sum_{n=0}^{\infty} \frac{i^n}{n!} \tau_{Z_1 \dots Z_n}^{(n)}(x_1 \dots x_n|a) L_{Z_1' Z_1}(\Lambda) \dots L_{Z_n' Z_n}(\Lambda) \\
&\quad \times L_{Z_1'' Z_1'}^{-1} \mathcal{V} j_{Z_1''}(x_1)\mathcal{V}^{-1} \dots L_{Z_n'' Z_n'}^{-1} \mathcal{V} j_{Z_n''}(x_n)\mathcal{V}^{-1}|0\rangle_F
\end{aligned}$$

with $\mathcal{V} \equiv \mathcal{V}(\Lambda, b)$. The products $\mathcal{V}^{-1}\mathcal{V}$ and LL^{-1} cancel out and we obtain

$$\begin{aligned}
|\mathcal{T}(j;a')\rangle &= \sum_{n=0}^{\infty} \frac{i^n}{n!} \tau^{(n)}(I_1 \dots I_n|a)\mathcal{V}(\Lambda, b) j_{I_1} \dots j_{I_n}|0\rangle_F \qquad (3.29) \\
&= \mathcal{V}(\Lambda, b)|\mathcal{T}(j;a)\rangle. \qquad \qquad \Diamond
\end{aligned}$$

In order to derive the functional counterpart of equations (2.117) we first prove the following proposition:

- **Proposition 3.4:** The effect of the Poincaré generators $P_\mu(x)$ and $M_{\mu\nu}(x)$ in function space on matrix elements of time ordered products of field operators (3.1) is given by the relations

$$\text{i)} \qquad \sum_{i=1}^{n} P_\mu(x_i)\langle 0|T(\psi_{I_1} \dots \psi_{I_n})|a\rangle \qquad (3.30)$$

$$= \langle 0|T \sum_{i=1}^{n} P_\mu(x_i)(\psi_{I_1} \dots \psi_{I_n})|a\rangle,$$

$$\text{ii)} \qquad \sum_{i=1}^{n} M_{\mu\nu}(x_i) \langle 0| T \left(\psi_{I_1} \dots \psi_{I_n} \right) |a\rangle \tag{3.31}$$

$$= \langle 0| T \sum_{i=1}^{n} M_{\mu\nu}(x_i) \left(\psi_{I_1} \dots \psi_{I_n} \right) |a\rangle .$$

Proof: i) We have to show that $\sum_{i=1}^{n} P_\mu(x_i)$ commutes with time ordering. This is nontrivial only for $\mu = 0$. With $P_0(x_i) := -i\hbar \partial_0(x_i) \equiv -i\frac{\hbar}{c} \partial/\partial t_i$ and (3.2) we obtain

$$\sum_{i=1}^{n} \partial_0(x_i) \sum_{p \in S_n} \text{sgn}(p) \Theta(t_{p_1}, \dots, t_{p_n}) \langle 0| \psi_{I_{p_1}} \dots \psi_{I_{p_n}} |a\rangle \tag{3.32}$$

$$= \sum_{p \in S_n} \text{sgn}(p) \langle 0| \psi_{I_{p_1}} \dots \psi_{I_{p_n}} |a\rangle \sum_{i=1}^{n} \partial_0(x_{p_i}) \Theta(t_{p_1}, \dots, t_{p_n})$$

$$+ \langle 0| T \left[\sum_{i=1}^{n} \partial_0(x_i) \psi_{I_1} \dots \psi_{I_n} \right] |a\rangle .$$

Furthermore, with (3.3) we have

$$\left[\sum_{i=1}^{n} \partial_0(x_{p_i}) \right] \Theta(t_{p_1}, \dots t_{p_n}) \tag{3.33}$$

$$= \sum_{\alpha=1}^{n-1} \left\{ \Theta(t_{p_1} - t_{p_2}) \dots \left[\sum_{i=1}^{n} \partial_0(x_{p_i}) \Theta(t_{p_\alpha} - t_{p_{\alpha+1}}) \right] \dots \Theta(t_{p_{n-1}} - t_{p_n}) \right\},$$

where the differentiation is applied to any factor in the product. Because of $d\Theta(t)/dt = \delta(t)$ it follows that

$$\left[\sum_{i=1}^{n} \partial_0(x_{p_i}) \right] \Theta(t_{p_k} - t_{p_{k+1}}) = 0 \qquad \text{for} \quad k = 1, \dots n-1, \tag{3.34}$$

so relation (3.30) is proved.

ii) We have to show that $\sum_{i=1}^{n} M_{\mu\nu}(x_i)$ commutes with time ordering. With (2.97) the only nontrivial part of the commutator comes from the term $\sum_{i=1}^{n} x_\mu^i \partial_0(x_i)$. The proof runs along the same lines as for i). We replace in (3.32) $\sum_{i=1}^{n} \partial_0(x_{p_i})$ by $\sum_{i=1}^{n} x_\mu^{p_i} \partial_0(x_{p_i})$. Then the analogue to (3.34) is

$$\left[\sum_{i=1}^{n} x_\mu^{p_i} \partial_0(x_{p_i}) \right] \Theta(t_{p_k} - t_{p_{k+1}}) = 0 \qquad \text{for} \quad k = 1, \dots n-1, \tag{3.35}$$

which is valid because we have restricted the definition of time ordered products to unequal times $t_i \neq t_j$, $\forall\, i \neq j$. $\diamond$

By means of Proposition 3.4 we can now derive the symmetry conditions for $\tau^{(n)}$ functions (3.1).

- **Proposition 3.5:** For all states $|p\rangle \in \mathbb{H}$ being eigenstates of equations (2.117) the corresponding $\tau^{(n)}$ functions satisfy the symmetry conditions

i) $$\left[\sum_{i=1}^{n} P_\mu(x_i) + p_\mu\right] \tau^{(n)}(I_1 \ldots I_n|p) = 0, \tag{3.36}$$

ii) $$\left[\sum_{i=1}^{n} P_\mu(x_i) \sum_{j=1}^{n} P^\mu(x_j) - (mc)^2\right] \tau^{(n)}(I_1 \ldots I_n|p) = 0, \tag{3.37}$$

iii) $$\left[\frac{1}{4m^2} \epsilon_{\mu\nu\rho\sigma} \epsilon^{\mu\nu'\rho'\sigma'} \sum_{i,j,k,l=1}^{n} P^\nu(x_i) M^{\rho\sigma}(x_j) P_{\nu'}(x_k) M_{\rho'\sigma'}(x_l)\right.$$
$$\left. - \hbar^2 s(s+1)\right] \tau^{(n)}(I_1 \ldots I_n|p) = 0, \tag{3.38}$$

iv) $$\left[\frac{1}{2m} \epsilon_{3\nu\rho\sigma} \sum_{i,j=1}^{n} P^\nu(x_i) M^{\rho\sigma}(x_j) - \hbar s_3\right] \tau^{(n)}(I_1 \ldots I_n|p) = 0. \tag{3.39}$$

Proof: According to (2.117) there holds $\mathbf{P}_\mu|p\rangle = p_\mu|p\rangle$ and therefore

$$\langle 0|T(\psi_{I_1} \ldots \psi_{I_n})\mathbf{P}_\mu|p\rangle = p_\mu\langle 0|T(\psi_{I_1} \ldots \psi_{I_n})|p\rangle. \tag{3.40}$$

Owing to the commutator of $\mathbf{P}_\mu$ and ψ_I given by (2.94) we can derive the commutator of $\mathbf{P}_\mu$ with $T(\psi_{I_1} \ldots \psi_{I_n})$ and obtain

$$T(\psi_{I_1} \ldots \psi_{I_n})\mathbf{P}_\mu = \mathbf{P}_\mu T(\psi_{I_1} \ldots \psi_{I_n}) - \sum_{i=1}^{n} P_\mu(x_i)T(\psi_{I_1} \ldots \psi_{I_n}). \tag{3.41}$$

Substitution of (3.41) into (3.40) and observing $\langle 0|\mathbf{P}_\mu = 0$ we obtain equation (3.36). Taking into account that the commutators of the rotation operators $\mathbf{M}_{\mu\nu}$ with ψ_I are given by (2.96), we can derive equations (3.37)–(3.39) in a way analogous to (3.36). $\diamond$

In the next step we formulate the symmetry conditions for the generating functional states (3.17).

- **Proposition 3.6:** For all states $|p\rangle \in \mathbb{H}$ being eigenstates of equations (2.117) the corresponding generating functional states $|\mathcal{T}(j;p)\rangle$ satisfy the symmetry conditions

$$\mathcal{P}_\mu|\mathcal{T}(j;p)\rangle = p_\mu|\mathcal{T}(j;p)\rangle, \tag{3.42}$$
$$\mathcal{P}_\mu\mathcal{P}^\mu|\mathcal{T}(j;p)\rangle = (mc)^2|\mathcal{T}(j;p)\rangle, \tag{3.43}$$
$$\mathcal{W}_\mu\mathcal{W}^\mu|\mathcal{T}(j;p)\rangle = \hbar^2 s(s+1)|\mathcal{T}(j;p)\rangle, \tag{3.44}$$
$$\mathcal{W}_3|\mathcal{T}(j;p)\rangle = \hbar s_3|\mathcal{T}(j;p)\rangle \tag{3.45}$$

with

$$\mathcal{P}_\mu := -\int j_Z(x) P_\mu(x) \partial_Z(x) d^4x, \tag{3.46}$$

$$\mathcal{W}_\mu := \frac{1}{2mc} \epsilon_{\mu\nu\rho\sigma} \mathcal{P}^\nu \mathcal{M}^{\rho\sigma}, \tag{3.47}$$

$$\mathcal{M}^{\rho\sigma} := -\int j_Z(x) M^{\rho\sigma}(x) \partial_Z(x) d^4x. \tag{3.48}$$

Proof: We apply (3.21) to evaluate $\mathcal{P}_\mu |T(j;p)\rangle$. This gives

$$\int d^4x j_Z(x) P_\mu(x) \partial_Z(x) |T(j;p)\rangle \tag{3.49}$$

$$= -i\hbar \int d^4x j_Z(x) \partial_\mu(x)$$

$$\times \sum_{n=0}^\infty \frac{i^n}{n!} \sum_{l=1}^n (-1)^{l-1} \delta_{ZZ_l} \delta(x - x_l)$$

$$\times \langle 0|T(\psi_{I_1} \dots \psi_{I_n})|p\rangle j_{I_1} \dots \overset{o}{j}_{I_l} \dots j_{I_n}|0\rangle_\mathrm{F}$$

$$= -i\hbar \sum_{n=0}^\infty \frac{i^n}{n!} \sum_{l=1}^n (-1)^l j_Z(x_l) \delta_{ZZ_l} \partial_\mu(x_l) \langle 0|T(\psi_{I_1} \dots \psi_{I_n})|p\rangle$$

$$\times j_{I_1} \dots \overset{o}{j}_{I_l} \dots j_{I_n}|0\rangle_\mathrm{F}$$

$$= \sum_{n=0}^\infty \frac{i^n}{n!} \sum_{l=1}^n \left[P_\mu(x_l) \langle 0|T(\psi_{I_1} \dots \psi_{I_n})|p\rangle \right] j_{I_1} \dots j_{I_n}|0\rangle_\mathrm{F}.$$

Combination with equation (3.36) yields (3.42).

In the same way one can proceed to prove the remaining equations (3.43), (3.44) and (3.45). $\diamond$

- **Proposition 3.7:** The operators $\mathcal{P}_\mu$ and $\mathcal{M}_{\mu\nu}$ defined by (3.46) and (3.48) satisfy the Lie algebra relations

$$[\mathcal{P}_\mu, \mathcal{P}_\nu]_- = 0, \tag{3.50}$$

$$[\mathcal{P}_\lambda, \mathcal{M}_{\mu\nu}]_- = -i\hbar (\eta_{\lambda\nu} \mathcal{P}_\mu - \eta_{\lambda\mu} \mathcal{P}_\nu), \tag{3.51}$$

$$[\mathcal{M}_{\mu\nu}, \mathcal{M}_{\rho\sigma}]_- = -i\hbar (\eta_{\mu\rho} \mathcal{M}_{\nu\sigma} + \eta_{\nu\sigma} \mathcal{M}_{\mu\rho} - \eta_{\mu\sigma} \mathcal{M}_{\nu\rho} - \eta_{\nu\rho} \mathcal{M}_{\mu\sigma}) \tag{3.52}$$

of the Poincaré group in functional space.

Proof: Let $\{G_a(x)\}$ be a set of generators satisfying the Lie algebraic relations

$$[G_a(x), G_b(x)]_- = f^c{}_{ab} G_c(x) \tag{3.53}$$

in function space. Then the corresponding representation in functional space is given by $\mathcal{G}_a = -\int j_Z(x) G_a(x) \partial_Z(x) d^4x$. Application of the anticommutation relations (3.10) and (3.11) yields for $\mathcal{G}_a \mathcal{G}_b$ the expression

$$\mathcal{G}_a\mathcal{G}_b = \int j_Z(x)G_a(x)\delta_{ZZ'}\delta(x-x')G_b(x')\partial_{Z'}(x')d^4xd^4x' \tag{3.54}$$

$$-\int j_Z(x)j_{Z'}(x')G_a(x)G_b(x')\partial_Z(x)\partial_{Z'}(x')d^4x'd^4x.$$

In the last term of equation (3.54) the operators $G_a(x)$ and $G_b(x')$ are referred to different coordinates and thus commute. Hence we obtain

$$[\mathcal{G}_a,\mathcal{G}_b]_- = \int j_Z(x)G_a(x)G_b(x)\partial_Z(x)d^4x \tag{3.55}$$

$$-\int j_Z(x)G_b(x)G_a(x)\partial_Z(x)d^4x$$

$$= \int j_Z(x)\,[G_a(x),G_b(x)]_-\,\partial_Z(x)d^4x,$$

and with (3.53) this yields

$$[\mathcal{G}_a,\mathcal{G}_b]_- = -f^c{}_{ab}\mathcal{G}_c. \qquad\qquad \diamond$$

Therefore a representation of the Poincaré group in functional space is generated by the infinitesimal transformation

$$\mathcal{V}(\omega,\epsilon) := 1 + \frac{i}{\hbar}\epsilon_\mu\mathcal{P}^\mu + \frac{i}{2\hbar}\omega_{\mu\nu}\mathcal{M}^{\mu\nu}. \tag{3.56}$$

- **Proposition 3.8:** In the restsystem with $\mathbf{p} = 0$ the symmetry conditions (3.42)–(3.45) go over into the equations

$$\mathcal{P}_0|\mathcal{T}(j;p)\rangle = mc|\mathcal{T}(j;p)\rangle, \tag{3.57}$$

$$\mathcal{P}_k|\mathcal{T}(j;p)\rangle = 0, \tag{3.58}$$

$$\mathcal{P}_\mu\mathcal{P}^\mu|\mathcal{T}(j;p)\rangle = (mc)^2|\mathcal{T}(j;p)\rangle, \tag{3.59}$$

$$\left(\mathcal{M}_{23}^2 + \mathcal{M}_{31}^2 + \mathcal{M}_{12}^2\right)|\mathcal{T}(j;p)\rangle = \hbar^2 s(s+1)|\mathcal{T}(j;p)\rangle, \tag{3.60}$$

$$\mathcal{M}_{12}|\mathcal{T}(j;p)\rangle = \hbar s_3|\mathcal{T}(j;p)\rangle. \tag{3.61}$$

Proof: The definition of the restsystem is given by (3.57) and (3.58) which immediately yield (3.59). By means of this definition we obtain from (3.47) for $\mu = 3$

$$\mathcal{W}_3|\mathcal{T}(j;p)\rangle = \frac{1}{2cm}\mathcal{P}_0\left(\mathcal{M}^{12} - \mathcal{M}^{21}\right)|\mathcal{T}(j;p)\rangle \tag{3.62}$$

$$= \mathcal{M}^{12}|\mathcal{T}(j;p)\rangle$$

$$= \hbar s_3|\mathcal{T}(j;p)\rangle.$$

In the same way (3.60) can be derived from (3.44). $\diamond$

The functional formulation of the global $SU(2) \times U(1)$ gauge groups runs along the same lines and for brevity we will directly discuss it in the context of the applications.

3.4   Functional Field Equations

So far we have discussed the functional formulation of symmetry constraints. To complete the map from $\mathbb{H}$ to $\mathbb{K}_F$ we have to represent the field equations in functional space. As a first step we prove a proposition, see Stumpf [Stu 70a], concerning time ordering and time derivatives.

- **Proposition 3.9:** The time derivative of time ordered products is given by

$$\partial_t \left[T(\psi_J \psi_{I_1} \ldots \psi_{I_n}) \right] \tag{3.63}$$

$$= T \left[(\partial_t \psi_J) \psi_{I_1} \ldots \psi_{I_n} \right] - \sum_{l=1}^{n} (-1)^l A_{J I_l} \delta(t - t_l) T(\psi_{I_1} \ldots \overset{o}{\psi}_{I_l} \ldots \psi_{I_n})$$

with

$$A_{I_1 I_2} = [\psi_{I_1}, \psi_{I_2}]_+^{t=t'}$$

and $J = (Z, \mathbf{r}, t)$, $I_i = (Z_i, \mathbf{r}_i, t_i)$.

Proof: For brevity we replace ψ_{I_i} by I_i for $i = 1 \ldots n$ and ψ_J by J. Then for $n \geq 2$ there holds

$$T(J I_1 \ldots I_n) \tag{3.64}$$

$$= \sum_{p \in S_n} \operatorname{sgn}(p) J I_{p_1} \ldots I_{p_n} \Theta(t, t_{p_1} \ldots, t_{p_n})$$

$$+ \sum_{p \in S_n} \sum_{k=1}^{n-1} \operatorname{sgn}(p)(-1)^k I_{p_1} \ldots I_{p_k} J I_{p_{k+1}} \ldots I_{p_n} \Theta(t_{p_1} \ldots, t_{p_k}, t, t_{p_{k+1}} \ldots, t_{p_n})$$

$$+ \sum_{p \in S_n} \operatorname{sgn}(p)(-1)^n I_{p_1} \ldots I_{p_n} J \Theta(t_{p_1} \ldots, t_{p_n}, t).$$

With

$$\Theta(t_{p_k} - t)\Theta(t - t_{p_{k+1}}) = \Theta(t_{p_k} - t_{p_{k+1}}) \left[\theta(t - t_{p_{k+1}}) - \theta(t - t_{p_k}) \right] \tag{3.65}$$

and

$$\theta(t - t_i) := \frac{1}{2} \left[\Theta(t - t_i) - \Theta(t_i - t) \right] \tag{3.66}$$

one obtains

$$T(J I_1 \ldots I_n) = \sum_{p \in S_n} \operatorname{sgn}(p) \Theta(t_{p_1} \ldots, t_{p_n}) \tag{3.67}$$

$$\times \left[J I_{p_1} \ldots I_{p_n} \Theta(t - t_{p_1}) \right.$$

$$\left. + \sum_{k=1}^{n-1} (-1)^k I_{p_1} \ldots I_{p_k} J I_{p_{k+1}} \ldots I_{p_n} \theta(t - t_{p_{k+1}}) \right.$$

$$-\sum_{k=1}^{n-1}(-1)^k I_{p_1}\ldots I_{p_k} JI_{p_{k+1}}\ldots I_{p_n}\theta(t-t_{p_k})$$

$$+(-1)^n I_{p_1}\ldots I_{p_n} J\Theta(t_{p_n}-t)\Big]$$

$$=\sum_{p\in S_n}\mathrm{sgn}(p)\Theta(t_{p_1}\ldots,t_{p_n})$$

$$\times\Big\{[JI_{p_1}\Theta(t-t_{p_1})+I_{p_1}J\theta(t-t_{p_1})]I_{p_2}\ldots,I_{p_n}$$

$$+(-1)^n I_{p_1}\ldots I_{p_{n-1}}[I_{p_n}J\Theta(t_{p_n}-t)-JI_{p_n}\theta(t-t_{p_n})]$$

$$+\sum_{k=2}^{n-1}(-1)^{k+1}I_{p_1}\ldots I_{p_{k-1}}[I_{p_k},J]_+I_{p_{k+1}}\ldots I_{p_n}\theta(t-t_{p_k})\Big\},$$

and with

$$\partial_t\theta(t-t_i)=\delta(t-t_i)\qquad\qquad(3.68)$$

it follows

$$\partial_t T(JI_1\ldots I_n)\qquad\qquad(3.69)$$

$$=\ T(\dot{J}I_1\ldots I_n)$$

$$-\sum_{p\in S_n}\mathrm{sgn}(p)\Theta(t_{p_1}\ldots,t_{p_n})\sum_{k=1}^{n}(-1)^k I_{p_1}\ldots I_{p_{k-1}}[I_{p_k},J]_+I_{p_{k+1}}\ldots I_{p_n}\delta(t-t_{p_k})$$

$$=\ T(\dot{J}I_1\ldots I_n)$$

$$-\sum_{p\in S_n}\mathrm{sgn}(p)\Theta(t_{p_1}\ldots,t_{p_n})\sum_{k=1}^{n}(-1)^k I_{p_1}\ldots \overset{o}{I}_{p_k}\ldots I_{p_n}A_{I_{p_k}}J\delta(t-t_{p_k}).$$

If one transforms the sequence of indices $(1,2,\ldots k-1,k,k+1,\ldots n)$ in the last term of (3.69) into the sequence $(2,3,\ldots k,1,k+1,\ldots n)$, then (3.69) goes over into

$$\partial_t T(JI_1\ldots I_n)\qquad\qquad(3.70)$$

$$=\ T(\dot{J}I_1\ldots I_n)$$

$$+\sum_{p\in S_n}\mathrm{sgn}(p)\sum_{k=1}^{n}A_{I_{p_1}}J\delta(t-t_{p_1})I_{p_2}\ldots I_{p_n}\Theta(t_{p_2},t_{p_3},\ldots t_{p_k},t_{p_1},t_{p_{k+1}},\ldots t_{p_n}).$$

Renewed permutation of $(1,2,\ldots l,l+1,\ldots n)$ into $(l,1,2,\ldots l-1,l+1,\ldots n)$ for $l=1,\ldots,n$ results in

$$\partial_t T(JI_1\ldots I_n)\qquad\qquad(3.71)$$

$$=\ T(\dot{J}I_1\ldots I_n)+\sum_{p\in S_n}\mathrm{sgn}(p)\sum_{l=1}^{n}(-1)^{l+1}A_{JI_l}\delta(t-t_l)(I_{p_1}\ldots \overset{o}{I}_l\ldots I_{p_n})$$

$$\times\Big[\sum_{k=1}^{l}\Theta(t_{p_1}\ldots t_{p_{k-1}},t_l,t_{p_k}\ldots \overset{o}{t}_l\ldots t_{p_n})$$

$$+\sum_{k=l+1}^{n}\Theta(t_{p_1}\ldots \overset{o}{t}_l\ldots t_{p_k},t_l,t_{p_{k+1}}\ldots t_{p_n})\Big].$$

For any $l, 1 \leq l \leq n$ and for any $p \in S_{n-1}$ there exists just one $k, 1 \leq k \leq n$, for which the sum over k does not vanish. Thus we have

$$\partial_t T(JI_1 \ldots I_n) \tag{3.72}$$

$$= T(\dot{J}I_1 \ldots I_n)$$

$$+ \sum_{p \in S_{n-1}} \text{sgn}(p) \sum_{l=1}^{n} (-1)^{l+1} A_{JI_l} \delta(t - t_l) I_{p_1} \ldots \overset{o}{I}_l \ldots I_{p_n} \Theta(t_{p_1}, \ldots \overset{o}{t}_l, \ldots t_{p_n})$$

$$= T(\dot{J}I_1 \ldots I_n) - \sum_{l=1}^{n} (-1)^l A_{JI_l} \delta(t - t_l) T(I_{p_1} \ldots \overset{o}{I}_l \ldots I_{p_n}).$$

The case $n = 1$ can easily be verified by direct calculation. $\Diamond$

Proposition 3.9 enables us to connect the field equations (2.69) with time ordered products (3.1). Multiplying (2.69) from the right by $\psi_{I'_1} \ldots \psi_{I'_n}$ and applying the time ordering operator T from the left, we obtain by means of (3.63) and with $I' = (Z', x')$

$$T\left[\left(D^{\mu}_{Z_1 Z_2} \partial_\mu - m_{Z_1 Z_2}\right) \psi_{Z_2}(x) \psi_{I'_1} \ldots \psi_{I'_n}\right] \tag{3.73}$$

$$= \left(D^{\mu}_{Z_1 Z_2} \partial_\mu - m_{Z_1 Z_2}\right) T\left[\psi_{Z_2}(x) \psi_{I'_1} \ldots \psi_{I'_n}\right]$$

$$+ D^0_{Z_1 Z_2} \sum_{l=1}^{n} (-1)^l A_{Z_2 Z'_l} \delta(x - x'_l) T(\psi_{I'_1} \ldots \overset{o}{\psi}_{I'_l} \ldots \psi_{I'_n})$$

$$= \hat{U}_{Z_1 Z_2 Z_3 Z_4} T\left[\psi_{Z_2}(x) \psi_{Z_3}(x) \psi_{Z_4}(x) \psi_{I'_1} \ldots \psi_{I'_n}\right].$$

The intention of this procedure is to transform equations (2.69) into a system of equations for $\tau^{(n)}$ functions which again have to be written as a functional equation. But this intention conflicts with the interaction term of (3.73), where the operator product $\psi_{Z_2}(x) \psi_{Z_3}(x) \psi_{Z_4}(x)$ cannot be expressed as a time ordered product. We circumvent this difficulty by a formal extension of the interaction term to a multi-time expression, cf. Pfister, [Pfi 87].

- **Proposition 3.10:** With the definition of the limit

$$\lim_U := \lim_{\substack{t_1, t_2, t_3 \to t \\ t_1 > t_2 > t_3}} \tag{3.74}$$

and for $t \neq t'_i$, $i = 1, \ldots n$ the interaction term of (3.73) can be expressed by

$$\lim_U \hat{U}_{Z_1 Z_2 Z_3 Z_4} T\left[\psi_{Z_2}(\mathbf{r}, t_1) \psi_{Z_3}(\mathbf{r}, t_2) \psi_{Z_4}(\mathbf{r}, t_3) \psi_{I'_1} \ldots \psi_{I'_n}\right] \tag{3.75}$$

$$= \hat{U}_{Z_1 Z_2 Z_3 Z_4} T\left[\psi_{Z_2}(\mathbf{r}, t) \psi_{Z_3}(\mathbf{r}, t) \psi_{Z_4}(\mathbf{r}, t) \psi_{I'_1} \ldots \psi_{I'_n}\right],$$

where $I' = (Z', \mathbf{r}', t')$.

Proof: We assume that the time ordered product $T[\psi_I \psi_{I'_1} \ldots \psi_{I'_n}]$ with $I = (Z, x)$ is well defined. This includes that for any admitted $(n + 1)$-tuple $(II'_1 \ldots I'_n)$ the corresponding time coordinates can be arranged in a sequence with increasing times $t'_{p_1} < \ldots t < \ldots t'_{p_n}$. Now consider the same set of arguments $t'_{p_1} \ldots t'_{p_n}$, but in time ordering with t_1, t_2, t_3. For finite differences between t_1, t_2, t_3 we obtain an arrangement $t'_{p_1} < \ldots t_1 < \ldots t_2 < \ldots t_3 < \ldots t'_{p_n}$, where in the intervals $[t_1, t_2]$ and $[t_2, t_3]$ other times t'_{p_i} may be included. But in the limit $t_1, t_2, t_3 \to t$ these intervals are null sequences and this enforces the limit $t'_{p_i} \to t$, which contradicts the assumption $t \neq t'_i$. So no t'_{p_i} can occur in the intervals $[t_1, t_2]$, etc..

Thus in the U limit all terms of the time ordered product $T[\psi_{I_1} \psi_{I_2} \psi_{I_3} \psi_{I'_1} \ldots \psi_{I'_n}]$ except for the terms containing the expression $\Theta\left(t'_{p_1} \ldots t_1, t_2, t_3 \ldots t'_{p_n}\right)$ vanish. With the definition of the Θ functions this limit yields (3.75). $\Diamond$

By means of this proposition we can derive from equations (3.73) a system of equations for $\tau^{(n)}$ functions:

$$\left(D^\mu_{Z_1 Z_2} \partial_\mu - m_{Z_1 Z_2}\right) \langle 0|T\left[\psi_{Z_2}(x)\psi_{I'_1} \ldots \psi_{I'_n}\right]|a\rangle \tag{3.76}$$

$$+ D^0_{Z_1 Z_2} \sum_{l=1}^{n} (-1)^l A_{Z_2 Z'_l} \delta(x - x'_l) \langle 0|T(\psi_{I'_1} \ldots \overset{\circ}{\psi}_{I'_l} \ldots \psi_{I'_n})|a\rangle$$

$$= \lim_U \hat{U}_{Z_1 Z_2 Z_3 Z_4} \langle 0|T\left[\psi_{Z_2}(\mathbf{r}, t_2)\psi_{Z_3}(\mathbf{r}, t_3)\psi_{Z_4}(\mathbf{r}, t_4)\psi_{I'_1} \ldots \psi_{I'_n}\right]|a\rangle.$$

In a last step we express equations (3.76) in functional space as an equation for functional states (3.17) by means of

- **Proposition 3.11:** The map of the field equations (2.69) into functional space yields the equation

$$\left[-\left(D^\mu_{Z_1 Z_2} \partial_\mu - m_{Z_1 Z_2}\right) \partial_{Z_2}(x) - D^0_{Z_1 Z_2} A_{Z_2 Z_3} j_{Z_3}(x)\right]|T(j; a)\rangle \tag{3.77}$$

$$= \lim_U \hat{U}_{Z_1 Z_2 Z_3 Z_4} \partial_{Z_4}(\mathbf{r}, t_4)\partial_{Z_3}(\mathbf{r}, t_3)\partial_{Z_2}(\mathbf{r}, t_2)|T(j; a)\rangle$$

for the determination of the time ordered functional states (3.17).

Proof: We multiply equation (3.76) with $i^n (n!)^{-1} j_{I'_1} \ldots j_{I'_n}|0\rangle_F$ and sum over $I'_1 \ldots I'_n$ and n. The first and the second term on the left hand side of (3.76) can be mapped into functional space by means of formulas (3.18) and (3.19). The interaction term of (3.76) needs a repeated application of (3.18). We define the functional state

$$|\hat{\mathcal{T}}_{K_1}(j; a)\rangle := i^{-1}\partial_{K_1}|T(j; a)\rangle. \tag{3.78}$$

For

$$|\hat{\mathcal{T}}_{K_1}(j; a)\rangle = \sum_{n=0}^{\infty} \frac{i^n}{n!} \hat{\tau}^{(n)}_{K_1}(I_1 \ldots I_n|a) j_{I_1} \ldots j_{I_n}|0\rangle_F$$

and $|T(j; a)\rangle$ given by (3.17) it holds

$$\hat{\tau}^{(n)}_{K_1}(I_1 \dots I_n | a) = \tau^{(n+1)}(K_1, I_1 \dots I_n | a). \tag{3.79}$$

By iteration of this procedure one can derive

$$i^{-3} \partial_{K_3} \partial_{K_2} \partial_{K_1} |\mathcal{T}(j; a)\rangle \tag{3.80}$$

$$= \sum_{n=0}^{\infty} \frac{i^n}{n!} \tau^{(n+3)}(K_3, K_2, K_1, I_1 \dots I_n | a) j_{I_1} \dots j_{I_n} |0\rangle_{\mathrm{F}}.$$

With (3.80) the interaction term of (3.76) can be transformed into functional space to yield the right hand side of (3.77). $\diamond$

If (3.20) is applied to (3.77) we can regain equations (3.76).

Equation (3.77) corresponds to the functional version of the Schwinger–Dyson–Freese equations of the spinor field if $|\mathcal{T}(j; a)\rangle$ is replaced by the state functional $\mathcal{T}(j)$ for $|a\rangle \equiv |0\rangle$. Originally such equations were derived by perturbation theory (see, e.g., Lurié [Luri 68]) or by a direct calculation of the time derivative of the generating functional, cf. Symanzik [Sym 54], which is a dubious operation. In contrast the derivation by means of Proposition 3.9 and the embedding of these equations into functional space are completely transparent mathematical operations.

Finally, we derive a functional energy eigenvalue equation from equation (3.77). Energy equations in the covariant formalism were formally derived by Zimmermann [Zim 54] and Symanzik [Sym 54]. But as we shall see in Chapter 4 the real problem consists in performing the limit to equal times. It will turn out that in order to give the energy equation a definite meaning it has to be considered in a single time formalism; see also the model discussion for the covariantly formulated anharmonic oscillator by Maison and Stumpf [Mai 66].

- **Proposition 3.12:** The functional energy eigenvalue equation of the spinor theory reads

$$E_{p0} |T(j; p)\rangle = \frac{ic}{\hbar} \int d^4x j_{Z_0}(x) D^0_{Z_0 Z_1} \Big[(D^k \partial_k - m)_{Z_1 Z_2} \partial_{Z_2}(x) \tag{3.81}$$

$$+ \lim_U \hat{U}_{Z_1 Z_2 Z_3 Z_4} \partial_{Z_4}(\mathbf{r}, t_4) \partial_{Z_3}(\mathbf{r}, t_3) \partial_{Z_2}(\mathbf{r}, t_2) \Big] |T(j; p)\rangle,$$

where $E_{p0} := E_p - E_0$ is the renormalized energy eigenvalue with E_p energy value of $|p\rangle$ and E_0 energy value of $|0\rangle$.

Proof: We combine the constraint equation (3.42) for $\mu = 0$ with equation (3.77). The symmetry condition (3.42) reads for $\mu = 0$

$$i\hbar c \int d^4x j_Z(x) \partial_0(x) \partial_Z(x) |\mathcal{T}(j; p)\rangle = E_{p0} |\mathcal{T}(j; p)\rangle \tag{3.82}$$

if we assume $\mathbf{P}_0 |0\rangle = E_0 |0\rangle$. Then by construction equation (3.82) explicitly shows the renormalization shift.

We multiply (3.77) from the left by $j_{Z_0}(x) D^0_{Z_0 Z_1}$ and integrate over x. Observing $D^0_{Z_0 Z_1} D^0_{Z_1 Z_2} = -\hbar^2 \delta_{Z_0 Z_2}$ we can replace the ∂_0 term of (3.77) by (3.82) with the result

$$E_{p0}|T(j;p)\rangle \tag{3.83}$$

$$= \frac{ic}{\hbar} \int d^4x j_{Z_0}(x) \Big[D^0_{Z_0 Z_1}(D^k \partial_k - m)_{Z_1 Z_2} \partial_{Z_2}(x) - \hbar^2 A_{Z_0 Z_1} j_{Z_1}(x)$$

$$+ \lim_U D^0_{Z_0 Z_1} \hat{U}_{Z_1 Z_2 Z_3 Z_4} \partial_{Z_4}(\mathbf{r}, t_4) \partial_{Z_3}(\mathbf{r}, t_3) \partial_{Z_2}(\mathbf{r}, t_2) \Big] |T(j;p)\rangle.$$

Furthermore, there holds

$$j_{Z_0}(x) A_{Z_0 Z_1} j_{Z_1}(x) = -j_{Z_1}(x) A_{Z_1 Z_0} j_{Z_0}(x) = 0 \tag{3.84}$$

owing to the properties of the anticommuting sources and the symmetry $A_{ZZ'} = A_{Z'Z}$. Thus the A term in (3.83) vanishes and (3.81) holds. $\diamond$

3.5 Nonperturbative Normal Ordering

In conventional quantum field theory normal ordering is an operator rearrangement which enables one to evaluate the S matrix in the interaction picture, *i.e.*, in perturbation theory with free-field representation. In our presentation of quantum field theory we work strictly outside perturbation theory; in particular we do not fix the representation of the field operator algebra *a priori*. Therefore, if we introduce 'normal ordering' in this context, this rearrangement needs a new physical and mathematical interpretation. First we shortly discuss the physical aspect, whilst the mathematical aspect will be treated at the end of this section.

It is a peculiarity of quantum field theory and statistical mechanics that the field or system dynamics expressed by the functional states (3.17), *etc.*, contain so called 'uncorrelated' parts. These uncorrelated parts are vacuum expectation values appearing in the decomposition of matrix elements (3.1) by intermediate states. If we try to establish a composite particle theory, such vacuum expectation values prevent a consistent composite particle interpretation, and thus partially have to be removed. This elimination serves as the definition of nonperturbative 'normal ordering'.

If, for instance, we consider transition matrix elements $\langle 0|\psi_{I_1}\psi_{I_2}|a\rangle$ as representatives of two-fermion composite particle states, then in the composite particle interpretation these states are mapped onto one-particle states of a phenomenological boson field $\chi(x)$ expressed by the set $\{\langle 0|\chi(x)|b\rangle\}$. But the vacuum expectation values $\langle 0|\psi_{I_1}\psi_{I_2}|0\rangle$ have no counterparts in the set of one-boson states $\langle 0|\chi(x)|b\rangle$ if symmetry breaking is excluded. Thus, in order to provide a bijective map between the original ψ field dynamics and the phenomenological χ field dynamics, the quantities $\langle 0|\psi_{I_1}\psi_{I_2}|0\rangle$ have to be removed from the original ψ field theory.

On the other hand, higher vacuum expectation values like $\langle 0|\psi_{I_1}\psi_{I_2}\psi_{I_3}\psi_{I_4}|0\rangle$ find their counterparts in terms like $\langle 0|\chi_{k_1}\chi_{k_2}|0\rangle$ and therefore need not be eliminated from the original ψ field theory. This example shows that nonperturbative normal ordering depends on the kind of composite particle dynamics under consideration.

In the following we discuss the elimination of two-point functions $\langle 0|T(\psi_{I_1}\psi_{I_2})|0\rangle$ from the original ψ field theory. The corresponding elimination procedure by normal ordering is in accordance with the general structure of matrix element decomposition by intermediate states and holds independently of perturbation theory.

We introduce normal ordering for two-point functions by the transformation

$$|\mathcal{F}(j;a)\rangle := \exp\left(\frac{1}{2}j_{I_1}F_{I_1 I_2}j_{I_2}\right)|T(j;a)\rangle \tag{3.85}$$

with the exponential function defined by its series and $F_{I_1 I_2}$ given by

$$F_{I_1 I_2} \equiv F_{Z_1 Z_2}(x_1, x_2) := \langle 0|T[\psi_{Z_1}(x_1)\psi_{Z_2}(x_2)]|0\rangle. \tag{3.86}$$

With (3.17) and the power series expansion of the exponential it follows that $|\mathcal{F}(j;a)\rangle$ can be represented in the form

$$|\mathcal{F}(j;a)\rangle = \sum_n \frac{i^n}{n!}\varphi^{(n)}(I_1\ldots I_n|a)j_{I_1}\ldots j_{I_n}|0\rangle_{\mathrm{F}}, \tag{3.87}$$

where the $\varphi^{(n)}$ functions are completely antisymmetric in their arguments $I_1\ldots I_n$.

For the special case $\psi_I \equiv \psi_I^{\mathrm{int}}$, i.e., in the interaction picture with free-field representation, $F_{I_1 I_2}$ is the free fermion propagator and (3.85) goes over into the functional formulation of the Wick rule, see Rampacher, Stumpf and Wagner [Ram 65]. But, independently of any representation, (3.87) is a functional state from which all propagator parts (3.86) are extracted. This can be explicitly seen by projecting (3.85) in coordinate space according to (3.20).

We introduce the notation

$$\varphi^{(n)}(I_1\ldots I_n|a) \equiv \langle 0|N(\psi_{I_1}\ldots\psi_{I_n})|a\rangle \tag{3.88}$$

for all states $|a\rangle \in \mathbb{H}$ with inclusion of the vacuum state $|0\rangle$, where N indicates that all two-point functions are extracted from the original time ordered product $T(\psi_{I_1}\ldots\psi_{I_n})$.

As a consequence of the normal ordering of functional states it is necessary also to transform the corresponding functional equations and the additional constraints. This can be achieved by means of the following proposition:

- **Proposition 3.13:** The commutator of the (inverse) normal transformation (3.85) with the functional differential operator ∂_K is given by

$$\left[\partial_K, \exp\left(-\frac{1}{2}j_{I_1}F_{I_1 I_2}j_{I_2}\right)\right]_- = -\left[\exp\left(-\frac{1}{2}j_{I_1}F_{I_1 I_2}j_{I_2}\right)\right]F_{K I_3}j_{I_3}. \tag{3.89}$$

Proof: Application of (3.11) yields

$$\partial_K \exp(-\frac{1}{2}j_{I_1}F_{I_1I_2}j_{I_2}) \tag{3.90}$$

$$= \sum_{n=0}^{\infty} \frac{(-1)^n}{n!2^n} F_{I_1I_2}F_{I_3I_4}\ldots F_{I_{2-1}I_{2n}} \partial_K j_{I_1}\ldots j_{I_{2n}}$$

$$= \sum_{n=0}^{\infty} \frac{(-1)^n}{n!2^n} F_{I_1I_2}\ldots F_{I_{2-1}I_{2n}} \left[j_{I_1}\ldots j_{I_{2n}}\partial_K + \sum_{l=1}^{2n}(-1)^{l+1}\delta_{I_lK}j_{I_1}\ldots \overset{o}{j}_{I_l}\ldots j_{I_{2n}} \right]$$

$$= \exp(-\frac{1}{2}j_{I_1}F_{I_1I_2}j_{I_2})\partial_K$$

$$+ \sum_{n=0}^{\infty} \frac{(-1)^n}{n!2^n} F_{I_1I_2}\ldots F_{I_{2-1}I_{2n}}$$

$$\times \left[\sum_{l=1}^{n}(-1)^{2l}\delta_{KI_{2l-1}}j_{I_1}\ldots \overset{o}{j}_{I_{2l-1}}\ldots j_{I_{2n}} \right.$$

$$\left. + \sum_{l=1}^{n}(-1)^{2l+1}\delta_{KI_{2l}}j_{I_1}\ldots \overset{o}{j}_{I_{2l}}\ldots j_{I_{2n}} \right]$$

$$= \exp(-\frac{1}{2}j_{I_1}F_{I_1I_2}j_{I_2})\partial_K$$

$$+ \sum_{n=0}^{\infty} \frac{(-1)^n}{n!2^n} \sum_{l=1}^{n} j_{I_1}F_{I_1I_2}j_{I_2}j_{I_3}F_{I_3I_4}j_{I_4}\ldots j_{I_{2l-2}}F_{KI_{2l}}j_{I_{2l}}\ldots j_{I_{2n-1}}F_{I_{2n-1}I_{2n}}j_{I_{2n}}$$

$$+ \sum_{n=0}^{\infty} \frac{(-1)^n}{n!2^n} \sum_{l=1}^{n} j_{I_1}F_{I_1I_2}\ldots j_{I_{2l-2}}F_{KI_{2l-1}}j_{I_{2l-1}}j_{I_{2l+1}}F_{I_{2l+1}I_{2l+2}}j_{I_{2l+2}}\ldots j_{I_{2n}},$$

where in the last step the antisymmetry of $F_{I_1I_2}$ was used.

By re-indexing $(2l - 1 \rightarrow 2l)$ of the last term, moving of $F_{KI_{2l}}j_{I_{2l}}$ to the last place and renewed indexing one obtains from (3.90)

$$\partial_K \exp(-\frac{1}{2}j_{I_1}F_{I_1I_2}j_{I_2}) \tag{3.91}$$

$$= \exp(-\frac{1}{2}j_{I_1}F_{I_1I_2}j_{I_2})\partial_K$$

$$+ \sum_{n=0}^{\infty} \frac{(-1)^n}{n!2^n}2\sum_{l=1}^{n} j_{I_1}F_{I_1I_2}j_{I_2}\ldots j_{I_{2l-2}}F_{KI_{2l}}j_{I_{2l}}j_{I_{2l+1}}\ldots F_{I_{2n-1}I_{2n}}j_{I_{2n}}$$

$$= \exp(-\frac{1}{2}j_{I_1}F_{I_1I_2}j_{I_2})\partial_K$$

$$+ \sum_{n=0}^{\infty} \frac{(-1)^n}{n!2^{n-1}}n j_{I_1}F_{I_1I_2}j_{I_2}\ldots j_{I_{2n-3}}F_{I_{2n-3}I_{2n-2}}j_{I_{2n-2}}F_{KK'}j_{K'}$$

$$= \exp(-\frac{1}{2}j_{I_1}F_{I_1I_2}j_{I_2})(\partial_K - F_{KK'}j_{K'}). \qquad \Diamond$$

By means of Proposition 3.13 we can derive the normal transformed functional equation (3.77).

- **Proposition 3.14:** The normal transformed functional equation (3.77) reads

$$\left[(D^\mu \partial_\mu - m)_{Z_1 Z_2} d_{Z_2}(x) + D^0_{Z_1 Z_2} A_{Z_2 Z_3} j_{Z_3}(x)\right] |\mathcal{F}(j;a)\rangle \tag{3.92}$$
$$= -\lim_U \hat{U}_{Z_1 Z_2 Z_3 Z_4} d_{Z_4}(\mathbf{r}, t_4) d_{Z_3}(\mathbf{r}, t_3) d_{Z_2}(\mathbf{r}, t_2) |\mathcal{F}(j;a)\rangle$$

with

$$d_Z(x) \equiv d_I := \partial_I - F_{IK} j_K. \tag{3.93}$$

Proof: The relation $e^{-z} e^z = 1$ can be proven by a power series expansion of the exponentials and thus it holds also for $z = j_{I_1} F_{I_1 I_2} j_{I_2}$. Therefore relation (3.85) can be resolved with respect to $|\mathcal{T}(j;a)\rangle$ and equivalently reads

$$|\mathcal{T}(j;a)\rangle = \exp(-\tfrac{1}{2} j_{I_1} F_{I_1 I_2} j_{I_2}) |\mathcal{F}(j;a)\rangle. \tag{3.94}$$

We substitute (3.94) into equation (3.77) and shift $\exp(-\tfrac{1}{2} j_{I_1} F_{I_1 I_2} j_{I_2})$ to the left hand side of the equation. This can be achieved by use of (3.89) and by means of

$$\left[j_K, \exp(-\tfrac{1}{2} j_{I_1} F_{I_1 I_2} j_{I_2})\right]_- = 0. \tag{3.95}$$

The resulting equation is given by

$$\exp(-\tfrac{1}{2} j_{K_1} F_{K_1 K_2} j_{K_2})$$
$$\times \left[(D^\mu \partial_\mu - m)_{Z_1 Z_2} d_{Z_2}(x) + D^0_{Z_1 Z_2} A_{Z_2 Z_3} j_{Z_3}(x)\right] |\mathcal{F}(j;a)\rangle$$
$$= -\exp(-\tfrac{1}{2} j_{K_1} F_{K_1 K_2} j_{K_2})$$
$$\times \lim_U \hat{U}_{Z_1 Z_2 Z_3 Z_4} d_{Z_4}(\mathbf{r}, t_4) d_{Z_3}(\mathbf{r}, t_3) d_{Z_2}(\mathbf{r}, t_2) |\mathcal{F}(j;a)\rangle, \tag{3.96}$$

and as the exponential function is a nondegenerate operator, equation (3.92) must hold.   ◇

We add that in (3.92) the $\lim_U$ has to be slightly modified in order to provide for a well defined limit $t_2, t_2, t_4 \to t$ in all expressions contained in the vertex term of (3.92). In the same way we can transform the symmetry constraints:

- **Proposition 3.15:** The constraint equations (3.42)–(3.45) are invariant under normal transformation, *i.e.*, it holds

$$\mathcal{P}_\mu |\mathcal{F}(j;p)\rangle = p_\mu |\mathcal{F}(j;p)\rangle, \tag{3.97}$$
$$\mathcal{P}_\mu \mathcal{P}^\mu |\mathcal{F}(j;p)\rangle = (mc)^2 |\mathcal{F}(j;p)\rangle, \tag{3.98}$$
$$\mathcal{W}_\mu \mathcal{W}^\mu |\mathcal{F}(j;p)\rangle = \hbar^2 s(s+1) |\mathcal{F}(j;p)\rangle, \tag{3.99}$$
$$\mathcal{W}_3 |\mathcal{F}(j;p)\rangle = \hbar s_3 |\mathcal{F}(j;p)\rangle, \tag{3.100}$$

where $\mathcal{P}_\mu$ and $\mathcal{W}_\mu$ are defined by (3.46) and (3.47).

Proof: Substitution of (3.94) into (3.42) gives

$$\mathcal{P}_\mu \exp(-\frac{1}{2}j_{I_1}F_{I_1I_2}j_{I_2})|\mathcal{F}(j;p)\rangle = \exp(-\frac{1}{2}j_{I_1}F_{I_1I_2}j_{I_2})p_\mu|\mathcal{F}(j;p)\rangle. \tag{3.101}$$

Observing (3.46) and using (3.89) and (3.95) we can commute the exponential function with $\mathcal{P}_\mu$ and obtain from (3.101)

$$p_\mu|\mathcal{F}(j;p)\rangle \tag{3.102}$$
$$= -\int d^4x_1 d^4x_2 j_{Z_1}(x_1)P_\mu(x_1)\Big[\partial_{Z_1}(x_1) - F_{Z_1Z_2}(x_1,x_2)j_{Z_2}(x_2)\Big]|\mathcal{F}(j;p)\rangle.$$

Owing to the translational invariance of $F_{I_1I_2}$ there holds

$$F_{Z_1Z_2}(x_1,x_2) = F_{Z_1Z_2}(x_1 - x_2), \tag{3.103}$$

and therefore

$$\partial_\mu(x_1)F_{Z_1Z_2}(x_1,x_2) = -\partial_\mu(x_2)F_{Z_1Z_2}(x_1,x_2). \tag{3.104}$$

Furthermore, $F_{I_1I_2}$ is antisymmetric, *i.e.*,

$$F_{Z_1Z_2}(x_1,x_2) = -F_{Z_2Z_1}(x_2,x_1). \tag{3.105}$$

Combining these properties and commuting the j_I sources, we obtain

$$\int d^4x_1 d^4x_2 j_{Z_1}(x_1)\partial_\mu(x_1)F_{Z_1Z_2}(x_1,x_2)j_{Z_2}(x_2) \tag{3.106}$$
$$= -\int d^4x_1 d^4x_2 j_{Z_2}(x_2)\partial_\mu(x_2)F_{Z_2Z_1}(x_2,x_1)j_{Z_1}(x_1).$$

If we interchange (Z_1, x_1) and (Z_2, x_2) on the right hand side of (3.106), the right hand side equals the negative of the left hand side and thus has to vanish. Hence from (3.102) there follows (3.97). By the same arguments (3.98), (3.99) and (3.100) can be proved. $\diamond$

Equation (3.92) can be simplified if for $F_{I_1I_2}$ the free fermion propagator is used. This free fermion propagator is defined by

$$F^f_{I_1I_2} := \langle 0|T\left[\psi^f_{Z_1}(x_1)\psi^f_{Z_2}(x_2)\right]|0\rangle, \tag{3.107}$$

where ψ^f is a solution of the fermion equation

$$\left(D^\mu_{Z_1Z_2}\partial_\mu - m_{Z_1Z_2}\right)\psi^f_{Z_2}(x) = 0. \tag{3.108}$$

As we work with superspinors, we cannot directly take over the propagator equation for Dirac spinors. We thus give a derivation of the properties of (3.107).

- **Proposition 3.16:** The free superspinor propagator $F_{I_1 I_2}^f$ satisfies the equation

$$[D^\mu \partial_\mu(x_1) - m]_{Z_0 Z_1} F_{Z_1 Z_2}^f(x_1, x_2) = D_{Z_0 Z_1}^0 A_{Z_1 Z_2} \delta(x_1 - x_2) \qquad (3.109)$$

with

$$\left[\psi_{Z_1}^f(x_1), \psi_{Z_2}^f(x_2)\right]_+^{t_1 = t_2} = A_{Z_1 Z_2} \delta(\mathbf{r}_1 - \mathbf{r}_2), \qquad (3.110)$$

and $A_{Z_1 Z_2}$ given by (2.73).

Proof: With (3.63) there holds

$$\partial_{t_1} T\left[\psi_{Z_1}^f(x_1)\psi_{Z_2}^f(x_2)\right] = T\left[\dot{\psi}_{Z_1}^f(x_1)\psi_{Z_2}^f(x_2)\right] + A_{Z_1 Z_2}\delta(x_1 - x_2). \qquad (3.111)$$

Multiplication of (3.111) from the left with $D_{Z_0 Z_1}^0$ and use of (3.108) yields

$$D_{Z_0 Z_1}^0 \partial_{t_1} T\left[\psi_{Z_1}^f(x_1)\psi_{Z_2}^f(x_2)\right] \qquad (3.112)$$

$$= T\left[\left(-D^k\partial_k(x_1) + m\right)_{Z_0 Z_1}\psi_{Z_1}^f(x_1)\psi_{Z_2}^f(x_2)\right] + D_{Z_0 Z_1}^0 A_{Z_1 Z_2}\delta(x_1 - x_2).$$

Formation of the vacuum expectation value of (3.112) yields (3.109).   ◇

We remark that equation (3.109) is of course identical with (3.76) for $n = 2$ and $U \equiv 0$.

Hence, choosing $F_{I_1 I_2} \equiv F_{I_1 I_2}^f$ and taking into account (3.109) we obtain for equation (3.92)

$$(D^\mu \partial_\mu - m)_{Z_1 Z_2}\partial_{Z_2}(x)|\mathcal{F}(j; a)\rangle \qquad (3.113)$$

$$= -\lim_U \hat{U}_{Z_1 Z_2 Z_3 Z_4} dz_4(\mathbf{r}, t_4) dz_3(\mathbf{r}, t_3) dz_2(\mathbf{r}, t_2)|\mathcal{F}(j; a)\rangle.$$

Resolution of (3.113) with respect to the time derivative by multiplication with $D_{Z_0 Z_1}^0$, subsequent multiplication with $j_{Z_0}(x)$ and integration over x in combination with (3.97) for $\mu = 0$ gives the normal transformed energy equation

$$E_{p0}|\mathcal{F}(j; p)\rangle = \frac{ic}{\hbar}\int d^4x j_{Z_0}(x)D_{Z_0 Z_1}^0\Big[(D^k\partial_k - m)_{Z_1 Z_2}\partial_{Z_2}(x) \qquad (3.114)$$

$$+ \lim_U \hat{U}_{Z_1 Z_2 Z_3 Z_4} dz_4(\mathbf{r}, t_4) dz_3(\mathbf{r}, t_3) dz_2(\mathbf{r}, t_2)\Big]|\mathcal{F}(j; p)\rangle.$$

However, equation (3.92), as well as equations (3.113) and (3.114), need a further discussion, as in general the introduction of F functions is connected with the occurrence of singularities.

By nonperturbative Pauli–Villars regularization the singularities can be removed, but nevertheless the choice of such propagators has to be justified (or rejected). This problem leads to the mathematical aspect of normal ordering: As we have pointed out in Section 3.1 the basis of nonperturbative quantum field theory is the GNS construction. The set of GNS basis vectors defines the representation of the corresponding field operators.

On the other hand, this set of GNS vectors can equally well be characterized by the algebraic state $\omega(A) = \langle\Omega|\pi_\omega(A)|\Omega\rangle$. The two-point function (3.86) (or more precisely its single time limit) is a part of $\omega(A)$, i.e., this quantity is representation dependent. Therefore if this quantity is introduced into the system dynamics, the dynamical law becomes representation dependent, too. Hence whilst the original dynamical law (3.77) is invariant with respect to representations, its normal transformed version (3.92) depends on them, and this is just what one has to expect from a physical point of view.

In addition the algebraic states themselves are objects of the system dynamics, i.e., the system itself defines its representation under suitable initial and boundary conditions. Therefore the propagators have to be self-consistently calculated and equation (3.109) can be considered only as a free-field approximation of the self-consistency condition. This will be discussed in more detail in Chapter 5.

With respect to applications the self-consistency has been studied so far only in the case of the BCS model, see Stumpf and Borne [Stu 94] and Grebe [Gre 98]. Therefore in the following we will use the free propagators of the auxiliary fields.

For later applications we explicitly give a solution of (3.109) by

$$F_{Z_1 Z_2}(x_1, x_2) \qquad (3.115)$$

$$= -i\hbar^{-3}(2\pi)^{-4}\lambda_{i_1}\delta_{i_1 i_2}\gamma^5_{\kappa_1\kappa_2}\int d^4p \left(\frac{\gamma^\mu p_\mu + cm_{i_1}}{p^2 - c^2 m_{i_1}^2 + i\epsilon}C\right)_{\alpha_1\alpha_2} e^{-ip(x_1-x_2)/\hbar}.$$

This solution represents the causal Feynman propagator and it is chosen in accordance with the application of the Wick Theorem to the transformation of time ordered into normal ordered field operator products, see Schweber, Bethe and de Hoffmann [Schwe 55]. It should, however, be noted that this is not identical with the introduction of a Fock space representation of the field dynamics.

In Fock space the spinor field operators had to be decomposed into positive and negative frequency parts, i.e., $\psi = \psi^{(+)} + \psi^{(-)}$ and had to satisfy the Fock space conditions $\psi^{(-)}|0\rangle = \bar{\psi}^{(+)}|0\rangle = 0$. At no stage of our investigation is such a transformation performed and no such additional conditions with respect to the vacuum are postulated. Hence, although referred to a free fermion propagator, we work with a state space beyond the interaction representation, i.e., beyond the Fock space structure as will be demonstrated in the following chapters.

3.6 Vertex Renormalization

To reveal the difficulties which are connected with the definition of the interaction term in (3.113) and (3.114) we explicitly evaluate it. As it stands, in this term — owing to the definiton (3.93) of d_K — the sources j_I and their duals ∂_K appear in mixed polynomial expressions. We disentangle these terms by giving them a 'functional normal ordered' form with all duals shifted to the right hand side of the sources. This gives

$$d_{I_4}d_{I_3}d_{I_2} \quad = \quad \partial_{I_4}\partial_{I_3}\partial_{I_2} \tag{3.116}$$

$$-F_{I_2I_3}\partial_{I_4} + F_{I_2I_4}\partial_{I_3} - F_{I_3I_4}\partial_{I_2}$$

$$-F_{I_2J_2}j_{J_2}\partial_{I_4}\partial_{I_3} - F_{I_3J_3}j_{J_3}\partial_{I_2}\partial_{I_4} - F_{I_4J_4}j_{J_4}\partial_{I_3}\partial_{I_2}$$

$$+F_{I_2I_3}F_{I_4J_4}j_{J_4} + F_{I_3I_4}F_{I_2J_2}j_{J_2} - F_{I_2I_4}F_{I_3J_3}j_{J_3}$$

$$+F_{I_4J_4}F_{I_2J_2}j_{J_2}j_{J_4}\partial_{I_3} + F_{I_3J_3}F_{I_2J_2}j_{J_3}j_{J_2}\partial_{I_4} + F_{I_4J_4}F_{I_3J_3}j_{J_4}j_{J_3}\partial_{I_2}$$

$$-F_{I_4J_4}F_{I_3J_3}F_{I_2J_2}j_{J_4}j_{J_3}j_{J_2}\,.$$

We consider the second and the fourth line of (3.116). If these lines are inserted in the vertex term of (3.113) and (or) (3.114) the $\lim_U$, together with the spatial locality of the vertex, leads to terms with $F_{Z_2Z_3}(0)$, $F_{Z_2Z_4}(0)$ and $F_{Z_3Z_4}(0)$. At the origin these propagators are singular terms and it is necessary, or at least convenient, to remove these terms by additive renormalization. This additive renormalization is commonly referred to as vertex renormalization, and it is defined by

$$: d_{Z_4}(x_4)d_{Z_3}(x_3)d_{Z_2}(x_2) : \tag{3.117}$$

$$:= \quad d_{Z_4}(x_4)d_{Z_3}(x_3)d_{Z_2}(x_2) + F_{Z_3Z_4}(x_3, x_4)d_{Z_2}(x_2)$$

$$-F_{Z_2Z_4}(x_2, x_4)d_{Z_3}(x_3) + F_{Z_2Z_3}(x_2, x_3)d_{Z_4}(x_4)\,.$$

If this operation is applied to (3.116), the unwanted terms of (3.116) are compensated by the additive renormalization terms of (3.117) and thus drop out.

The vertex renormalization can be transferred to the dynamical equation (2.69) itself and leads to

$$(D^\mu\partial_\mu - m)_{Z_1Z_2}\psi_{Z_2}(x) = \hat{U}_{Z_1Z_2Z_3Z_4} : \psi_{Z_2}(x)\psi_{Z_3}(x)\psi_{Z_4}(x) : \tag{3.118}$$

with

$$: \psi_{Z_2}(x)\psi_{Z_3}(x)\psi_{Z_4}(x) : \quad := \quad \psi_{Z_2}(x)\psi_{Z_3}(x)\psi_{Z_4}(x) + F_{Z_3Z_4}(0)\psi_{Z_2}(x) \tag{3.119}$$

$$-F_{Z_2Z_4}(0)\psi_{Z_3}(x) + F_{Z_2Z_3}(0)\psi_{Z_4}(x).$$

For instance with (3.117) one obtains the corresponding functional energy equation

$$E_{p0}|\mathcal{F}(j;p)\rangle \quad = \quad \frac{ic}{\hbar}\int d^4x j_{Z_0}(x)D^0_{Z_0Z_1} \tag{3.120}$$

$$\times \left[(D^k\partial_k - m)_{Z_1Z_2}\partial_{Z_2}(x)\right.$$

$$\left. + \lim_U \hat{U}_{Z_1Z_2Z_3Z_4} : d_{Z_4}(\mathbf{r}, t_4)d_{Z_3}(\mathbf{r}, t_3)d_{Z_2}(\mathbf{r}, t_2) :\right]|\mathcal{F}(j;p)\rangle.$$

The latter equation is used for a further investigation in Chapter 4.

3.7 Limits of Covariant Formalism

Time ordered products (3.1) and normal ordering of time ordered products originally stem from perturbation theory. In perturbation theory the representation of the field operator algebra is fixed to be the interaction picture with free-field representation, and as its evaluation is carried out according to fixed rules, there is little room for further investigations. Beyond perturbation theory the situation is quite different: the genuine field representation is influenced, or even completely determined, by the field dynamics and its constraints themselves, and one needs the full apparatus of formal quantum theory for the investigation of nonperturbative phenomena. Thus one is faced with the question: is the covariant formalism the suitable starting point for such investigations?

Apart from the covariant formalism formal quantum theory is governed by the concept of state evolution in time. This means algebraically that the algebra of a quantum system is completely determined already on a space-like hyperplane and that time evolution is a continuous group the action of which generates a sequence of inner automorphisms of the algebra in the course of time. With respect to representations this means that the corresponding quantum state space must be complete on any space-like hyperplane.

In a preliminary way in (3.1) the states $\langle 0|T(\psi_{I_1} \ldots \psi_{I_n})$ were referred to as basis states of the covariant representation of a state vector $|a\rangle$. These states are assumed to exist for all values of their time coordinates $t_1 \neq \ldots \neq t_n$. However, the time evolution gives rise to doubts about this concept because already for any fixed $t = t_1 = \ldots = t_n$ there must exist a complete set of basis vectors. Hence we are forced to suppose that the set of time ordered 'basis' states $\langle 0|T(\psi_{I_1} \ldots \psi_{I_n})$ is over complete.

Mathematically over completeness is an unpleasant property because it prevents the application of orthonormality and completeness relations, and with respect to the time ordered basis states their over completeness mixes past and future in an obscure way which admits physical interpretation only by asymptotic behaviour of the S matrix.

Apart from this difficulty the time ordered products are unpleasant in yet another respect. If one wants to perform explicit calculations one is faced with the difficulty that these products are themselves products of c number and operator distributions which, for instance, are not well defined for equal times, see Bogoliubov, Logunov and Todorov [Bogo 75]. As a consequence the Schwinger–Dyson–Freese equations cannot be considered as differential equations for continuous functions.

These difficulties already occur in field-theoretic perturbation theory and enforce a very careful distributional analysis, cf. Scharf [Schar 89]. But whilst in perturbation theory one has to deal with well known distributions, in the general case of nonperturbative phenomena one would have to solve Schwinger–Dyson–Freese equations by a careful distributional analysis for yet unknown distributions — at present a nearly insurmountable task.

As a way out of these difficulties path integrals were derived as formal solutions of the Schwinger–Dyson–Freese equations. At present, path integrals are commonly considered as an ideal means for the treatment of quantum field theories. However, a closer inspection reveals many weak points concerning the definition and evaluation of such integrals, see Rivers [Riv 88], Cheng [Che 90], Roepstorff [Roep 91] etc..

In particular, their evaluation with respect to composite particle dynamics shows deficiencies which were extensively discussed by Stumpf and Borne in [Stu 94]. For instance, composite particle dynamics for $n > 2$ constituents cannot be derived by path integrals. But apart from the deficiences of path integrals concerning the derivation of effective dynamical laws, the objections against their existence altogether are still more serious (see also Section 2.1). Therefore owing to the mathematical fallacies and defects which are inherent in these integrals and which cannot be cured, they cannot be used as solutions of the Schwinger–Dyson–Freese equations beyond perturbation theory.

Summarizing this discussion we conclude that the complete covariant formalism cannot be used as a suitable starting point for nonperturbative calculations. Therefore in the next chapter we will develop a quantum field-theoretic formalism which is based on state space descriptions on space-like hyperplanes and thus avoids over completeness and the difficulties connected with it.

In the following chapters we shall demonstrate that this formalism allows to successfully treat the problems mentioned in Section 2.1. In doing so we do not completely reject the covariant formalism. Rather we use it as an auxiliary constraint if relativistic transformations of the states on the hyperplanes have to be performed. For details we refer to the following chapters.

4 Algebraic Schrödinger Representation

4.1 Introduction

In the preceding chapter we pointed out our aim of developing a functional quantum theory which allows a successful treatment of quantum fields and which is governed by two ideas: that of taking into account inequivalent representations; and that of explicit state space construction. To this end we proposed a map from a state space $\mathbb{H}$ onto a functional space $\mathbb{K}_F$ which has to leave metric and quantum numbers invariant.

In Chapter 3 we discussed such a map within a covariant formalism. With respect to the quantum numbers we succeeded in constructing a functional space which can carry representations of the Poincaré group and in formulating symmetry conditions which characterize these representations. But with respect to the explicit construction of the state space we found the assumed time ordered basis states $T(\psi_{I_1} \dots \psi_{I_n})|0\rangle$ to be over complete. This prevented us from examining the problem of the existence and of the metrical structure of the state space.

As the over completeness stems from this choice of the basis states ignoring the concept of causal time evolution, it suggests itself to abandon explicit covariance and to look for a Hamiltonian formalism where time ordered, *i.e.*, multi-time basis vectors are avoided in favour of a single time formulation. We follow this line of argument and start with a more detailed analysis of the state spaces of quantum fields.

Compared with quantum mechanics, for quantum fields the discussion of the state space, *i.e.*, the representation space of the corresponding field operator algebra, is more complicated in a twofold way: on the one hand, quantum fields as systems with an infinite number of degrees of freedom do not *a priori* possess representations, rather the conditions have to be studied under which such representations do exist. On the other hand, if representations exist quantum fields are confronted with an infinity of inequivalent representations, see Friedrichs [Frie 53], Bratteli and Robinson [Brat 79], Haag [Haag 92].

Obviously the first problem is more fundamental. It immediately leads to nonperturbative Pauli–Villars regularization, which is considered as a means of securing the existence of field representations as a whole. By definition the nonperturbative Pauli–Villars regularization is connected with the use of a higher order spinor field equation (2.19), (2.23) instead of a nonrenormalizable first order equation. In classical physics for higher order field equations, for instance in electrodynamics, the existence of solutions and the finiteness of self-energies, *etc.*, can be demonstrated, see Bopp [Bopp 40], Podolski [Pod 41], *etc.*, and in a review of Iwanenko and Sokolow [Iwa 53].

But what happens in quantum field theory? Because of to the canonical quantization of the auxiliary fields the original Ψ field has vanishing anticommutators (2.62),

(2.63) and is therefore unsuitable for the algebraic treatment. Thus the nontrivial canonical field operator dynamics must be formulated by means of the auxiliary fields (2.44) and (2.45). But can one detect the action of the nonperturbative Pauli–Villars regularization on the level of auxiliary fields, and does this lead to the existence of quantum solutions? With respect to this problem the knowledge of the conventional Pauli–Villars regularization by auxiliary fields in perturbation theory is of no help, as this method is not based on a systematic algebraic calculation and is not nonperturbative.

In addition: the alternating signs in the anticommutators (2.62), (2.63) enforce an indefinite metric in state space. How can one guarantee the probabilistic interpretation in this case? In the following we first introduce a general formalism and afterwards try to answer the fundamental questions mentioned above, see Section 6.8.

In the subsequent discussion of a Hamiltonian formalism with respect to inequivalent representations we follow a general line of algebraic quantum theory, and guided by the GNS construction we consider antisymmetrized basis states $\mathcal{A}(\psi_{I_1} \ldots \psi_{I_n})_t|0\rangle$, where the arguments $I_1 \ldots I_n$ are taken for equal times $t_1 = \ldots = t_n = t$. With respect to the indefiniteness of the representation spaces we assume that these spaces are Krein spaces, see Bognar [Bogn 74].

This assumption is in accordance with the evaluation of axiomatic quantum field theory in indefinite metric spaces. In order to enlarge the formalism to include such theories, in particular gauge theories, Morchio and Strocchi [Mor 80] abandoned the positive definiteness of the Wightman functions, i.e., of algebraic states. Within this context Jakobczyk [Jak 84] proved that (algebraic) Strocchi–Wightman states lead to a reconstruction theorem in Krein space, and Jakobczyk and Strocchi [Jak 88] studied the Euclidean continuation of Strocchi–Wightman states in Krein space. Krein spaces exclude the more difficult cases of dipole fields (ghosts) or dipole regularizations, which, however, were also studied by some authors, e.g. by Moschella and Strocchi [Mos 90]. A group-theoretical and algebraic classification of the various possibilities of indefinite metric was aimed at by Saller [Sall 89,91,92,93].

In conventional algebraic theory, cf., e.g., Bratteli and Robinson [Brat 79], as well as in original axiomatic quantum field theory, cf. Völkel [Völ 77], Glimm and Jaffe [Gli 87], Haag [Haag 92], one deals only with positive definite metrical spaces, i.e., with Hilbert spaces. The extension of these treatments of quantum fields with positive metric to the case of quantum fields with indefinite metric was performed by the abovementioned authors, and led to general statements about the structure of such theories.

However, as with Hilbert space theories, in this case also these efforts did not lead to a final result allowing a straightforward treatment of specific realistic models. Hence we shall be content with adopting general features of both approaches, in particular of algebraic representation theory to the problem of constructing a suitable Hamiltonian formalism, and with developing our own approach which is directly concerned with the foundation of a composite particle dynamics. In the following we will demonstrate that a single time formalism is the suitable framework to formulate our functional theory. As this formalism yields the field-theoretic analogue of ordinary Schrödinger states we may call it the Algebraic Schrödinger Representation.

The essential of the Algebraic Schrödinger Representation is the explicit state space construction which is, however, different from the usual approach in quantum mechanics. Projection of the abovementioned basis states onto physical states $|a\rangle$ leads to transition matrix elements, and these matrix elements are considered as representatives of this physical state $|a\rangle$. Thus, for the construction of the state space one has to work with these quantities.

As far as the calculation of transition matrix elements for equal times is concerned, the New Tamm–Dancoff Method was proposed by Dyson [Dys 53a,b]. Dyson replaced the quantum mechanical Schrödinger equation by Heisenberg's time evolution equation for products of field operators, and projecting with states from both sides of this equation leads to equations for transition matrix elements. But only an insufficient attempt was made to derive an appropriate normalization of these transition matrix elements, and no attempt at all to formulate a corresponding theory of states.

Apart from some applications of the single time New Tamm–Dancoff Method in solid state physics, cf. Wahl, Duscher, Göbel and Maichle [Wahl 84], and in nuclear physics, cf. Friedrich, Gerling and Bleuler [Frie 75], Friedrich and Gerling [Frie 76], Swift and Marrero [Swi 84], Fetter and Walecka [Fett 71], this formalism was not further pursued owing to difficulties with renormalization, see Bogoliubov and Shirkov [Bogo 59]. Hence there was no further progress with respect to its relation to an explicit state space construction.

Apart from Heisenberg's fundamental evolution equation the treatment given in the following has nothing in common with the New Tamm–Dancoff Method. Rather this treatment is the field-theoretic counterpart of quantum mechanical state representations. However, it will be seen in the following that the transfer of the method of state representations from quantum mechanics to quantum field theory is absolutely nontrivial and leads to new specific problems which are connected with the infinite number of degrees of freedom of a quantum field.

4.2 Subfermion State Spaces

In general a quantum theory is assumed to possess a representation in a Hilbert space, i.e., a vector space with a positive definite metric. As already mentioned in the introduction there are, however, prominent examples of theories which enlarge the concept of Hilbert space. Without referring to specific physical systems we first discuss the formal aspects of such generalizations.

The properties of representations and corresponding representation spaces of CAR or CCR algebras, respectively, depend on the sign of the anticommutators or commutators defining the algebra. If all basic anticommutators or commutators have the same sign, then there exist representation spaces with positive definite metric. If these quantities have alternating signs, then the corresponding representation spaces are indefinite. For instance, vector fields in covariant quantization possess basic commutators with alternating sign owing to their dependence on the metric tensor $\eta_{\mu\nu}$.

Another example are the spinorial anticommutators (2.60), and, as announced in the introduction, we will treat just these kinds of systems. Hence the state spaces under consideration must be vector spaces $\mathbb{H}$ with indefinite metric, $i.e.$, vector spaces with a scalarproduct $\langle \cdot, \cdot \rangle : \mathbb{H} \times \mathbb{H} \to \mathbb{C}$, and with the properties

i) $\langle \alpha_1 x_1 + \alpha_2 x_2, y \rangle = \alpha_1 \langle x_1, y \rangle + \alpha_2 \langle x_2, y \rangle$ for all $\alpha_1, \alpha_2 \in \mathbb{C}$, $x_1, x_2, y \in \mathbb{H}$,

ii) $\langle y, x \rangle = \langle x, y \rangle^*$ for all $x, y \in \mathbb{H}$,

but with $\langle x, x, \rangle \leq 0$ for at least one $x \in \mathbb{H}$.

Furthermore, to be able to make nontrivial statements about the state space, in particular to introduce a basis in $\mathbb{H}$, we assume $\mathbb{H}$ to be a Krein space, $i.e.$, $\mathbb{H}$ is nondegenerate and has the decomposition

$$\mathbb{H} = \mathbb{H}_+ \oplus \mathbb{H}_-$$

with $\mathbb{H}_+ \subset P^{++}$, $\mathbb{H}_- \subset P^{--}$, where

$$P^{++} := \{x \in \mathbb{H} , \langle x, x \rangle > 0 \text{ or } x = 0\},$$

$$P^{--} := \{x \in \mathbb{H} , \langle x, x \rangle < 0 \text{ or } x = 0\},$$

and where $\mathbb{H}_+$ and $\mathbb{H}_-$ are assumed to be intrinsic complete, whilst $\oplus$ denotes the orthogonal direct sum. The decomposition is referred to as fundamental decomposition.

We shortly explain the notions used in the definition of the Krein space: $\mathbb{H}$ is nondegenerate if for $\langle x, x \rangle = 0$ there holds $x = 0$. Intrinsic completeness of the spaces $\mathbb{H}_+$ and $\mathbb{H}_-$ means the completeness with respect to their topologies which are generated by $\| x \|_+ := \langle x, x \rangle^{\frac{1}{2}}$ for $\mathbb{H}_+$ and $\| x \|_- := |\langle x, x \rangle|^{\frac{1}{2}}$ for $\mathbb{H}_-$ and for all $x \in \mathbb{H}_+$ or all $x \in \mathbb{H}_-$, respectively. A metric space is complete if any Cauchy sequence of elements in the space has a limit which is contained in this space. The direct orthogonal sum implies that for all $x \in \mathbb{H}_+$ and for all $y \in \mathbb{H}_-$ it is $\langle x, y \rangle = 0$.

We assume that the Krein spaces are spanned by a set of generalized eigenstates for a set of commuting self-adjoint observables. With respect to the latter the following statements can be verified for Krein spaces: The adjoint of an operator with dense domain of definition in $\mathbb{H}$ exists and is unique; the adjunction is an involution on the representation of the algebra, $cf.$ Bognar [Bogn 74]. Furthermore, in such spaces suitable representations of the physical transformation groups exist, see, $e.g.$, Bracci, Morchio and Strocchi [Brac 76] or Schlieder [Schli 60a,b,c].

In general the eigenstates of a set of commuting observables on an indefinite state space need not constitute a basis of this space. However, in Krein space one can find generalized eigenstates enlarging the set of eigenstates to a set of basis states of $\mathbb{H}$, $cf.$ Bognar [Bogn 74]. Hence all the physics and 'non-physics' can be covered by this construction and one can base the probability interpretation on it.

As the physical subspace $\mathbb{H}_{\text{ph}}$ of $\mathbb{H}$ we define the closed linear hull of all eigenstates with positive norm in $\mathbb{H}$ with respect to a suitable topology, see Bognar [Bogn 74].

Another complication in indefinite state spaces is that the eigenvalues of self-adjoint operators, in our case the momentum and spin parameters p and s from (2.117), may take complex values. However, it is easy to show that in Krein space the eigenvalues of self-adjoint operators for eigenstates with a nonvanishing norm are real. Owing to the non-degeneracy of the Krein space in particular the eigenvalues of the states spanning the physical subspace $\mathbb{H}_{ph}$ are real. Hence quantum theory in $\mathbb{H}_{ph}$ can be developed as usual.

According to definition any physical state which can be realized must be a super-position of eigenstates of with positive norm. This statement is trivial in conventional quantum mechanics. *But in Krein space the elements of $\mathbb{H}_{ph}$ in general form no complete basis set and, owing to the restriction to the physical part of the state space, one has to accept that some information from the whole state space gets lost. One cannot say what is lost without considering special models. In any case this loss will effect restrictions of measurement. Then one can decide by experiment whether such restrictions can be found or not. From a theoretical point of view this definition of $\mathbb{H}_{ph}$ is sufficient to allow a probability interpretation of the theory,* cf. also Feynman [Feyn 87].

The system which is assumed to be treated along these lines is the regularized spinor field of Chapter 2. It is characterized by the Hamiltonian operator for the auxiliary fields, *cf.* (2.57):

$$\mathbf{H}: \quad = \quad c\sum_{i=1}^{3}\lambda_i^{-1}\int \bar{\psi}_{A\alpha i}(\mathbf{r})\left(-i\hbar\gamma^k\partial_k + cm_i\right)_{\alpha\beta}\delta_{AB}\psi_{B\beta i}(\mathbf{r})d^3r \qquad (4.1)$$

$$+c\frac{g}{2}V_{ABCD}\sum_{\substack{\alpha\beta\gamma\delta \\ i,j,k,l=1}}^{3}\int \bar{\psi}_{A\alpha i}(\mathbf{r})\psi_{B\beta j}(\mathbf{r})\bar{\psi}_{C\gamma k}(\mathbf{r})\psi_{D\delta l}(\mathbf{r})d^3r,$$

with $\psi_{A\alpha i}(\mathbf{r}) \equiv \psi_{A\alpha i}(\mathbf{r}, 0)$.

In the algebraic treatment of the theory the Hamiltonian (4.1) is considered as the generator of a one-parameter group of automorphisms, and the dynamical law connected with these automorphisms is expressed by Heisenberg's equation of motion

$$i\hbar\dot{\mathbf{O}} = [\mathbf{O}, \mathbf{H}]_- , \qquad (4.2)$$

where $\mathbf{O}$ can be any element of the field operator algebra.

In order to avoid the complications connected with the transition to an infinite system contained in $\mathbb{R}^3$, we first assume the systems to be contained in an arbitrarily large but simply connected compact volume V. In this volume we assume that (4.1) possesses a complete set of eigenstates and eigenvalues which constitute the elements of a corresponding Krein space. In the limit $V \to \mathbb{R}^3$ one has to apply the GNS construction and we assume that the results of this construction, which can be also applied to finite volumes, converge toward a well defined state space, *etc.*, for $V \to \mathbb{R}^3$.

Without investigating this limit process we use the results for an arbitrarily large but finite volume in our discussion as long as we perform the general evaluation of our formalism, *i.e.*, the introduction of the Algebraic Schrödinger Representation. For such finite systems it is reasonable to consider the embedding into Krein spaces as a suitable theoretical framework. But as the GNS construction is conceived to exceed this framework in the limit $V \to \mathbb{R}^3$ and to replace it, in all practical calculations we consider the system to be defined in $\mathbb{R}^3$.

Of course, this formalism should also apply to gauge theories. But in this case the most simple way of solving the metrical problem consists in eliminating the gauge constraints from the beginning in order to work only in the physically admitted subspace. This leads in Quantum Electrodynamics to the Coulomb gauge, see Stumpf, Fauser and Pfister [Stu 93a], Stumpf and Pfister [Stu 96], Fauser and Stumpf [Fau 97], whilst for Quantum Chromodynamics the subsidiary conditions are properly formulated in temporal gauge, see Stumpf and Pfister [Stu 97], where these theories were treated in the Algebraic Schrödinger Representation. For the discussion of gauge theories we also refer to Chapters 7 and 8.

In the super-renormalizable spinor field theory, however, the indefinite states cannot be characterized by subsidiary conditions and one has to work with the above given criterion, i.e., with norm calculations. After having expounded the algebraic formalism in detail, in Section 6.8 following a paper by Stumpf [Stu 00] we propose a representation of the spinor field which guarantees positive definiteness of the state norms. For this representation the loss of information refers to the measurability of the auxiliary fields, which for physical reasons are anyway unobservable.

4.3 Nonorthogonal Basis Sets

Although we will work along the lines of conventional quantum theory, in some respect quantum fields and Krein spaces need an extension. Whilst in conventional quantum theory in principle orthonormal basis sets are used and nonorthogonal basis sets merely appear in applications, for quantum fields nonorthogonal basis sets have to be assumed from the outset. In this section we shall treat this extension which also covers Krein spaces.

For brevity we use a symbolic notation. Such a notation allows a transparent description of the general algebraic formalism. For concrete calculations we return to the full set of indices.

Let $\mathcal{J}$ be a set of indices and

$$\mathbb{B} := \{|e_n\rangle, \ n \in \mathcal{J}\} \tag{4.3}$$

be a basis of a linear space $\mathbb{H}$ with an inner product $\langle . | . \rangle$. The metric of $\mathbb{H}$ is defined by the scalar product

$$G_{mn} := \langle e_m | e_n \rangle. \tag{4.4}$$

Owing to the assumed nonorthonormality of the basis vectors and the indefiniteness of $\mathbb{H}$ there holds $G_{mn} \neq \delta_{mn}$.

For a suitable domain of definition a linear operator $\mathbf{T}$ on $\mathbb{H}$ is Hermitian (self-adjoint) if

$$\langle e_n | \mathbf{T} | e_m \rangle = \langle e_m | \mathbf{T} | e_n \rangle^*, \quad \forall\, n, m \in \mathcal{J}. \tag{4.5}$$

In particular in accordance with Section 4.2, G_{mn} is Hermitian. The nondiagonal form of G_{mn} forces us to introduce a complete set of dual basis vectors $|e^m\rangle$ which are defined by the relation

$$\langle e^m | e_n \rangle = \delta_n^m \tag{4.6}$$

and additionally satisfy the completeness relation

$$|e^m\rangle\langle e_m| = \mathbf{1}. \tag{4.7}$$

The dual basis vectors are collected in a dual basis

$$\mathbb{B}^d := \{|e^m\rangle, \quad m \in \mathcal{J}\}. \tag{4.8}$$

Provided that $\mathbb{B}$ is nondegenerate, *i.e.*, that $(G_{nm})^{-1}$ exists, the dual basis vectors are unambiguously determined.

- **Proposition 4.1:** Provided that G_{mn} is invertible, then $G^{mn} := \langle e^m | e^n \rangle$ equals $(G_{mn})^{-1}$, *i.e.*, it holds

$$G^{mk} G_{kn} = \delta_n^m, \qquad G_{mk} G^{kn} = \delta_n^m. \tag{4.9}$$

Proof: We expand the dual basis vectors in the original basis $\mathbb{B}$ by means of (4.7):

$$|e^m\rangle = (G^{mk})^* |e_k\rangle. \tag{4.10}$$

From (4.6) it follows $\langle e_n | e^m \rangle = \delta_n^m$. Multiplying (4.10) with $\langle e_n|$ we obtain

$$\langle e_n | e^m \rangle = (G^{mk})^* \langle e_n | e_k \rangle = (G^{mk})^* G_{kn}^* = \delta_n^m, \tag{4.11}$$

thus it holds the first part of (4.9). The second part of (4.9) can be proved along the same lines. $\diamond$

For the following we assume that $\mathbb{B}$ is nondegenerate. Then any element $|a\rangle \in \mathbb{H}$ can be represented by a linear combination of basis vectors. We assume that a Krein space is isomorphic to its dual, so either the set $\mathbb{B}$ or the set $\mathbb{B}^d$ can be used for this representation of $|a\rangle$. This gives

$$|a\rangle = \sum_{n \in \mathcal{J}} \sigma^n(a) |e_n\rangle \tag{4.12}$$

or

$$|a\rangle = \sum_{n \in \mathcal{J}} \tau_n(a) |e^n\rangle, \tag{4.13}$$

where the summation over $n \in \mathcal{J}$ includes summation over the discrete parts of $\mathcal{J}$ and integration over the continuous parts of $\mathcal{J}$.

Projection of (4.12) and (4.13) with $\langle e^m|$ or $\langle e_m|$ yields

$$\sigma^m(a) = \langle e^m | a \rangle, \quad m \in \mathcal{J}, \tag{4.14}$$

and

$$\tau_m(a) = \langle e_m | a \rangle, \quad m \in \mathcal{J}. \tag{4.15}$$

By (4.14) and (4.15) two bijective maps are defined:

$$|a\rangle \to \{\tau_n(a), \ n \in \mathcal{J}\} \tag{4.16}$$

and

$$|a\rangle \to \{\sigma^n(a), \ n \in \mathcal{J}\}, \tag{4.17}$$

and the set $\{\tau_n(a), \ n \in \mathcal{J}\}$ is referred to as the covariant representation of $|a\rangle$, whilst $\{\sigma^n(a), \ n \in \mathcal{J}\}$ is referred to as the contravariant representation of $|a\rangle$, see Lagally and Franz [Lag 64].

In the same way as vectors also linear operators $\mathbf{T}$ on $\mathbb{H}$ can be represented in the basis $\mathbb{B}$ or (and) $\mathbb{B}^d$.

- **Proposition 4.2:** For a linear operator $\mathbf{T}$ on $\mathbb{H}$ the following representations exist:

$$T_{mn} := \langle e_n | \mathbf{T} | e_m \rangle, \tag{4.18}$$

$$T^{mn} := \langle e^n | \mathbf{T} | e^m \rangle, \tag{4.19}$$

$$T_n^{\ m} := \langle e_n | \mathbf{T} | e^m \rangle, \tag{4.20}$$

$$T_{\cdot\cdot m}^n := \langle e^n | \mathbf{T} | e_m \rangle. \tag{4.21}$$

Proof: Owing to (4.4), (4.7) and (4.9) the unit operator $\mathbf{1}$ on $\mathbb{H}$ can be represented by

$$\mathbf{1} = \sum_{m,n} |e^m\rangle G_{mn} \langle e^n| = \sum_{m,n} |e_m\rangle G^{mn} \langle e_n| \tag{4.22}$$

$$= \sum_{m,n} |e^m\rangle G_m^{\ n} \langle e_n| = \sum_m |e^m\rangle \langle e_m|.$$

With these representations one obtains

$$\mathbf{T} = \sum_{m,n} |e^m\rangle T_{mn} \langle e^n| = \sum_{m,n} |e_m\rangle T^{mn} \langle e_n| \tag{4.23}$$

$$= \sum_{m,n} |e^m\rangle T_m^{\ n} \langle e_n| = \sum_{m,n} |e_n\rangle T_{\cdot\cdot m}^n \langle e^m|,$$

where the coefficients are given by (4.18)–(4.21). So by (4.18)–(4.21) a map from linear operators onto $\mathbb{H}$ on the space of matrix representations is defined. One easily verifies the representation properties of (4.18)–(4.21).   ◊

The matrix representations (4.18)–(4.21) are referred to as double covariant, double contravariant or mixed co–contravariant representations.

By means of the representations (4.22) the scalar product for any elements $|a\rangle, |b\rangle \in \mathbb{H}$ can be reduced to the scalar products of the basis vectors of $\mathbb{B}$ and $\mathbb{B}^d$, and we obtain with (4.12), (4.13) and (4.22)

$$g_{ab} := \langle a|b \rangle \tag{4.24}$$
$$= \sum_n \sigma^n(a)^* \tau_n(b)$$
$$= \sum_{m,n} \tau_m(a)^* G^{mn} \tau_n(b)$$
$$= \sum_{m,n} \sigma^m(a)^* G_{mn} \sigma^n(b).$$

Of special interest are eigenvalue equations for self-adjoint operators S on $\mathbb{H}$ in the matrix representations (4.18)–(4.21).

- **Proposition 4.3:** Let S be a self-adjoint operator on $\mathbb{H}$ and $|p\rangle \in \mathbb{H}$ an eigenstate of S with $S|p\rangle = s_p|p\rangle$. Then there exists a mixed co–contravariant matrix representation of S, $S_m^{\cdot n} := \langle e_m|S|e^n \rangle$, and the co- and contravariant components $\tau_n(p)$ and $\sigma^n(p)$ of $|p\rangle$ are right hand side or left hand side solutions of the eigenvalue equations

$$\sum_n S_m^{\cdot n} \tau_n(p) = s_p \tau_m(p) \tag{4.25}$$

and

$$\sum_m \sigma^m(p)^* S_m^{\cdot n} = s_p^* \sigma^n(p)^*. \tag{4.26}$$

Proof: We substitute (4.13) into the eigenvalue equation of S and project from the left with $\langle e_m|$. This gives

$$\sum_n S_m^{\cdot n} \tau_n(p) = s_p \tau_m(p)$$

with $S_m^{\cdot n} = \langle e_m|S|e^n \rangle$. Because of the self-adjointness of S from $S|p\rangle = s_p|p\rangle$ it follows by Hermitian conjugation

$$\langle p|S = s_p^* \langle p|. \tag{4.27}$$

Substitution of the Hermitian conjugate of (4.12) into (4.27) and projection with $|e^n\rangle$ yields equation (4.26). $\Diamond$

Finally we observe that in spite of the self-adjointness of S the mixed co–contravariant matrix representation $S_m^{\cdot n}$ in general is non-Hermitian. This can be seen from matrix conjugation:

$$(S_n^{\cdot m})^+ = (\langle e_m|S|e^n\rangle)^* = \langle e^n|S|e_m \rangle = S_{\cdot m}^n.$$

But $\langle e^n|S|e_m\rangle$ is unequal $\langle e_n|S|e^m\rangle$ as long as $\langle e^m| \neq \langle e_m|$. Hence $(S_n^{\cdot m})^+ \neq S_n^{\cdot m}$. However, in any case we have $S_{mn} = S_{mn}^+$ etc., i.e., the double co- and contravariant matrix representations are Hermitian.

The latter property leads to the following conclusion:

- **Proposition 4.4:** If the matrix representations S^{mn} and G^{mn} are Hermitian, then the eigenvectors $\tau_n(a)$ and $\tau_n(b)$ of the eigenvalue equation (4.25) with eigenvalues s_a and s_b, $s_a \neq s_b$ are orthogonal with respect to the scalar product (4.24).

Proof: For nonorthogonal basis systems the proof goes in analogy to the corresponding proof for orthonormal basis systems in quantum mechanics.   ◊

This proposition is of special interest for inequivalent representations. In this case there is no universally valid state spectrum of the abstract operator **S**. Rather the state spectrum depends on the chosen representation. But if this representation leads to Hermitian matrix operators the orthogonality of eigenvectors is secured as in conventional quantum mechanics. The two properties are important results of the algebraic line of approach to quantum field theory. They generalize and extend the formalism of perturbation theory and conventional quantum mechanics, but at the same time conserve the valuable properties of eigenvectors.

4.4 Cyclic Basis Vector Representations

The explicit evaluation of eigenvalue equations, *etc.*, depends on the ability to define appropriate basis vector systems. We emphasized that in quantum field theory this must be done in accordance with general ideas of algebraic representation theory and axiomatic quantum field theory. The corresponding construction of basis vector systems in Algebraic Quantum Theory is the GNS construction (Gelfand–Naimark–Segal construction, *cf.* Section 3.1), a generalization of the Fock space construction.

The GNS construction provides genuine representations of the basic field operator algebra. As already mentioned this algebra is already complete (or even over complete) at any space-like hyperplane. In our case this is expressed by the equal time anticommutation relations (2.61). For simplicity we will only consider space-like hyperplanes defined by $t = $ const..

The GNS construction is based on the existence of cyclic vectors. Excluding the ground states of thermo-field dynamics we assume that a suitable cyclic vector is given by the physical ground state $|0\rangle \in$ IH. Then the set of basis vectors is generated by the application of products of field operators to the ground state $|0\rangle$, and according to the completeness, or even over completeness of the algebra on any space-like hyperplane we confine these products to equal times in their coordinates.

In the following we use the abbreviation

$$(\psi_{I_1} \ldots \psi_{I_n})_t := \psi_{Z_1}(\mathbf{r}_1, t) \ldots \psi_{Z_n}(\mathbf{r}_n, t) \qquad (4.28)$$

for field operator products at equal times which are the generating elements of our GNS construction.

However, these operator products have to be subjected to further restrictions with respect to their domain of definition. Owing to the equal time anticommutators

$$[\psi_{I_1}, \psi_{I_2}]_+^{t_1=t_2} = A_{I_1 I_2}, \qquad (4.29)$$

we can rearrange any product (4.28) into a series of products obeying some ordering relation, see Grimm, Hailer and Stumpf [Gri 91]. We choose a lexicographic ordering relation which enables us to decide for two general indices I_1, I_2 whether $I_1 < I_2$ or $I_2 < I_1$ apart from $I_1 = I_2$. Then any product (4.28) can be expressed by a series of products in lexicographic order, or equivalently: an arbitrary product (4.28) is linearly dependent on a set of representative products with a definite denumeration. For brevity we do not discuss this rearrangement in detail. Rather we immediately define the generating GNS products by the set

$$\left\{ (\psi_{I_1} \ldots \psi_{I_n})_t, \; I_1 < I_2 < \ldots < I_n, \; n \in \mathbb{N} \right\}, \tag{4.30}$$

where here and in the following we have $I_m \in \mathcal{I}, \; \forall \, 1 \leq m \leq n$, and where a lexicographic ordering relation for $I_1, \ldots I_n$ is assumed. In this set, owing to their continuum part the points $I_k = I_l$, etc., have measure zero and thus can be omitted. On this domain of the indices we now can define a set of GNS basis vectors

$$\mathbb{B}_1 := \left\{ (\psi_{I_1} \ldots \psi_{I_n})_t \, |0\rangle, \; I_1 < I_2 < \ldots < I_n, \; n \in \mathbb{N} \right\}, \tag{4.31}$$

and this set is assumed to be complete and linear independent. As a consequence of this any state vector $|a\rangle \in \mathbb{H}$ can uniquely be represented by the expansion

$$|a\rangle = \sum_n \sum_{I_1 < \ldots < I_n} \rho_t^{(n)}(I_1 \ldots I_n | a) \, (\psi_{I_1} \ldots \psi_{I_n})_t \, |0\rangle, \tag{4.32}$$

where the coefficient functions $\rho_t^{(n)}(I_1 \ldots I_n | a)$ are only defined for ordered sequences $I_1 \ldots I_n$ at equal times $t_1 = \ldots = t_n = t$. In this way we have obtained the starting point for field theoretic state space theory. The only flaw in this representation is the restriction of the summation (integration) imposed by the ordering relation. As is familiar from statistical mechanics, one can avoid this restriction by the introduction of symmetric or antisymmetric basis vectors. In that case the restricted summation can be enlarged to comprise all equivalence classes of the original domain, and this allows integration over the whole 'phase space' of $I_1 \ldots I_n$ (for equal times).

For spinor fields the symmetric products can be excluded and we define for anti-symmetric products the corresponding GNS basis vectors by the set

$$\mathbb{B}_2 := \left\{ \mathcal{A} \, (\psi_{I_1} \ldots \psi_{I_n})_t \, |0\rangle, \; n \in \mathbb{N} \right\}, \tag{4.33}$$

where $\mathcal{A}$ means antisymmetrization in $I_1 \ldots I_n$, i.e.,

$$\mathcal{A} \, (\psi_{I_1} \ldots \psi_{I_n})_t := \sum_{p \in S_n} \frac{1}{n!} \mathrm{sgn}(p) \psi_{I_{p_1}} \ldots \psi_{I_{p_n}}. \tag{4.34}$$

With respect to the set $\mathbb{B}_2$ any state $|a\rangle \in \mathbb{H}$ can be represented by

$$|a\rangle = \sum_n \sum_{I_1 \ldots I_n} \frac{1}{n!} \hat{\rho}_t^{(n)}(I_1 \ldots I_n | a) \mathcal{A} \, (\psi_{I_1} \ldots \psi_{I_n})_t \, |0\rangle, \tag{4.35}$$

where $\hat{\rho}_n^t(I_1 \ldots I_n | a)$ is antisymmetric in $I_1 \ldots I_n$ for $t_1 = \ldots = t_n = t$.

The proof of equivalence of (4.32) and (4.35) makes use of equivalence classes with respect to the various domains of $I_1 \ldots I_n$ and the antisymmetry properties of the constituents of (4.35). For its explicit discussion we refer to statistical mechanics.

For technical reasons we still slightly modify our basis set $\mathbb{B}_2$. As the adjunction is an inner automorphism of the field algebra we can equivalently use the adjoints of the vectors of $\mathbb{B}_2$ defined by

$$\mathbb{B}_0 := \left\{ [\mathcal{A}(\psi_{I_1} \ldots \psi_{I_n})_t]^+ |0\rangle, \ n \in \mathbb{N} \right\}, \tag{4.36}$$

and with respect to $\mathbb{B}_0$ the expansion of $|a\rangle$ reads

$$|a\rangle = \sum_n \sum_{I_1 \ldots I_n} \sigma_t^{(n)}(I_1 \ldots I_n | a) [\mathcal{A}(\psi_{I_1} \ldots \psi_{I_n})_t]^+ |0\rangle, \tag{4.37}$$

where we have incorporated the factor $(n!)^{-1}$ in the definition of $\sigma_t^{(n)}$.

We can now relate expansion (4.37) to our formalism of Section 4.3 by defining the operator $\mathbf{A}_n$ by

$$\mathbf{A}_n := \mathcal{A}(\psi_{I_1} \ldots \psi_{I_n})_t \tag{4.38}$$

and the basis vectors $|e_n\rangle$ by

$$|e_n\rangle := \mathbf{A}_n^+ |0\rangle, \tag{4.39}$$

where in this symbolic notation we have $n \in \mathcal{J}$, $\mathcal{J} = \{I_1, \ldots I_n\}$. Thus we can write

$$|a\rangle = \sum_n \sigma^n(a)|e_n\rangle = \sum_n \tau_n(a)|e^n\rangle, \tag{4.40}$$

and according to (4.15)

$$\tau_n(a) = \langle e_n | a \rangle = \langle 0 | \mathcal{A}(\psi_{I_1} \ldots \psi_{I_n})_t | a \rangle. \tag{4.41}$$

In Fock space such objects like (4.41) are referred to as 'wave functions'. In GNS spaces we can maintain this designation, but have to pay attention to the fact that the state norms are given by (4.24), i.e., they are not as simple as in Fock space.

In order to complete our discussion of the construction of state space representations by cyclic basis vector systems, we have to derive the metric tensor with respect to these basis vectors. This can be performed by the following proposition, Stumpf [Stu 89]:

• **Proposition 4.5:** For $G_{mn} := \langle e_m | e_n \rangle$ with $|e_m\rangle, |e_n\rangle \in \mathbb{B}_0$ the representation

$$G_{mn} = \sum_{l,k} T_m^l C_{ln}^k \tau_k(0) \tag{4.42}$$

holds, where the expansion coefficients T_m^l and C_{ln}^k are well defined.

Proof: With (4.38) and (4.39) there holds

$$\langle e_m | e_n \rangle = \langle 0 | \mathcal{A} \left(\psi_{I_1} \dots \psi_{I_m} \right)_t \left[\mathcal{A} \left(\psi_{K_1} \dots \psi_{K_n} \right)_t \right]^+ | 0 \rangle. \tag{4.43}$$

As the adjunction is an automorphism of the field operator algebra, the relation

$$\mathbf{A}_n^+ \equiv \mathcal{A} \left(\psi_{K_n}^+ \dots \psi_{K_1}^+ \right)_t = \sum_m \sum_{I_1 \dots I_m} T_{K_1 \dots K_n}^{I_1 \dots I_m} \mathcal{A} \left(\psi_{I_1} \dots \psi_{I_m} \right)_t \tag{4.44}$$

holds with well defined expansion coefficients. In symbolic notation, with (4.44) equation (4.43) goes over into

$$\langle e_m | e_n \rangle = \sum_l T_n^l \langle 0 | \mathbf{A}_m \mathbf{A}_l | 0 \rangle. \tag{4.45}$$

By means of the anticommutator (4.29) the product $\mathbf{A}_m \mathbf{A}_l$ can be rearranged into a series of antisymmetric products, see Proposition 4.11:

$$\mathbf{A}_m \mathbf{A}_l = \sum_k C_{ml}^k \mathbf{A}_k. \tag{4.46}$$

If we substitute (4.46) into (4.45) and observe that owing to the definiton (4.41) there holds

$$\langle 0 | \mathbf{A}_k | 0 \rangle = \langle e_k | 0 \rangle = \tau_k(0),$$

we obtain equation (4.42). $\diamond$

The advantage of this representation stems from the fact that the calculation of G_{mn} is reduced to the calculation of the set $\{\tau_k(0)\}$, which is a solution of the corresponding functional equation for antisymmetrized vacuum expectation values.

Concerning the normalization of states in an indefinite metric space we have the following proposition:

- **Proposition 4.6:** Any eigenstate $|p\rangle$ of a set of self-adjoint operators can be normalized to $+1$ or -1, provided $\|p\| \neq 0$.

Proof: For eigenstates $|p\rangle$ of a set of self-adjoint operators the wave functions $\tau_n(p)$ are solutions of homogeneous eigenvalue equations. They are determined up to a common complex factor $\lambda(p)$. This factor can be used to transform $g_{pp'}$, given by (4.24), into its normal form, which is given by $|g_{pp}| = 1$ and $g_{pp'} = 0$ for $p \neq p'$. If we replace $\tau_n(p)$ by $\lambda(p)\tau_n(p)$ then (4.24) goes over into

$$\|p\|^2 = g'_{pp} = \sum_{m,n} \lambda(p)^* \tau_m(p)^* G^{mn} \tau_n(p) \lambda(p) \tag{4.47}$$

$$= |\lambda(p)|^2 \sum_{m,n} \tau_m(p)^* G^{mn} \tau_n(p)$$

$$= |\lambda(p)|^2 g_{pp},$$

and thus the eigenstate $|p\rangle$ can be normalized to $+1$ or -1, provided $\|p\| \neq 0$. $\diamond$

Furthermore, this normalization factor is transferred to $\sigma^n(p)$, since from (4.24) we can read off

$$\sigma^n(p)^* = \sum_m \tau_m(p)^* G^{mn} . \tag{4.48}$$

This shows that the calculation of left hand side solutions $\sigma^{(n)}$ does not lead to a new freedom of normalization. For a continuous spectrum of $|p\rangle$ this statement can be generalized to the normalization of distributions.

4.5 Renormalized Eigenvalue Equations

Poincaré invariant field theories describe systems with infinite spatial and temporal extension. This implies that at least some of the eigenvalues of extensive operators as, for instance, total energy, particle number, *etc.*, have infinite values, and if these systems are to describe an idealized 'world', their eigenvalues are not measurable, because one can only observe differences between a state of reference (ground state) and excited states of the system. Mathematically this leads to an additive renormalization replacing the absolute values of observables by difference values. We elaborate this concept in the cyclic representation referring to Grimm, Hailer and Stumpf, [Gri 91].

Let $\mathbf{S}$ be a self-adjoint operator on $\mathbb{H}$ with $\mathbf{S}|p\rangle = s_p|p\rangle$. Then we introduce the operator $\mathbf{S}_0$ on $\mathbb{H}$ by

$$\mathbf{S}_0|a\rangle = s_0|a\rangle , \quad \forall \, |a\rangle \in \mathbb{H} \tag{4.49}$$

and define the renormalized operator $\mathbf{S}^r$ by

$$\mathbf{S}^r := \mathbf{S} - \mathbf{S}_0 , \tag{4.50}$$

where renormalization refers to the physical ground state $|0\rangle$. The explicit form of $\mathbf{S}_0$ is given by $\mathbf{S}_0 = s_0 \mathbf{1}$, from which we gather that $\mathbf{S}_0$ is self-adjoint, too, and hence also $\mathbf{S}^r$ is self-adjoint.

- **Proposition 4.7:** Let $\mathbf{S}^r$ be the corresponding renormalized operator (4.50) of a self-adjoint operator $\mathbf{S}$ on $\mathbb{H}$ with

$$\mathbf{S}^r|p\rangle = (s_p - s_0)|p\rangle \tag{4.51}$$

for all elements $|p\rangle \in \mathbb{B}_{\mathrm{ph}}$ with $\| \, p \, \| = 1$. Then there exist 'matrix' representations of equation (4.51) in the cyclic basis $\mathbb{B}_0$ given by

$$\sum_n (S^r)_m{}^{\cdot n} \tau_n(p) = (s_p - s_0)\tau_m(p), \tag{4.52}$$

$$\sum_m \sigma^m(p)^* (S^r)_m{}^{\cdot n} = (s_p^* - s_0^*)\sigma^n(p)^* \tag{4.53}$$

with

$$(S^r)_m{}^{\cdot n} := \langle e_m|\mathbf{S}^r|e^n\rangle = \langle e_m|\mathbf{S} - \mathbf{A}_m^{-1}\mathbf{S}\mathbf{A}_m|e^n\rangle . \tag{4.54}$$

Proof: We substitute (4.13) into the eigenvalue equation (4.49) and project from the left by $\langle e_m|$. This gives

$$\sum_n \langle e_m|S_0|e^n\rangle \tau_n(p) \;=\; s_0 \sum_n \langle e_m|e^n\rangle \tau_n(p) \tag{4.55}$$

$$=\; s_0 \tau_m(p).$$

On the other hand, for self-adjoint S we have $\langle 0|S = s_0\langle 0|$ and therefore

$$s_0 \sum_n \langle e_m|e^n\rangle \tau_n(p) \;=\; \sum_n \langle 0|SA_m|e^n\rangle \tau_n(p) \tag{4.56}$$

$$=\; s_0 \tau_m(p).$$

Comparison of equations (4.55) and (4.56) yields

$$\sum_n \langle e_m|S_0|e^n\rangle \tau_n(p) = \sum_n \langle 0|SA_m|e^n\rangle \tau_n(p). \tag{4.57}$$

If we formally write $\langle 0| \equiv \langle 0|\, A_m A_m^{-1}$ this is equivalent to

$$\sum_n \langle e_m|S_0|e^n\rangle \tau_n(p) = \sum_n \langle e_m|A_m^{-1}SA_m|e^n\rangle \tau_n(p), \tag{4.58}$$

which has to be satisfied for all $\langle e_m| \in \mathbb{B}_0^d$. Hence, if we represent (4.51) in the cyclic basis $\mathbb{B}_0$ and observe (4.58) we obtain equation (4.52) with (4.54). In the same way (4.53) can be proved.

To make sure that the replacement of $\langle 0|$ by $\langle e_m|A_m^{-1}$ is possible, we represent A_m and A_m^{-1} in accordance with Proposition 4.3 by

$$A_m \;=\; \sum_{k,l} |e^k\rangle (A_m)_k^{\;\cdot\, l}\langle e_l|, \tag{4.59}$$

$$A_m^{-1} \;=\; \sum_{h,j} |e^h\rangle \left(A_m^{-1}\right)_h^{\;\cdot\, j}\langle e_j|. \tag{4.60}$$

In this representation the condition $\langle 0| = \langle e_m|A_m^{-1}$ goes over into

$$\sum_l (A_m)_0^{\;\cdot\, l}\left(A_m^{-1}\right)_l^{\;\cdot\, j} = \delta_0^j, \tag{4.61}$$

where owing to the duality we used $\langle 0|e_j\rangle = \delta_j^0$. If we consider the matrices $(A_m^{-1})_l^{\;\cdot\, j}$ as a set of j vectors with components $l \in \mathcal{J}$, then (4.61) means that all vectors $j \neq 0$ must be orthogonal to a given vector $(A_m)_0^{\;\cdot\, l}$ with components l. This condition can be satisfied by direct construction. $\diamond$

For the practical evaluation of renormalized eigenvalue equations Proposition 4.7 is not very well suited; so according to Grimm, Hailer and Stumpf [Gri 91] we give another version of this proposition which later on allows an elegant functional evaluation method.

- **Proposition 4.8:** If the operator S is a power series functional of the field operators on a space-like hyperplane $t = $ const., then the commutator of S with A_n can be expressed by the expansion

$$[A_n, S]_- = \sum_m C_n^{\cdot\ m} A_m \qquad (4.62)$$

with

$$C_n^{\cdot\ m} := (S^r)_n^{\cdot\ m}, \qquad (4.63)$$

where $(S^r)_n^{\cdot\ m}$ is given by (4.54).

Proof: Owing to the assumption about S the commutator $[A_n, S]_-$ can be directly evaluated by means of (4.29) and leads to a power series expansion in the field operators. If some of these power series terms are not antisymmetrized, they can be rearranged into an antisymmetrized product expansion. Hence (4.62) holds true.

To prove (4.63) we project (4.62) from the left with $\langle 0|$ and obtain

$$\langle 0| [A_n, S]_- = \sum_k C_n^{\cdot\ k} \langle e_k|. \qquad (4.64)$$

Furthermore, we have with $\langle 0| \equiv \langle 0|A_m A_m^{-1}$, $m \in \mathcal{J}$:

$$\langle 0| (A_n S - S A_n) = \langle e_n| \left(S - A_n^{-1} S A_n \right), \qquad (4.65)$$

and combination of (4.64) with (4.65) yields

$$\sum_k C_n^{\cdot\ k} \langle e_k| = \langle e_n| \left(S - A_n^{-1} S A_n \right), \qquad (4.66)$$

which by projection with $|e^m\rangle$ from the right leads to (4.63). $\Diamond$

If we project equation (4.65) from the right with an eigenvector $|p\rangle$ of S and use expansion (4.40), it follows

$$\langle 0| [A_n, S]_- |p\rangle = \sum_k \langle e_n| \left(S - A_n^{-1} S A_n \right) |e^k\rangle \tau_k(p), \qquad (4.67)$$

and with (4.54) and (4.52) this yields

$$\langle 0| [A_n, S]_- |p\rangle = (s_p - s_0) \tau_n(p). \qquad (4.68)$$

In particular for $S \equiv H$ we obtain

$$\langle 0| [A_n, H]_- |p\rangle = (E_p - E_0) \tau_n(p), \qquad (4.69)$$

the transition matrix element of Heisenberg's formula $i\hbar \dot{O} = [O, H]_-$ for $O \equiv A_n$ and energy eigenstates $\langle 0|$ and $|p\rangle$. The latter equation is the starting point of the New Tamm–Dancoff Method, see Silin and Fainberg [Si 56].

Summarizing our statements we see that they end up with a renormalized energy equation, but in addition this approach establishes a relation of this equation with an explicit state construction which so far has not been considered in conventional quantum field theory. It is just this relation which is crucial for composite particle theory.

4.6   Functional Eigenvalue Equations

In order to derive the renormalized eigenvalue equation (4.68) for a self-adjoint operator S or S^r, respectively, we assumed S to be a power series functional of the field operators on a space-like hyperplane $t = $ const.. As far as the spinor field is concerned, this assumption applies to the generators of the Poincaré group $P_\mu, M_{\mu\nu}$ given by (2.99) and (2.109). Therefore, by means of (4.68) the eigenvalue equations (2.117) can be formulated in the cyclic representation. In the following we concentrate on the evaluation of (4.68) for $P_0 \equiv H$, because the resulting renormalized energy eigenvalue equation is fundamental for composite particle theory. The evaluation of (4.68) for the other Poincaré generators appearing in (2.117) runs along the same lines, and for brevity we do not explicitly discuss their eigenvalue equations.

In analogy with the covariant formalism the appropriate means for the evaluation of (4.68) is the mapping into functional space. But in contrast to the covariant formalism, in this case we do not define the functional CAR algebra on the whole IM^4, rather we confine this algebra to a space-like hyperplane $t = $ const.. To this end we introduce the generators of a CAR algebra on the hyperplane $t = $ const. by

$$j_I^t := j_Z^t(\mathbf{r}) \quad \text{and} \quad \partial_I^t := \partial_Z^t(\mathbf{r}),\tag{4.70}$$

and mark them with the superscript t. *In this case we understand the symbol I to be an abbreviation for $I = (Z, \mathbf{r}, t)$ with the fixed parameter t.* In particular the summation convention does not include the parameter t in the following.

The anticommutators of the CAR generators on the hyperplane are then given by

$$\left[j_I^t, j_{I'}^t \right]_+ = \left[\partial_I^t, \partial_{I'}^t \right]_+ = 0,\tag{4.71}$$

$$\left[j_I^t, \partial_{I'}^t \right]_+ = \delta_{ZZ'}\delta(\mathbf{r} - \mathbf{r}'),\tag{4.72}$$

and their Fock space representation results from $(j_I^t)^+ |0\rangle_F = \partial_I^t|0\rangle_F = 0$. The state $|0\rangle_F$ is again assumed to be a cyclic state of a corresponding functional space IK_F^t. For brevity we do not explicitly introduce the transformation properties of the generators (4.70). They are analogous to (3.12) and (3.13) but restricted to the rotation group O_3 and the additional symmetry groups.

We define functional states on the hyperplane by

$$|\mathcal{A}(j^t; a)\rangle := \sum_{n=1}^{\infty} \frac{i^n}{n!} \sum_{I_1 \ldots I_n} \tau_t^{(n)}(I_1 \ldots I_n|a) j_{I_1}^t \ldots j_{I_n}^t |0\rangle_F\tag{4.73}$$

with

$$\tau_t^{(n)}(I_1 \ldots I_n|a) := \langle 0|\mathcal{A}(\psi_{I_1} \ldots \psi_{I_n})_t|a\rangle,\tag{4.74}$$

and in this way we associate to every state $|a\rangle \in IH$ a functional state $|\mathcal{A}(j^t; a)\rangle \in IK_F^t$.

The discussion of the 'physical' metric in IH which is based on the cyclic representation and which was given in the preceding sections enables us to establish an isometric map between IH and IK_F^t by postulating the isometry condition

$$\langle \mathcal{D}(\partial^t; a) | \mathcal{A}(j^t; b)\rangle = \langle a|b\rangle, \tag{4.75}$$

where the dual state $\langle \mathcal{D}(\partial^t; a)|$ is defined by

$$\langle \mathcal{D}(\partial^t, a)| := \sum_{n=1}^{\infty} (-i)^n \sigma_t^{(n)} (I_1 \ldots I_n | a)^* {}_{\mathrm{F}}\langle 0| \partial_{I_n}^t \ldots \partial_{I_1}^t. \tag{4.76}$$

Then the following proposition holds:

- **Proposition 4.9:** If the expansion coefficients $\sigma_t^{(n)}(I_1 \ldots I_n | a)$ of the dual functional state (4.76) are the contravariant components of the state $|a\rangle$, *cf.* (4.14), then (4.75) is valid and an isometric map from $\mathbb{H}$ to $\mathbb{K}_{\mathrm{F}}^t$ is established.

Proof: Owing to the basic commutator relations (4.71), (4.72) the dual basis vectors $(-i)_{\mathrm{F}}^n \langle 0| \partial_{I_n}^t \ldots \partial_{I_1}^t$ are orthonormal to the original basis vectors $i^n (n!)^{-1} j_{I_1}^t \ldots j_{I_n}^t |0\rangle_{\mathrm{F}}$. By means of this property we can calculate the left hand side of (4.75). This gives

$$\langle \mathcal{D}(\partial^t; a) | \mathcal{A}(j^t; b)\rangle \tag{4.77}$$

$$= \sum_{m,n=1}^{\infty} (-i)^m \frac{i^n}{n!} \delta_{mn} \sigma_t^{(m)} (K_1 \ldots K_m | a)^* \tau_t^{(n)} (I_1 \ldots I_n | b)$$

$$\times \sum_{p \in S_m} \mathrm{sgn}(p) \delta_{I_1 K_{p_1}} \ldots \delta_{I_n K_{p_m}}$$

$$= \sum_{p \in S_n} \sum_{n=1}^{\infty} \mathrm{sgn}(p) \frac{1}{n!} \sigma_t^{(n)} (K_1 \ldots K_n | a)^* \tau_t^{(n)} (K_{p_1} \ldots K_{p_n} | b)$$

$$= \sum_{n=1}^{\infty} \sigma_t^{(n)} (K_1 \ldots K_n | a)^* \tau_t^{(n)} (K_1 \ldots K_n | b)$$

$$= \langle a|b\rangle,$$

if one observes (4.12), (4.13) and (4.24).   ◇

Having provided the map from the state space $\mathbb{H}$ into functional space $\mathbb{K}_{\mathrm{F}}^t$ we can now proceed to the derivation of eigenvalue equations in $\mathbb{K}_{\mathrm{F}}^t$. We concentrate on equation (4.69). There are several possibilities to perform its explicit evaluation:

i) direct or indirect calculation of the commutator in (4.69);

ii) the equal time limit of the multitime covariant energy equation (3.81);

iii) application of the Hausdorff formula.

We shall treat the methods i) and ii) but omit method iii) which needs a too specialized mathematical preparation. In order to gain confidence in these methods they were first tested for the anharmonic oscillator by Maison and Stumpf [Mai 66], and later on extended to the treatment of spinor fields by Pfister [Pfi 87]. In this section we apply method i) with a direct calculation of the commutator in (4.69), which seems to be the simplest way to solve our problem.

We observe that the Heisenberg formula $i\hbar\dot{O} = [O, H]_{-}$ can be applied to any observable O, which is given as a power series of field operators, in particular to the antisymmetrized products A_n. This leads to

$$i\hbar\frac{\partial}{\partial t}A(\psi_{I_1}\ldots\psi_{I_n})_t = \left[A(\psi_{I_1}\ldots\psi_{I_n})_t, H\right]_{-}, \tag{4.78}$$

or in combination with equation (4.69) to

$$i\hbar\frac{\partial}{\partial t}\langle 0|A(\psi_{I_1}\ldots\psi_{I_n})_t|p\rangle = (E_p - E_0)\langle 0|A(\psi_{I_1}\ldots\psi_{I_n})_t|p\rangle. \tag{4.79}$$

The evaluation of these relations can be performed by bringing the field dynamics into play. In Stumpf, Fauser and Pfister [Stu 93a] and in Stumpf and Borne [Stu 94] this evaluation was done with the replacement of the time derivatives in (4.79) by the corresponding equations of motion for the ψ field and subsequent antisymmetrization. But the most simple way is the direct evaluation of the right hand side of (4.78) in functional space as was demonstrated by Stumpf and Pfister [Stu 96].

In order to keep the formulae to a minimum we write instead of ψ_I only the argument I and omit the explicit notation of the states $|0\rangle$ and $|p\rangle$. Furthermore, we set $\hbar = c = 1$ and remember, that the spinor field Hamiltonian (4.1) is polynomial in ψ. The task is now to rearrange partially disordered polynomials of ψ in (4.78) into antisymmetrical ones.

For the rearrangment of the standard form (4.78) into the final form with antisymmetrized transition matrix elements we need a proposition which was proved by Pfister [Pfi 87], Stumpf, Fauser and Pfister [Stu 93a] (see also Fauser [Fau 98]):

- **Proposition 4.10:** The recursion formula

$$\sum_{p\in S_n}\frac{1}{n!}\text{sgn}(p)\langle 0|\psi_J\psi_{I_{p_1}}\ldots\psi_{I_{p_n}}|a\rangle \tag{4.80}$$

$$=: \sum_{p\in S_n}\frac{1}{n!}\text{sgn}(p)\langle JI_{p_1}\ldots I_{p_n}\rangle$$

$$= \sum_{p\in S_n}\frac{1}{(n+1)!}\text{sgn}(p)\left[\sum_{\ell=1}^{n}(-1)^{\ell}\langle I_{p_1}\ldots I_{p_\ell}JI_{p_{\ell+1}}\ldots I_{p_n}\rangle\right.$$

$$\left.+\frac{1}{2}n(n+1)A_{JI_{p_1}}\langle I_{p_2}\ldots I_{p_n}\rangle\right], \quad n\in\mathbb{N},$$

generates the relation between non-antisymmetric standard forms and antisymmetric matrix elements of field operators. With the definition

$$|A_{J_1\ldots J_m}(j^t; a))\rangle := \sum_{n=0}^{\infty}\frac{i^n}{n!}\langle 0|\psi_{J_1}\ldots\psi_{J_m}A(\psi_{I_1}\ldots\psi_{I_n})|a\rangle j_{I_1}^t\ldots j_{I_n}^t|0\rangle_F \tag{4.81}$$

equation (4.80) can be expressed by the functional relation

$$|A_I(j^t; a)\rangle = \left(\frac{1}{i}\partial_I^t + \frac{i}{2}A_{IJ}j_J^t\right)|A(j^t; a)\rangle. \tag{4.82}$$

Proof: We use the identity

$$\langle JI_{p_1} \ldots I_{p_n} \rangle \equiv \frac{1}{(n+1)} \sum_{k=0}^{n} \langle JI_{p_1} \ldots I_{p_n} \rangle, \tag{4.83}$$

and in the summand $k = \ell$ on the right hand side of (4.83) we shift the operator $\psi_J \equiv J$ by repeated anticommutations to the position ℓ, i.e., to $\langle I_{p_1} \ldots I_{p_\ell} J I_{p_{\ell+1}} \ldots I_{p_n} \rangle$. Reindexing of the permutations in the terms with the anticommutator A_{JI} yields (4.80).

Next we multiply (4.80) from the right with $i^n (n!)^{-1} j_{I_1}^t \ldots j_{I_n}^t |0\rangle_F$, sum over n and obtain

$$\sum_n \frac{i^n}{n!} \sum_{p \in S_n} \frac{1}{n!} \mathrm{sgn}(p) \langle JI_{p_1} \ldots I_{p_n} \rangle j_{I_1}^t \ldots j_{I_n}^t |0\rangle_F \tag{4.84}$$

$$= \sum_n \frac{i^n}{n!} \left[\sum_{p \in S_n} \frac{1}{(n+1)!} \mathrm{sgn}(p) \sum_{\ell=0}^{n} (-1)^\ell \langle I_{p_1} \ldots I_{p_\ell} J I_{p_{\ell+1}} \ldots I_{p_n} \rangle \right.$$

$$\left. + \frac{1}{2} \sum_{p \in S_n} \frac{1}{(n-1)!} \mathrm{sgn}(p) A_{JI_{p_1}} \langle I_{p_2} \ldots I_{p_n} \rangle \right] j_{I_1}^t \ldots j_{I_n}^t |0\rangle_F.$$

By means of the single time version of (3.18) the first term on the right hand side of (4.84) can equivalently be written in the form

$$\sum_n \frac{i^n}{n!} \langle \mathcal{A}(JI_1 \ldots I_n) \rangle j_{I_1}^t \ldots j_{I_n}^t |0\rangle_F = \frac{1}{i} \partial_J^t |\mathcal{A}(j^t; a)\rangle, \tag{4.85}$$

whilst the second term can be rearranged into

$$\frac{1}{2} \sum_n \frac{i^n}{(n-1)!} A_{JI_1} \langle I_2 \ldots I_n \rangle j_{I_1}^t \ldots j_{I_n}^t |0\rangle_F \tag{4.86}$$

$$= \frac{i}{2} A_{JI} j_I^t \sum_n \frac{i^{n-1}}{(n-1)!} \langle \mathcal{A}(I_2 \ldots I_n) \rangle j_{I_1}^t \ldots j_{I_n}^t |0\rangle_F$$

$$= \frac{i}{2} A_{JI} j_I^t |\mathcal{A}(j^t; a)\rangle.$$

Substitution of (4.85) and (4.86) into (4.84) and observance of (4.81) for $m = 1$ results in (4.82).   ◇

Repeated application of (4.82) leads to the general formula

$$|\mathcal{A}_{J_m \ldots J_1}(j^t; a)\rangle = \prod_{k=1}^{m} \left(\frac{1}{i} \partial_{J_k}^t + \frac{i}{2} A_{J_k I_k} j_{I_k}^t \right) |\mathcal{A}(j^t; a)\rangle. \tag{4.87}$$

In addition to Proposition 4.10 the following proposition holds:

- **Proposition 4.11:** With the definition

$$|\overline{\mathcal{A}}_{J_1 \ldots J_m}(j^t; a)\rangle := \sum_{n=0}^{\infty} \frac{i^n}{n!} \langle 0| \mathcal{A}(\psi_{I_1} \ldots \psi_{I_n}) \psi_{J_1} \ldots \psi_{J_m} |a\rangle j_{I_1}^t \ldots j_{I_n}^t |0\rangle_F \tag{4.88}$$

this quantity can be expressed by the functional relation

$$|\mathcal{A}_{J_m \ldots J_1}(j^t; a)\rangle = \prod_{k=1}^{m} \left(\frac{1}{i} \partial_{J_k}^t - \frac{i}{2} A_{J_k I_k} j_{I_k}^t \right) |\mathcal{A}(j^t; a)\rangle. \tag{4.89}$$

Proof:
The proof runs along the same lines as that of Proposition 4.10 and of (4.87). ◊

We now proceed to the derivation of the functional energy equation for equal times.

• **Proposition 4.12:** The set of equations (4.78) for $n = 1, 2, \ldots \infty$ can be equivalently expressed by the functional equation

$$E_{p0}|\mathcal{A}(j^t; p)\rangle \tag{4.90}$$

$$= \left[\widehat{K}_{I_1 I_2} j_{I_1}^t \partial_{I_2}^t - \widehat{W}_{I_1 I_2 I_3 I_4} j_{I_1}^t \left(\partial_{I_4}^t \partial_{I_3}^t \partial_{I_2}^t + \frac{1}{4} A_{I_4 J_1} A_{I_3 J_2} j_{J_1}^t j_{J_2}^t \partial_{I_2}^t \right) \right] |\mathcal{A}(j^t; p)\rangle.$$

Proof: Written in superindices the Hamiltonian **H** is a polynomial in ψ_I. Hence with $\mathbf{H} = \mathbf{H}[\psi_I]$ defined by (4.1) the application of Propositions 4.10 and 4.11 to equations (4.78) and (4.79) yields

$$E_{p0}|\mathcal{A}(j^t, p)\rangle = \left[H\left(\frac{1}{i}\partial^t + \frac{i}{2}Aj^t \right) - H\left(\frac{1}{i}\partial^t - \frac{i}{2}Aj^t \right) \right] |\mathcal{A}(j^t, p)\rangle, \tag{4.91}$$

where the field operators ψ_I in H are replaced by the corresponding functional expressions in accordance with Propositions 4.10 and 4.11. Explicit evaluation of (4.91) leads to (4.90). In the following we shall only use this short cut calculation. ◊

Equation (4.90) can be rewritten from the symbolic notation into the original notation. With $I_k \equiv (Z_k, \mathbf{r}_k, t)$ and equations (2.82), (2.83), equation (4.90) reads with reconstructed $\hbar$ and c:

$$E_{p0}|\mathcal{A}(j^t; p)\rangle \tag{4.92}$$

$$= \left\{ \frac{ic}{\hbar} \int d^3 r j_{Z_0}^t(\mathbf{r}) D_{Z_0 Z_1}^0 (D^k \partial_k - m)_{Z_1 Z_2} \partial_{Z_2}^t(\mathbf{r}) \right.$$

$$+ \frac{ic}{\hbar} \int d^3 r j_{Z_0}^t(\mathbf{r}) D_{Z_0 Z_1}^0 \widehat{U}_{Z_1 Z_2 Z_3 Z_4}$$

$$\left. \times \left[\partial_{Z_4}^t(\mathbf{r}) \partial_{Z_3}^t(\mathbf{r}) \partial_{Z_2}^t(\mathbf{r}) + \frac{1}{4} A_{Z_4 Z_5} j_{Z_5}^t(\mathbf{r}) A_{Z_3 Z_6} j_{Z_6}^t(\mathbf{r}) \partial_{Z_2}^t(\mathbf{r}) \right] \right\} |\mathcal{A}(j^t; p)\rangle$$

or concisely

$$E_{p0}|\mathcal{A}(j^t; p)\rangle = \mathcal{H}_F(j^t, \partial^t)|\mathcal{A}(j^t; p)\rangle, \tag{4.93}$$

where $\mathcal{H}_F$ is defined by the functional operator on the right hand side of (4.92).

In this way we have obtained a functional map from the renormalized energy equation (4.69) onto the functional equation (4.93). If we want to regain equation (4.69) from equation (4.93), we write in a symbolic notation for (4.73)

$$|\mathcal{A}(j^t; p)\rangle = \sum_n \tau_n(p)|d^n\rangle. \tag{4.94}$$

Then, by the definition of the projection states

$$\langle d_m| := {}_F\langle 0|i^{-m}\partial^t_{I_m} \ldots \partial^t_{I_1} \tag{4.95}$$

from (4.93) we can derive the matrix equation

$$\sum_n \langle d_m|\mathcal{H}_F|d^n\rangle \tau_n(p) = E_{p0}\tau_m(p), \tag{4.96}$$

and by comparison with (4.67) and (4.69) it follows that

$$\sum_n \langle d_m|\mathcal{H}_F|d^n\rangle \tau_n(p) = \sum_n \langle e_m|\mathbf{H} - \mathbf{A}_m^{-1}\mathbf{H}\mathbf{A}_m|e^n\rangle \tau_n(p) \tag{4.97}$$

for $m = 0, 1, 2 \ldots \infty$. Hence with (4.96) and (4.97) we have reproduced the renormalized energy equation (4.69). In all following chapters we shall use the functional equation (4.93) instead of equation (4.69) as the starting point of our discussions.

The derivation of the eigenvalue equations (4.90) or (4.92) respectively, clearly shows the physical content of this eigenvalue problem: in contrast to ordinary quantum mechanical Schrödinger equations which directly diagonalize the energy operators, the renormalization of infinite systems leads to eigenvalue equations which are based on the corresponding field equations. This results from the equivalence of (4.78) and (4.79), where (4.79) brings the field equations into play. Thus the functional operator $\mathcal{H}_F$ in (4.93) describes field equations and not Hamiltonians. This fact is essential for the interpretation of effective functional equations for composite particles which will be derived in the following chapters.

4.7 Normal Ordering

For the covariant theory the need for normal ordering was discussed in great detail in Section 3.5. The same arguments apply to the formulation of the theory on a spacelike hyperplane, and therefore normal ordering must be introduced in this case, too. This will be done in this section.

In Section 3.5 we applied a free fermion propagator for normal ordering. This choice will be justified in Chapter 6 in detail. In a preliminary way we only mention that the choice of a free fermion propagator means that in the basic NJL model no condensation phenomena and no thermo-states are considered in the present investigations. For this reason we assume that equation (3.109) is a suitable approximation of the self-consistency condition which was mentioned in Section 3.5.

Thus for any state $|a) \in \mathbb{IH}$ we define normal ordered functional states on a spacelike hyperplane $t = $ const. by

$$|\mathcal{F}(j^t; a)\rangle \ := \ \exp\left[\frac{1}{2}\int j_{Z_1}^t(\mathbf{r}_1)F_{Z_1 Z_2}^t(\mathbf{r}_1, \mathbf{r}_2)j_{Z_2}^t(\mathbf{r}_2)d^3r_1 d^3r_2\right]|\mathcal{A}(j^t; a)\rangle$$

$$=: \ \sum_{n=1}^{\infty}\frac{i^n}{n!}\varphi_t^{(n)}(I_1\ldots I_n|a)j_{I_1}^t\ldots j_{I_n}^t|0\rangle_{\mathrm{F}} \tag{4.98}$$

with the propagator

$$F_{Z_1 Z_2}^t(\mathbf{r}_1, \mathbf{r}_2) := \langle 0|\mathcal{A}\left[\psi_{Z_1}^f(\mathbf{r}_1, t)\psi_{Z_2}^f(\mathbf{r}_2, t)\right]|0\rangle, \tag{4.99}$$

where $\psi_Z^f(\mathbf{r}, t)$ is a solution of equation (3.108).

- **Proposition 4.13:** The normal transform of equation (4.92) reads

$$E_{p0}|\mathcal{F}(j^t; p)\rangle = \frac{ic}{\hbar}\int d^3r\, j_{Z_0}^t(\mathbf{r})D_{Z_0 Z_1}^0$$

$$\times\Bigg\{(D^k\partial_k - m)_{Z_1 Z_2}\partial_{Z_2}^t(\mathbf{r})$$

$$+\hat{U}_{Z_1 Z_2 Z_3 Z_4}\Big[d_{Z_4}^t(\mathbf{r})d_{Z_3}^t(\mathbf{r})d_{Z_2}^t(\mathbf{r})$$

$$+\frac{1}{4}A_{Z_4 Z_5}j_{Z_5}^t(\mathbf{r})A_{Z_3 Z_6}j_{Z_6}^t(\mathbf{r})d_{Z_2}^t(\mathbf{r})\Big]\Bigg\}|\mathcal{F}(j^t; p)\rangle \tag{4.100}$$

with

$$d_{Z_1}^t(\mathbf{r}_1) := \partial_{Z_1}^t(\mathbf{r}_1) - \int F_{Z_1 Z_2}^t(\mathbf{r}_1, \mathbf{r}_2)j_{Z_2}^t(\mathbf{r}_2)d^3r_2. \tag{4.101}$$

Proof: From (4.98) we obtain by inversion of the exponential function

$$|\mathcal{A}(j^t; a)\rangle = \exp(-\frac{1}{2}j_{I_1}^t F_{I_1 I_2}^t j_{I_2}^t)|\mathcal{F}(j^t; a)\rangle. \tag{4.102}$$

We substitute (4.102) into equation (4.92) and commute the exponential factor with $\mathcal{H}_{\mathrm{F}}$. Afterwards we cancel it by division. The commutation can be achieved by means of formula (3.89) which holds for functionals on the hyperplane, too, and for which with (4.101) we simply write

$$\partial_K^t \exp(-\frac{1}{2}j_{I_1}^t F_{I_1 I_2}^t j_{I_2}^t) = \exp(-\frac{1}{2}j_{I_1}^t F_{I_1 I_2}^t j_{I_2}^t)d_K^t. \tag{4.103}$$

By repeated application of (4.103) we can derive the interaction terms of (4.100) from those of (4.92). With respect to the kinetic energy term we observe that due to (3.108) and (4.99) there holds

$$\left[D^k\partial_k(\mathbf{r}_1) - m\right]_{Z_0 Z_1}F_{Z_1 Z_2}^t(\mathbf{r}_1, \mathbf{r}_2) \tag{4.104}$$

$$= \frac{1}{2}D_{Z_0 Z_1}^0\langle 0|\psi_{Z_2}^f(\mathbf{r}_2, t)\dot{\psi}_{Z_1}^f(\mathbf{r}_1, t) - \dot{\psi}_{Z_1}^f(\mathbf{r}_1, t)\psi_{Z_2}^f(\mathbf{r}_2, t)|0\rangle.$$

If $\mathbf{H}^0$ is the Hamiltonian of the free field we have $\dot{\psi}_Z^f(\mathbf{r},t) = \frac{i}{\hbar}[\mathbf{H}^0, \psi_Z^f(\mathbf{r},t)]_-$, and (4.104) can be rewritten as

$$\left[D^k \partial_k(\mathbf{r}_1) - m\right]_{Z_0 Z_1} F_{Z_1 Z_2}^t(\mathbf{r}_1, \mathbf{r}_2) = \frac{i}{2} D_{Z_0 Z_1}^0 O_{Z_1 Z_2}(\mathbf{r}_1, \mathbf{r}_2, t), \qquad (4.105)$$

where

$$O_{Z_1 Z_2}(\mathbf{r}_1, \mathbf{r}_2, t) := \langle 0| \psi_{Z_1}^f(\mathbf{r}_1, t) \mathbf{H}^0 \psi_{Z_2}^f(\mathbf{r}_2, t) + \psi_{Z_2}^f(\mathbf{r}_2, t) \mathbf{H}^0 \psi_{Z_1}^f(\mathbf{r}_1, t)|0\rangle, \qquad (4.106)$$

and thus

$$O_{Z_1 Z_2}(\mathbf{r}_1, \mathbf{r}_2, t) = O_{Z_2 Z_1}(\mathbf{r}_2, \mathbf{r}_1, t). \qquad (4.107)$$

If (4.105) is substituted in the transformed kinetic energy term, due to (4.107) the F term vanishes and hence only the kinetic energy term in (4.100) remains.   ◇

Finally we mention that the vertex renormalization, as was discussed in Section 3.6 for the covariant formalism, can be transferred to the energy equation on a space-like hyperplane without any alteration. In particular equation (3.117), which describes a normal ordering of the sources, can be taken over to the latter case; *i.e.*, by the effect of the counter-terms the sources and their duals formally anticommute within the vertex expression, and in this way divergencies by $F(0)$ terms are avoided.

4.8   Covariant Equations on the Hyperplane

We shortly discuss the question whether in the limit of equal times the covariant formalism leads to the Hamiltonian formalism on the hyperplane, referring to Maison and Stumpf [Mai 66], Pfister [Pfi 87], Borne [Bor 91].

The starting point of our investigation is the multi-time energy equation (3.81). By means of formulas (3.18)–(3.20) we project (3.81) into configuration space and obtain a system of equations between various time ordered products. This system reads

$$E_{p0}\langle 0|T\left(\psi_{I_1} \ldots \psi_{I_N}\right)|p\rangle \qquad (4.108)$$

$$= \frac{ic}{\hbar} \sum_{k=1}^{N} \Big\{ \langle 0|T\left(\psi_{I_1} \ldots \left[D_{Z_k X_1}^0 \left(D^m \partial_m(x_k) - m\right)_{X_1 X_2} \psi_{X_2}(x_k)\right] \ldots \psi_{I_N}\right)|p\rangle$$

$$- \langle 0|T\left(\psi_{I_1} \ldots \left[D_{Z_k X_1}^0 \hat{U}_{X_1 X_2 X_3 X_4} \psi_{X_2}(x_k) \psi_{X_3}(x_k) \psi_{X_4}(x_k)\right] \ldots \psi_{I_N}\right)|p\rangle \Big\},$$

where we have to observe $I_k = (Z_k, x_k)$ for $1 \leq k \leq N$. Equation (4.108) corresponds to the projection of equation (4.90) into configuration space with the only difference that time ordered products are used instead of antisymmetrized single time expressions. The connection between both kinds of products can be established by a limit procedure.

We define the following limiting process for time ordered products:

$$\lim_T T(\psi_{I_1} \ldots \psi_{I_N}) := \frac{1}{N!} \sum_{p \in S_N} \lim_{\substack{t_{p_1} > \ldots > t_{p_N} \\ t_{p_1} \ldots t_{p_N} \to t}} T(\psi_{I_{p_1}} \ldots \psi_{I_{p_N}}). \qquad (4.109)$$

By means of this limiting process T products can be related to A products.

- **Proposition 4.14:** The limiting process of (4.109) yields

$$\lim_T T(\psi_{I_1} \ldots \psi_{I_N}) = A(\psi_{I_1} \ldots \psi_{I_N})_t. \qquad (4.110)$$

Proof: By definition we have

$$T(\psi_{I_1} \ldots \psi_{I_N}) = \sum_{p \in S_N} \operatorname{sgn}(p) \psi_{I_{p_1}} \ldots \psi_{I_{p_N}} \Theta(t_{p_1} \ldots t_{p_N}), \qquad (4.111)$$

and thus

$$\lim_T T(\psi_{I_1} \ldots \psi_{I_N}) = \frac{1}{N!} \sum_{p,p' \in S_N} \lim_{\substack{t_{p_1} > \ldots > t_{p_N} \\ t_{p_1} \ldots t_{p_N} \to t}} \operatorname{sgn}(p') \psi_{I_{p'_1}} \ldots \psi_{I_{p'_N}} \Theta(t_{p'_1} \ldots t_{p'_N}). \qquad (4.112)$$

For any ordering $t_{p_1} > \ldots > t_{p_N}$ there exists only one $p' \in S_N$, namely $p' = p$, for which $\Theta(t_{p'_1}, \ldots t_{p'_N})$ is unequal zero, i.e., we obtain

$$\lim_T T(\psi_{I_1} \ldots \psi_{I_N}) = \frac{1}{N!} \sum_{p \in S_N} \lim_{\substack{t_{p_1} > \ldots > t_{p_N} \\ t_{p_1} \ldots t_{p_N} \to t}} \operatorname{sgn}(p) \psi_{I_{p_1}} \ldots \psi_{I_{p_N}} \Theta(t_{p_1} \ldots t_{p_N}); \qquad (4.113)$$

if the limiting process for Θ is performed, we obtain (4.110). $\Diamond$

As equations (4.108) are linear equations with respect to the T products we can immediately apply the limiting process (4.109) to (4.108). With (4.110) this limiting process leads to the projection of equation (4.90) into configuration space.

We showed in this way that the covariant formalism is indeed compatible with the Hamiltonian formalism. Conversely this formalism can be used to demonstrate the relativistic covariance of the Hamiltonian formalism or of the Algebraic Schrödinger Representation, respectively. As the essential relativistic effects result from the homogeneous Lorentz transformations we confine our discussion to this subgroup of the Poincaré group.

By construction the basic nonlinear spinor field equations (2.48) and (2.49) are form invariant under orthochronous Lorentz transformations. Although the anticommutation relations (2.61) are not manifestly covariant, in combination with time ordered products one regains complete form invariance under the Lorentz group for the functional operator of the functional equation (3.77), whilst the time ordered state functionals (3.17) transform in accordance with (3.29) as elements of functional representation spaces of the Lorentz group.

Hence if we describe the Lorentz transformed quantities by primed coordinates the derivation of the transformed many-time energy equation runs along the same lines as that of (4.108) and we therefore obtain

$$E_{p'0}\langle 0|T\left(\psi_{I_1'}\ldots\psi_{I_N'}\right)|p'\rangle \tag{4.114}$$

$$= \frac{ic}{\hbar}\sum_{k=1}^{N}\Big\{\langle 0|T\left(\psi_{I_1'}\ldots\left[D^0_{Z_kX_1}\left(D^m\partial_m(x_k')-m\right)_{X_1X_2}\psi_{X_2}(x_k')\right]\ldots\psi_{I_N'}\right)|p'\rangle$$

$$-\langle 0|T\left(\psi_{I_1'}\ldots\left[D^0_{Z_kX_1}\hat{U}_{X_1X_2X_3X_4}\psi_{X_2}(x_k')\psi_{X_3}(x_k')\psi_{X_4}(x_k')\right]\ldots\psi_{I_N'}\right)|p'\rangle\Big\}.$$

Apart from the single time limit having to be performed for the primed quantities the Algebraic Schrödinger operator resulting from (4.114) is formally the same, as that resulting from (4.108). Thus the only quantities which are nontrivially transformed are the functional state vectors themselves. They can be defined in a relativistic covariant way, if we generalize the definition of the T products. Instead of (4.111) we define

$$T_n(\psi_{I_1}\ldots\psi_{I_N}) = \sum_{p\in S_N}\text{sgn}(p)\psi_{I_{p_1}}\ldots\psi_{I_{p_N}}\Theta(nx_{p_1}\ldots nx_{p_N}), \tag{4.115}$$

where n is any time-like, but fixed unit vector. With this modification we have demonstrated the relativistic covariance of the functional many-time energy equation.

Another point in the discussion of the covariance of our Algebraic Schrödinger Representation is the investigation of the scalar products. Such scalar products cannot be studied without specifying the state space involved. It is our aim to derive effective theories resulting from the dynamics of composite particles. Then the corresponding state spaces have to describe composite particle (field) configurations which clearly are part of nonperturbative quantum field theory.

The theoretical basis for the description of composite particles and composite particle reactions is developed in Chapter 5. Hence in order to discuss the respective states we have to anticipate the content of this chapter with respect to state formation. We demonstrate this for the special case of composite particles which are composed of two subfermions.

Let $\left\{C_k^{II'}\right\}$ be a complete set of two-particle states, where k runs through all relevant quantum numbers. Then in the auxiliary field formalism the single time normal ordered state amplitudes $\varphi^{(2n)}$ are assumed to be representable by expansions into series of the two-particle functions $C_k^{II'}$.

Such expansions can (to some extent) be justified by inserting intermediate states into the state amplitudes and read

$$\varphi^{(2n)}(I_1\ldots I_{2n}|a) = C_{k_1}^{[I_1I_2}\ldots C_{k_n}^{I_{2n-1}I_{2n}]}\rho^{(n)}(k_1\ldots k_n|a), \tag{4.116}$$

where the set of basis functions

$$D_{2n}\begin{pmatrix}I_1\ldots I_{2n}\\k_1\ldots k_n\end{pmatrix} := C_{k_1}^{[I_1I_2}\ldots C_{k_n}^{I_{2n-1}I_{2n}]} \tag{4.117}$$

is defined by the determinants of two-subfermion states and describe the whole space–time dependence and associated symmetries of the normal ordered matrix elements. Hence we can confine ourselves to study the properties of this set.

The set (4.117) starts with the two-particle functions $C_k^{II'}$ themselves. Via their quantum numbers k they define state amplitudes, *i.e.*, matrix elements, for themselves and in this way induce image states in a corresponding Hilbert (Krein) space. The scalar product of states from this Hilbert (Krein) space then reads

$$\langle k'|k\rangle = \int d\mathbf{r}_1 d\mathbf{r}_2 \, \Theta_{Z_1 Z_2}(\mathbf{r}_1, \mathbf{r}_2|k')^* C_{Z_1 Z_2}(\mathbf{r}_1, \mathbf{r}_2|k) \tag{4.118}$$

in the Algebraic Schrödinger Representation.

Suppose now that the corresponding covariant state amplitude $C_{Z_1 Z_2}(x_1, x_2|k)$ exists and is an element of a representation space of the inhomogeneous Lorentz group. Furthermore, also $\Theta_{Z_1 Z_2}(x_1, x_2|k')$ is assumed to exist and to transform contragredient to $C_{Z_1 Z_2}(x_1, x_2|k)$.

Then by definition the many-time function $C_{Z_1 Z_2}(x_1, x_2|k)$ is related to the single time function $C_{Z_1 Z_2}(\mathbf{r}_1, \mathbf{r}_2|k)$ by the limiting process (4.109) and we can rearrange (4.118) as follows:

$$\begin{aligned}
\langle k'|k\rangle &= \int d\mathbf{r}_1 d\mathbf{r}_2 dt_1 dt_2 \, \Theta_{Z_1 Z_2}(\mathbf{r}_1, t_1, \mathbf{r}_2, t_2|k')^* \tag{4.119}\\
&\quad \times \frac{1}{2} \sum_{p \in S_2} \operatorname{sgn}(p) \lim_{\substack{\epsilon_{p_1} > \epsilon_{p_2} \\ \epsilon_{p_1}, \epsilon_{p_2} \to 0}} \delta(t_{p_1} - \epsilon_{p_1})\delta(t_{p_2} - \epsilon_{p_2}) C_{Z_{p_1} Z_{p_2}}(\mathbf{r}_{p_1}, t_{p_1}, \mathbf{r}_{p_2}, t_{p_2}|k)\\
&= \int dx_1 dx_2 \, \Theta_{Z_1 Z_2}(x_1, x_2|k')^*\\
&\quad \times \frac{1}{2} \sum_{p \in S_2} \operatorname{sgn}(p) \lim_{\substack{\epsilon_{p_1} > \epsilon_{p_2} \\ \epsilon_{p_1}, \epsilon_{p_2} \to 0}} \delta(nx_{p_1} - \epsilon_{p_1})\delta(nx_{p_2} - \epsilon_{p_2}) C_{Z_{p_1} Z_{p_2}}(x_{p_1}, x_{p_2}|k)
\end{aligned}$$

for $n = (0, 0, 0, 1)$. This expression is completely relativistically invariant, provided it is independent of n.

We now show that the latter expression is the same for all time-like vectors n, see Stumpf [Stu 85a]. For brevity we specify the quantum numbers k and k' of the corresponding representations C_k and $\Theta_{k'}$ only by their four-momenta p and q and omit the other quantum numbers resulting from Proposition 3.15.

- **Proposition 4.15:** Let $|p\rangle$ and $|q\rangle$ be two eigenstates of the four-momentum operator $\mathbf{P}_\mu$. Then the scalar product $\langle p|q\rangle$ is given by the relativistically invariant expression

$$\langle p|q\rangle = \int d^4 x_1 d^4 x_2 \, \Theta_{Z_1 Z_2}(x_1, x_2|p)^* \lim_{T,n} \delta(nx_1)\delta(nx_2) C_{Z_1 Z_2}(x_1, x_2|q), \tag{4.120}$$

  where n is an arbitrary but fixed time-like unit vector and $\lim_{T,n}$ is defined by the last formula of (4.119).

Proof: We compare (4.120) for two different n-vectors n and n'. In an abbreviated notation we have

$$\langle p|q\rangle_n = \int \Theta(p)^* \lim_{T,n} \delta(nx)C(q)\, d\tau, \qquad (4.121)$$

$$\langle p|q\rangle_{n'} = \int \Theta(p)^* \lim_{T,n'} \delta(n'x)C(q)\, d\tau. \qquad (4.122)$$

As both n vectors are time-like we can find a Lorentz transformation Λ which transforms n into n' by means of $n' = \Lambda n$. Hence if we write $n'_\mu = \Lambda_\mu^\nu n_\nu$ we can substitute this into $n'_\mu x^\mu$ and obtain $n'_\mu x^\mu = \Lambda_\mu^\nu n_\nu x^\mu = n_\nu x'^\nu$ with $x'^\nu = \Lambda_\mu^\nu x^\mu$.

For the further discussion we use the symbolic notation $x' = \Lambda x$. Now we replace in (4.122) x_i by $\Lambda^{-1}x'_i$, i.e., we apply the transformation $x_i = \Lambda^{-1}x'_i$, $i = 1, 2$. This transformation leaves the volume element invariant, i.e., $d^4x_1 d^4x_2 = d^4x'_1 d^4x'_2$. Then (4.122) goes over into

$$\langle p|q\rangle_{n'} = \int d^4x'_1 d^4x'_2\, \Theta_{Z_1 Z_2}(\Lambda^{-1}x'_1, \Lambda^{-1}x'_2|p)^* \qquad (4.123)$$
$$\times \lim_{T,n} \delta(nx'_1)\delta(nx'_2)C_{Z_1 Z_2}(\Lambda^{-1}x'_1, \Lambda^{-1}x'_2|q).$$

From the explicit representation of C functions in Chapter 6 it follows that

$$C_{Z_1 Z_2}(x_1, x_2|q) = C_{Z_1 Z_2}(\Lambda^{-1}x'_1, \Lambda^{-1}x'_2|q) \qquad (4.124)$$
$$= L_{Z_1 Z'_1}(\Lambda)^{-1} L_{Z_2 Z'_2}(\Lambda)^{-1} C_{Z'_1 Z'_2}(x'_1, x'_2|\Lambda^{-1}q),$$

which expresses the fact that a Lorentz transformation induces a transformation in the representation space of C functions.

As the Θ functions span a dual representation space, similar formulas hold for them. If these relations are substituted in (4.123), owing to duality the spinor parts L cancel out and one obtains

$$\langle p|q\rangle_{n'} = \int d^4x'_1 d^4x'_2\, \Theta_{Z'_1 Z'_2}(x'_1, x'_2|\Lambda^{-1}p)^* \qquad (4.125)$$
$$\times \lim_{T,n} \delta(nx'_1)\delta(nx'_2)C_{Z'_1 Z'_2}(x'_1, x'_2|\Lambda^{-1}q)$$
$$= \langle \Lambda^{-1}p|\Lambda^{-1}q\rangle_n.$$

As by definition the C_k amplitudes are assumed to describe real physical particles, the corresponding representations must be unitary. Then with $|\Lambda^{-1}q\rangle = U(\Lambda^{-1})|q\rangle$ and $\langle\Lambda^{-1}p| = \langle p|U^{-1}(\Lambda^{-1})$ we have $\langle\Lambda^{-1}p|\Lambda^{-1}q\rangle_n = \langle p|q\rangle_n$ and thus $\langle p|q\rangle_{n'} = \langle p|q\rangle_n$. ◇

Having derived a relativistically invariant scalar product for two-particle (field) states one can proceed to prove the relativistic invariance of scalar products for the general basis functions (4.117) and their contragredient counterparts.

Such proofs run along the lines of Proposition 4.15 and are of course rather lengthy. So we will not further pursue this subject. But it is obvious that in this procedure the single composite particle states play a central role. Can one find single composite particle states and their contragredient counterparts which constitute elements of representation spaces of the inhomogeneous Lorentz group and how do they depend on special field representations?

Such questions can only be answered by a detailed study of the construction of these states which we postpone to Chapter 6, where the foundations of state construction will be developed.

So far we have been concerned with composite particle state construction. Finally, one should say something about their constituents, the subfermions themselves. These subfermions are described by Dirac fields, and one of the most important representations in elementary particle physics is generated by the Dirac vacuum which is inequivalent to the Fock vacuum of the original Dirac spinors.

In the interaction representation the Dirac vacuum directly leads to the existence of particles and antiparticles. But subfermions are unobservable and subjected to interactions. So the purpose of the Dirac vacuum to shift indefiniteness of the energy of free Dirac fermions to their density and to interpret the effect of this as opposite charges is rather meaningless on the level of unobservable subfermions; thus the Dirac vacuum seems to be necessary only in the interaction representation of effective theories.

5   Weak Mapping Theorems

5.1   Introduction

What are Weak Mappings? Weak Mappings are defined by mappings between quantum field-theoretic functional states which were introduced in Chapter 4. The purpose of a Weak Mapping is to provide a suitable means for deriving effective quantum field dynamics. The attribute 'weak' is borrowed from the concept of the weak topology and means that these mappings are concerned with matrix elements of operators, which constitute the functional states.

In contrast to this method are those methods which use operator products for the description of composite particles. These methods are based on the assumption that the group-theoretical decomposition of local operator products into their irreducible parts allows a unique characterization of corresponding particles. That this conception is a too narrow one shows the simple example of the states of a hydrogen atom which in its rest system has infinitely many singlet states of the rotation group and cannot be described by a simple operator product. For an extensive discussion see Stumpf and Borne [Stu 94] and Section 2.1. We thus reject the operator methods for the derivation of effective composite particle dynamics in favor of the state concept which according to the preceding chapters is expressed by transition matrix elements.

For illustrating the role of Weak Mappings we briefly summarize the steps we have already made towards a composite particle theory: in Chapter 4 we pointed out that antisymmetrized expectation values of field operators on a space-like hyperplane are regarded as the basic quantities of the algebraic theory. From the algebraic point of view these matrix elements are algebraic 'states'. Guided by the GNS construction of Algebraic Quantum Theory we assumed that for each algebraic state on the field algebra there exists at least one regular cyclic representation in Hilbert space.

The 'GNS basis vectors' of this representation were obtained by the application of symmetrized or antisymmetrized monomials of field operators on this cyclic Hilbert space vector. We then defined functional states generating the set of projections of an arbitrary but fixed state in Hilbert space onto these basis vectors, see (4.73), (4.74). Furthermore, from the original Heisenberg dynamics of the quantum fields for these functional states corresponding functional equations could be derived, see (4.92), (4.100).

Weak Mapping is now introduced by a reordering of such generating functional states according to certain bound state structures. As a result one obtains another functional state which by definition is referred to these bound states. The functional field equations may be viewed as being generated by operators acting on the set of generating functional states. In a second step, by Weak Mapping these operators are simultaneously transformed with the states, and this yields new functional field equations which are considered as the functional description of the effective dynamics of composite particles in quantum field theory.

The interesting applications in physics are studies of maps between functional equations for basic particles and functional equations for composite particles, where the latter may be called 'effective' functional equations, in analogy with the frequently used notion 'effective' Lagrangian.

In this chapter we will study the formal evaluation of exact Weak Mapping theorems which we will derive for the case of the super-renormalizable spinor field model of Chapter 2. But Weak Mapping is not restricted to this model. Rather it can also be applied to renormalizable theories like quantum electrodynamics, see Stumpf, Fauser and Pfister [Stu 93a], Fauser and Stumpf [Fau 97], provided a canonical quantization is uniquely defined. From these considerations it follows that nonrelativistic quantum field theories can also be treated in this way. Physical applications and interpretations will be given in the following chapters.

Originally Weak Mappings were performed by means of functional 'chain rules' in the transformation of elementary particle functional states to composite particle functional states, see Stumpf [Stu 85a,b,c]. This method was the starting point for the investigation of exact Weak Mapping theorems, and it can be justified as a borderline case for negligible exchange forces of the exact theory. As the exact mapping theorems are rather unwieldy, we will also discuss the justification of the chain rule mappings allowing a very elegant (and manageable) treatment of composite particle dynamics.

Before we start with the derivation of the Weak Mapping theorems it is necessary to give some general comments upon composite particles, *i.e.*, bound states of various elementary or basic particles.

Bound states are the elements and the starting point of any composite particle dynamics. However, in contrast to quantum mechanics in quantum field theory no unique definition of bound states is possible. Rather, one has to distinguish two kinds of bound states: hard core bound states and dressed particle states. The latter consist of a hard core surrounded by a polarization cloud, which in nontrivial quantum field theories is generated by the interaction of the physical vacuum with hard core state constituents.

Whilst the hard core state is the analogue of the quantum mechanical bound states which, in general, exist for themselves, the polarization cloud is a specific quantum field theoretic contribution to the formation of bound states, and sometimes it even stabilizes hard core bound states which otherwise would be unstable. Therefore, apart from the latter possibility, for the formulation of a composite particle dynamics one can either consider only the hard core of a bound state as the basic quantity, or one considers the hard core with inclusion of its polarization cloud as the basic quantity, what then yields a 'dressed' composite particle.

The Weak Mapping of dressed particle states is, however, much more complicated than that of hard core states and so far it has not yet been satisfactorily explored. Thus for deriving the exact mapping theorems we concentrate on the mappings of hard core states and finally add some remarks about dressed composite particle states in

Sections 5.7 and 5.8.

A peculiarity of quantum field theory is the occurrence of uncorrelated parts in the definition of the original antisymmetrized functional states. In order to prepare composite particle states and their dynamics these uncorrelated parts must be removed from the theory. This can be achieved by applying normal transformations to the original functional states (see Chapter 3). For the case of hard core bound states of two, three, and four subfermions it is sufficient to apply the normal transform (4.98). For the treatment of bound states of more than four constituent subfermions it is necessary to extract higher order vacuum expectation values. This also results in considerable complications which are not fully explored.

Thus in the following we restrict ourselves to the normal transform (4.98) which allows the bound state calculation for two, three and four subfermions and the treatment of the full dynamics of two- and three-fermion bound states including special cases of four-fermion bound state dynamics. This comprises a wide range of applications in elementary particle and in solid state physics. In particular, within this framework de Broglie's original fusion idea can be further pursued.

5.2   Hard Core States

As just mentioned, we first consider Weak Mapping on the basis of hard core bound states. Hard core states are related to the minimal number of constituents which are necessary to obtain the correct quantum numbers of the corresponding physical particle, apart from its mass and charge which undergo renormalization. In this way they are not different from dressed particle states, because in general dressed particles also need this final renormalization. The only difference between hard core and dressed particle states is the exclusion or inclusion of a polarization cloud.

Weak Mapping with hard core states, however, does not mean omission of the polarization cloud. As the exact mapping theorems provide bi-unique maps between the basic theory and effective theories, in the case of Weak Mapping with hard core states the polarization cloud manifests itself in effective self-interactions and mutual interactions of the effective hard core states. So nothing is lost as the dressing problem is only shifted from the original theory into the effective theory.

In order to derive a definition for hard core states we consider the normal ordered functional energy equation (4.100). If we return to the symbolic notation of (2.81)–(2.83) and observe the vertex renormalization (3.117), then equation (4.100) can be equivalently written in the following form:

$$E_{p0}|\mathcal{F}(j^t;p)\rangle = \mathcal{H}_{\mathrm{F}}(j^t,\partial^t)|\mathcal{F}(j^t;p)\rangle \tag{5.1}$$

with the functional Hamiltonian operator

$$\mathcal{H}_{\mathrm{F}}(j^t,\partial^t) := j^t_{I_1}\widehat{K}_{I_1I_2}\partial^t_{I_2}$$

$$+\widehat{W}_{I_1I_2I_3I_4}\left[j^t_{I_1}\partial^t_{I_4}\partial^t_{I_3}\partial^t_{I_2} - 3F^t_{I_4K}j^t_{I_1}j^t_K\partial^t_{I_3}\partial^t_{I_2}\right] \tag{5.2}$$

$$+(3F^t_{I_4K_1}F^t_{I_3K_2} + \frac{1}{4}A_{I_4K_1}A_{I_3K_2})j^t_{I_1}j^t_{K_1}j^t_{K_2}\partial^t_{I_2}$$

$$-(F^t_{I_4K_1}F^t_{I_3K_2} + \frac{1}{4}A_{I_4K_1}A_{I_3K_2})F^t_{I_2K_3}j^t_{I_1}j^t_{K_1}j^t_{K_2}j^t_{K_3} \Big].$$

To recognize the difference between hard core and dressed particle states in quantum field theory, we consider quantum mechanical bound states, such as atoms, nuclei, *etc.*. These bound states contain only their constituent particles and no polarization cloud. Therefore they are hard core states, and if they are described in Fock space they are characterized by the appearance of only diagonal part operators, *i.e.*, operators which, after normal ordering, contain equal numbers of creation and annihilation operators.

Although owing to the GNS construction and the embedding in functional space the basic energy equation (5.1) is not directly comparable with bound state equations in Fock space, we take over the hard core criterion from Fock space to functional space. We therefore assume that the nondiagonal part of equation (5.1) is related to the polarization cloud and characterize hard core states by the following definition, see [Stu 83b]:

- **Definition 5.1:** Hard core states are defined as solutions of the diagonal part of equation (5.1) with corresponding functional states $|\mathcal{F}(j^t; a))\rangle^d$.

Thus in our case the defining equation for hard core bound states is the diagonal part of (5.1) and reads

$$E^d_{p0}|\mathcal{F}(j^t;p))\rangle^d = \left(j^t_{I_1}\widehat{K}_{I_1I_2}\partial^t_{I_2} - 3\widehat{W}_{I_1I_2I_3I_4}F^t_{I_4K}j^t_{I_1}j^t_K\partial^t_{I_3}\partial^t_{I_2}\right)|\mathcal{F}(j^t;p))\rangle^d. \tag{5.3}$$

To solve equation (5.3) we project it into configuration space by applying $\langle d_m|$ of (4.95) from the left for $m = 1, 2 \ldots \infty$. Owing to the appearance of only diagonal part operators the various sectors $m = 1, 2 \ldots \infty$ completely decouple and thus can be treated separately. We denote the solutions of (5.3) in configuration space by $C^{I_1 \ldots I_m}_k$, $m = 1, 2 \ldots \infty$, where k are the quantum numbers of the state for fixed m, and assume that the solutions in any sector form a complete set of functions.

For the formulation of Weak Mapping theorems we treat the simplest example of a bound state dynamics. We restrict ourselves to the two-particle sector given by the set of functional states $|\mathcal{F}(j^t; 2, k))\rangle^d := -\frac{1}{2}C^{I_1I_2}_k j^t_{I_1} j^t_{I_2}|0\rangle_F$. The corresponding equations arise from (5.3) by projection with $\langle d_2| \equiv -_F\langle 0|\partial^t_{K_2}\partial^t_{K_1}$ and by substitution of $|\mathcal{F}(j^t; 2, k))\rangle^d$; they read

$$E^d_{k0}C^{K_1K_2}_k = \left[\widehat{K}_{K_1I_1}C^{I_1K_2}_k - 3\widehat{W}_{K_1I_2I_3I_4}F^t_{I_4K_2}C^{I_2I_3}_k\right]_{\mathrm{as}[K_1,K_2]}. \tag{5.4}$$

For the following treatment a comment about the set of solutions of (5.4) is necessary. This set contains bound states as well as scattering states. How to deal with the latter will be discussed in the applications of Weak Mapping.

Furthermore, in spite of its assumed completeness the set of solutions is in general *not* an orthogonal set. This is because the eigenvalue equation (5.3) or (5.4), respectively, is *not* a Schrödinger equation, and the solutions in general have *not* the meaning of quantum mechanical probability amplitudes. This non-Schrödinger property is a consequence of the quantum field-theoretic state construction given in Chapter 4, and it also manifests itself in the hard core equations (5.3). How to proceed in this case was discussed in detail in Chapter 4.

But for the Weak Mapping theorems the non-orthogonality of the $C_k^{I_1...I_m}$ is of no relevance. It suffices to introduce in any sector dual functions $R_{I_1...I_m}^k$, $m = 1, 2 ... \infty$, which are defined by orthogonality and completeness relations. For the two-particle sector we introduce the duals $R_{I_1 I_2}^k$ by the orthogonality conditions

$$\sum_{I_1, I_2} R_{I_1 I_2}^k C_{k'}^{I_1 I_2} = \delta_{kk'} , \tag{5.5}$$

and assuming the completeness of the original set, the completeness relations

$$\sum_k R_{I_1 I_2}^k C_k^{K_1 K_2} = \frac{1}{2} \left(\delta_{I_1 K_1} \delta_{I_2 K_2} - \delta_{I_2 K_1} \delta_{I_1 K_2} \right) \tag{5.6}$$

have to hold, where in (5.6) the antisymmetries $C_k^{I_1 I_2} = -C_k^{I_2 I_1}$ and $R_{I_1 I_2}^k = -R_{I_2 I_1}^k$ were taken into account. In Section 5.7 we will give a method for calculating the duals explicitly.

5.3 Self-consistent Propagators

The hard core state equation (5.3) shows that the hard core states themselves essentially depend on the single time propagator $F_{I_1 I_2}^t$ which is the relativistic analogue to the quantum mechanical interaction potential. Hence a thorough investigation of $F_{I_1 I_2}^t$ is necessary.

From a formal point of view $F_{I_1 I_2}^t$ is a vacuum expectation value and hence an uncorrelated part of the state functional. It was emphasized in Section 4.8 that Weak Mapping theorems are based on an expansion of matrix elements and that uncorrelated parts of these matrix elements must be compatible with the basis functions of the expansion. In particular, if the basis functions themselves are (exact) solutions of an (unperturbed) functional Hamiltonian the corresponding uncorrelated parts must be vacuum expectation values of the same functional equations. For hard core states the functional Hamiltonian is defined by the operator equation (5.3):

$$\mathcal{H}_F^d := j_{I_1}^t \widehat{K}_{I_1 I_2} \partial_{I_2}^t - 3 \widehat{W}_{I_1 I_2 I_3 I_4} F_{I_4 K}^t j_{I_1}^t j_K^t \partial_{I_3}^t \partial_{I_2}^t . \tag{5.7}$$

The propagator $F^t_{I_1 I_2}$ can be embedded into this formalism if one observes that vacuum expectation values have vanishing energy values, as one can easily verify by means of equation (4.79) for $|p\rangle \equiv |0\rangle$. Therefore, if we interpret the functions $C^{I_1 I_2}_k$ as two-particle matrix elements, the solution $C^{I_1 I_2}_0$ of equation (5.4) for $E_{p0} = 0$ must be a vacuum expectation value, *i.e.*, we have $C^{I_1 I_2}_0 \equiv F^t_{I_1 I_2}$. Thus $F^t_{I_1 I_2}$ has to satisfy the equation

$$\widehat{K}_{K_1 I} F^t_{I K_2} - \widehat{K}_{K_2 I} F^t_{I K_1} = 3 \widehat{W}_{K_1 I_2 I_3 I_4} F^t_{I_4 K_2} F^t_{I_3 I_2} - 3 \widehat{W}_{K_2 I_2 I_3 I_4} F^t_{I_4 K_1} F^t_{I_3 I_2}, \qquad (5.8)$$

which can be considered as a self-consistency condition for the determination of $F^t_{I_1 I_2}$. In any case from these considerations it follows that the determination of $F^t_{I_1 I_2}$ is closely related to the dynamical fixing of the quantum field-theoretic representations, and further investigations are necessary.

For our investigations in *relativistic* quantum field theory we use an elementary argument to derive $F^t_{I_1 I_2}$. According to the Lehmann–Källen spectral theorem propagators can be decomposed into products of one-particle matrix elements $\langle 0|\psi_I|a\rangle$. For equal times $t = t'$ one obtains for $F^t_{I_1 I_2}$

$$F^t_{I_1 I_2} = \langle 0|\mathcal{A}(\psi_{I_1} \psi_{I_2})_t|0\rangle = \sum_{a,b} \left\{ \langle 0|\psi_{I_1}|a\rangle g^{ab} \langle b|\psi_{I_2}|0\rangle \right\}_{\text{as}[I_1, I_2], t_1 = t_2}, \qquad (5.9)$$

where g^{ab} is the dual to the metric tensor $g_{ab} = \langle a|b\rangle$. Hence the properties of $F^t_{I_1 I_2}$ depend on the properties of these matrix elements. If the states $|a\rangle$ are dressed particle states apart from symmetry considerations no general statements can be made about the resulting one-particle matrix elements. For hard core states $|k\rangle$, however, we can directly calculate the corresponding one-particle matrix elements from equation (5.3) by substitution of $|\mathcal{F}(j^t; 1, k)\rangle^d = iC^I_k j^t_I|0\rangle_F$ into (5.3). Projecting (5.3) with $\langle d_1|$ from the left, we obtain the sector equation

$$E^d_{k0} C^{I_1}_k = \widehat{K}_{I_1 I_2} C^{I_2}_k. \qquad (5.10)$$

From this equation it follows that the matrix elements $\langle 0|\psi_I|k\rangle \equiv C^I_k$ for hard core states $|k\rangle$ must be free-field matrix elements, as (5.10) is the free Dirac equation, *i.e.*, we have $C^I_k \equiv \langle 0|\psi^f_I|k\rangle$. For free fields the metric tensor $g^{kk'}$ can be directly calculated from the corresponding anticommutation relations (2.60), see [Stu 94]. Therefore, in this case we obtain for (5.9)

$$F^t_{I_1 I_2} = \sum_{k_1, k_2} \left\{ \langle 0|\psi^f_{I_1}|k_1\rangle g^{k_1 k_2} \langle k_2|\psi^f_{I_2}|0\rangle \right\}_{\text{as}[I_1, I_2], t_1 = t_2}, \qquad (5.11)$$

and this expression is equal to the spectral decomposition on a space-like hyperplane of $F^f_{I_1 I_2}$ given by Proposition 3.16. Hence from (5.10), together with (2.60) we have $F^t_{I_1 I_2} = F^f_{I_1 I_2}$ on the hyperplane.

The statement $F_{I_1 I_2}^t = F_{I_1 I_2}^f$ is, however, in general not compatible with equation (5.8). The reason for this discrepancy is the fact that $\mathcal{H}_F^d$ of (5.7) does not really arise from a corresponding field theory with state space Hamiltonian $\mathbf{H}^d$ and corresponding field operators ψ_I^d. As a consequence in the decomposition (5.9) the one-particle functions cannot be regarded as solutions of a corresponding eigenvalue equation resulting from a $\mathbf{H}^d$. In spite of this discrepancy we use $F_{I_1 I_2}^t = F_{I_1 I_2}^f$ as a first approximation for $F_{I_1 I_2}^t$.

5.4   Effective Boson Dynamics

A general proof of a Weak Mapping theorem which does not rely upon the kind of basic quantum fields and their bound states cannot be given. Although in any case the general reasoning runs along the same lines, each quantum field or quantum field combination and their bound states need an individual treatment.

In order to make the deduction as transparent as possible we demonstrate such mappings for the spinor field of Chapter 2, *i.e.*, for a single type of basic fields. Furthermore, we confine ourselves to two-fermion bound states, *i.e.*, to the special mapping of the spinor field equation onto effective boson field equations. Other cases of Weak Mapping will be treated in the following chapters.

For convenience we list various symbols which will be used in the following:

$$j_I^t :$$ fermion sources on the hyperplane $t = \text{const.}$,

$$b_k :$$ boson sources on the hyperplane $t = \text{const.}$,

$$\partial_I^t := \delta/\delta j_I^t :$$ fermion 'duals' on the hyperplane $t = \text{const.}$,

$$\partial_k := \delta/\delta b_k :$$ boson 'duals' on the hyperplane $t = \text{const.}$,

$$\partial_\mu := \partial/\partial x_\mu :$$ for Greek small letter index.

In particular, capital Latin letters refer to the indices of fermionic functional operators, whilst small Latin letters refer to bosonic quantities. Furthermore, as long as no misunderstanding is to be expected we omit the index t which indicates the hyperplane $t = \text{const.}$.

We assume that the boson dynamics can be formulated by functional boson states and corresponding functional boson equations. To this end we define a functional space $\mathbb{K}_B$ analogously to the fermion case in Chapter 4 as a representation space of the bosonic functional Fock states which are generated by the boson sources b_k and supplied by their 'duals' ∂_k, where k is the collection of quantum numbers characterizing the corresponding boson states. The Fock representation of the corresponding CCR algebra is then generated by

$$\partial_k |0\rangle_B = 0, \qquad b_k^+ |0\rangle_B = 0, \tag{5.12}$$

where $|0\rangle_B$ is the bosonic functional Fock 'vacuum'.

As the functional spaces $\mathbb{K}_F$ and $\mathbb{K}_B$ are in principle only auxiliary spaces allowing a compact formulation of the field dynamics, they have to be independent of each other. Therefore sources and their 'duals' for fermions and bosons are assumed to commute completely. Later on, by working with the chain rule we shall abandon this postulate in favour of an approximate treatment, but for the present investigation it strictly holds. The commutation rules for the boson operators are

$$[\partial_k, \partial_{k'}]_- = [b_k, b_{k'}]_- = 0, \tag{5.13}$$
$$[\partial_k, b_{k'}]_- = \delta_{kk'},$$

and owing to our assumption any boson field state $|a\rangle$ can be represented by a functional state $|\mathcal{B}(b; a)\rangle \in \mathbb{K}_B$ with the expansion

$$|\mathcal{B}(b; a)\rangle = \sum_{n=0}^{\infty} \frac{1}{n!} \rho^{(n)}(k_1 \ldots k_n | a) b_{k_1} \ldots b_{k_n} |0\rangle_B, \tag{5.14}$$

where $\rho^{(n)}(k_1 \ldots k_n | a)$ is symmetric in $k_1 \ldots k_n$.

The calculation rules for boson functionals are analogous to the fermion case (3.18)–(3.20) and are given by

- **Proposition 5.1:** For $|\mathcal{B}(b; a)\rangle \in \mathbb{K}_B$ with representation (5.14) the following relations hold:

(i) $_B\langle 0| \partial_{k_m} \ldots \partial_{k_1} |\mathcal{B}(b; a)\rangle = \rho^{(m)}(k_1 \ldots k_m | a),$ \hfill (5.15)

(ii) $\partial_j |\mathcal{B}(b; a)\rangle = \sum_{n=0}^{\infty} \frac{1}{n!} \rho^{(n+1)}(j, k_1 \ldots k_n | a) b_{k_1} \ldots b_{k_n} |0\rangle_B,$ \hfill (5.16)

(iii) $b_j |\mathcal{B}(b; a)\rangle = \sum_{n=1}^{\infty} \frac{1}{n!} \sum_{l=1}^{n} \delta_{jk_l} \rho^{(n-1)}(k_1 \ldots \overset{o}{k_l} \ldots k_n | a) b_{k_1} \ldots b_{k_n} |0\rangle_B$ (5.17)

Proof: With (5.12) and (5.13) there holds

$$_B\langle 0| \partial_{k_m} \ldots \partial_{k_1} b_{j_1} \ldots b_{j_n} |0\rangle_B = \delta_{mn} \sum_{p \in S_n} \delta_{j_1 k_{p_1}} \ldots \delta_{j_n k_{p_n}}. \tag{5.18}$$

With (5.18) and the symmetry of the coefficients $\rho^{(n)}(k_1 \ldots k_n | a)$ one obtains

$$_B\langle 0| \partial_{k_m} \ldots \partial_{k_1} |\mathcal{B}(b; a)\rangle = \frac{1}{m!} \rho^{(m)}(k_1 \ldots k_m | a) \sum_{p \in S_n} \delta_{j_1 k_{p_1}} \ldots \delta_{j_m k_{p_m}} \tag{5.19}$$
$$= \rho^{(m)}(k_1 \ldots k_m | a).$$

The proof of ii) and iii) runs along the same lines as that for formulas (3.18) and (3.19). $\diamond$

The description of bosons as bound states of subfermions and the derivation of a corresponding effective dynamics has to start from the normal ordered energy equation on the hyperplane (4.100). This was justified in detail in Section 4.7. Hence the intended map has to be applied to (4.100). An inspection of this equation shows that the set of solutions can be subdivided into two classes, which completely decouple: those with all coeffcent functions of $|\mathcal{F}(j^t; a)\rangle$ belonging to an even number of arguments, and those with an odd number of arguments. With respect to the mapping onto a boson theory for two-particle bound boson states only the first class of solutions is of interest. Thus in Sections 5.4–5.7 we confine our considerations to the functional states

$$|\mathcal{F}(j; a)\rangle = \sum_{n=0}^{\infty} \frac{(-1)^n}{(2n)!} \varphi^{(2n)}(I_1 \ldots I_{2n}|a) j_{I_1} \ldots j_{I_{2n}} |0\rangle_{\mathrm{F}} \qquad (5.20)$$

on the hyperplane $t = \text{const.}$.

Without any regard to functional space we define the Weak Mapping of a fermion theory onto a boson theory only by means of the coefficient functions $\varphi^{(2n)}(I_1 \ldots I_{2n})$. In particular with respect to the map by two-particle boson functions we assume the set of solutions $C_k^{I_1 I_2}$ of equation (5.4) to be complete. Then we define the antisymmetrized generalized determinants by

$$D_{2n} \begin{pmatrix} I_1 \ldots I_{2n} \\ k_1 \ldots k_n \end{pmatrix} := C_{k_1}^{[I_1 I_2} \ldots C_{k_n}^{I_{2n-1} I_{2n}]}, \qquad (5.21)$$

where $[I_1 \ldots I_{2n}]$ means antisymmetrization in $I_1 \ldots I_{2n}$ with a corresponding factor $1/(2n)!$. By means of the completeness relations (5.6) it can be easily shown that the set of all D_{2n}, where $k_1 \ldots k_n$ run through all quantum numbers of the two-particle states, is complete in the space of antisymmetrized functions $f(I_1 \ldots I_{2n})$. Therefore we can expand any coefficient function $\varphi^{(2n)}(I_1 \ldots I_{2n})$ in terms of the determinantal functions (5.21); *i.e.*,

$$\varphi^{(2n)}(I_1 \ldots I_{2n}|a) = \sum_{k_1 \ldots k_n} D_{2n} \begin{pmatrix} I_1 \ldots I_{2n} \\ k_1 \ldots k_n \end{pmatrix} f(k_1 \ldots k_n|a) \qquad (5.22)$$

with the expansion coefficients $f(k_1 \ldots k_n|a)$.

We now introduce Weak Mapping of a fermion theory onto a boson theory by the following definition, see Stumpf [Stu 85a,b,c], Kerschner and Stumpf [Ker 91]:

- **Definition 5.2:** Weak Mapping is defined by the identification of the expansion coefficient functions $f(k_1 \ldots k_n|a)$ of (5.22) with the coefficient functions $\rho^{(n)}$ of the boson functional state (5.14), *i.e.*,

$$f(k_1 \ldots k_n|a) \equiv \rho^{(n)}(k_1 \ldots k_n|a) \qquad (5.23)$$

  or

$$\varphi^{(2n)}(I_1 \ldots I_{2n}|a) = C_{k_1}^{[I_1 I_2} \ldots C_{k_n}^{I_{2n-1} I_{2n}]} \rho^{(n)}(k_1 \ldots k_n|a). \qquad (5.24)$$

Equation (5.24) establishes a map $\rho^{(n)} \to \varphi^{(2n)}$. If the set (5.21) is complete the inverse mapping $\varphi^{(2n)} \to \rho^{(n)}$ exists, *i.e.*, the map is bijective, provided only symmetrized $\rho^{(n)}(k_1 \ldots k_n)$ are admitted. The connection to effective field equations is established by interpreting the boson coefficient functions $\rho^{(n)}$ as symmetrized expectation values of effective boson field operators; for a detailed discussion we refer to the applications in the following chapters.

Equation (5.24) provides additional evidence for the use of equation (4.100) as the starting point of Weak Mapping. One can expect the set of functions $C_k^{I_1 I_2}$ to be complete only on a hyperplane $t = \text{const.}$, if one rejects Euclidean field theories as unphysical and compares the functions $C_k^{I_1 I_2}$ with ordinary quantum mechanical wave functions. Only if due to completeness relations the map (5.24) is bijective, (5.24) allows a faithful mapping of the basic field dynamics and prevents an uncontrollable loss of information of the basic theory. Thus the need for completeness enforces the use of the noncovariant formalism for the meaningfull definition of Weak Mapping.

In spite of the independence of Weak Mapping from the functional space, we want to formulate the *results* of Weak Mapping by functional equations. This was done in the following proposition derived by Pfister and Stumpf [Pfi 91]. This proposition is an alternative version to a theorem by Kerschner and Stumpf [Ker 91].

- **Proposition 5.2:** By the transformation (5.24) the functional energy equation (4.100) of the fermionic spinor field (2.68) is mapped onto the following functional boson energy equation:

$$E_{p0}|\tilde{B}(b;p)\rangle = \tag{5.25}$$

$$2R_{I_1 K}^k \widehat{K}_{I_1 I_2} C_{k'}^{I_2 K} b_k \partial_{k'} |\tilde{B}(b;p)\rangle$$

$$-\widehat{W}_{K_1 I_2 I_3 I_4} \Big\{ 6 F_{I_4 K_2} R_{K_1 K_2}^k C_{k'}^{I_2 I_3} b_k \partial_{k'}$$

$$-36 F_{I_4 K_2} R_{K_1 K_3}^{k_1} R_{K_2 K_4}^{k_2} C_{k_1'}^{[I_2 I_3} C_{k_2'}^{K_3 K_4]} b_{k_1} b_{k_2} \partial_{k_1'} \partial_{k_2'}$$

$$-6 R_{K_1 K_2}^k C_{k_1'}^{[I_2 I_3} C_{k_2'}^{I_4 K_2]} b_k \partial_{k_1'} \partial_{k_2'}$$

$$-\Big(3 F_{I_4 K_2} F_{I_3 K_3} + \frac{1}{4} A_{I_4 K_2} A_{I_3 K_3}\Big)$$

$$\times \sum_{p \in S_3} \text{sgn}(p) \frac{1}{3!}$$

$$\times \Big[12 R_{K_{p_1} K_{p_2}}^{k_1} R_{K_{p_3} K_4}^{k_2} C_{k_1'}^{I_2 K_4} b_{k_1} b_{k_2} \partial_{k'}$$

$$+24 R_{K_{p_1} K_4}^{k_1} R_{K_{p_2} K_5}^{k_2} R_{K_{p_3} K_6}^{k_3} C_{k_1'}^{[I_2 K_6} C_{k_2'}^{K_5 K_4]} b_{k_1} b_{k_2} b_{k_3} \partial_{k_1'} \partial_{k_2'} \Big]$$

$$+\Big(F_{I_4 K_2} F_{I_3 K_3} + \frac{1}{4} A_{I_4 K_2} A_{I_3 K_3}\Big) F_{I_2 K_4}$$

$$\times \sum_{p \in S_4} \text{sgn}(p) \frac{1}{4!}$$

$$\times \left[12 R^{k_1}_{K_{p_1} K_{p_2}} R^{k_2}_{K_{p_3} K_{p_4}} b_{k_1} b_{k_2} \right.$$

$$+ 48 R^{k_1}_{K_{p_1} K_5} R^{k_2}_{K_{p_2} K_6} R^{k_3}_{K_{p_3} K_{p_4}} C^{K_5 K_6}_{k'_1} b_{k_1} b_{k_2} b_{k_3} \partial_{k'}$$

$$\left. \left. + 48 R^{k_1}_{K_{p_1} K_5} \cdots R^{k_4}_{K_{p_4} K_8} C^{[K_5 K_6}_{k'_1} C^{K_7 K_8]}_{k'_2} b_{k_1} \cdots b_{k_4} \partial_{k'_1} \partial_{k'_2} \right] \right\} |\tilde{\mathcal{B}}(b; p)\rangle$$

with

$$|\tilde{\mathcal{B}}(b; p)\rangle := \mathcal{U} |\mathcal{B}(b; p)\rangle \qquad (5.26)$$

and

$$\mathcal{U} := \sum_{n=0}^{\infty} \frac{1}{n!} b_{k'_1} \cdots b_{k'_n} |0\rangle_{\mathrm{B}} R^{k'_1}_{K_1 K_2} \cdots R^{k'_n}_{K_{2n-1} K_{2n}} \qquad (5.27)$$

$$\times C^{[K_1 K_2}_{k_1} \cdots C^{K_{2n-1} K_{2n}]}_{k_n} {}_{\mathrm{B}}\langle 0 | \partial_{k_n} \cdots \partial_{k_1}.$$

For the rather extensive proof of Proposition 5.2 we refer to [Stu 94]. Without proof we remark that for comparison with the results by [Pfi 91] and [Stu 94] the following identities have to be used:

$$\mathcal{V} \begin{pmatrix} I_1 I_2 \\ k \end{pmatrix} \partial_k |\mathcal{B}(b; p)\rangle = \sum_l C^{I_1 I_2}_l \partial_l |\tilde{\mathcal{B}}(b; p)\rangle \qquad (5.28)$$

and

$$\mathcal{W} \begin{pmatrix} I_1 I_2 & I_3 I_4 \\ k & k' \end{pmatrix} \partial_k \partial_{k'} |\mathcal{B}(b; p)\rangle = \sum_{l l'} C^{I_1 I_2}_l C^{I_3 I_4}_{l'} \partial_l \partial_{l'} |\tilde{\mathcal{B}}(b; p)\rangle, \qquad (5.29)$$

where $\mathcal{V}$ and $\mathcal{W}$ are defined in [Pfi 91] and [Stu 94].

Equation (5.25) represents a faithful mapping of the spinor field dynamics onto a boson field dynamics and it contains all subtleties of the interactions of the fermionic constituents of the boson states under consideration. In order to extract the main physical effects from this equation we have to simplify and to approximate it in a suitable way.

5.5   Direct and Exchange Forces

The most suitable way of reducing the complicated structure of equation (5.25) and to obtain an insight into its physical meaning is the decomposition of the interactions into direct and exchange force interactions.

In quantum mechanics direct forces are characterized by the exclusive occurrence of probability densities in the interaction terms, whilst all other bilinear combinations of probability amplitudes in the interaction terms are attributed to exchange forces. In the quantum field theoretic energy equation (5.25) the expressions $R^k_{I_1 I_2} C^{I_1 I_2}_k$ (no summation) take over the part of quantum mechanical probability densities and thus are connected with direct forces, whilst all other R/C combinations have to be attributed to exchange forces. We expect the direct forces to overbalance the exchange forces; thus in a first step we separate the former from the exchange forces by rewriting the expression (5.27).

If we define the dual generalized determinant by

$$D^{2n}\begin{pmatrix} k_1 \ldots k_n \\ I_1 \ldots I_{2n} \end{pmatrix} := R^{k_1}_{[I_1 I_2} \ldots R^{k_n}_{I_{2n-1} I_{2n}]}, \qquad (5.30)$$

then with (5.21) the definition (5.27) can be rewritten as

$$\mathcal{U} = \sum_{n=0}^{\infty} \frac{1}{n!} b_{k'_1} \ldots b_{k'_n} |0\rangle_B \left\langle D^{2n}\begin{pmatrix} k'_1 \ldots k'_n \\ I_1 \ldots I_{2n} \end{pmatrix}, D_{2n}\begin{pmatrix} I_1 \ldots I_{2n} \\ k_1 \ldots k_n \end{pmatrix} \right\rangle {}_B\langle 0|\partial_{k_n} \ldots \partial_{k_1}, \quad (5.31)$$

where $\langle \, , \, \rangle$ denotes the scalar product of these determinants in configuration space, i.e., the summation over $I_1, \ldots I_{2n}$. In contrast to Hartree–Fock determinants with orthogonal one-particle states, the determinants (5.21) and (5.30) are not orthogonal. This leads to nontrivial exchange forces, formulated by the following proposition, derived by Pfister and Stumpf [Pfi 91]:

• **Proposition 5.3:** Let $\lambda := (\lambda_1 \ldots \lambda_j)$ be a partition of n with

$$\lambda_i \in \mathbb{N}; \quad \lambda_1 \geq \lambda_2 \geq \ldots \geq \lambda_j; \quad \sum_{i=1}^{j} \lambda_i = n$$

and $s_1, \ldots s_j$ the multiplicities of identical values of the $\lambda_1 \ldots \lambda_j$. Then the following relation holds:

$$R^{k'_1}_{K_1 K_2} \ldots R^{k'_n}_{K_{2n-1} K_{2n}} C^{[K_1 K_2}_{k_1} \ldots C^{K_{2n-1} K_{2n}]}_{k_n} \qquad (5.32)$$

$$= \frac{(n!)^2}{(2n)!} \sum_{\text{part } \lambda} (-1)^j \frac{2^{2n-j}}{s_1! \ldots s_j! \, \lambda_1 \ldots \lambda_j} \left\{ \begin{matrix} k'_1 & \ldots & k'_n \\ k_1 & \ldots & k_n \\ \lambda_1 & \ldots & \lambda_j \end{matrix} \right\}_{\text{sym}},$$

where

$$\left\{ \begin{matrix} k'_1 & \ldots & k'_n \\ k_1 & \ldots & k_n \\ \lambda_1 & \ldots & \lambda_j \end{matrix} \right\} \qquad (5.33)$$

$$:= \text{tr}\left[R^{k'_1} C_{k_1} \ldots R^{k'_{\lambda_1}} C_{k_{\lambda_1}} \right] \ldots \text{tr}\left[R^{k'_{n-\lambda_j+1}} C_{k_{n-\lambda_j+1}} \ldots R^{k'_n} C_{k_n} \right],$$

and the suffix 'sym' means separate symmetrization in $k_1, \ldots k_n$ and $k'_1, \ldots k'_n$, i.e.,

$$\left\{ \begin{matrix} k_1' & \cdots & k_n' \\ k_1 & \cdots & k_n \\ \lambda_1 & \cdots & \lambda_j \end{matrix} \right\}_{\text{sym}} := \frac{1}{n!} \sum_{p \in S_n} \frac{1}{n!} \sum_{p' \in S_n} \left\{ \begin{matrix} k_{p_1}' & \cdots & k_{p_n}' \\ k_{p_1'} & \cdots & k_{p_n'} \\ \lambda_1 & \cdots & \lambda_j \end{matrix} \right\} . \tag{5.34}$$

Proof: For the rather extensive proof we refer to the original paper [Pfi 91] or to [Stu 94]. ◊

By means of this proposition the direct forces can be separated from the exchange forces. According to our definition the direct forces are characterized by the exclusive appearance of density expressions, which in our case are given by $\text{tr}[RC]$. Obviously these expressions result from all partitions with $\lambda_1 = \ldots = \lambda_j = 1$. Hence only these partitions lead to direct forces whilst all other partitions have to be identified with the exchange forces.

In Section 8.8 we will show that the exchange forces are weaker than the direct forces, and thus the former may be neglected in a low energy evaluation of the effective boson dynamics (5.25).

5.6 Weak Mapping in Functional Space

In order to accentuate the physical and mathematical foundations of Weak Mappings we have strictly separated the definition of the Weak Mapping from the embedding of its results into functional space. In this section we soften up this point of view in favour of an elegant functional treatment of Weak Mapping which neglects the exchange forces contained in (5.27) from the beginning and which was developed by Stumpf [Stu 85a,b,c].

To introduce this method we define bosonic source operators by a bilinear form of fermion sources in functional space:

$$b_k := \frac{1}{2} C_k^{I_1 I_2} j_{I_1} j_{I_2} , \tag{5.35}$$

where $C_k^{I_1 I_2}$ are the two-particle solutions of (5.4), k denotes the boson quantum numbers, and where in contrast to Section 5.4 the boson operators b_k depend on the fermionic sources. It is just this property which now allows to directly transform equation (5.1) in functional space.

Let $|\mathcal{F}(j; p)\rangle$ be the functional eigenstate (5.20) of equation (5.1). Then in accordance with the theory of linear vector spaces we consider the boson functional state as the same vector in another frame of reference and postulate the invariance of the functional state $|\mathcal{F}(j; p)\rangle$ under the boson transformation (5.35) by

$$|\mathcal{F}(j; p)\rangle = |\widehat{B}(b; p)\rangle \tag{5.36}$$

with the boson state functional $|\widehat{B}(b; p)\rangle$, where according to (5.35) $|\widehat{B}(b; p)\rangle$ is an abbreviation for $|\widehat{B}(b[j]; p)\rangle$ and denotes the transformed functional state which is assumed to have a power series expansion in the b operators. With (5.36) equation (5.1) can be equivalently written as

$$E_{p0}|\widehat{B}(b;p)\rangle = \mathcal{H}_{\mathrm{F}}(j,\partial)|\widehat{B}(b;p)\rangle. \tag{5.37}$$

If we introduce functional boson 'duals' ∂^b and transform $\mathcal{H}_{\mathrm{F}}(j,\partial)$ into $\mathcal{H}_{\mathrm{B}}(b,\partial^b) \equiv \mathcal{H}_{\mathrm{F}}(j[b],\partial[\partial^b])$ and subsequently take the b and ∂^b operators as independent from the j and ∂ operators we arrive at an effective boson theory given by

$$E_{p0}|\widehat{B}(b;p)\rangle = \mathcal{H}_{\mathrm{B}}(b,\partial^b)|\widehat{B}(b;p)\rangle. \tag{5.38}$$

Thus we first have to calculate the effect of fermion duals ∂_I on $|\widehat{B}(b[j];p)\rangle$. Analogously to Section 5.4 we introduce the dual ∂_k^b to b_k by the commutation relation

$$\left[\partial_{k'}^b, b_k\right]_- = \delta_{kk'} \tag{5.39}$$

and apply the functional chain rule

$$\partial_I|\widehat{B}(b[j];a)\rangle = \sum_k (\partial_I b_k)\partial_k^b|\widehat{B}(b;a)\rangle \equiv \sum_k \frac{\delta b_k[j]}{\delta j_I}\frac{\delta}{\delta b_k}|\widehat{B}(b[j];a)\rangle. \tag{5.40}$$

With (5.35) we obtain $\partial_{I_1} b_k = C_k^{I_1 I_2} j_{I_2}$ and therefore

$$\partial_{I_1}|\widehat{B}(b;a)\rangle = \sum_k C_k^{I_1 I_2} j_{I_2}\partial_k^b|\widehat{B}(b;a)\rangle. \tag{5.41}$$

Repeated application of the chain rule yields

$$\partial_{I_1}\partial_{I_2}|\widehat{B}(b;a)\rangle = \sum_k C_k^{I_2 I_1}\partial_k^b|\widehat{B}(b;a)\rangle - \sum_{k_1,k_2} 2C_{k_1}^{I_2 K_1}C_{k_2}^{I_1 K_2} j_{K_1} j_{K_2}\partial_{k_2}^b\partial_{k_1}^b|\widehat{B}(b;a)\rangle \tag{5.42}$$

and

$$\begin{aligned}
\partial_{I_1}\partial_{I_2}\partial_{I_3}&|\widehat{B}(b;a)\rangle \\
&= 3\sum_{k_1,k_2} C_{k_1}^{I_2 I_1}C_{k_2}^{I_3 K_1} j_{K_1}\partial_{k_2}^b\partial_{k_1}^b|\widehat{B}(b;a)\rangle \\
&\quad + \sum_{k_1,k_2,k_3} C_{k_1}^{I_1 K_1}C_{k_2}^{I_2 K_2}C_{k_3}^{I_3 K_3} j_{K_1} j_{K_2} j_{K_3}\partial_{k_1}^b\partial_{k_2}^b\partial_{k_3}^b|\widehat{B}(b;a)\rangle_{|\mathrm{as}[I_1,I_2,I_3]}.
\end{aligned} \tag{5.43}$$

In this way the fermionic duals can be transformed into bosonic duals. As is obvious from $\mathcal{H}_{\mathrm{F}}$ and from (5.41), (5.42) and (5.43) the transformation has to be completed by giving a transformation rule for the fermionic sources themselves. By observing (5.5) and (5.6) we can resolve (5.35) with respect to the fermionic sources in the form

$$j_{I_1} j_{I_2} = 2R_{I_1 I_2}^k b_k. \tag{5.44}$$

The transformation of products of more than two fermion sources can be consistently performed only in connection with dressed particle states. This will be shown in Section 5.7. Here we only mention that it is necessary to observe antisymmetries of these products; in particular we assume

$$j_{I_1} j_{I_2} j_{I_3} j_{I_4} = 4R_{[I_1 I_2}^{k_1} R_{I_3 I_4]}^{k_2} b_{k_1} b_{k_2}. \tag{5.45}$$

The five equations (5.41)–(5.45) allow a direct transformation of $\mathcal{H}_F$ defined in (5.2). As a result of this transformation one obtains the effective boson functional equation

$$E_{p0}|\widehat{B}(b;p)\rangle \tag{5.46}$$

$$
\begin{aligned}
= \ & 2\widehat{K}_{I_1 I_2} R^k_{I_1 K} C^{I_2 K}_{k'} b_k \partial^b_{k'} |\widehat{B}(b;p)\rangle \\
& + \widehat{W}_{I_1 I_2 I_3 I_4} \Big\{ -6 F_{I_4 K} R^k_{I_1 K} C^{I_2 I_3}_{k'} b_k \partial^b_{k'} \\
& \qquad + 6 R^k_{I_1 K} C^{[I_2 I_3}_{k'_1} C^{I_4 K]}_{k'_2} b_k \partial^b_{k_1} \partial^b_{k_2} \\
& \qquad + 4 R^{k_1}_{I_1 K_1} R^{k_2}_{K_2 K_3} C^{I_2 K_1}_{k'_1} C^{I_3 K_2}_{k'_2} C^{I_4 K_3}_{k'_3} b_{k_1} b_{k_2} \partial^b_{k'_1} \partial^b_{k'_2} \partial^b_{k'_3} \\
& \qquad + 12 F_{I_4 K} R^{k_1}_{[I_1 K_1} R^{k_2}_{K_2 K_3]} C^{I_2 K_2}_{k'_1} C^{I_3 K_3}_{k'_2} b_{k_1} b_{k_2} \partial^b_{k'_1} \partial^b_{k'_2} \Big) \\
& + 12 \Big(3 F_{I_4 K_1} F_{I_3 K_2} + \frac{1}{4} A_{I_4 K_1} A_{I_3 K_2} \Big) R^{k_1}_{[I_1 K_1} R^{k_2}_{K_2 K_3]} C^{I_2 K_3}_{k'} b_{k_1} b_{k_2} \partial^b_{k'} \\
& - 12 \Big(F_{I_4 K_1} F_{I_3 K_2} + \frac{1}{4} A_{I_4 K_1} A_{I_3 K_2} \Big) F_{I_2 K_3} R^{k_1}_{[I_1 K_1} R^{k_2}_{K_2 K_3]} b_{k_1} b_{k_2} \Big\} |\widehat{B}(b;p)\rangle
\end{aligned}
$$

We add that we can 'forget' the relation (5.35) between boson and fermion sources as in (5.46) no reference is made to (5.35) and because the duals ∂^b_k were introduced without reference to fermionic duals. Thus we have obtained a genuine boson functional equation, *i.e.*, a dynamical law for bosons without reference to fermions.

In Sections 8.7 and 8.8 we will identify the last four terms of (5.46) as residual interactions and will show that in a low energy limit the exchange forces and the residual interactions can be neglected. In this limit we will demonstrate that equation (5.46) is equivalent to equation (5.25), which was obtained by the exact Weak Mapping procedure. So in this way the method of the functional chain rule is justified as a method which automatically reproduces the direct forces and the 'physical' interactions of the corresponding effective theories. In the following we take over this result also for other mappings and work solely with the chain rule mapping.

The successful use of the chain rule suggests to formulate Weak Mapping completely in a functional version. This was performed by Kerschner [Ker 94]. For brevity we do not explicitly discuss this approach.

5.7 Dressed Particle States

Hard core bound states are characterized by a definite number of coordinates in their wave functions, the minimal number of subfermions, and they are described by the state functionals of Section 5.2. Corresponding dressed particle states start with the same number of coordinates in their wave functions, but in addition they contain an infinite set of wave functions with higher coordinate numbers than the minimal hard core number. Thus we can describe dressed particle states by the functional states

$$|\mathcal{F}(j;f,k)\rangle^{\mathrm{dr}} = \sum_{n=0}^{\infty} \frac{i^{2n+f}}{(2n+f)!} C_{f,k}^{I_1\ldots I_{2n+f}} j_{I_1}\ldots j_{I_{2n+f}}|0\rangle_{\mathrm{F}}, \qquad (5.47)$$

where k denotes the full set of quantum numbers which characterize the dressed particle, whilst f is the minimal number of fermionic subcomponents which is needed for the formation of the hard core of a specific dressed particle state. In contrast to the hard core states which are assumed to be solutions of the diagonal part (5.3) of the energy equation, the dressed particle states are assumed to be solutions of the full energy equation (5.1). As in (5.1) the sectors of even and odd fermion numbers completely decouple, the specific form of (5.47) is justified.

However, it should be noted that although (5.47) is assumed to be a solution of (5.1), the direct product of (5.47) with itself, i.e., a state of two dressed particles, is in general not a solution of (5.1). Rather equation (5.1) enforces an interaction between two (or more) dressed particles, i.e., a dressed particle dynamics, a situation which is completely analogous to the hard core bound state dynamics.

In order to derive an effective dressed particle dynamics we need orthonormality and completeness relations analogously to those for hard core states, cf. Stumpf [Stu 87]. By definition we introduce dual sets $\left\{R_{I_1\ldots I_{2n+f}}^{f,k}, n=0,1,\ldots,\infty\right\}$, where f and k run through all dressed particle state quantum numbers. These dual sets are characterized by the orthonormality relations

$$\sum_{I_1\ldots I_{2n+f}} \sum_{n,n'} R_{I_1\ldots I_{2n+f}}^{f,k} C_{f',k'}^{I_1\ldots I_{2n'+f'}} \delta_{2n+f,2n'+f'} = \delta_{kk'}\delta_{ff'}. \qquad (5.48)$$

These relations are an appropriate extension of the hard core state orthonormality relations (5.5).

From (5.48) it follows that orthonormality or duality, respectively, is connected with a selection rule. Owing to (5.47) either even or odd numbers $2n+f$ appear in (5.48), i.e., boson states with even numbers and fermion states with odd numbers are completely separated. Therefore, if we generalize Weak Mapping to dressed particle states, it is convenient to distinguish between fermions and bosons from the beginning. In full generality the analogue of the hard core boson transformation (5.35) for dressed particles then has to be defined by the transformations

$$f_{2j+1,l} = \sum_{n=0}^{\infty} C_{2j+1,l}^{I_1\ldots I_{2j+1+2n}} j_{I_1}\ldots j_{I_{2j+1+2n}}, \qquad j=0,1,\ldots\infty, \qquad (5.49)$$

$$b_{2j,k} = \sum_{n=0}^{\infty} C_{2j,k}^{I_1\ldots I_{2j+2n}} j_{I_1}\ldots j_{I_{2j+2n}}, \qquad j=1,2\ldots\infty, \qquad (5.50)$$

for dressed fermion operators $f_{2j+1,l}$ and dressed boson operators $b_{2j,k}$, where $2j$ or $2j+1$ are the minimal subfermion numbers, whilst k or l are the quantum numbers which characterize the state of the respective particle. With respect to Weak Mapping the inverse relations of (5.49), (5.50) are needed.

- **Proposition 5.4:** The right inverses of (5.49) and (5.50) are given by the relations

$$j_{I_1} \cdots j_{I_{2r+1}} = \sum_l \sum_{j=0}^r R_{I_1 \ldots I_{2r+1}}^{2j+1,l} f_{2j+1,l} , \tag{5.51}$$

$$j_{I_1} \cdots j_{I_{2r}} = \sum_k \sum_{j=1}^r R_{I_1 \ldots I_{2r}}^{2j,k} b_{2j,k} , \tag{5.52}$$

where $k \neq k'$ for all $R^{f,k}$ and $R^{f',k'}$ with $f \neq f'$.

Proof: We assume $j < j'$, but $k = k'$. Then we have a state $(2j+1, k)$ with $C_{2j+1,k}^{I_1 \ldots I_{2j+1}} \neq 0$ and a state $(2j' + 1, k)$ with $C_{2j+1,k}^{I_1 \ldots I_{2j'+1}} \equiv 0$, as by definition $C_{2j'+1,k}^{I_1 \ldots I_{2j'+1}}$ is the first nonvanishing amplitude for the state $(2j' + 1, k)$. This is a contradiction, and thus it follows $k \neq k'$.

To prove (5.51) we write it in the form

$$j_{I_1} \cdots j_{I_{2j+1+2n}} = \sum_{l'} \sum_{j'=0}^{j+n} R_{I_1 \ldots I_{2j+1+2n}}^{2j'+1,l'} f_{2j'+1,l'} \tag{5.53}$$

and substitute (5.53) into (5.49). This gives

$$f_{2j+1,l} = \sum_{n=0}^{\infty} C_{2j+1,l}^{I_1 \ldots I_{2j+1+2n}} \sum_{l'} \sum_{j'=0}^{j+n} R_{I_1 \ldots I_{2j+1+2n}}^{2j'+1,l'} f_{2j'+1,l'} , \tag{5.54}$$

and if we apply (5.48) we obtain

$$f_{2j+1,l} = \sum_{l'} \sum_{j'=0}^{j+n} \delta_{2j+1,2j'+1} \delta_{l,l'} f_{2j'+1,l'} = f_{2j+1,l} . \tag{5.55}$$

The same considerations can be applied to (5.52). If we symbolically write $f = U \circ j$ we have $j = U^{-1} \circ f$. By relation (5.55) we have thus shown $U \circ U^{-1} = \mathbb{I}$, i.e., that U^{-1} given by (5.51) and (5.52) is a right inverse to U. $\Diamond$

On the other hand, one should also show $j = U^{-1} \circ Uj = j$, i.e., that U^{-1} is simultaneously a left and a right inverse. The proof of the latter property depends on completeness relations for dressed particle states. For the simple applications of dressed particle states which we will discuss in the following this proof is not needed.

Finally we have to derive the duals $R_{I_1 \ldots I_{2n+f}}^{f,k}$ for themselves. There is no way to simply guess these functions; rather they have to be calculated. We assume to have calculated a complete set of left hand side dressed particle solutions of equation (5.1) which are given by their coefficient functions $L_{I_1 \ldots I_{2n+f}}^{f,k}$. Then according to (4.24) we obtain for the scalar product of right hand and left hand side solutions with the same subfermion number $f = f'$:

$$g_{kk'} = \sum_n \sum_{I_1 \ldots I_{2n+f}} \left(L_{I_1 \ldots I_{2n+f}}^{f,k} \right)^* C_{f,k'}^{I_1 \ldots I_{2n+f}} . \tag{5.56}$$

If we assume that $g_{kk'}$ is non-degenerate, it follows from (5.56)

$$\sum_{n,k''}\sum_{I_1\ldots I_{2n+f}}\left(g^{-1}\right)_{kk''}\left(L_{I_1\ldots I_{2n+f}}^{f,k''}\right)^* C_{f,k'}^{I_1\ldots I_{2n+f}} = \delta_{kk'},\tag{5.57}$$

and by comparison with (5.48) we obtain

$$R_{I_1\ldots I_{2n+f}}^{f,k} \equiv \sum_{k'}\left(g^{-1}\right)_{kk'}\left(L_{I_1\ldots I_{2n+f}}^{f,k'}\right)^*.\tag{5.58}$$

It remains the problem of the calculation of the left hand side solutions $L^{f,k}$. We demonstrate that this problem can be reduced to the problem of determining the metric tensor G^{mn} of the subfermion state space.

Besides (5.56) the scalar product (4.24) is given by

$$g_{kk'} = \sum_{m,n}\left(C_{f,k}^{I_1\ldots I_{2m+f}}\right)^* G_{I_1\ldots I_{2m+f},I_1'\ldots I_{2n+f}'}^{-1} C_{f,k'}^{I_1'\ldots I_{2n+f}'}.\tag{5.59}$$

Comparing (5.59) with (5.56) we obtain the general formula

$$\left(L_{I_1\ldots I_{2n+f}}^{f,k}\right)^* = \sum_m C_{f,k}^{*I_1'\ldots I_{2m+f}'} G_{I_1'\ldots I_{2m+f}',I_1\ldots I_{2n+f}}^{-1}.\tag{5.60}$$

For normal ordered basis states the metric tensor G reads according to Sections 4.4 and 4.5

$$G_{I_1\ldots I_{2m+f},I_1'\ldots I_{2n+f}'} = \langle 0|\mathcal{N}(\psi_{I_1}\ldots\psi_{I_{2m+f}})\mathcal{N}(\psi_{I_1'}\ldots\psi_{I_{2n+f}'})^+|0\rangle.\tag{5.61}$$

However, owing to the fermion number conservation of the original subfermion theory the nondiagonal elements vanish for a hard core metric. We extend this to the polarization cloud, so that only the diagonal elements are left:

$$G_{I_1\ldots I_{2m+f},I_1'\ldots I_{2n+f}'} = \delta_{m,n}\langle 0|\mathcal{N}(\psi_{I_1}\ldots\psi_{I_{2n+f}})\mathcal{N}(\psi_{I_1'}\ldots\psi_{I_{2n+f}'})^+|0\rangle.\tag{5.62}$$

In (5.62) we explicitly see that the metric tensor G depends on the representation of the subfermion operator algebra; it has to be calculated in accordance with the representation of the corresponding matrix elements. In [Stu 94] the diagonal elements of G were given in an approximation which is consistent with the free fermion propagator. We cite the result:

$$G_{I_1\ldots I_{2n},I_1'\ldots I_{2n}'} = 2^{-n}\lambda_{i_1}\ldots\lambda_{i_n}\delta_{I_1 I_1'}\ldots\delta_{I_{2n}I_{2n}'}|_{as[I_1\ldots I_{2n}]}.\tag{5.63}$$

Thus, with (5.60) we obtain for the duals (5.58)

$$R_{I_1\ldots I_{2n+f}}^{f,k} = 2^{2n+f}\sum_{k'}\left(g^{-1}\right)_{kk'}\lambda_{i_1}^{-1}\ldots\lambda_{i_{2n+f}}^{-1}\left(C_{f,k'}^{I_1\ldots I_{2n+f}}\right)^*.\tag{5.64}$$

The scalar products $g_{kk'}$ can be calculated by inserting explicit functions C into (5.59).

5.8   Effective Boson and Composite Fermion Dynamics

In Section 5.4 the Weak Mapping of quantum fields was exemplified by the proof of a Weak Mapping theorem for bosons as two-fermion composites. In physical reality, however, bosons and fermions occur simultaneously, and often both kinds of particles are assumed to be non-elementary. The most prominent example is offered by quantum chromodynamics, where mesons and nucleons are considered to be composed of two or three quarks, respectively. Another example is provided by the standard model. This model contains at least 61 elementary bosons and fermions and more than 20 parameters, *cf.*, for instance, Nachtmann [Nach 86].

This strongly supports the assumption that this model describes composite particles and effective forces rather than elementary ones. In a subfermion model for the explanation of the standard model the standard fermions should be composed of three subfermions, whilst the weak bosons should be composed of two subfermions. Hence in both cases one has to study the effective dynamics for two-fermion and three-fermion bound states.

With respect to the quark model this problem is more involved than for the subfermion model, as the quark model is based on quantum chromodynamics, *i.e.*, on a complicated non-abelian coupling theory whilst the subfermion model simply rests on a spinor field with nonlinear self-interaction. It is, however, interesting to note that the attempts to evaluate quark–gluon dynamics are mainly based on various effective NJL quark models, see for instance Vogl and Weise [Vogl 91], Alkofer and Reinhard [Alk 95]. Therefore, as a first step for treating these problems we concentrate on the spinor field model.

Within this model Kerschner [Ker 94] has derived a Weak Mapping theorem for the combined dynamics of two-fermion and three-fermion hard core states. Whilst this theorem demonstrates the ability of the Weak Mapping technique to derive effective dynamics for complicated systems, *there are indications that the effective dynamics for combined boson–fermion hard core states does not correctly reflect the physical situation.*

Without going into details we merely state that there are compelling reasons to treat, instead of a hard core dynamics, a dressed particle dynamics, because only in the latter case one can achieve a correct coupling of the bosons to the composite fermion current. An exact treatment for the combined dynamics of two-fermion and three-fermion *dressed* particle states has not yet been given. Hence for a first exploration we discuss this problem by means of the chain rule formalism. This formalism was already applied by Stumpf [Stu 87], [Stu 88] to the problem under consideration, but the improved discussion of dressed particle states also allows an improved application of the chain rule which we will formally discuss in this section.

The relation of the chain rule to the exact mapping theorems was discussed in Section 5.7. Of course, these arguments are still valid if we generalize the chain rule to comprise dressed particles. So without further comments we rewrite the formalism of Section 5.7 for dressed particles, *i.e.*, in the case under consideration, for dressed two-fermion and dressed three-fermion states. In accordance with (5.49) and (5.50) we define boson and fermion source operators by

$$b_{2,k} := \sum_{n=0}^{\infty} C_{2,k}^{I_1 \ldots I_{2n+2}} j_{I_1} \ldots j_{I_{2n+2}}, \tag{5.65}$$

$$f_{3,l} := \sum_{n=0}^{\infty} C_{3,l}^{I_1 \ldots I_{2n+3}} j_{I_1} \ldots j_{I_{2n+3}}, \tag{5.66}$$

whilst the reciprocal relations are given by (5.51) and (5.52). In defining (5.65) and (5.66) we have restricted ourselves to $j = 1$ for fermions in (5.49) and $j = 1$ for bosons in (5.50).

In order to study the consequences of this restriction we consider the reciprocal relations (5.51) and (5.52). From these formulae it follows that $j_{I_1} \ldots j_{I_{2r+1}}$ is correlated with $f_{1,l}, \ldots f_{2r+1,l}$ and $j_{I_1} \ldots j_{I_{2r}}$ is correlated with $b_{2,k}, \ldots b_{2r,k}$. So if we suppress $f_{2r+1,l}$, $r = 2, \ldots \infty$, and $b_{2r,k}$, $r = 2, \ldots \infty$, we are not able to correctly express higher order products of the j sources.

On the other hand, owing to the chain rule such higher order products appear in the transformation formulae if we insist on the full expansion in (5.65) and (5.66). From this we conclude that full expansion in (5.65) and (5.66) and restriction to $b_{2,k}$ and $f_{3,l}$ are not compatible. Therefore, if we want to exclusively study the $(b_{2,k}, f_{3,l})$ dynamics we are forced to start with the lowest order of the expansions (5.65) and (5.66), i.e., with the corresponding hard core states. The dressing property of these states then solely appears in the reciprocal relations (5.51) and (5.52). Thus we define the hard core sources (without using new letters)

$$b_{2,k} := C_{2,k}^{I_1 I_2} j_{I_1} j_{I_2}, \tag{5.67}$$

$$f_{3,l} := C_{3,l}^{I_1 I_2 I_3} j_{I_1} j_{I_2} j_{I_3}, \tag{5.68}$$

and assume that the original functional state $|\mathcal{F}(j; a)\rangle$, which is an eigenstate of equation (5.1), is transformed into a state $|\mathcal{G}(b, f; a)\rangle$, which is defined by the invariance relation $|\mathcal{F}(j; a)\rangle = |\mathcal{G}(b, f; a)\rangle$. Thus equation (5.1) can equivalently be written as

$$E_{p0}|\mathcal{G}(b, f; p)\rangle = \mathcal{H}_{\mathrm{F}}(j, \partial)|\mathcal{G}(b, f; p)\rangle, \tag{5.69}$$

and the chain rule now reads

$$\begin{aligned} \partial_I |\mathcal{G}\rangle &= \left[(\partial_I b_k)\partial_k^b + (\partial_I f_l)\partial_l^f \right] |\mathcal{G}\rangle \\ &= \left[2C_{2,k}^{IK} j_K \partial_k^b + 3C_{3,l}^{IK_1 K_2} j_{K_1} j_{K_2} \partial_l^f \right] |\mathcal{G}\rangle \end{aligned} \tag{5.70}$$

with $b_{2,k} \equiv b_k$, $f_{3,l} \equiv f_l$ and $\partial_{2,k}^b \equiv \partial_k^b$, $\partial_{3,l}^f \equiv \partial_l^f$. By repeated application of this chain rule all derivatives $\partial_I \equiv \delta/\delta j_I$ of $\mathcal{H}_{\mathrm{F}}$ can be expressed by ∂^b and ∂^f derivatives together with products of sources j_I. The latter can be replaced by linear combinations of b and f sources if one substitutes for these products the relations (5.51) and (5.52). If we restrict ourselves to $b_{2,k}$ and $f_{3,l}$ sources, these relations can be expressed by

$$j_{I_1} \ldots j_{I_{2r+1}} = \sum_l R^{3,l}_{I_1 \ldots I_{2r+1}} f_l, \tag{5.71}$$

$$j_{I_1} \ldots j_{I_{2r}} = \sum_k R^{2,k}_{I_1 \ldots I_{2r}} b_k, \tag{5.72}$$

which means that we exclude from our calculation any admixture of hard core states (and their dressing) with fermion number $n \geq 4$. Furthermore, from the beginning we have suppressed all contributions of the elementary subfermions with fermion number $n = 1$, because due to the decoupling theorem these heavy particles do not contribute to the low energy dynamics.

The corresponding calculations are straightforward, so we can directly give the result. By application of the chain rule and substitution of (5.71) and (5.72) the following equation results from (5.69):

$$E_{p_0}|\mathcal{G}(b, f; p)\rangle = \mathcal{H}_C(b, f, \partial^b, \partial^f)|\mathcal{G}(b, f; p)\rangle \tag{5.73}$$

with the functional Hamiltonian operator

$$
\begin{aligned}
\mathcal{H}_C(&b, f, \partial^b, \partial^f) \\
= \ & 2K_{I_1 I_2} C^{2,k'}_{I_2 I_3} R^{2,k}_{I_1 I_3} b_k \partial^b_{k'} \\
&+ 3K_{I_1 I_2} C^{3,\ell'}_{I_2 K_1 K_2} R^{3,\ell}_{I_1 K_1 K_2} f_\ell \partial^f_{\ell'} \\
&+ W_{I_4 I_1 I_2 I_3} \Big[-6 F_{I_3 K} C^{2,k'}_{I_1 I_2} R^{2,k}_{I_4 K} b_k \partial^b_{k'} \\
&\quad - 18 F_{I_3 K_1} C^{3,\ell'}_{I_1 I_2 K_2} R^{3,\ell}_{I_4 K_1 K_2} f_\ell \partial^f_{\ell'} \\
&\quad + 12 C^{2,k'_1}_{I_1 I_2} C^{2,k'_2}_{I_3 K} R^{2,k}_{I_4 K} b_k \partial^b_{k'_1} \partial^b_{k'_2} \\
&\quad + 12 F_{I_3 K_1} C^{2,k'_1}_{I_1 K_2} C^{2,k'_2}_{I_2 K_3} R^{2,k}_{I_4 K_1 K_2 K_3} b_k \partial^b_{k'_1} \partial^b_{k'_2} \\
&\quad - 8 C^{2,k'_1}_{I_1 K_1} C^{2,k'_2}_{I_2 K_2} C^{2,k'_3}_{I_3 K_3} R^{2,k}_{I_4 K_1 K_2 K_3} b_k \partial^b_{k'_1} \partial^b_{k'_2} \partial^b_{k'_3} \\
&\quad + 27 C^{3,\ell'_1}_{I_1 K_1 K_2} C^{3,\ell'_2}_{I_2 K_3 K_4} C^{3,\ell'_3}_{I_3 K_5 K_6} R^{3,\ell}_{I_4 K_1 K_2 K_3 K_4 K_5 K_6} f_\ell \partial^f_{\ell'_1} \partial^f_{\ell'_2} \partial^f_{\ell'_3} \\
&\quad + 36 F_{I_3 K_1} C^{3,\ell'}_{I_2 K_3 K_4} C^{2,k'}_{I_1 K_2} R^{3,\ell}_{I_4 K_1 K_2 K_3 K_4} f_\ell \partial^b_{k'} \partial^f_{\ell'} \\
&\quad + \left(18 C^{3,\ell'}_{I_3 K_1 K_2} C^{2,k}_{I_1 I_2} + 36 C^{3,\ell'}_{I_2 I_3 K_2} C^{2,k'}_{I_1 K_1} \right) R^{3,\ell}_{I_4 K_1 K_2} f_\ell \partial^b_{k'} \partial^f_{\ell'} \\
&\quad + 27 F_{I_3 K_1} C^{3,\ell'_1}_{I_1 K_2 K_3} C^{3,\ell'_2}_{I_2 K_4 K_5} R^{2,k}_{I_4 K_1 K_2 K_3 K_4 K_5} b_k \partial^f_{\ell'_1} \partial^f_{\ell'_2} \\
&\quad - 36 C^{3,\ell'_1}_{I_1 I_2 K_1} C^{3,\ell'_2}_{I_3 K_2 K_3} R^{2,k}_{I_4 K_1 K_2 K_3} b_k \partial^f_{\ell'_1} \partial^f_{\ell'_2} \\
&\quad + 54 C^{2,k'}_{I_1 K_1} C^{3,\ell'_1}_{I_2 K_2 K_3} C^{3,\ell'_2}_{I_3 K_4 K_5} R^{2,k}_{I_4 K_1 K_2 K_3 K_4 K_5} b_k \partial^b_{k'} \partial^f_{\ell'_1} \partial^f_{\ell'_2} \\
&\quad - 36 C^{2,k'_1}_{I_1 K_1} C^{2,k'_2}_{I_2 K_2} C^{3,\ell'}_{I_3 K_3 K_4} R^{3,\ell}_{I_4 K_1 K_2 K_3 K_4} f_\ell \partial^b_{k'_1} \partial^b_{k'_2} \partial^f_{\ell'} \\
&\quad - \left(3 F_{I_1 K_1} F_{I_2 K_2} + \frac{1}{4} A_{I_1 K_1} A_{I_2 K_2} \right) \\
&\qquad \times \left(2 C^{2,k'}_{I_3 K_3} R^{2,k}_{I_4 K_1 K_2 K_3} b_k \partial^b_{k'} + 3 C^{3,\ell'}_{I_3 K_3 K_4} R^{3,\ell}_{I_4 K_1 K_2 K_3 K_4} f_\ell \partial^f_{\ell'} \right) \\
&\quad + \left(F_{I_1 K_1} F_{I_2 K_2} + \frac{1}{4} A_{I_1 K_1} A_{I_2 K_2} \right) F_{I_3 K_3} R^{2,k}_{I_4 K_1 K_2 K_3} b_k \Big].
\end{aligned}
$$
(5.74)

We can use the results from Stumpf [Stu 85a,b,c], [Stu 86a,b] and Chapter 7 in order to draw some general conclusions.

First of all, the various terms in (5.74) allow a distinct physical interpretation, which we will explain in more detail later on. In a preliminary way one can say the following: apart from time derivatives the expressions in (5.73) and (5.74) are the functional version of the field equations. The first two terms of (5.74) contain the kinetic terms of the boson and fermion field equations, respectively. The third and fourth terms yield bosonic and fermionic mass corrections in the field equations. The next three terms represent the (non-abelian) bosonic self-coupling, the eighth term a fermionic self-coupling. The following terms represent the coupling of bosons and fermions; we expect them to contain the coupling of the boson fields to the fermionic current in the field equations as well as the covariant coupling of fermions to bosons. The last two terms of (5.74) are quantization terms which contain the subfermion anticommutator and propagator.

In order to justify these assumptions we divide up $\mathcal{H}_C$ into a physical part $\mathcal{H}_C^{\text{ph}}$ and a rest term $\mathcal{H}_r$ which contains residual forces, and we show in Chapter 8 that the latter term can be neglected in the low energy limit. Chapters 7 and 8 are dedicated to the evaluation of the physical content of (5.73) and (5.74).

6 Generalized de Broglie–Bargmann–Wigner Equations

6.1 Introduction

The aim of our investigations is the derivation of a composite particle dynamics by means of Weak Mapping. As this procedure is referred to corresponding bound states we have to define and to calculate them in the framework of Functional Quantum Theory.

According to Section 5.1 there are two possibilities of formulating a composite particle theory: either in terms of hard core bound states or in terms of dressed composite particle states. At least in lowest order the dressing of hard core states is indispensable for obtaining a physically meaningful effective dynamics of composite particles. But for dressing the knowledge of the corresponding hard core states is required. Thus in this chapter we concentrate on the calculation of hard core bound states.

We will discuss composite particle states which correspond to vector bosons and which are generated by the fusion of two subfermions. By means of this example the general technique of composite particle state calculations will be illustrated. After this discussion we will give a short review of hard core calulations for three- and four-subfermion bound states which are to represent leptons and quarks or gravitons, respectively.

For decades in quantum field theory the calculation of bound states was dominated by the use of Bethe–Salpeter equations, see Salpeter and Bethe [Sal 51]. However, with increasing familiarity with this type of equations, strange and unphysical properties of their solutions were discovered, see, *e.g.*, Itzykson and Zuber [Itz 80]. Bethe–Salpeter equations were derived within the covariant formalism which has inherent defects as was discussed in Section 3.7. As we work in the Algebraic Schrödinger Representation (*cf.* Chapter 4) the (hard core) bound state equations (6.1) are of another type than Bethe–Salpeter equations. Structurally they can be considered as the energy representation of generalized de Broglie equations.

Indeed, beyond quantum field theory there was a continuous interest in the discussion of the original de Broglie equations, which are usually referred to as Bargmann–Wigner equations. A review concerning this topic can be found in a paper by Pfister, Rosa and Stumpf [Pfi 89]. It should be noted that de Broglie–Bargmann–Wigner equations are also primarily covariant equations. But in contrast to the Bethe–Salpeter equations, de Broglie–Bargmann–Wigner equations possess an energy representation and admit in their covariant generalized form the transition from the multi-time version to equations for equal times. Obviously this difference is responsible for the fact that in the latter case the covariance does not lead to difficulties.

In this chapter we describe the general way of deriving exact bound state solutions of equation (6.1) for the case of the subfermion numbers $n = 2$. The explicit evaluation of the corresponding wave functions is partly postponed to subsequent chapters, where these functions are used for Weak Mapping calculations. In deriving solutions for $n = 2$ we use as an intermediate step covariant equations, but the close connection of these equations with the energy representation of corresponding single time equations is explicitly demonstrated.

Furthermore, we intend to derive three-subfermion states, which are to be interpreted as quark and lepton states of the standard model, as well as four-fermion graviton states. Concerning these three- and four-particle states, we have to make some qualifying statements.

Exact solutions of three-body, four-body equations, *etc.*, can only be given for special choices of interactions, see, *e.g.*, Gerjuoy and Adhikari [Ger 85]. In general exact solutions for such systems cannot be achieved. This holds *a fortiori* for the calculations of dressed particle states. Therefore also for our generalized de Broglie–Bargmann–Wigner equations one can expect to obtain solutions only by suitable approximations. This implies the fundamental question: does the inability to derive exact solutions of generalized de Broglie–Bargmann–Wigner equations prevent the application of the Weak Mapping theory?

The answer is clearly negative. The exact Weak Mapping theorems describe an invertible mapping between the respective fundamental and effective theories, see Chapter 5. Thus as long as one works with these exact theorems nothing of the original dynamics is lost if one performs the mapping procedure with wave functions which are not exact solutions of generalized de Broglie–Bargmann–Wigner equations.

Therefore these equations are more or less to be considered as a guideline for the derivation of wave functions with correct relativistic transformation properties and optimal adaption to the physical spectrum. We will demonstrate our handling of the generalized de Broglie–Bargmann–Wigner equations for three- and four-particle bound states in Sections 6.5 and 6.6.

The calculations for vector bosons were performed by Pfister, Rosa and Stumpf [Pfi 89], for vector and scalar bosons by Grimm [Gri 94d,95] and for spin 2 bosons by Stumpf, Borne and Kaus [Stu 93b], [Kaus 94]. Three-subfermion bound states were treated by Pfister [Pfi 94,95c].

As we will calculate massless vector bosons, we have to make a comment upon these states: the so called Weinberg–Witten theorem [Wein 80] excludes massless charged vector bosons as being composite. However, this statement does not apply to our case, because its presuppositions are too restrictive. Hence from this theorem no contradiction can be established to our approach.

Finally, we mention that the solution method for (6.1) via covariant equations is not imperative. The definition of the hard core equation (6.1) and a direct integration method for $n = 2$ was proposed by Stumpf [Stu 83b] and further investigated by Hailer [Hai 85], Hornung [Hor 86], and for $n = 3$ by Lauxmann [Lau 86]. For an illustrative treatment of bound states see also Grosser [Gro 87]. As the extension to $n = 4$ has not been performed so far we will not discuss this method.

Furthermore we mention, that for abbreviation we work with 'natural units' from now on, *i.e.*, we set $\hbar = c = 1$.

6.2 Covariant Bound State Equations

According to Section 5.2 we define hard core bound states for fermion numbers $n = 2, 3, \ldots$ as solutions of equation (5.3):

$$E_{p0}|\mathcal{F}(j^t; p)\rangle^{\mathrm{d}} = (j^t_{I_1}\widehat{K}_{I_1 I_2}\partial^t_{I_2} - 3\widehat{W}_{I_1 I_2 I_3 I_4}F^t_{I_4 K}j^t_{I_1}j^t_K\partial^t_{I_3}\partial^t_{I_2})|\mathcal{F}(j^t; p)\rangle^{\mathrm{d}}. \tag{6.1}$$

Owing to the algebraic degrees of freedom which are contained in the symbolic indices I, equation (6.1) is rather complicated even for the low fermion numbers $n = 2, 3, 4$. However, these complications can be reduced if (6.1) is related to its covariant counterpart equation. This is a special feature of quantum field theory and it may be used to obtain solutions of (6.1) via the solution of the associated covariant equation

$$K_{I_1 I_2}\partial_{I_2}|\mathcal{F}(j; a)\rangle^{\mathrm{d}} = 3W_{I_1 I_2 I_3 I_4}F_{I_4 K}j_K\partial_{I_3}\partial_{I_2}|\mathcal{F}(j; a)\rangle^{\mathrm{d}}, \tag{6.2}$$

where according to (2.77)–(2.80) and (3.115) there holds

$$K_{I_1 I_2} = \left[D^\mu_{Z_1 Z_2}\partial_\mu(x_1) - m_{Z_1 Z_2}\right]\delta(x_1 - x_2), \tag{6.3}$$

$$W_{I_1 I_2 I_3 I_4} = \widehat{U}_{Z_1[Z_2 Z_3 Z_4]}\delta(x_1 - x_2)\delta(x_1 - x_3)\delta(x_1 - x_4), \tag{6.4}$$

$$F_{I_1 I_2} = -i\lambda_{i_1}\delta_{i_1 i_2}\gamma^5_{\kappa_1 \kappa_2}\left[(i\gamma^\mu\partial_\mu(x_1) + m_{i_1})C\right]_{\alpha_1 \alpha_2}\Delta(x_1 - x_2, m_{i_1}) \tag{6.5}$$

with the super-index κ defined by (6.7), and where $\Delta(x_1 - x_2, m_{i_1})$ is the Feynman propagator

$$\Delta(x, m) := (2\pi)^{-4}\int \frac{e^{-ipx}}{p^2 - m^2 + i\epsilon}d^4 p.$$

Equation (6.2) is the hard core part of equation (3.92) with the regularization (3.117). We will explicitly discuss the relation between equations (6.2) and (6.1) for the case $n = 2$.

Equation (6.2) admits separate solutions in any fermion number sector. Thus for the bound state functionals we apply the following ansatz:

$$|\mathcal{F}(j; a, n)\rangle^{\mathrm{d}} := \frac{i^n}{n!}\varphi^{(n)}(I_1 \ldots I_n|a)j_{I_1}\ldots j_{I_n}|0\rangle_{\mathrm{F}}. \tag{6.6}$$

The bound state equation (6.2) can now be reduced by a technique developed for the 'covariant' treatment of the anharmonic oscillator by Schuler and Stumpf [Schu 67]. For that purpose we introduce some useful definitions.

First we combine the isospin index A and the superspin index Λ (*cf.* Section 2.6) into one index κ which is given by the following correspondence:

$$\kappa = \begin{cases} 1 & \text{for} \quad \Lambda = 1, A = 1 \\ 2 & \text{for} \quad \Lambda = 1, A = 2 \\ 3 & \text{for} \quad \Lambda = 2, A = 1 \\ 4 & \text{for} \quad \Lambda = 2, A = 2. \end{cases} \tag{6.7}$$

Then we have for the vertex (6.4)

$$W_{I_1 I_2 I_3 I_4} = \lambda_{i_1} B_{i_2 i_3 i_4} V_{\alpha_1 \alpha_2 \alpha_3 \alpha_4}^{\kappa_1 \kappa_2 \kappa_3 \kappa_4} \delta(x_1 - x_2)\delta(x_1 - x_3)\delta(x_1 - x_4), \tag{6.8}$$

where $B_{i_2 i_3 i_4}$ indicates the summation over the auxiliary field indices and

$$V_{\alpha_1 \alpha_2 \alpha_3 \alpha_4}^{\kappa_1 \kappa_2 \kappa_3 \kappa_4} := \frac{g}{2} \sum_{h=1}^{2} \left\{ v_{\alpha_1 \alpha_2}^h (v^h C)_{\alpha_3 \alpha_4} \delta_{\kappa_1 \kappa_2} [\gamma^5(1 - \gamma^0)]_{\kappa_3 \kappa_4} \right\}_{\text{as}[2,3,4]}. \tag{6.9}$$

The Green's function G is defined by $GK\psi = KG\psi = \psi$, the averages over the auxiliary fields are defined by

$$\varphi_{a_1 \ldots a_n}^{(n)} := \sum_{i_1 \ldots i_n} \varphi_{a_1 \ldots a_n}^{(n)}, \tag{6.10}$$

$$F_{a_1 a_2} := \sum_{i_1, i_2} F_{\substack{i_1 i_2 \\ a_1 a_2}}, \tag{6.11}$$

$$G_{a_1 a_2} := \sum_{i_1, i_2} \lambda_{i_1} G_{\substack{i_1 i_2 \\ a_1 a_2}}, \tag{6.12}$$

and the Fourier transforms of $\varphi^{(n)}$, F and G are denoted by $\tilde{\varphi}^{(n)}$, $\tilde{F}$ and $\tilde{G}$ with coordinates q, p instead of x, y.

One easily verifies the translational invariance of the hard core equation (6.2). Therefore $|\mathcal{F}(j; a, k, 2)\rangle^{\text{d}}$ has to satisfy (3.97) with the momentum quantum number k; this is equivalent to

$$\varphi_{Z_1 Z_2}^{(2)}(x_1, x_2 | a, k) = e^{-ik(x_1 + x_2)/2} \chi_{Z_1 Z_2}(x_1 - x_2 | a, k), \tag{6.13}$$

where a is a set of quantum numbers in addition to the four-momentum k. We interpret k as the center of mass momentum of the two-particle state.

The Fourier transform of (6.13) reads

$$\tilde{\varphi}_{Z_1 Z_2}^{(2)}(q_1, q_2 | a, k) = \delta(q_1 + q_2 - k)\tilde{\chi}_{Z_1 Z_2}\left(\frac{q_1 - q_2}{2}\bigg| a, k\right). \tag{6.14}$$

Then for $n = 2$ equation (6.2) can be reduced to a purely algebraic problem.

- **Proposition 6.** : Let $\varphi_{I_1 I_2}^{(2)}$ be the wave function of the hard core functional state (6.6) for $n = 2$. Then $\varphi_{I_1 I_2}^{(2)}$ is a solution of equation (6.2), if it is given by

$$\varphi_{\substack{i_1 i_2 \\ \kappa_1 \kappa_2 \\ \alpha_1 \alpha_2}}^{(2)}(x_1, x_2) \tag{6.15}$$

$$= 3 \int G_{\substack{i_1 j_1 \\ \kappa_1 \nu_1 \\ \alpha_1 \beta_1}}(x_1 - y)\lambda_{j_1} V_{\beta_1 \beta_2 \beta_3 \beta_4}^{\nu_1 \nu_2 \nu_3 \nu_4} \sum_{\substack{j_2 \\ \beta_2 \alpha_2}} F_{\substack{j_2 i_2 \\ \nu_2 \kappa_2 \\ \beta_2 \alpha_2}}(y - x_2)\varphi_{\substack{\nu_3 \nu_4 \\ \beta_3 \beta_4}}^{(2)}(x_1, x_1)d^4y$$

with

$$\varphi^{(2)}_{\substack{\kappa_1\kappa_2 \\ \alpha_1\alpha_2}}(x_1, x_2) = \sum_{i_1,i_2} \varphi^{(2)}_{\substack{i_1 i_2 \\ \kappa_1\kappa_2 \\ \alpha_1\alpha_2}}(x_1, x_2).$$

The Fourier transform of $\varphi^{(2)}_{\substack{\kappa_1\kappa_2 \\ \alpha_1\alpha_2}}(x_1, x_2)$ is determined by

$$\tilde{\varphi}^{(2)}_{\substack{\kappa_1\kappa_2 \\ \alpha_1\alpha_2}}(q_1, q_2) \tag{6.16}$$

$$= -3(2\pi)^4 \tilde{G}_{\substack{\kappa_1\nu_1 \\ \alpha_1\beta_1}}(q_1) V_{\substack{\nu_1\nu_2\nu_3\nu_4 \\ \beta_1\beta_2\beta_3\beta_4}} \tilde{F}_{\substack{\kappa_2\nu_2 \\ \alpha_2\beta_2}}(q_2)\delta(q_1 + q_2 - k)\tilde{\chi}_{\substack{\nu_3\nu_4 \\ \beta_3\beta_4}},$$

where $\tilde{\chi}$ has to satisfy the eigenvalue equation

$$\tilde{\chi}_{\substack{\kappa_1\kappa_2 \\ \alpha_1\alpha_2}} = -3(2\pi)^4 \int d^4\eta \; \tilde{G}_{\substack{\kappa_1\nu_1 \\ \alpha_1\beta_1}}(\eta) V_{\substack{\nu_1\nu_2\nu_3\nu_4 \\ \beta_1\beta_2\beta_3\beta_4}} \tilde{F}_{\substack{\kappa_2\nu_2 \\ \alpha_2\beta_2}}(k - \eta)\tilde{\chi}_{\substack{\nu_3\nu_4 \\ \beta_3\beta_4}}, \tag{6.17}$$

and the eigenvalue k is the four-momentum of the state.

Proof: We project equation (6.2) for $n = 2$ into coordinate space and apply the Green's function G. This gives (6.15), where $\varphi^{(2)}_{I_1 I_2}$ has to be antisymmetric in I_1, I_2. Averaging (6.15) over the auxiliary indices and Fourier transforming the resulting equation we obtain

$$\tilde{\varphi}^{(2)}_{\substack{\kappa_1\kappa_2 \\ \alpha_1\alpha_2}}(q_1, q_2) = -3(2\pi)^4 \int d^4u \; \tilde{G}_{\substack{\kappa_1\nu_1 \\ \alpha_1\beta_1}}(q_1) V_{\substack{\nu_1\nu_2\nu_3\nu_4 \\ \beta_1\beta_2\beta_3\beta_4}} \tilde{F}_{\substack{\kappa_2\nu_2 \\ \alpha_2\beta_2}}(q_2)\tilde{\varphi}^{(2)}_{\substack{\nu_3\nu_4 \\ \beta_3\beta_4}}(q_1+q_2-u, u). \tag{6.18}$$

We define the contracted function

$$\tilde{\varphi}^{(2)}(q_1|) := \int \tilde{\varphi}^{(2)}(\eta, q_1 - \eta)d^4\eta. \tag{6.19}$$

Contraction of (6.18) according to (6.19) yields

$$\tilde{\varphi}^{(2)}_{\substack{\kappa_1\kappa_2 \\ \alpha_1\alpha_2}}(q_1|) = -3(2\pi)^4 \int d^4\eta \; \tilde{G}_{\substack{\kappa_1\nu_1 \\ \alpha_1\beta_1}}(\eta) V_{\substack{\nu_1\nu_2\nu_3\nu_4 \\ \beta_1\beta_2\beta_3\beta_4}} \tilde{F}_{\substack{\kappa_2\nu_2 \\ \alpha_2\beta_2}}(q_1 - \eta)\tilde{\varphi}^{(2)}_{\substack{\nu_3\nu_4 \\ \beta_3\beta_4}}(q_1|). \tag{6.20}$$

Substitution of the ansatz (6.14) into (6.19) gives

$$\tilde{\varphi}^{(2)}_{\substack{\kappa_1\kappa_2 \\ \alpha_1\alpha_2}}(q_1|) = \delta(q_1 - k)\tilde{\chi}_{\substack{\kappa_1\kappa_2 \\ \alpha_1\alpha_2}} \tag{6.21}$$

with

$$\tilde{\chi}_{\substack{\kappa_1\kappa_2 \\ \alpha_1\alpha_2}} := \int d^4u \tilde{\chi}_{\substack{\kappa_1\kappa_2 \\ \alpha_1\alpha_2}}(u); \tag{6.22}$$

furthermore substituting (6.21) into (6.20) one obtains (6.17), and substitution of (6.21) into (6.18) gives (6.16). $\diamondsuit$

Thus, by means of Proposition 6.1 we have reduced the solution of equation (6.2) for $n = 2$ to the solution of an algebraic equation (6.17). In a similar way we can proceed with (6.2) for $n = 3$ and $n = 4$; for the corresponding calculations we refer to the sections following.

With respect to the eigenvalue k we add that owing to the relativistic invariance of (6.2) it is sufficient to calculate the solutions of (6.2) in the rest frame, whilst the solution for arbitrary k can be obtained by an appropriate Lorentz transformation of the corresponding wave function. The case of vanishing boson mass μ_B needs an extra discussion.

The simultaneous existence of the hard core energy equation (6.1) and its covariant counterpart equation (6.2) raises the question about the compatibility and the relationship of their solutions. For simplicity we discuss this problem for $n = 2$. For $n > 2$ the treatment runs along similar lines but becomes rather complicated.

- **Proposition 6.2:** Let $\varphi^{(2)}_{Z_1 Z_2}(x_1, x_2 | a, k)$ be the wave function of the hard core state functional (6.6) for $n = 2$ and (6.6) be a solution of equation (6.2). Then for

$$\varphi^{(2)}_{Z_1 Z_2}(\mathbf{r}_1, \mathbf{r}_2, t | a, k) := \frac{1}{2} \left(\lim_{t_1 - t_2 \to t + 0} + \lim_{t_1 - t_2 \to t - 0} \right) \varphi^{(2)}_{Z_1 Z_2}(x_1, x_2 | a, k) \qquad (6.23)$$

the corresponding state functional

$$|\mathcal{F}(j^t; a, k, 2)\rangle^{\mathrm{d}} := -\frac{1}{2} \int d^3 r_1 d^3 r_2 \; \varphi^{(2)}_{Z_1 Z_2}(\mathbf{r}_1, \mathbf{r}_2, t | a, k) j^t_{Z_1}(\mathbf{r}_1) j^t_{Z_2}(\mathbf{r}_2) |0\rangle_{\mathrm{F}} \qquad (6.24)$$

is a solution of equation (6.1).

Proof: We consider the function $\varphi^{(2)}_{Z_1 Z_2}(x_1, x_2 | a, k)$. By projection of equation (6.2) into configuration space due to the antisymmetry of $\varphi^{(2)}(I_1, I_2)$ we obtain the following equations:

$$K_{I_1 K_1} \varphi^{(2)}(K_1, I_2) = 3 W_{I_1 K_2 K_3 K_4} F_{K_4 I_2} \varphi^{(2)}(K_2, K_3), \qquad (6.25)$$

$$K_{I_2 K_1} \varphi^{(2)}(I_1, K_1) = -3 W_{I_2 K_2 K_3 K_4} F_{K_4 I_1} \varphi^{(2)}(K_2, K_3). \qquad (6.26)$$

With $I = (Z, x)$, (6.3) and (6.4), equations (6.25) and (6.26) can be rewritten as

$$\left[-\delta_{Z_1 X_1} \partial_0(x_1) + (D^0 D^k)_{Z_1 X_1} \partial_k(x_1) - (D^0 m)_{Z_1 X_1} \right] \varphi^{(2)}_{X_1 Z_2}(x_1, x_2) \qquad (6.27)$$

$$= 3 D^0_{Z_1 X_1} \hat{U}_{X_1 X_2 X_3 X_4} F_{X_4 Z_2}(x_1 - x_2) \varphi^{(2)}_{X_2 X_3}(x_1, x_1)$$

and

$$\left[-\delta_{Z_2 X_1} \partial_0(x_2) + (D^0 D^k)_{Z_2 X_1} \partial_k(x_2) - (D^0 m)_{Z_2 X_1} \right] \varphi^{(2)}_{Z_1 X_1}(x_1, x_2) \qquad (6.28)$$

$$= -3 D^0_{Z_2 X_1} \hat{U}_{X_1 X_2 X_3 X_4} F_{X_4 Z_1}(x_2 - x_1) \varphi^{(2)}_{X_2 X_3}(x_2, x_2).$$

We introduce new coordinates

$$x_s := \frac{1}{2}(x_1 + x_2), \qquad x_r := (x_1 - x_2) \qquad (6.29)$$

with

$$\partial_k(x_1) = \frac{1}{2}\partial_k(x_s) + \partial_k(x_r), \qquad \partial_k(x_2) = \frac{1}{2}\partial_k(x_s) - \partial_k(x_r) \qquad (6.30)$$

and observe that

$$\varphi^{(2)}_{Z_1 Z_2}(x_1, x_2) = \exp\left(-ikx_s\right)\chi_{Z_1 Z_2}(x_r | a, k). \tag{6.31}$$

Substitution of (6.30), (6.31) into (6.27), (6.28) and subsequent addition of the resulting equations yields after reintroduction of $\varphi^{(2)}$

$$
\begin{aligned}
k_0 \varphi^{(2)}_{Z_1 Z_2}(x_1, x_2) &= iD^0_{Z_1 X_1}\left[D^k_{X_1 X_2}\partial_k(x_1) - m_{X_1 X_2}\right]\varphi^{(2)}_{X_2 Z_2}(x_1, x_2) \\
&\quad + iD^0_{Z_2 X_1}\left[D^k_{X_1 X_2}\partial_k(x_2) - m_{X_1 X_2}\right]\varphi^{(2)}_{Z_1 X_2}(x_1, x_2) \\
&\quad - 3i\left[D^0_{Z_1 X_1}\hat{U}_{X_1 X_2 X_3 X_4}F_{X_4 Z_2}(x_1 - x_2)\varphi^{(2)}_{X_2 X_3}(x_1, x_1)\right. \\
&\quad \left. - D^0_{Z_2 X_1}\hat{U}_{X_1 X_2 X_3 X_4}F_{X_4 Z_1}(x_2 - x_1)\varphi^{(2)}_{X_2 X_3}(x_2, x_2)\right].
\end{aligned}
\tag{6.32}
$$

We mention that apart from energy renormalization (6.32) is the diagonal part of the multi-time energy equation (3.120), projected into coordinate space for $n = 2$.

Due to the normal transform (3.85) we have with $\langle a|0\rangle = \delta_{a0}$:

$$\varphi^{(2)}_{Z_1 Z_2}(x_1, x_2 | a, k) = \tau^{(2)}_{Z_1 Z_2}(x_1, x_2 | a, k) - \delta_{0a}\delta_{0k}F_{Z_1 Z_2}(x_1, x_2), \tag{6.33}$$

for the two-point functions, *i.e.*, for $a, k \neq 0$, $\varphi^{(2)}$ and $\tau^{(2)}$ are identical. Now according to Proposition 4.14 the limiting process (6.23) gives an equation which is identical with the coordinate space representation of (6.1) for $n = 2$. $\diamond$

The transition from the covariant representation (6.6) to the energy representation (6.24) of a state $|a, k\rangle$ does not imply any loss of information. This is secured by the following proposition, *cf.* Stumpf [Stu 85d]:

- **Proposition 6.3:** The covariant multi-time solution $\varphi^{(2)}_{Z_1 Z_2}(x_1, x_2 | a, k)$ is completely determined by the single time solution $\varphi^{(2)}_{Z_1 Z_2}(\mathbf{r}_1, \mathbf{r}_2, t | a, k)$.

Proof: We again consider the function $\varphi^{(2)}_{Z_1 Z_2}(x_1, x_2 | a)$. Substitution of (6.30), (6.31) into (6.27), (6.28) and subsequent subtraction of the resulting equations yields with $x_r \equiv (\mathbf{u}, t)$:

$$
\begin{aligned}
-2\partial_0(t)\chi_{Z_1 Z_2}(\mathbf{u}, t | a, k) & \tag{6.34} \\
= \quad -D^0_{Z_1 X_1}&\left[D^j_{X_1 X_2}\left(-\frac{i}{2}k + \partial_j(\mathbf{u})\right) - m_{X_1 X_2}\right]\chi_{X_2 Z_2}(\mathbf{u}, t | a, k) \\
+ D^0_{Z_2 X_1}&\left[D^j_{X_1 X_2}\left(-\frac{i}{2}k - \partial_j(\mathbf{u})\right) - m_{X_1 X_2}\right]\chi_{Z_1 X_2}(\mathbf{u}, t | a, k) \\
+ 3D^0_{Z_1 X_1}&\hat{U}_{X_1 X_2 X_3 X_4}F_{X_4 Z_2}(\mathbf{u}, t)\chi_{X_2 X_3}(0, 0 | a, k) \\
+ 3D^0_{Z_2 X_1}&\hat{U}_{X_1 X_2 X_3 X_4}F_{X_4 Z_1}(-\mathbf{u}, -t)\chi_{X_2 X_3}(0, 0 | a, k) \\
=: \quad H_1(\nabla_\mathbf{u})&\chi(\mathbf{u}, t) + H_2(\mathbf{u}, t)\chi(0, 0)
\end{aligned}
$$

or

$$\chi(\mathbf{u}, t) = -\frac{1}{2} \int_0^t H_1(\nabla_\mathbf{u}) \chi(\mathbf{u}, \tau) d\tau + f(\mathbf{u}, t) \tag{6.35}$$

with

$$f(\mathbf{u}, t) := -\frac{1}{2} \int_0^t H_2(\mathbf{u}, \tau) \chi(0, 0) d\tau + \chi(\mathbf{u}, 0).$$

Equation (6.35) is an integral equation of the Volterra type which allows convergent solutions by iteration for all values of its coupling constant λ, in particular for $\lambda = -\frac{1}{2}$. The iteration solution of (6.35) is completely fixed by the inhomogeneous term $f(\mathbf{u}, t)$. For $t = 0$ this term is given by $f(\mathbf{u}, 0) \equiv \chi(\mathbf{u}, 0)$, i.e., the single time solution.   ◇

Finally, it should be noted that although we will concentrate on hard core bound states, equations (6.1) and (6.2) are not restricted to the calculation of the latter, but can also be used for the calculation of hard core scattering states. For $n = 2$ such scattering states were calculated by Pfister, Rosa and Stumpf in [Pfi 89]. For brevity we do not discuss this topic, but only mention that the scattering states are essential for the construction of complete sets of functions which are applied in the Weak Mapping procedure.

6.3   Vector Boson States

The simplest bound states are two-fermion states. As it was shown in the preceding section these states, defined as solutions of (6.1) or (6.2), can be exactly calculated. With respect to their physical interpretation we assume that spin 1 two-fermion bound states are to be identified with gauge vector bosons of the weak and electromagnetic interactions before symmetry breaking. Hence in the following we will concentrate on the discussion of these spin 1 states by evaluating (6.17). After a slight rearrangement we obtain for this equation

$$\tilde{\chi}^{\kappa_1 \kappa_2}_{\alpha_1 \alpha_2}(k) = 3(2\pi)^4 \int d^4 q \, \tilde{G}^{\kappa_1 \rho_1}_{\alpha_1 \beta_1}(k - q) V_{\beta_1 \beta_2 \beta_3 \beta_4}^{\rho_1 \rho_2 \rho_3 \rho_4} \tilde{F}^{\rho_4 \kappa_2}_{\beta_4 \alpha_2}(-q) \tilde{\chi}^{\rho_2 \rho_3}_{\beta_2 \beta_3}(k). \tag{6.36}$$

Owing to Proposition 6.1 for the solution of (6.15) it is sufficient to solve (6.36). To do this we expand $\tilde{\chi}$ on the superspinor–isospinor algebra (index κ) and on the spinor algebra (index α). As both algebras have the same dimension we can use the same basis elements. Denoting the symmetric basis elements of the Dirac algebra by S^b and the antisymmetric elements by T^a we may represent $\tilde{\chi}$ by

$$\tilde{\chi}^{\kappa_1 \kappa_2}_{\alpha_1 \alpha_2} = \sum_{a,b} m^{ab} T^a_{\kappa_1 \kappa_2} S^b_{\alpha_1 \alpha_2} + \sum_{a,b} n^{ab} S^b_{\kappa_1 \kappa_2} T^a_{\alpha_1 \alpha_2} \tag{6.37}$$

with coefficients m^{ab} and n^{ab} in order to obtain a completely antisymmetric $\tilde{\chi}$. With the explicit form of $\tilde{G}$, $\tilde{F}$ and V, cf. (6.46), (6.47) and (6.9), one can immediately deduce from (6.36) that linear independent solutions of (6.36) are given by the following submanifolds of (6.37):

$$\tilde{\chi}^{(a,1)}_{\substack{\kappa_1 \kappa_2 \\ \alpha_1 \alpha_2}} := \sum_b m^b T^a_{\kappa_1 \kappa_2} S^b_{\alpha_1 \alpha_2} \tag{6.38}$$

and

$$\tilde{\chi}^{(2)}_{\substack{\kappa_1 \kappa_2 \\ \alpha_1 \alpha_2}} := \sum_{a,b} n^{ab} S^b_{\kappa_1 \kappa_2} T^a_{\alpha_1 \alpha_2} \tag{6.39}$$

with $a = 1, \ldots 6$ and $b = 1, \ldots 10$. By substitution of (6.38) into (6.36) it follows that the solutions (6.38) are degenerate with respect to any element of the superspinor-isospinor algebra. Differences appear with respect to the spinor algebra: whilst (6.38) describes vector boson states, solutions (6.39) describe scalar boson states (see Section 6.4). In the following we will discuss solutions of the type (6.38), *i.e.*, vector boson states.

Choosing the symmetric basis elements $S^b \in \{\gamma^\mu C, \Sigma^{\mu\nu} C\}$ of the Dirac algebra we can write the vector boson ansatz (6.38) explicitly as

$$\tilde{\chi}^{(a,1)}_{\substack{\kappa_1 \kappa_2 \\ \alpha_1 \alpha_2}} = T^a_{\kappa_1 \kappa_2} \left[A^\mu (\gamma_\mu C)_{\alpha_1 \alpha_2} + F^{\mu\nu} (\Sigma_{\mu\nu} C)_{\alpha_1 \alpha_2} \right] \tag{6.40}$$

with $A^\mu = A^\mu(k)$, $F^{\mu\nu} = F^{\mu\nu}(k)$ and obtain the wave functions for vector bosons by

- **Proposition 6.4:** Let $\varphi^{(2)}_{I_1 I_2}$ be a solution of (6.15). Then $\varphi^{(2)}_{I_1 I_2}$ describes a vector boson with momentum k, if it is given by

$$\varphi^{(2)}_{\substack{i_1 i_2 \\ \kappa_1 \kappa_2 \\ \alpha_1 \alpha_2}} (x_1, x_2) = T^a_{\kappa_1 \kappa_2} \exp\left[-i\frac{k}{2}(x_1 + x_2) \right] A_\mu \chi^\mu_{\substack{i_1 i_2 \\ \alpha_1 \alpha_2}} (x_1 - x_2 | k) \tag{6.41}$$

with the relative wave function

$$\chi^\mu_{\substack{i_1 i_2 \\ \alpha_1 \alpha_2}} (x|k) := \frac{2ig}{(2\pi)^4} \lambda_{i_1} \lambda_{i_2} \int d^4 p \, \exp\left(-ipx \right)$$
$$\times \left[S_F\left(p + \frac{k}{2}, m_{i_1} \right) \gamma^\mu S_F\left(p - \frac{k}{2}, m_{i_2} \right) C \right]_{\alpha_1 \alpha_2} \tag{6.42}$$

(no summation over i_1, i_2) and

$$S_F(p, m) := \frac{\gamma^\mu p_\mu + m}{p^2 - m^2 + i\epsilon}. \tag{6.43}$$

Proof: The general form of the solution in coordinate space is given by (6.13). By introducing center of mass coordinates (6.29) one verifies that (6.13) is a general solution of (6.15). Hence for the determination of $\varphi^{(2)}$ in (6.13) we have to calculate only $\chi_{Z_1 Z_2}(x_1 - x_2)$. It is convenient to consider instead of $\chi_{Z_1 Z_2}$ its Fourier transform $\tilde{\chi}_{Z_1 Z_2}$. We transform (6.15) into momentum space, introduce center of mass coordinates $q = q_1 + q_2$, $p = \frac{1}{2}(q_1 - q_2)$ and substitute (6.14) into the left hand side of the Fourier transform of (6.15). This gives

$$\delta(q-k)\tilde{\chi}_{\substack{i_1i_2\\\kappa_1\kappa_2\\\alpha_1\alpha_2}}(p|k) \tag{6.44}$$

$$= 3(2\pi)^4\,\tilde{G}_{\substack{i_1j_1\\\kappa_1\rho_1\\\alpha_1\beta_1}}(q_1)\lambda_{j_1}V_{\rho_1\rho_2\rho_3\rho_4}^{\beta_1\beta_2\beta_3\beta_4}\sum_{j_2}\tilde{F}_{\substack{j_2i_2\\\rho_2\kappa_2\\\beta_2\alpha_2}}(-q_2)\delta(q-k)\tilde{\chi}_{\substack{\rho_3\rho_4\\\beta_3\beta_4}}(k),$$

where $\tilde{\chi}$ is defined by (6.22). Then owing to the δ distribution on the right hand side of (6.44) we can replace $q_1 = \frac{1}{2}q+p$, $q_2 = \frac{1}{2}q-p$ by $q_1 = \frac{1}{2}k+p$, $q_2 = \frac{1}{2}k-p$. Subsequent integration over q yields

$$\tilde{\chi}_{\substack{i_1i_2\\\kappa_1\kappa_2\\\alpha_1\alpha_2}}(p|k) = 3(2\pi)^4\,\tilde{G}_{\substack{i_1j_1\\\kappa_1\rho_1\\\alpha_1\beta_1}}(\frac{k}{2}+p)\lambda_{j_1}V_{\rho_1\rho_2\rho_3\rho_4}^{\beta_1\beta_2\beta_3\beta_4}\sum_{j_2}\tilde{F}_{\substack{j_2i_2\\\rho_2\kappa_3\\\beta_2\alpha_3}}(p-\frac{k}{2})\tilde{\chi}_{\substack{\rho_3\rho_4\\\beta_3\beta_4}}(k). \tag{6.45}$$

In the latter equation we substitute V given by (6.9), $\tilde{F}$ as the Fourier transform of (6.5), and $\tilde{\chi}$ given by (6.40). Furthermore, there holds $\tilde{F} = -i\lambda\gamma^5\tilde{G}C$. Direct evaluation of (6.45) then leads to the representation (6.41), (6.42). $\diamond$

A detailed calculation with respect to this proposition and the subsequent discussion of eigenvalue equations is contained in the paper by Sand [Sand 91].

The above given representation of the boson wave functions holds for all k, as the calculation was performed in a strictly relativistic invariant way. The integral in (6.42) can be evaluated by standard methods and leads, of course, to a singular behavior on the light cone. Performing the equal time limit one explicitly sees that $\varphi^{(2)}$ vanishes for large relative distances $\mathbf{r}_1 - \mathbf{r}_2$, so (6.41), (6.42) really describe bound states; we will discuss these functions in more detail in Chapter 8. Furthermore, one can verify the antisymmetry of (6.41) in the indices I_1, I_2. Finally, we mention that (6.41) is only determined up to a normalization constant.

To discuss the role of A^μ and $F^{\mu\nu}$ in the expansion (6.40) we observe that on the right hand side of (6.45) the term $F^{\mu\nu}(\Sigma_{\mu\nu})C$ drops out owing to $v^h\Sigma_{\mu\nu}v^h = 0$. This means: $F^{\mu\nu}$ cannot be an arbitrary parameter for an exact solution of (6.15). To demonstrate this we consider the secular equation (6.36). On both sides of (6.36) we substitute (6.40) and introduce the explicit representations

$$\tilde{G}_{\substack{\alpha_1\alpha_2\\\kappa_1\kappa_2}}(p) = \delta_{\kappa_1\kappa_2}\big[R_\mu(p)\gamma^\mu_{\alpha_1\alpha_2} + S(p)\delta_{\alpha_1\alpha_2}\big], \tag{6.46}$$

$$\tilde{F}_{\substack{\alpha_1\alpha_2\\\kappa_1\kappa_2}}(p) = -i\gamma^5_{\kappa_1\kappa_2}\big[R_\mu(p)(\gamma^\mu C)_{\alpha_1\alpha_2} + S(p)C_{\alpha_1\alpha_2}\big] \tag{6.47}$$

with

$$R_\mu(p) := \sum_{i=0}^{2}\lambda_i\Delta(p,m_i)p_\mu, \tag{6.48}$$

$$S(p) := \sum_{i=0}^{2}\lambda_i m_i\Delta(p,m_i), \tag{6.49}$$

which follow from (6.9), (6.11) and (6.12). Next step we multiply (6.36) one after the other by $(S^b)^+$ for all S^b of the symmetric set, form the trace and obtain the equations

$$A^\mu = 2i(2\pi)^4 g \int d^4q \big\{ [S(k-q)S(q) + R^\rho(k-q)R_\rho(q)]A^\mu \tag{6.50}$$
$$-2R^\mu(k-q)R_\rho(q)A^\rho \big\}$$

and

$$F^{\mu\nu} = 4(2\pi)^4 g \int d^4q \, S(q) \, [R^\nu(k-q)A^\mu]_{\mathrm{as}[\mu,\nu]} . \tag{6.51}$$

From equations (6.50) a secular equation for vector bosons can be derived. This secular equation constitutes a relation between subfermion coupling constant g, the auxiliary subfermion masses m_1, m_2 and m_3 and the four-momentum k of the vector boson. Owing to the strict relativistic invariance of our calculation this secular equation must respect the relation $k^2 = \mu_B^2$. For brevity we do not specify the corresponding expressions but refer to [Sand 91].

We remark, however, that the effective *dynamical* equations for composite vector bosons contain a mass renormalization term. Thus in these effective equations a boson mass $\mu_B' = \mu_B + \Delta$ occurs; massless vector bosons are then described by the condition $\mu_B' = 0$. For the moment we do not further discuss these subtleties, which are connected with the determination of the masses and coupling constants of the effective theory and refer to Chapter 8.

As far as A^μ and $F^{\mu\nu}$ are concerned, equations (6.51) show that the 'field strengths' $F^{\mu\nu}$ are determined by the 'vector potentials' A^μ, whilst the A^μ themselves are eigenvectors of the homogeneous eigenvalue equations (6.50) for the corresponding eigenvalue. Furthermore, it can be shown that equations (6.50) are compatible with the transversality condition $k^\rho A_\rho = 0$ and from (6.51) it follows that $F^{\mu\nu}$ must have the form $c(k)k^{[\mu}A^{\nu]}$. But attention must be paid to these relations holding only for (6.40), *i.e.*, for $\chi^\mu(0|k)$. In the full exact solutions (6.41) only the vector potentials A^μ occur.

With (6.42) we can discuss the transformation properties of $\chi^\mu(x|k)$. If $\varphi^{(2)}_{I_1 I_2}$ undergoes an orthochronous Lorentz transformation, then according to Proposition 3.1 it transforms as a direct product of Dirac spinors. Therefore $\chi^\mu(x|k)$ also transforms in this way and we have, see (3.6),

$$\chi^\mu_{\substack{i_1 i_2 \\ \alpha_1 \alpha_2}} (x'|k') = S_{\alpha_1\beta_1}(\Lambda)\chi^\mu_{\substack{i_1 i_2 \\ \beta_1\beta_2}} (x|k)S_{\alpha_2\beta_2}(\Lambda) \tag{6.52}$$

with $x' = \Lambda x$ and $k' = \Lambda k$, where Λ is the ordinary Lorentz matrix and $S(\Lambda)$ the representation of the Lorentz transformation in classical spinor space.

Defining

$$\chi^\mu_{\substack{i_1 i_2 \\ \alpha_1 \alpha_2}} (x|k)' := \chi^\mu_{\substack{i_1 i_2 \\ \alpha_1 \alpha_2}} (x'|k') \tag{6.53}$$

for $x' = \Lambda x$ and $k' = \Lambda k$, with (6.42) one easily verifies that (6.52) can be rewritten as

$$\chi^\mu_{\substack{i_1 i_2 \\ \alpha_1 \alpha_2}} (x|k)' = \Lambda^\mu_\rho \chi^\rho_{\substack{i_1 i_2 \\ \alpha_1 \alpha_2}} (x|k). \tag{6.54}$$

Equation (6.54) shows that $\chi^\mu(x|k)$ indeed transforms as a vector boson. Note that $A^\mu k_\mu = 0$ implies that A transforms as a four-vector, which exhibits the form invariance property

$$A_\mu \chi^\mu_{\substack{i_1 i_2 \\ \alpha_1 \alpha_2}} (x|k) = A'_\mu \chi^\mu_{\substack{i_1 i_2 \\ \alpha_1 \alpha_2}} (x|k)' \qquad (6.55)$$

with $A' = \Lambda A$.

The functions (6.41), (6.42) are exact solutions of generalized de Broglie–Bargmann–Wigner equations (6.25), (6.26), the energy representation of which coincides with (6.1). As the energy representation (6.1) is referred to the state description in the equal time limit which is defined by (6.23), this holds also for the special state representation (6.41). Thus the proper vector boson functions are given by the equal time limit of the functions (6.41), (6.42).

But with respect to the subsequent applications of Weak Mapping the exact solutions (6.41) are no suitable starting point. Weak Mapping is conceived to describe the effect of interactions between composite particles. A theoretical consequence of such interactions is that composite particles can be no longer described by the exact solutions of their corresponding eigenvalue equations because the interactions influence and disturb their motion and their internal structure. For instance, this can be expressed by considering the vector potential A^μ and the field strength $F^{\mu\nu}$ as independent new dynamical variables which allow to describe the influence of interactions between vector bosons on their center of mass motion.

Naturally one can also think of the inclusion of inner deformations, but such deformations of the inner structure of composite particles can be covered by the formalism of the corresponding effective theories. Hence the variables which describe the center of mass motion are the most elementary ones which we will use in the following.

We can discover both kinds of variables by identifying terms $k^{[\mu} A^{\nu]}$ in (6.41) with $F^{\mu\nu}$ as a heuristic principle. Already by this identification we lift the relations (6.50), (6.51) in favour of a more flexible state description and introduce a new interpretation of (6.41).

This will be discussed in more detail in Section 8.5, together with a thorough discussion of the resulting vector boson functions. Here we anticipate the general form of these single time functions for symmetric s states. In this case we obtain from the bound state functions (6.41) in the equal time limit:

$$\varphi^{(2)}_{\substack{i_1 i_2 \\ \kappa_1 \kappa_2 \\ \alpha_1 \alpha_2}} (\mathbf{z}, \mathbf{u}, t | a, \mathbf{k}) \qquad (6.56)$$

$$= T^a_{\kappa_1 \kappa_2} e^{-ik_0 t} e^{i\mathbf{k}\mathbf{z}} \left[A_\mu f^A_{i_1 i_2}(\mathbf{u}) (\gamma^\mu C)_{\alpha_1 \alpha_2} + F_{\mu\nu} f^F_{i_1 i_2}(\mathbf{u}) (\Sigma^{\mu\nu} C)_{\alpha_1 \alpha_2} \right]$$

with the new coordinates $\mathbf{z} := (\mathbf{r}_1 + \mathbf{r}_2)/2$ and $\mathbf{u} := \mathbf{r}_1 - \mathbf{r}_2$. In Section 8.5 we will show that f^A and f^F are even functions, i.e., that $f^{A/F}(\mathbf{u}) = f^{A/F}(|\mathbf{u}|)$.

Finally, a physical interpretation of the results is needed. With regard to the spin content of the functions (6.56) we remember that we restricted ourselves to s states with vanishing orbital angular momentum. In this case the spin analysis according to Section 6.4 can easily be performed and it is demonstrated that in temporal gauge, i.e., for $A^0 = 0$, we indeed obtain spin 1 states.

Thus only the superspin–isospin parts $T^a_{\kappa_1 \kappa_2}$ need a further discussion. The full set of antisymmetric T^a matrices explicitly reads

$$\left\{ T^a, a = 1, \ldots 6 \right\} \tag{6.57}$$

$$= \left\{ \begin{pmatrix} \sigma^2 & 0 \\ 0 & \sigma^2 \end{pmatrix}, \begin{pmatrix} \sigma^2 & 0 \\ 0 & -\sigma^2 \end{pmatrix}, \begin{pmatrix} 0 & \mathbb{I} \\ -\mathbb{I} & 0 \end{pmatrix}, \begin{pmatrix} 0 & \sigma^t \\ (-1)^t \sigma^t & 0 \end{pmatrix}, t = 1, 2, 3 \right\}.$$

In Section 6.4 it will be shown that the triplet

$$T_{SU(2)}^t \in \left\{ \begin{pmatrix} 0 & \sigma^t \\ (-1)^t \sigma^t & 0 \end{pmatrix}, \; t = 1, 2, 3 \right\} \tag{6.58}$$

is an isospin 1 triplet with fermion number $f = 0$, whilst

$$T_{U(1)} := \begin{pmatrix} 0 & \mathbb{I} \\ -\mathbb{I} & 0 \end{pmatrix} \tag{6.59}$$

is an isospin 0 representation with fermion number $f = 0$, too.

In Chapter 7 it will be shown that in connection with a Weak Mapping the triplet $T_{SU(2)}^t$ has to be identified with $SU(2)$ gauge vector boson states and the $T_{U(1)}$ state gives the electro–weak $U(1)$ gauge boson. The states (6.58) and (6.59) lead to the effective $SU(2) \times U(1)$ electro–weak dynamics, whilst the first two states in (6.57) constitute eigenstates with $f = \pm 2$ and isospin 0 and completely decouple from this dynamics.

6.4   Bound State Quantum Numbers

Regardless of whether we consider hard core bound states or dressed particle (bound) states, these states must have definite quantum numbers in order to admit a unique classification and characterization of the corresponding effective fields which are subject to the effective dynamics. Hence in accordance with this condition we postulate that the functional composite particle states, for instance the vector boson hard core states (6.41), the three-particle states (6.93) and the four-subfermion graviton states (6.124) are eigenstates of the set of commuting Poincaré generators (3.97)–(3.100) and in addition, of course, have definite quantum numbers with respect to the discrete symmetry operations and global gauge groups.

For complete dressed particle states this postulate is in accordance with the corresponding defining equations, as the functional equations are invariant under the above mentioned symmetry groups. With respect to hard core states we observe that any term in the basic functional equation is invariant for itself. Hence the hard core equation (6.2) admits the full invariance group of the original functional equation and hence also hard core states can be classified in the postulated way, provided that the propagator is invariant.

In this context we emphasize that it is convenient to discuss the covariant states, although for Weak Mapping we only need the equal time limits of these states. The use of covariant states allows to define the equal time limits in any frame of reference and hence is a prerequisite for a correct transformation property of the Algebraic Schrödinger Representation.

According to the results of Section 3.3 the eigenvalue equations for corresponding matrix elements are given by suitable sums of one-particle representations of the associated generators (see Proposition 3.5). We start our investigations by evaluating these equations for two-particle bound states.

i) **Momentum conditions**

Projection of (3.97) into configuration space with (6.6) for $n = 2$ yields according to Proposition 3.5:

$$
\begin{aligned}
{}^{(2)}P_\mu\varphi^{(2)}_{Z_1Z_2}(x_1, x_2) &:= \left[i\partial_\mu(x_1) + i\partial_\mu(x_2)\right]\varphi^{(2)}_{Z_1Z_2}(x_1, x_2) \\
&= p_\mu\varphi^{(2)}_{Z_1Z_2}(x_1, x_2).
\end{aligned}
\tag{6.60}
$$

This condition is satisfied by

$$
\varphi^{(2)}_{Z_1Z_2}(x_1, x_2) = \exp\left(ip\frac{x_1 + x_2}{2}\right)\chi_{Z_1Z_2}(x_1 - x_2)
\tag{6.61}
$$

which in addition satisfies

$$
{}^{(2)}P_\mu{}^{(2)}P_\mu\varphi^{(2)}_{Z_1Z_2}(x_1, x_2) = m^2\varphi^{(2)}_{Z_1Z_2}(x_1, x_2),
\tag{6.62}
$$

as for bound states p_μ is on the mass shell. The generalization to n-particle states is evident.

ii) **Angular momentum conditions**

Without loss of generality we consider the massive bound state $\varphi^{(2)}_{Z_1Z_2}(x_1, x_2)$ with $m \neq 0$ in its rest system $p_\mu = (p_0, 0, 0, 0)$. In this case (and for momentum eigenvectors) the Pauli–Lubanski spin vector (2.116) reads

$$
W_\mu = \frac{1}{2p_0}\varepsilon_{\mu\nu\rho0}M^{\nu\rho}P^0,
\tag{6.63}
$$

and in this expression the representation of the generators P^μ and $M^{\mu\nu}$ depends on the particle number or the dimension of the coordinate space, respectively.

With $\varepsilon_{ijk0} = \varepsilon^{0ijk} = -\varepsilon_{ijk}$ we obtain

$$
W_0 = 0,
\tag{6.64}
$$

$$
W_i = -\frac{1}{2}\varepsilon_{ijk}M^{jk} = -J_i,
$$

where J_i are the nonrelativistic angular momentum operators. According to Proposition 3.5 the representation of the operator W_μ in the two-particle configuration space reads

$$
{}^{(2)}W_i\varphi^{(2)}(u, z) = -\left({}^{(2)}L^i + {}^{(2)}S^i\right)\varphi^{(2)}(u, z)
\tag{6.65}
$$

with the orbital angular momentum operator

$$
{}^{(2)}L^i\varphi^{(2)}(u, z) = \frac{i}{2}\varepsilon_{ijk}\left[(u_j\partial_k(u) - u_k\partial_j(u)\right]\varphi^{(2)}(u, z)
\tag{6.66}
$$

and the two-particle spin operator

$$^{(2)}S^i\varphi^{(2)}(u,z) = \left(^{(2)}S_1^i + {}^{(2)}S_2^i\right)\varphi^{(2)}(u,z) \tag{6.67}$$

with

$$^{(2)}S_1^i := \frac{1}{2}\Sigma_{\alpha_1\alpha_1'}^i \delta_{\alpha_2\alpha_2'} \qquad \text{etc.,} \tag{6.68}$$

and the Dirac algebra elements Σ^i, where $u = x_1 - x_2$, $\partial(u) = \frac{1}{2}[\partial(x_1) - \partial(x_2)]$ and $z = \frac{1}{2}(x_1 + x_2)$.

Thus the eigenvalue condition (3.100) for $\varphi^{(2)}$ is given by

$$\begin{aligned}
W_3\varphi^{(2)}(u,z) &= \frac{1}{2}\Sigma^3\varphi^{(2)}(u,z) + \frac{1}{2}\varphi^{(2)}(u,z)(\Sigma^3)^T \\
&\quad + i\left[u_1\partial_2(u) - u_2\partial_1(u)\right]\varphi^{(2)}(u,z) \\
&= s_3\varphi^{(2)}(u,z).
\end{aligned} \tag{6.69}$$

In application to $\varphi^{(2)}$ the square of W_i reads in the rest frame in accordance with (6.64)

$$^{(2)}W^2\varphi^{(2)}(x_1,x_2) = J^2\varphi^{(2)}(x_1,x_2) = (J_1^2 + J_2^2 + J_3^2)\varphi^{(2)}(x_1,x_2) = s(s+1)\varphi^{(2)}(x_1,x_2). \tag{6.70}$$

For massless quanta we choose the frame $p_\mu = (p_0, 0, 0, p_0)$ and obtain in this case the same formula (6.69) as in the massive case.

Having derived the eigenvalues of J^2 in the rest system, owing to the relativistic invariance there holds $J^2 \equiv W_\mu W^\mu$, and we have a general statement about the spin values in any frame of reference.

The generalization to three-particle states is obvious. We obtain angular momentum operators

$$^{(3)}J^i = {}^{(3)}S^i + {}^{(3)}L^i = {}^{(3)}S^i + {}^{(3)}L_{u_1}^i + {}^{(3)}L_{u_2}^i, \tag{6.71}$$

where the spin part is given by

$$^{(3)}S^i = {}^{(3)}S_1^i + {}^{(3)}S_2^i + {}^{(3)}S_3^i \tag{6.72}$$

with

$$^{(3)}S_1^i\begin{pmatrix} \alpha_1 & \alpha_2 & \alpha_3 \\ \alpha_1' & \alpha_2' & \alpha_3' \end{pmatrix} := \frac{1}{2}(\Sigma^i)_{\alpha_1\alpha_1'}\delta_{\alpha_2\alpha_2'}\delta_{\alpha_3\alpha_3'} \qquad \text{etc.,} \tag{6.73}$$

and the orbital angular momentum is given by

$$^{(3)}L_{u_r}^i = \frac{1}{i}\,\varepsilon_{ijk}\,u_r^j\frac{\partial}{\partial u_r^k}, \qquad r = 1, 2 \tag{6.74}$$

with u_r, $r = 1, 2$ as the relative coordinates of the three-particle system.

In the two- and the four-particle sector we consider s states without orbital angular momentum. In this case the corresponding symmetry equations reduce to purely algebraic equations for the spin which can be explicitly solved.

Expanding the two-particle bound state functions on the symmetric basis elements $\hat{S}^a$ of the Dirac algebra one obtains the eigenvalue equations of $\varphi^{(2)}$ for $^{(2)}W_3 = {}^{(2)}S_3$ and $(^{(2)}W)^2 = (^{(2)}S)^2 = \sum_{i,h} {}^{(2)}S^i_h {}^{(2)}S^i_h$ in the reduced form

$$\frac{1}{2}\left[\Sigma^3 \hat{S}^a + \hat{S}^a \left(\Sigma^3\right)^T\right]_{\alpha_1\alpha_2} = s_3 \hat{S}^a_{\alpha_1\alpha_2}, \tag{6.75}$$

$$\frac{1}{2}\left[3\hat{S}^a + \Sigma^k \hat{S}^a \left(\Sigma^k\right)^T\right]_{\alpha_1\alpha_2} = s(s+1)\hat{S}^a_{\alpha_1\alpha_2}. \tag{6.76}$$

With $\hat{S}^0 := \gamma^0 C$ and $\hat{S}^k \in \{\gamma^k C, \Sigma^{0k} C, \Sigma^k C\}$ a complete classification of the eigenvectors of massive two-particle states is given by the following table:

	$\hat{S}^0$	$\hat{S}^1 + i\hat{S}^2$	$\hat{S}^1 - i\hat{S}^2$	$\hat{S}^3$
s	0	1	1	1
s_3	0	1	−1	0

There exist three spin 1 'generations', namely either $\hat{S}^k = \gamma^k C$ or $\hat{S}^k = \Sigma^{0k} C$ or $\hat{S}^k = \Sigma^k C$. As we shall see, all these spin 1 representations are needed for the formulation of the dynamics of vector bosons, and by the gauge condition $A_0 = 0$ it is possible to exclude spin 0 parts from the two-particle bound state functions.

With respect to the subsequent discussion it is also of interest to analyze the antisymmetric basis elements C, $\gamma^5 C$ and $\gamma^\mu \gamma^5 C$ of the Dirac algebra. Application of the spin operators (6.75) and (6.76) to these elements yields the following table:

	C	$\gamma^5 C$	$\gamma^0 \gamma^5 C$	$\gamma^5(\gamma^1 + i\gamma^2)C$	$\gamma^5(\gamma^1 - i\gamma^2)C$	$\gamma^5 \gamma^3 C$
s	0	0	0	1	1	1
s_3	0	0	0	1	−1	0

For massless vector boson states the expansion (6.40) can be maintained as was already stated in Section 6.3; for massless scalar states and discrete symmetry operations we refer to [Gri 94].

For the four-particle bound states we observe that we assume them to be composed of two spin 1 states; i.e., they have the structure

$$\varphi^{(4)}_{\alpha_1\alpha_2\alpha_3\alpha_4} = \varphi^{(2)}_{\alpha_1\alpha_2} \varphi^{(2)}_{\alpha_3\alpha_4} \tag{6.77}$$

with respect to the spinor indices.

In this case the symmetry equations can be reduced to corresponding equations for two-particle functions. For s states the corresponding spin eigenvalue equation reads

$$\left[^{(4)}S^k \varphi^{(4)}\right]_{\alpha_1\alpha_2\alpha_3\alpha_4} = \left[^{(2)}S^k \varphi^{(2)}\right]_{\alpha_1\alpha_2} \varphi^{(2)}_{\alpha_3\alpha_4} + \varphi^{(2)}_{\alpha_1\alpha_2} \left[^{(2)}S^k \varphi^{(2)}\right]_{\alpha_3\alpha_4} \tag{6.78}$$

$$= s_k \varphi^{(4)}_{\alpha_1\alpha_2\alpha_3\alpha_4}$$

with the two-particle spin operators $^{(2)}S^k$ from (6.67). In an analogous manner the square of the four-particle operators can be reduced to expressions containing only two-particle operators; for brevity we do not explicitly give these expressions.

The result of the analysis of the spin eigenvalue equations for the massive four-particle states, the spin dependence of which is given by

$$\hat{S}^\mu \otimes \hat{S}'^\nu \equiv \hat{S}^\mu_{\alpha_1\alpha_2} \hat{S}'^\nu_{\alpha_3\alpha_4} \tag{6.79}$$

with $\hat{S}^\mu, \hat{S}'^\mu \in \{\gamma^0 C, \gamma^k C, \Sigma^{0k} C, \Sigma^k C\}$, may be represented in the following table:

s	s_3	eigenstate
2	2	$(\hat{S}^1 + i\hat{S}^2) \otimes (\hat{S}'^1 + i\hat{S}'^2)$
2	-2	$(\hat{S}^1 - i\hat{S}^2) \otimes (\hat{S}'^1 - i\hat{S}'^2)$
2	1	$(\hat{S}^1 + i\hat{S}^2) \otimes \hat{S}'^3 + \hat{S}^3 \otimes (\hat{S}'^1 + i\hat{S}'^2)$
2	-1	$(\hat{S}^1 - i\hat{S}^2) \otimes \hat{S}'^3 + \hat{S}^3 \otimes (\hat{S}'^1 - i\hat{S}'^2)$
2	0	$\hat{S}^1 \otimes \hat{S}'^1 + \hat{S}^2 \otimes \hat{S}'^2 - 2\hat{S}^3 \otimes \hat{S}'^3$
1	1	$\hat{S}^0 \otimes (\hat{S}'^1 + i\hat{S}'^2), (\hat{S}^1 + i\hat{S}^2) \otimes \hat{S}'^0,$ $(\hat{S}^1 + i\hat{S}^2) \otimes \hat{S}'^3 - \hat{S}^3 \otimes (\hat{S}'^1 + i\hat{S}'^2)$
1	-1	$\hat{S}^0 \otimes (\hat{S}'^1 - i\hat{S}'^2), (\hat{S}^1 - i\hat{S}^2) \otimes \hat{S}'^0,$ $(\hat{S}^1 - i\hat{S}^2) \otimes \hat{S}'^3 - \hat{S}^3 \otimes (\hat{S}'^1 - i\hat{S}'^2)$
1	0	$\hat{S}^0 \otimes \hat{S}'^3, \hat{S}^3 \otimes \hat{S}'^0, \hat{S}^1 \otimes \hat{S}'^2 - \hat{S}^2 \otimes \hat{S}'^1$
0	0	$\hat{S}^0 \otimes \hat{S}'^0, \sum_k \hat{S}^k \otimes \hat{S}'^k$

The table demonstrates the mechanism of the fusion of two-particle spin eigenstates to four-particle eigenstates. In particular one obtains for $m \neq 0$ nine complete sets of eigenstates for $s = 2$, as $\hat{S}^k$ and $\hat{S}'^k$ can independently be chosen from the set $\{\gamma^k C, \Sigma^{0k} C, \Sigma^k C\}$.

We turn to the discussion of the internal symmetries of our theory. Because of the $SU(2)$ isospin and the $U(1)$ invariance of our initial spinor field equation (2.20) we have isospin and fermion number as good quantum numbers of our states. These quantum numbers are given by the symmetry equations

$$\begin{aligned}
(\mathbf{G}_T)^2 |p\rangle &= t(t+1)|p\rangle, \\
\mathbf{G}_T^3 |p\rangle &= t_3|p\rangle, \\
\mathbf{G}_F |p\rangle &= f|p\rangle,
\end{aligned} \tag{6.80}$$

where $\mathbf{G}_T^k$, $k = 1, 2, 3$, and $\mathbf{G}_F$ are the $SU(2)$ and $U(1)$ generators in the the corresponding representation space.

The generators of the fundamental representation of the $SU(2)$ are the Pauli matrices σ^k:

$$\left(G_T^k\right)_{A_1 A_2} = \sigma^k_{A_1 A_2}, \qquad k = 1, 2, 3, \tag{6.81}$$

acting on the isospin indices of the spinor fields ψ_I, whilst the $U(1)$ distinguishes fermion numbers and acts with its generator $G_F = \sigma^3$ on the super-indices Λ, see Section 2.6, (2.120). In (6.7) we combined isospin and super-index into one index κ with $\kappa = 1, 2, 3, 4$. Hence we have to formulate these transformations for our superspinors.

A corresponding four-dimensional representation of the $SU(2)_T \times U(1)_F$ generators is given by

$$\left(G_T^k\right)_{\kappa_1\kappa_2} = \frac{1}{2}\begin{pmatrix} \sigma^k & 0 \\ 0 & (-1)^k\sigma^k \end{pmatrix}_{\kappa_1\kappa_2}, \tag{6.82}$$

$$\left(G_F\right)_{\kappa_1\kappa_2} = \begin{pmatrix} \mathbb{1} & 0 \\ 0 & -\mathbb{1} \end{pmatrix}_{\kappa_1\kappa_2}. \tag{6.83}$$

Owing to the transformation laws (2.122) and (2.123) of the field operators for infinitesimal transformations the Hilbert space generators $\mathbf{G}$ and their representation in the configuration space are connected by

$$[\mathbf{G}, \psi_I] = -G_{II'}\psi_{I'}, \tag{6.84}$$

where $G \in \{G_T^k, G_F\}$ is defined by

$$G_{II'} := G_{\kappa\kappa'}\delta_{\alpha\alpha'}\delta_{ii'}\delta(\mathbf{r} - \mathbf{r}'). \tag{6.85}$$

If we project the state $\mathbf{G}|p\rangle$ onto the set of GNS basis vectors $\langle 0|\mathcal{A}(\psi_{I_1}\dots\psi_{I_n})$, with (6.84) we obtain

$$\langle 0|\mathcal{A}(\psi_{I_1}\dots\psi_{I_n})\mathbf{G}|p\rangle = \sum_{l=1}^{n}G_{I_lI'}\langle 0|\psi_{[I_1}\dots\psi_{I_{l-1}}\psi_{I'}\psi_{I_{l+1}}\dots\psi_{I_n]}|p\rangle. \tag{6.86}$$

In Proposition 3.6 the Hilbert space generators were mapped into functional space. In application to the present case we obtain with

$$\mathcal{G} := j_I G_{II'}\partial_{I'}, \tag{6.87}$$

where $\mathcal{G} \in \{\mathcal{G}_T^k, \mathcal{G}_F\}$, and with (6.86) the relation

$$i^{-n}\langle 0|\partial_{I_n}\dots\partial_{I_1}\mathcal{G}|\mathcal{A}(j;a)\rangle = -\langle 0|\mathcal{A}(\psi_{I_1}\dots\psi_{I_n})\mathbf{G}|a\rangle. \tag{6.88}$$

Therefore the eigenvalue equations (6.80) read in functional space

$$\begin{aligned}
(\mathcal{G}_T)^2|\mathcal{A}(j;a)\rangle &= t(t+1)|\mathcal{A}(j;a)\rangle, \\
\mathcal{G}_T^3|\mathcal{A}(j;a)\rangle &= t_3|\mathcal{A}(j;a)\rangle, \\
\mathcal{G}_F|\mathcal{A}(j;a)\rangle &= f|\mathcal{A}(j;a)\rangle.
\end{aligned} \tag{6.89}$$

For Weak Mapping calculations normal transformed functional states $|\mathcal{F}(j;a)\rangle$ have to be introduced according to definition (4.98). With respect to these normal ordered states the following proposition holds:

- **Proposition 6.5:** The normal ordered state functionals $|\mathcal{F}(j;a)\rangle$ satisfy the same eigenvalue equations (6.89) as the states $|\mathcal{A}(j;a)\rangle$ provided the propagator $F_{II'}$ is invariant under the group $SU(2)_T \times U(1)_F$.

Proof: The proof runs along the same lines as that of Proposition 3.15.   ◇

We now calculate the quantum numbers of the vector boson states (6.41). According to (6.41) the isospin/superspin part of our two-particle solutions is constituted by the antisymmetric part of the Dirac algebra. A basis for this part is given by the matrices

$$\hat{T}^1 := \frac{i}{2}\left[\begin{pmatrix} \sigma^2 & 0 \\ 0 & \sigma^2 \end{pmatrix} + \begin{pmatrix} \sigma^2 & 0 \\ 0 & -\sigma^2 \end{pmatrix}\right], \tag{6.90}$$

$$\hat{T}^2 := \frac{i}{2}\left[\begin{pmatrix} \sigma^2 & 0 \\ 0 & \sigma^2 \end{pmatrix} + \begin{pmatrix} -\sigma^2 & 0 \\ 0 & \sigma^2 \end{pmatrix}\right],$$

$$\hat{T}^3 := \begin{pmatrix} 0 & \mathbb{1} \\ -\mathbb{1} & 0 \end{pmatrix},$$

$$\hat{T}^4 := \frac{1}{2}\left[\begin{pmatrix} 0 & \sigma^1 \\ -\sigma^1 & 0 \end{pmatrix} + i\begin{pmatrix} 0 & \sigma^2 \\ \sigma^2 & 0 \end{pmatrix}\right],$$

$$\hat{T}^5 := \frac{1}{2}\left[\begin{pmatrix} 0 & \sigma^1 \\ -\sigma^1 & 0 \end{pmatrix} - i\begin{pmatrix} 0 & \sigma^2 \\ \sigma^2 & 0 \end{pmatrix}\right],$$

$$\hat{T}^6 := \begin{pmatrix} 0 & \sigma^3 \\ -\sigma^3 & 0 \end{pmatrix},$$

where the set of matrices (6.57) can be regained by the linear combinations

$$\begin{aligned}
T^1 &= i^{-1}(\hat{T}^1 + \hat{T}^2), \tag{6.91}\\
T^2 &= i^{-1}(\hat{T}^1 - \hat{T}^2),\\
T^3 &= \hat{T}^3,\\
T^4 &= \hat{T}^4 + \hat{T}^5,\\
T^5 &= i^{-1}(\hat{T}^4 - \hat{T}^5),\\
T^6 &= \hat{T}^6.
\end{aligned}$$

As the eigenvalue equation (6.1) for vector bosons admits any linear combination of superspin/isospin matrices, we can equally well use the set (6.90) instead of the set (6.57). The quantum conditions (6.89) then read for $\varphi^{(2)}_{II'} \equiv \langle 0|\mathcal{A}(\psi_I\psi_{I'})|p\rangle$:

$$\frac{3}{2}\varphi^{(2)}_{I_1I_2} + 2(G^k_{\mathrm{T}})_{I_1K_1}(G^k_{\mathrm{T}})_{I_2K_2}\varphi^{(2)}_{K_1K_2} = t(t+1)\varphi^{(2)}_{I_1I_2}, \tag{6.92}$$

$$(G^3_{\mathrm{T}})_{I_1K}\varphi^{(2)}_{KI_2} + (G^3_{\mathrm{T}})_{I_2K}\varphi^{(2)}_{I_1K} = t_3\varphi^{(2)}_{I_1I_2},$$

$$(G^3_{\mathrm{F}})_{I_1K}\varphi^{(2)}_{KI_2} + (G^3_{\mathrm{F}})_{I_2K}\varphi^{(2)}_{I_1K} = f\varphi^{(2)}_{I_1I_2}.$$

By explicit evaluation of these equations the eigenvectors can be determined to give

	$\hat{T}^1$	$\hat{T}^2$	$\hat{T}^3$	$\hat{T}^4$	$\hat{T}^5$	$\hat{T}^6$
t	0	0	0	1	1	1
t_3	0	0	0	1	-1	0
f	2	-2	0	0	0	0

In addition to vector boson states, equations (6.1) and (6.2) also admit scalar boson states. So the problem arises of which criteria states of the complete manifold of solutions of (6.1) or (6.2), respectively, are to be included into the physical considerations or have to be excluded from it.

We postpone the discussion of this problem to Chapter 8 and consider only the relevant quantum numbers in this section. Scalar bosons are to be described by the antisymmetric spin algebra elements C, $\gamma^5 C$ and $\gamma^\mu \gamma^5 C$. The spin properties of this set were discussed above.

Owing to the complete antisymmetry of the wave functions the superspin/isospin algebra has to be symmetric in this case. The corresponding basis elements are $S^a \in \{\gamma^\mu C, \Sigma^{\mu\nu} C\}$. As equations (6.1) and (6.2) are invariant under superspin/isospin transformations the solutions can be classified by irreducible representations of these groups. For brevity we consider only $f = 0$ representations. This yields the table

	$(\gamma^0 C)_{\kappa\kappa'}$	$-(\Sigma^1 + i\Sigma^2)C$	$-(\Sigma^1 - i\Sigma^2)C$	$-\Sigma^3 C$
t	0	1	1	1
t_3	0	1	-1	0
f	0	0	0	0

One can easily verify this table by observing that the symmetric elements $S_{\kappa\kappa'}$ can equally well be written as $S_{\kappa\kappa'} \equiv S_{\Lambda A, \Lambda' A'}$, i.e., as direct products of isospin and superspin matrices, the representations of which are well known.

The extension of the superspin/isospin analysis to the three- and four-particle case is again straightforward; the three-particle case will be explicitly discussed in Section 6.5.

6.5   Three-Particle Bound States

In various field theoretic models three-fermion bound states are assumed to play an important role. The most prominent example is the Nambu–Jona–Lasinio (NJL) quark model, where nucleons are described by three-quark bound states. In the context of our fusion program the constituent fermions of the standard model, namely quarks and leptons, are considered as bound states of three subfermions.

This assumption is shared with many other preon models, in particular with the rishon model, compare [Lyo 83]. But whilst (apart from quark theory) the model building in the literature is done only on a combinatorial level, for Weak Mapping analytical expressions of the whole bound state wave functions are needed.

With respect to three-subfermion bound states Pfister [Pfi 94] has solved the energy equation (6.1) in an ultra-local limit, i.e., by neglect of the kinetic energy for very large subfermion masses. In this limit the algebraic structure of the three-particle bound states can be exactly derived. However, as the energy equation (6.1) is a single time equation, the orbital parts of its solutions cannot show the correct transformation properties under the Lorentz group. As we already mentioned, Lorentz covariant wave functions can only be obtained by the solution of generalized de Broglie–Bargmann–Wigner equations, see Section 6.1.

Unfortunately the latter equations have not been treated so far for the three-particle case. Therefore, for a first inspection of the quark and lepton states we restrict ourselves to the discussion of the algebraic properties of these states which we analyze in accordance with the investigations of Pfister, see [Pfi 94].

We emphasize that in the following the algebraic parts of all Weak Mapping calculations are exactly evaluated. With respect to the orbital parts, however, we shall apply nonrelativistic approximations for the internal coordinates, which corresponds to the low energy limit of the effective theory. In Section 6.1 we have already stated the general compatibility of approximated wave functions with Weak Mapping. So, what are the consequences of such approximations?

In the low energy limit the effective interactions which arise in the course of the Weak Mapping calculations are pointlike, *i.e.*, local, and one has no chance to discover deviations from these interactions resulting from approximate orbital wave functions. This means that as long as one does not try to investigate high energy processes (with particle energies in the vicinity of the Planck mass?), where the subfermionic orbital bound state structure becomes relevant, one is justified to use the approximate functions as pointed out above. Thus we will concentrate on the analysis of the algebraic structure of three-subfermion bound states without insisting on their space–time relativistic covariance.

In the preceding sections we have calculated the vector boson bound states by evaluating the nonperturbative subfermion regularization from Chapter 2. Thus for this case we have demonstrated the applicability of this regularization. For the three-particle solutions, however, for simplicity we deviate from this procedure by suppressing all auxiliary field indices i as well as the regularization parameters λ_i. This means, that in this section the collective index I is given by $I = (Z, \mathbf{r}) = (\alpha, \kappa, \mathbf{r})$.

Without specifying the kind of regularization we assume the theory to be regularized by means of a regularized subfermion propagator so that $F^t|_{\mathbf{r}_1 = \mathbf{r}_2} < \infty$. As the regularization is connected with the calculation of mass parameters and coupling constants this means that for a first examination we give up the possibility of a reliable determination of the parameters of the effective coupling theory.

The results of [Pfi 94] motivate the following ansatz for the lepton and quark states:

$$\varphi^{(3)}_{I_1 I_2 I_3}(a, t) \tag{6.93}$$

$$= e^{-ik_0 t} e^{i\mathbf{k}(\mathbf{r}_1 + \mathbf{r}_2 + \mathbf{r}_3)/3} \Theta^l_{\kappa_1 \kappa_2 \kappa_3} \left[\chi^s_{\alpha_1 \alpha_2 \alpha_3} \zeta^\rho \left(\mathbf{r}_2 - \mathbf{r}_3, \mathbf{r}_1 - \frac{1}{2}(\mathbf{r}_2 + \mathbf{r}_3) \right) \right]_{\text{as}[1,2,3]}$$

with $a = (\mathbf{k}, s, \rho)$ and with the symmetric isospinor/superspinor part Θ^l. The orbital functions ζ^ρ will be discussed in Section 7.6 in connection with shell model states, and the spin tensors are given by

$$\chi^s_{\alpha_1 \alpha_2 \alpha_3} := (\gamma^5 C)_{\alpha_1 \alpha_2} \chi^s_{\alpha_3} . \tag{6.94}$$

The quantum number s of the elementary Dirac spinors χ_α^s characterizes the set of basis vectors in spin space, which is defined by

$$\chi_\alpha^s \in \left\{ \begin{pmatrix} 1 \\ 0 \\ 0 \\ 0 \end{pmatrix}, \begin{pmatrix} 0 \\ 1 \\ 0 \\ 0 \end{pmatrix}, \begin{pmatrix} 0 \\ 0 \\ 1 \\ 0 \end{pmatrix}, \begin{pmatrix} 0 \\ 0 \\ 0 \\ 1 \end{pmatrix} \right\}. \tag{6.95}$$

From the calculations of vector boson states in Sections 6.2 and 6.3 and from the calculations by Pfister [Pfi 94] with the energy equation (6.1) it follows that complete sets of bound state functions (and if necessary scattering state functions) contain in a symmetric way positive and negative energy solutions. *This is not an unwanted accident*. Rather, Weak Mapping works correctly only if complete state systems are used. Such state systems necessarily have to include negative energies, which themselves depend on the complete subfermion spectrum.

I.e., *at the level of unobservable subfermions no Fock–Dirac vacuum is introduced*, and this in turn gives rise to formulate our terminology more precisely: in no stage of our considerations do genuine subfermion *particles* appear. Rather, the whole formalism is concerned with subfermion *fields* and effective *fields*. Particle interpretation can be introduced in the evaluation of effective theories by the introduction of a specific algebraic representation, namely the interaction representation, corresponding to the Fock–Dirac vacuum. Thus in our theory the Fock–Dirac vacuum plays no fundamental role, although in the following to simplify matters we speak of particles.

We turn to the determination of the superspin/isospin part Θ^l of the three-particle solutions. According to the ansatz (6.93), Θ^l completely decouples from the spin part. In [Pfi 94] it was shown that a condition for this decoupling is given by

$$\gamma^5_{\kappa_1 \kappa_2} \Theta^l_{\kappa_1 \kappa_2 \kappa_3} = 0. \tag{6.96}$$

Together with the symmetry of Θ^l this gives sixteen possible superspin/isospin states Θ^l.

As superspin and isospin are connected with the global $SU(2) \times U(1)$ form invariance of the spinor theory, we can choose the quantities Θ^l to be eigenstates of isospin and fermion number generators. Thereby we classify the three-particle states according to $SU(2)$ and $U(1)$ quantum numbers, ignoring the possibility of symmetry breaking which takes place afterwards.

The classification of the quantum numbers of the three-particle wave functions runs along the lines of Section 6.4. We require the three-particle state $|p\rangle$ to satisfy the equations (6.80)–(6.81):

$$\begin{aligned} \mathbf{G}_T^3 |p\rangle &= t_3 |p\rangle, \\ \left(\mathbf{G}_T\right)^2 |p\rangle &= t(t+1)|p\rangle, \\ \mathbf{G}_F |p\rangle &= f|p\rangle. \end{aligned} \tag{6.97}$$

From the spinor field transformation law (2.89) it follows that the corresponding generators satisfy the commutator relations

$$[\mathbf{G}, \psi_I] = -G_{II'}\psi_{I'},\qquad(6.98)$$

where $G_{II'}$ are the corresponding classical generators. They are given by $\{G_{I_1 I_2}\} \equiv \{T^k_{I_1 I_2}, F_{I_1 I_2}\}$, where

$$T^k_{I_1 I_2} = \frac{1}{2}\begin{pmatrix} \sigma^k & 0 \\ 0 & (-1)^k \sigma^k \end{pmatrix}_{\kappa_1 \kappa_2} \delta_{\alpha_1 \alpha_2}\delta(\mathbf{r}_1 - \mathbf{r}_2),\qquad(6.99)$$

$$F_{I_1 I_2} = \frac{1}{3}\begin{pmatrix} \mathbb{I} & 0 \\ 0 & -\mathbb{I} \end{pmatrix}_{\kappa_1 \kappa_2} \delta_{\alpha_1 \alpha_2}\delta(\mathbf{r}_1 - \mathbf{r}_2),\qquad(6.100)$$

with the commonly used notations T^k and F for the isospin and fermion number generators. *We add that the isospin generators should not be confused with the isospin/superspin matrices T^a which we used for our vector boson functions.*

With this choice of the symmetry group generators the elementary subfermions are associated to the isospin quantum number $t = 1/2$ and the fermion number $f = 1/3$. The fermion number of subfermions can be arbitrarily chosen because the subfermions are unobservable. Once the subfermion quantum number is fixed (the isospin is treated in the conventional manner) there is no freedom for further manipulations. That means, charge and hypercharge for the subfermion bound states, *i.e.*, for leptons and quarks are to be derived and have to coincide with the corresponding phenomenological values.

In order to reproduce the phenomenological charge we define a charge generator by

$$Q := T^3 + Y\qquad(6.101)$$

with the hypercharge generator

$$Y := \frac{1}{2}F.\qquad(6.102)$$

The corresponding quantum numbers are $q = t_3 + y$ and $y = f/2$, and the charge generator is explicitly given by

$$Q_{I_1 I_2} := \frac{1}{3}\begin{pmatrix} 2 & 0 & 0 & 0 \\ 0 & -1 & 0 & 0 \\ 0 & 0 & -2 & 0 \\ 0 & 0 & 0 & 1 \end{pmatrix}_{\kappa_1 \kappa_2} \delta_{\alpha_1 \alpha_2}\delta(\mathbf{r}_1 - \mathbf{r}_2).\qquad(6.103)$$

Next we have to derive symmetry conditions for the three-particle solutions $\varphi^{(3)}$ from the corresponding state space eigenvalue equations (6.97). By observing that the normal ordered matrix elements $\varphi(I_1 \ldots I_n|a)$ have the same transformation properties as the matrix elements $\langle 0|\psi_{I_1} \ldots \psi_{I_n}|a\rangle$ one can easily transfer the results from Section 6.4 to the present case.

In physical state space the eigenvalue equations for the gauge group generators are given by (6.97). The corresponding generators in functional space are defined by

$$\mathcal{G}^a := -\int j^t_{\kappa\alpha}(\mathbf{r})G^a_{\kappa\kappa'}\partial^t_{\kappa'\alpha}(\mathbf{r})\,d\mathbf{r}\qquad(6.104)$$

in accordance with the definition (6.87).

Then for $|\mathcal{F}(j^t;p)\rangle^{\mathrm{d}}$ the functional eigenvalue equations read

$$\mathcal{G}_{\mathrm{T}}^3|\mathcal{F}(j^t;p)\rangle^{\mathrm{d}} = t_3|\mathcal{F}(j^t;p)\rangle^{\mathrm{d}}, \tag{6.105}$$

$$(\mathcal{G}_{\mathrm{T}})^2|\mathcal{F}(j^t;p)\rangle^{\mathrm{d}} = t(t+1)|\mathcal{F}(j^t;p)\rangle^{\mathrm{d}},$$

$$\mathcal{G}_{\mathrm{F}}|\mathcal{F}(j^t;p)\rangle^{\mathrm{d}} = f|\mathcal{F}(j^t;p)\rangle^{\mathrm{d}}.$$

Projection onto the space of three-subfermion solutions and observing $(T^k)^2 = 1$ yields the equations

$$\frac{9}{4}\varphi^{(3)}_{I_1I_2I_3} + 2\left[T^k_{I_1K_1}T^k_{I_2K_2}\varphi^{(3)}_{K_1K_2I_3}\right. \tag{6.106}$$

$$\left.+T^k_{I_1K_1}T^k_{I_3K_2}\varphi^{(3)}_{K_1I_2K_2} + T^k_{I_2K_1}T^k_{I_3K_2}\varphi^{(3)}_{I_1K_1K_2}\right] = t(t+1)\varphi^{(3)}_{I_1I_2I_3},$$

$$T^3_{I_1K}\varphi^{(3)}_{KI_2I_3} + T^3_{I_2K}\varphi^{(3)}_{I_1KI_3} + T^3_{I_3K}\varphi^{(3)}_{I_1I_2K} = t_3\varphi^{(3)}_{I_1I_2I_3}, \tag{6.107}$$

and corresponding equations for the generators F, Y and Q.

If we substitute the ansatz (6.93) into these equations we obtain equations for the isospin–superspin part Θ^l of $\varphi^{(3)}$. We obtain four isospin quadruplets with $t = 3/2$, the quantum numbers of which are conveniently displayed in a table. Presenting only those states which satisfy (6.96), we have:

	Θ^1	Θ^2	Θ^3	Θ^4	Θ^5	Θ^6	Θ^7	Θ^8
t	3/2	3/2	3/2	3/2	3/2	3/2	3/2	3/2
t_3	3/2	1/2	-1/2	-3/2	3/2	1/2	-1/2	-3/2
f	-1	-1	-1	-1	1	1	1	1
$y = f/2$	-1/2	-1/2	-1/2	-1/2	1/2	1/2	1/2	1/2
$q = t_3 + y$	1	0	-1	-2	2	1	0	-1
interpretation:	?	ν	e	?	?	e^+	$\bar{\nu}$	?

	Θ^9	Θ^{10}	Θ^{11}	Θ^{12}	Θ^{13}	Θ^{14}	Θ^{15}	Θ^{16}
t	3/2	3/2	3/2	3/2	3/2	3/2	3/2	3/2
t_3	3/2	1/2	-1/2	-3/2	3/2	1/2	-1/2	-3/2
f	-1/3	-1/3	-1/3	-1/3	1/3	1/3	1/3	1/3
$y = f/2$	-1/6	-1/6	-1/6	-1/6	1/6	1/6	1/6	1/6
$q = t_3 + y$	4/3	1/3	-2/3	-5/3	5/3	2/3	-1/3	-4/3
interpretation:	?	$\bar{d}$	$\bar{u}$	?	?	u	d	?

The corresponding superspin/isospin tensors which allow a physical interpretation are explicitly given by

$$\Theta^2_{\kappa_1\kappa_2\kappa_3} = \frac{1}{\sqrt{3}}\left[\delta_{4\kappa_1}\delta_{4\kappa_2}\delta_{3\kappa_3} + \delta_{4\kappa_1}\delta_{3\kappa_2}\delta_{4\kappa_3} + \delta_{3\kappa_1}\delta_{4\kappa_2}\delta_{4\kappa_3}\right], \tag{6.108}$$

$$\Theta^3_{\kappa_1\kappa_2\kappa_3} = \frac{1}{\sqrt{3}} [\delta_{3\kappa_1}\delta_{3\kappa_2}\delta_{4\kappa_3} + \delta_{4\kappa_1}\delta_{3\kappa_2}\delta_{3\kappa_3} + \delta_{3\kappa_1}\delta_{4\kappa_2}\delta_{3\kappa_3}],$$

$$\Theta^6_{\kappa_1\kappa_2\kappa_3} = \frac{1}{\sqrt{3}} [\delta_{1\kappa_1}\delta_{1\kappa_2}\delta_{2\kappa_3} + \delta_{1\kappa_1}\delta_{2\kappa_2}\delta_{1\kappa_3} + \delta_{2\kappa_1}\delta_{1\kappa_2}\delta_{1\kappa_3}],$$

$$\Theta^7_{\kappa_1\kappa_2\kappa_3} = \frac{1}{\sqrt{3}} [\delta_{2\kappa_1}\delta_{2\kappa_2}\delta_{1\kappa_3} + \delta_{2\kappa_1}\delta_{1\kappa_2}\delta_{2\kappa_3} + \delta_{1\kappa_1}\delta_{2\kappa_2}\delta_{2\kappa_3}],$$

$$\Theta^{10}_{\kappa_1\kappa_2\kappa_3} = \frac{1}{3} [\delta_{4\kappa_1}\delta_{4\kappa_2}\delta_{2\kappa_3} + \delta_{4\kappa_1}\delta_{2\kappa_2}\delta_{4\kappa_3} + \delta_{2\kappa_1}\delta_{4\kappa_2}\delta_{4\kappa_3}$$
$$-\delta_{4\kappa_1}\delta_{3\kappa_2}\delta_{1\kappa_3} - \delta_{3\kappa_1}\delta_{4\kappa_2}\delta_{1\kappa_3} - \delta_{1\kappa_1}\delta_{4\kappa_2}\delta_{3\kappa_3} - \delta_{4\kappa_1}\delta_{1\kappa_2}\delta_{3\kappa_3}$$
$$-\delta_{3\kappa_1}\delta_{1\kappa_2}\delta_{4\kappa_3} - \delta_{1\kappa_1}\delta_{3\kappa_2}\delta_{4\kappa_3}],$$

$$\Theta^{11}_{\kappa_1\kappa_2\kappa_3} = -\frac{1}{3} [\delta_{3\kappa_1}\delta_{3\kappa_2}\delta_{1\kappa_3} + \delta_{3\kappa_1}\delta_{1\kappa_2}\delta_{3\kappa_3} + \delta_{1\kappa_1}\delta_{3\kappa_2}\delta_{3\kappa_3}$$
$$- \delta_{4\kappa_1}\delta_{3\kappa_2}\delta_{2\kappa_3} - \delta_{3\kappa_1}\delta_{4\kappa_2}\delta_{2\kappa_3} - \delta_{2\kappa_1}\delta_{4\kappa_2}\delta_{3\kappa_3} - \delta_{4\kappa_1}\delta_{2\kappa_2}\delta_{3\kappa_3}$$
$$-\delta_{3\kappa_1}\delta_{2\kappa_2}\delta_{4\kappa_3} - \delta_{2\kappa_1}\delta_{3\kappa_2}\delta_{4\kappa_3}],$$

$$\Theta^{14}_{\kappa_1\kappa_2\kappa_3} = -\frac{1}{3} [\delta_{1\kappa_1}\delta_{1\kappa_2}\delta_{3\kappa_3} + \delta_{3\kappa_1}\delta_{1\kappa_2}\delta_{1\kappa_3} + \delta_{1\kappa_1}\delta_{3\kappa_2}\delta_{1\kappa_3}$$
$$- \delta_{4\kappa_1}\delta_{2\kappa_2}\delta_{1\kappa_3} - \delta_{2\kappa_1}\delta_{4\kappa_2}\delta_{1\kappa_3} - \delta_{1\kappa_1}\delta_{4\kappa_2}\delta_{2\kappa_3} - \delta_{4\kappa_1}\delta_{1\kappa_2}\delta_{2\kappa_3}$$
$$-\delta_{1\kappa_1}\delta_{2\kappa_2}\delta_{4\kappa_3} - \delta_{2\kappa_1}\delta_{1\kappa_2}\delta_{4\kappa_3}],$$

$$\Theta^{15}_{\kappa_1\kappa_2\kappa_3} = \frac{1}{3} [\delta_{2\kappa_1}\delta_{2\kappa_2}\delta_{4\kappa_3} + \delta_{4\kappa_1}\delta_{2\kappa_2}\delta_{2\kappa_3} + \delta_{2\kappa_1}\delta_{4\kappa_2}\delta_{2\kappa_3}$$
$$-(\delta_{3\kappa_1}\delta_{2\kappa_2} + \delta_{2\kappa_1}\delta_{3\kappa_2})\delta_{1\kappa_3} - (\delta_{1\kappa_1}\delta_{3\kappa_2} + \delta_{3\kappa_1}\delta_{1\kappa_2})\delta_{2\kappa_3}$$
$$-(\delta_{1\kappa_1}\delta_{2\kappa_2} + \delta_{2\kappa_1}\delta_{1\kappa_2})\delta_{3\kappa_3}].$$

The interpretation of the various states in the table is based on the total agreement of the numbers of t_3, y and q with the quantum numbers of the corresponding phenomenological particles.

It should be emphasized that these quantum numbers are the result of an exact algebraic state calculation and that they are determined on the subfermion level by means of the fusion process. It is remarkable that one subfermion number f leads to the correct hypercharges for leptons as well as for quarks, as in phenomenological theory these hypercharges have to be separately defined in order to achieve agreement with experiments.

If we apply the formula $q = t_3 + f/2$ to the quantum numbers of the elementary subfermions, we obtain the charges $q_f = 2/3, -1/3$. Because the quantum numbers t_3 and y are additive, the values of q_f may serve as a book keeping charge which enable us to determine the charge of a three-particle bound state if the fermion constituents are known. However, as the subfermions are assumed to be unobservable, their quantum numbers have no direct physical meaning.

Of course, the above mentioned interpretation, which has been suggested by the analysis of the quantum numbers of the various fermion states, has to be justified by taking into account the mutual interactions of these states. This will be done in the next chapter by the evaluation of their effective dynamics.

The classification according to the above table has, however, some disadvantages. First of all the states are collected in isospin quadruplets instead of the phenomenological doublets, leading to superfluous states which cannot be interpreted. The second drawback is the lack of the three families of leptons and quarks.

With respect to these points Pfister [Pfi 95b,c] has performed an improved analysis of three-particle states by considering states of mixed symmetry, $i.e.$, states with a mixed symmetry in isospin/superspin space and in spin-orbit space. This modified three-particle state ansatz removes the above mentioned drawbacks: one obtains $three$ linear independent solution manifolds, which may be interpreted as the three families of leptons and quarks, and one obtains isospin $t = \frac{1}{2}$ doublets as is required from the phenomenology.

$Thus$ $with$ $these$ $mixed$ $symmetry$ $states$ all $quantum$ $numbers$ of the $phenomenological$ $leptons$ and $quarks$, $including$ $their$ $family$ $structure$, are $reproduced$ in our $fusion$ $model$. Of course these interpretations also have to be justified by an investigation of the effective equations including the interactions; this was done by Pfister in [Pfi 95a].

Because of the rather extensive formalism which is necessary for the discussion of mixed symmetry states we do not describe their explicit treatment but demonstrate the mechanism of the fusion of three-particle states and the derivation of their effective dynamics in the case of the symmetric superspin/isospin states.

Owing to the decoupling (6.93) the various isospin/superspin tensors Θ^l, $l = 1 \dots 16$, are connected with the same space function ζ and thus there would be no discrimination with respect to the space functions of the various quarks and leptons. This problem will be solved in the framework of a shell model in Chapter 7; as we have already mentioned in this chapter we will also justify the interpretation of our three-particle states as quarks and leptons by evaluating their effective dynamical equations.

6.6   Four-Fermion Bound States

Apart from the intention to derive the standard model of elementary particles it is obvious to try to explain the gravitational forces by an effective theory of elementary subfermions. For that purpose we assume the gravitation to be generated by the effective dynamics of four-subfermion bound states. Thus in this section we discuss four-particle states as 'graviton' states.

The idea to describe the gravitation by bound states of four spin $\frac{1}{2}$ subfermions was originated by de Broglie [Brog 54], see Section 1.8. However, his fusion theory and the de Broglie–Bargmann–Wigner equations [Barg 48] for spin 2 particles are too restrictive for the derivation of a gravitation theory with many-particle processes, see the comment at the beginning of Section 2.1. This fact again stresses the necessity of a generalized fusion concept, $i.e.$, the transition to generalized de Broglie–Bargmann–Wigner equations and Weak Mapping. As far as the relations between de Broglie–Bargmann–Wigner equations and spin 2 graviton representations are concerned, we refer to Niederer [Nie 75], Rodrigues and Lorente [Rod 81] [Rod 84], and Nous [Nous 83].

Our generalized fusion equations generate spin 2 states as well as spin 0 and spin 1 parts. As in principle we intend to derive a nonlinear gravitation theory we will not *a priori* restrict our graviton states to spin 2 but leave it to the full effective dynamics to select proper eigenstates.

The treatment of four-fermion bound states is similar to that of the two-fermion bound states, *i.e.*, by contraction of the corresponding integral equations. However, in contrast to the two-fermion case, in the four-fermion case even the contracted integral equations cannot be solved exactly. Thus we will introduce physically motivated approximations. The corresponding calculations were performed by Stumpf, Borne and Kaus, see [Stu 88], [Stu 93b] and [Bor 98]. For brevity we describe only the results and omit the intermediate steps of calculation.

The starting point of the calculation is the covariant bound state equation (6.2), which, as already remarked, admits separate solutions in any sector of the fermion number $n = 1, 2, \ldots$.

We assume that the bound state functional (6.6) is defined for $n = 4$. Substitution of (6.6) into (6.2), projection into coordinate space and subsequent application of the Green's function G yield the defining equation

$$\varphi^{(4)}(I_1, I_2, I_3, I_4|a) \tag{6.109}$$
$$= 3 G_{I_4 K_1} W_{K_1 K_2 K_3 K_4} \Big[F_{K_2 I_3} \varphi^{(4)}(I_1, I_2, K_3, K_4|a)$$
$$- F_{K_2 I_2} \varphi^{(4)}(I_1, I_3, K_3, K_4|a) + F_{K_2 I_1} \varphi^{(4)}(I_2, I_3, K_3, K_4|a) \Big]_{\mathrm{as}[I_1 \ldots I_4]}.$$

The analogue to Proposition 6.2 is given by

- **Proposition 6.6:** Let $\varphi^{(4)}(I_1, I_2, I_3, I_4)$ be the wave function of the hard core state functional (6.6) for $n = 4$. Then if $\varphi^{(4)}(I_1, I_2, I_3, I_4)$ is a solution of equation (6.2) for $n = 4$, it can be reconstructed from the contracted functions $\tilde{\varphi}^{(4)}(q_1, q_3|)$ and $\tilde{\varphi}^{(4)}(q_1|q_3, q_4)$, which satisfy the integral equation

$$\tilde{\varphi}^{(4)}_{\substack{\kappa_1 \kappa_2 \kappa_3 \kappa_4 \\ \alpha_1 \alpha_2 \alpha_3 \alpha_4}}(q_1, q_3|) \tag{6.110}$$
$$= 3(2\pi)^4 V_{\substack{\rho_1 \rho_2 \rho_3 \rho_4 \\ \beta_1 \beta_2 \beta_3 \beta_4}}$$
$$\times \Bigg\{ \int d^4\eta \; \tilde{G}_{\substack{\kappa_4 \rho_1 \\ \alpha_4 \beta_1}}(q_3 - \eta) \tilde{F}_{\substack{\rho_2 \kappa_3 \\ \beta_2 \alpha_3}}(-\eta) \tilde{\varphi}^{(4)}_{\substack{\kappa_1 \kappa_2 \rho_3 \rho_4 \\ \alpha_1 \alpha_2 \beta_3 \beta_4}}(q_3, q_1|)$$
$$- \int d^4\eta \, d^4\xi \; \tilde{G}_{\substack{\kappa_4 \rho_1 \\ \alpha_4 \beta_1}}(q_3 - \eta) \tilde{F}_{\substack{\rho_2 \kappa_2 \\ \beta_2 \alpha_2}}(\xi - q_1) \tilde{\varphi}^{(4)}_{\substack{\kappa_1 \kappa_3 \rho_3 \rho_4 \\ \alpha_1 \alpha_3 \beta_3 \beta_4}}(q_1 + q_3 - \xi - \eta|\xi, \eta)$$
$$+ \int d^4\eta \, d^4\xi \; \tilde{G}_{\substack{\kappa_4 \rho_1 \\ \alpha_4 \beta_1}}(q_3 - \eta) \tilde{F}_{\substack{\rho_2 \kappa_1 \\ \beta_2 \alpha_1}}(-\xi) \tilde{\varphi}^{(4)}_{\substack{\kappa_2 \kappa_3 \rho_3 \rho_4 \\ \alpha_2 \alpha_3 \beta_3 \beta_4}}(q_3 + \xi - \eta|q_1 - \xi, \eta) \Bigg\},$$

where $\tilde{\varphi}^{(4)}$ is the Fourier transform of $\varphi^{(4)}$, the contractions are defined by

$$\tilde{\varphi}^{(4)}(q_1|q_3, q_4) := \int d^4\xi \; \tilde{\varphi}^{(4)}(\xi, q_1 - \xi, q_3, q_4), \tag{6.111}$$

$$\varphi^{(4)}(q_1, q_3|) := \int d^4\eta \; \varphi^{(4)}(q_1|\eta, q_3 - \eta), \tag{6.112}$$

and the definitions (6.10)–(6.12), as well as the antisymmetrization in all indices, have to be observed.

Proof: We consider the defining equation (6.109). If we sum over the auxiliary field indices i_1, i_2, i_3, i_4, observe (6.8) and definitions (6.10)–(6.12) and eventually apply a Fourier transformation, equation (6.109) goes over into

$$\tilde{\varphi}^{(4)}_{\substack{\kappa_1\kappa_2\kappa_3\kappa_4 \\ \alpha_1\alpha_2\alpha_3\alpha_4}}(q_1, q_2, q_3, q_4) \tag{6.113}$$

$$= 3(2\pi)^4 \tilde{G}_{\substack{\kappa_4\rho_1 \\ \alpha_4\beta_1}}(q_4) V_{\substack{\rho_1\rho_2\rho_3\rho_4 \\ \beta_1\beta_2\beta_3\beta_4}}$$

$$\times \left\{ \tilde{F}_{\substack{\rho_2\kappa_3 \\ \beta_2\alpha_3}}(-q_3)\tilde{\varphi}^{(4)}_{\substack{\rho_3\rho_4\kappa_1\kappa_2 \\ \beta_3\beta_4\alpha_1\alpha_2}}(q_3 + q_4 | q_1, q_2) \right.$$

$$- \tilde{F}_{\substack{\rho_2\kappa_2 \\ \beta_2\alpha_2}}(-q_2)\tilde{\varphi}^{(4)}_{\substack{\rho_3\rho_4\kappa_1\kappa_3 \\ \beta_3\beta_4\alpha_1\alpha_3}}(q_2 + q_4 | q_1, q_3)$$

$$\left. + \tilde{F}_{\substack{\rho_2\kappa_1 \\ \beta_2\alpha_1}}(-q_1)\tilde{\varphi}^{(4)}_{\substack{\rho_3\rho_4\kappa_2\kappa_3 \\ \beta_3\beta_4\alpha_2\alpha_3}}(q_1 + q_4 | q_2, q_3) \right\}.$$

Application of the contractions (6.111), (6.112) to (6.113) gives equation (6.110).

Let $\tilde{\varphi}^{(4)}(q_1, q_3|)$ and $\tilde{\varphi}^{(4)}(q_1|q_3, q_4)$ be solutions of (6.110). Then we can substitute $\tilde{\varphi}_4(q_1|q_3, q_4)$ into equation (6.113) and obtain $\tilde{\varphi}^{(4)}(q_1, q_2, q_3, q_4)$. Hence in this way the full $\tilde{\varphi}^{(4)}$ or $\varphi^{(4)}$ can be reconstructed, and $\tilde{\varphi}^{(4)}$ is a solution of the uncontracted equation (6.113).   ◇

In contrast to the secular equation (6.17) for the two-particle case the integral equation (6.110) does not allow an immediate solution. As we are only interested in approximate solutions we do not speculate about exact methods for the solution of equation (6.110) in order to obtain four-fermion bound states.

To properly approximate the wave functions in (6.109) we only consider a sub- manifold of solutions. As we will investigate gravitational effects we are interested in four-fermion states with spin 2 parts which can be related to the tensor expressions of relativistic gravity theory.

The requirement of spin 2 states suggests to imagine these states as being composed of two spin 1 states, i.e., we consider the four-fermion state as a bound state of two vector bosons. We remark that already de Broglie discussed this idea in connection with the analysis of his fusion states, see [Brog 54] and Section 1.8.

In order to utilize this idea for an approximate calculation of the four-fermion states we further assume that in spite of their fusion the internal wave functions of the vector bosons are approximately conserved. In this case the remaining task is the determination of the part of the wave function for the center of mass motion of both bosons, and for their relative motion.

The covariant wave functions of the single vector bosons are given by equations (6.41) and (6.42), which can be written in the following form:

$$\varphi^{(2)}_{Z_1 Z_2}(x_1, x_2) \tag{6.114}$$

$$= N_B T^a_{\kappa_1\kappa_2} \exp\left[-\frac{i}{2}k(x_1 + x_2)\right] \left[A_\mu u^{\mu}_{\substack{i_1 i_2 \\ \alpha_1\alpha_2}}(x_1 - x_2) + F_{\mu\nu} v^{\mu\nu}_{\substack{i_1 i_2 \\ \alpha_1\alpha_2}}(x_1 - x_2)\right]$$

with $F_{\mu\nu} := iA_{[\mu}k_{\nu]}$; we remember that $I = (Z, x) = (\alpha, \kappa, i, x)$.

As a preliminary step we consider the direct product of two vector boson states:

$$\varphi^{(2)}_{Z_1 Z_2}(x_1, x_2) \otimes \tilde{\varphi}^{(2)}_{Z_3 Z_4}(x_3, x_4) \tag{6.115}$$

$$= N_G T^a_{\kappa_1 \kappa_2} \exp\left[-\frac{i}{2}k(x_1 + x_2)\right] \left[A_\mu u^\mu_{\underset{\alpha_1 \alpha_2}{i_1 i_2}}(x_1 - x_2) + F_{\mu\nu} v^{\mu\nu}_{\underset{\alpha_1 \alpha_2}{i_1 i_2}}(x_1 - x_2)\right]$$

$$\otimes T^b_{\kappa_3 \kappa_4} \exp\left[-\frac{i}{2}k(x_3 + x_4)\right] \left[A_\mu u^\mu_{\underset{\alpha_3 \alpha_4}{i_3 i_4}}(x_3 - x_4) + F_{\mu\nu} v^{\mu\nu}_{\underset{\alpha_3 \alpha_4}{i_3 i_4}}(x_3 - x_4)\right]$$

with a normalization constant N_G. This ansatz describes two bosons moving in directions k_1 and k_2 without any mutual interaction. If the bosons are assumed to constitute a bound state, according to our assumption their internal wave functions remain unchanged, but their center of mass coordinates become correlated.

We therefore modify (6.115) and make the following ansatz for the four-fermion spin 2 hard core state, which respects the symmetry conditions (3.97) and shows a definite four-momentum k:

$$\varphi^{(4,g)}_{Z_1 Z_2 Z_3 Z_4}(x_1, x_2, x_3, x_4|k) \tag{6.116}$$

$$= N_G t_{ab} T^a_{\kappa_1 \kappa_2} T^b_{\kappa_3 \kappa_4} \exp\left[-\frac{i}{4}k(x_1 + x_2 + x_3 + x_4)\right]$$

$$\times \left\{ u^\mu_{\underset{\alpha_1 \alpha_2}{i_1 i_2}}(x_1 - x_2) u^\nu_{\underset{\alpha_3 \alpha_4}{i_3 i_4}}(x_3 - x_4) X_{\mu\nu}[(x_1 + x_2) - (x_3 + x_4)|k] \right.$$

$$+ u^\mu_{\underset{\alpha_1 \alpha_2}{i_1 i_2}}(x_1 - x_2) v^{\rho\sigma}_{\underset{\alpha_3 \alpha_4}{i_3 i_4}}(x_3 - x_4) Y_{\mu\rho\sigma}[(x_1 + x_2) - (x_3 + x_4)|k]$$

$$+ v^{\rho\sigma}_{\underset{\alpha_1 \alpha_2}{i_1 i_2}}(x_1 - x_2) u^\mu_{\underset{\alpha_3 \alpha_4}{i_3 i_4}}(x_3 - x_4) \overline{Y}_{\rho\sigma\mu}[(x_1 + x_2) - (x_3 + x_4)|k]$$

$$\left. + v^{\mu\nu}_{\underset{\alpha_1 \alpha_2}{i_1 i_2}}(x_1 - x_2) v^{\rho\sigma}_{\underset{\alpha_3 \alpha_4}{i_3 i_4}}(x_3 - x_4) Z_{\mu\nu\rho\sigma}[(x_1 + x_2) - (x_3 + x_4)|k)] \right\}.$$

The normalization constant N_G and the (complex) coefficients t_{ab} will be determined later.

The tensors of the boson bonding $X_{\mu\nu}, Y_{\mu\rho\sigma}, \overline{Y}_{\rho\sigma\mu}, Z_{\mu\nu\rho\sigma}$ are unknown functions which have to be calculated by means of equation (6.110). The index k denotes the quantum numbers $\mathbf{k}, \ldots$ of the graviton state (6.117). Furthermore, we remark that the right hand side of (6.117) has to be simultaneously antisymmetrized in all indices $I_1 \ldots I_4$.

Analogously to the vector boson case of Section 6.3 from the covariant equation (6.110) one can derive secular equations for the determination of the bonding functions $X, Y, \overline{Y}, Z$. We will, however, not explicitly specify these lenghty equations but refer to [Stu 94].

As in the two-particle case we need the graviton single time functions. We observe that according to Section 6.3 the single time vector boson functions in a symmetric s wave approximation read

$$\varphi^{(2)}_{Z_1 Z_2}(\mathbf{r}_1, \mathbf{r}_2, t | a, k) \tag{6.117}$$

$$= N_B T^a_{\kappa_1 \kappa_2} e^{ik_0 t} e^{-ik(\mathbf{r}_1 - \mathbf{r}_2)/2}$$

$$\times \left[A_\mu f^A_{i_1 i_2}(\mathbf{r}_1 - \mathbf{r}_2)(\gamma^\mu C)_{\alpha_1 \alpha_2} + F_{\mu\nu} f^F_{i_1 i_2}(\mathbf{r}_1 - \mathbf{r}_2)(\Sigma^{\mu\nu} C)_{\alpha_1 \alpha_2} \right],$$

where $F_{\mu\nu} := iA_{[\mu} k_{\nu]}$.

The corresponding single time 'graviton' function is given by

$$\varphi^{(4,g)}_{Z_1 Z_2 Z_3 Z_4}(\mathbf{r}_1, \mathbf{r}_2, \mathbf{r}_3, \mathbf{r}_4, t | k) \tag{6.118}$$

$$= N_G t_{ab} T^a_{\kappa_1 \kappa_2} T^b_{\kappa_3 \kappa_4} e^{ik_0 t} e^{-ik(\mathbf{r}_1 + \mathbf{r}_2 + \mathbf{r}_3 + \mathbf{r}_4)/4}$$

$$\times \Big\{ f^A_{i_1 i_2}(\mathbf{r}_1 - \mathbf{r}_2) f^A_{i_3 i_4}(\mathbf{r}_3 - \mathbf{r}_4)(\gamma^\mu C)_{\alpha_1 \alpha_2}(\gamma^\nu C)_{\alpha_3 \alpha_4}$$

$$\times X_{\mu\nu}[(\mathbf{r}_1 + \mathbf{r}_2) - (\mathbf{r}_3 + \mathbf{r}_4)|k]$$

$$+ f^A_{i_1 i_2}(\mathbf{r}_1 - \mathbf{r}_2) f^F_{i_3 i_4}(\mathbf{r}_3 - \mathbf{r}_4)(\gamma^\mu C)_{\alpha_1 \alpha_2}(\Sigma^{\rho\sigma} C)_{\alpha_3 \alpha_4}$$

$$\times Y_{\mu\rho\sigma}[(\mathbf{r}_1 + \mathbf{r}_2) - (\mathbf{r}_3 + \mathbf{r}_4)|k]$$

$$+ f^F_{i_1 i_2}(\mathbf{r}_1 - \mathbf{r}_2) f^A_{i_3 i_4}(\mathbf{r}_3 - \mathbf{r}_4)(\Sigma^{\rho\sigma} C)_{\alpha_1 \alpha_2}(\gamma^\mu C)_{\alpha_3 \alpha_4}$$

$$\times \overline{Y}_{\rho\sigma\mu}[(\mathbf{r}_1 + \mathbf{r}_2) - (\mathbf{r}_3 + \mathbf{r}_4)|k]$$

$$+ f^F_{i_1 i_2}(\mathbf{r}_1 - \mathbf{r}_2) f^F_{i_3 i_4}(\mathbf{r}_3 - \mathbf{r}_4)(\Sigma^{\mu\nu} C)_{\alpha_1 \alpha_2}(\Sigma^{\rho\sigma} C)_{\alpha_3 \alpha_4}$$

$$\times Z_{\mu\nu\rho\sigma}[(\mathbf{r}_1 + \mathbf{r}_2) - (\mathbf{r}_3 + \mathbf{r}_4)|k] \Big\}$$

With the ansatz

$$X_{\mu\nu}(\mathbf{r}|k) = \theta^X(\mathbf{r}) X_{\mu\nu}(k), \tag{6.119}$$

$$Y_{\mu\rho\sigma}(\mathbf{r}|k) = \theta^Y(\mathbf{r}) Y_{\mu\rho\sigma}(k),$$

$$\overline{Y}_{\rho\sigma\mu}(\mathbf{r}|k) = \theta^{\overline{Y}}(\mathbf{r}) \overline{Y}_{\rho\sigma\mu}(k),$$

$$Z_{\mu\nu\rho\sigma}(\mathbf{r}|k) = \theta^Z(\mathbf{r}) Z_{\mu\nu\rho\sigma}(k)$$

for the tensor functions $X, Y, \overline{Y}, Z$ and by assuming the functions $\theta^X, \theta^Y, \theta^{\overline{Y}}, \theta^Z$ to be even functions (which corresponds to our restriction to s states with vanishing orbital momentum) we can approximately calculate these functions, see [Stu 93b], [Bor 98]; for the functions θ^Y and θ^Z these calculations yield

$$\theta^Y(\mathbf{r}) = -g \frac{1}{2^4 \pi^2} r^2 K_1 [2mr]^2, \tag{6.120}$$

$$\theta^Z(\mathbf{r}) = -g \frac{3^2 5^2}{2^5 \pi^2} (mr)^2 K_1 [2mr]^2 \tag{6.121}$$

with the subfermion coupling constant g and the mean subfermion mass m.

If by means of the ansatz (6.117) (approximate) solutions of (6.110) are to be derived, owing to the eigenvalue conditions the bonding functions $X, Y, \overline{Y}, Z$ are mutually correlated. On the other hand, these bonding functions are closely related to the effective gravitational field quantities, which are the dynamical variables of the effective theory.

Thus for the application of Weak Mapping we are compelled to assume the independence of these quantities, *i.e.*, we do not maintain their mutual correlations following from (6.110). This procedure is in complete analogy to the vector boson case with the bonding functions A_μ and $F_{\mu\nu}$.

In the present case these quantities are given by $X_{\mu\nu}(k)$, $Y_{\mu\rho\sigma}(k)$, $\overline{Y}_{\rho\sigma\mu}(k)$ and $Z_{\mu\nu\rho\sigma}(k)$. The introduction of such independent quantities will be discussed in detail in the following chapters.

The possible quantum numbers of the states are the energy–momentum, spin, isospin and the subfermion number. Owing to our ansatz we have an eigenstate of the center of mass momentum k. In addition we assume $k^2 = m_G^2$, *i.e.*, we assume our state to be on the mass shell m_G with a unique graviton mass m_G, and because of $k^2 = m_G^2$ we have $X_{\mu\nu}(k) = X_{\mu\nu}(\mathbf{k})$ etc.. We do not fix this graviton mass m_G for the moment as the effective dynamical equations induce a mass renormalization which enables us to derive equations for gravitons with an effective renormalized mass $m_G' = 0$.

An analysis of the spin content of the states (6.118) shows, that these states in general are not spin eigenstates but contain spin 0, spin 1 and spin 2 parts. We do not restrict this variety of spins in our bound states at this stage but leave it to the effective dynamics to select the corresponding spin properties.

Furthermore, the bound state equations (6.109) do not determine the isospin–superspin dependence of the bound states. Thus we have to make a suitable ansatz for this part of the graviton states. As the gravitons should have a universal coupling to matter, the graviton bound states must be formed just from the two elementary subfermion fields with isospin $A = 1$ and $A = 2$ and their conjugated fields. Thus the graviton states have to be eigenstates to isospin 0 and subfermion number 0; according to Section 6.4 this determines the states (6.118) to be given by

$$t_{ab}T^a_{\kappa_1\kappa_2}T^b_{\kappa_3\kappa_4} = T^{(1}_{\kappa_1\kappa_2}T^{2)}_{\kappa_3\kappa_4} \tag{6.122}$$

in their isospin–superspin part with

$$T^1_{\kappa_1\kappa_2} \quad := \quad -\frac{1}{2}(\gamma^0\gamma^5 C + \gamma^5 C)_{\kappa_1\kappa_2} \tag{6.123}$$

$$T^2_{\kappa_1\kappa_2} \quad := \quad \frac{1}{2}(\gamma^0\gamma^5 C - \gamma^5 C)_{\kappa_1\kappa_2}.$$

If we summarize our results, the 'graviton' states are given by

$$\varphi^{(4,g)}_{Z_1 Z_2 Z_3 Z_4}(\mathbf{r}_1, \mathbf{r}_2, \mathbf{r}_3, \mathbf{r}_4, t|\mathbf{k}) \tag{6.124}$$

$$= \; N_G T^{(1}_{\kappa_1\kappa_2} T^{2)}_{\kappa_3\kappa_4} e^{iE_k t} e^{-i\mathbf{k}(\mathbf{r}_1+\mathbf{r}_2+\mathbf{r}_3+\mathbf{r}_4)/4}$$

$$\times \Big[f^A_{i_1 i_2}(\mathbf{r}_1 - \mathbf{r}_2) f^A_{i_3 i_4}(\mathbf{r}_3 - \mathbf{r}_4)\theta^X[(\mathbf{r}_1 + \mathbf{r}_2) - (\mathbf{r}_3 + \mathbf{r}_4)]$$

$$\times (\gamma^\mu C)_{\alpha_1\alpha_2}(\gamma^\nu C)_{\alpha_3\alpha_4} X_{\mu\nu}(\mathbf{k})$$

$$+ f^A_{i_1 i_2}(\mathbf{r}_1 - \mathbf{r}_2) f^F_{i_3 i_4}(\mathbf{r}_3 - \mathbf{r}_4)\theta^Y[(\mathbf{r}_1 + \mathbf{r}_2) - (\mathbf{r}_3 + \mathbf{r}_4)]$$

$$\times (\gamma^\mu C)_{\alpha_1\alpha_2}(\Sigma^{\rho\sigma} C)_{\alpha_3\alpha_4} Y_{\mu\rho\sigma}(\mathbf{k})$$

$$+f^F_{i_1i_2}(\mathbf{r}_1 - \mathbf{r}_2)f^A_{i_3i_4}(\mathbf{r}_3 - \mathbf{r}_4)\theta^Y[(\mathbf{r}_1 + \mathbf{r}_2) - (\mathbf{r}_3 + \mathbf{r}_4)]$$
$$\times(\Sigma^{\rho\sigma}C)_{\alpha_1\alpha_2}(\gamma^\mu C)_{\alpha_3\alpha_4}\overline{Y}_{\rho\sigma\mu}(\mathbf{k})$$
$$+f^F_{i_1i_2}(\mathbf{r}_1 - \mathbf{r}_2)f^F_{i_3i_4}(\mathbf{r}_3 - \mathbf{r}_4)\theta^Z[(\mathbf{r}_1 + \mathbf{r}_2) - (\mathbf{r}_3 + \mathbf{r}_4)]$$
$$\times(\Sigma^{\mu\nu}C)_{\alpha_1\alpha_2}(\Sigma^{\rho\sigma}C)_{\alpha_3\alpha_4}Z_{\mu\nu\rho\sigma}(\mathbf{k})\Big]_{\text{as}[I_1,I_2,I_3,I_4]}$$

The bound states (6.124) are the starting point for the derivation of an effective gravitation theory in Chapter 9, and only by the result of this derivation it can be decided whether or not these graviton 'test' functions are a suitable starting point for Weak Mapping.

6.7 Bound State Stability

From the considerations of the preceding chapter it follows that the various single bound states, $i.e.$, single composite particle states, are the elements by means of which the whole state space of a composite particle theory can be generated, $i.e.$, the state space contains all possible many-particle configurations of free composite particles. In the absence of interactions such free composite many-particle states should be stable, $i.e.$, in the course of time their norms have to be conserved. This is the case if the norms of their single constituents are conserved. Hence we study norm conservation of single composite particles.

We start this investigation with the discussion of the norm conservation of single hard core subfermion states themselves. In order to obtain an idea how to proceed we first consider free Dirac spinors $\psi^f(x)$.

The free Dirac equation with mass m reads

$$(i\hbar\gamma^\mu\partial_\mu - cm)\,\psi^f(x) = 0, \tag{6.125}$$

and for its solutions the conservation law

$$\partial^\mu j_\mu(x) = 0 \tag{6.126}$$

with

$$j_\mu(x) := c\bar{\psi}^f(x)\gamma_\mu\psi^f(x) \tag{6.127}$$

can be derived. The density $\rho = j_0$ is given by

$$\rho(\mathbf{r}, t) := \psi^f(\mathbf{r}, t)^+\psi^f(\mathbf{r}, t) = \sum_\alpha \psi^f_\alpha(\mathbf{r}, t)^*\psi^f_\alpha(\mathbf{r}, t) \tag{6.128}$$

and is thus positive definite. Therefore the quantum mechanical Dirac theory supports a probability interpretation. Connected with this interpretation is the metric of the corresponding state space. In particular the norm expression is given by

$$\| \psi^f \|^2 := \sum_\alpha \int \psi^f_\alpha(\mathbf{r}, t)^*\psi^f_\alpha(\mathbf{r}, t)d^3r, \tag{6.129}$$

and as a result of (6.126) the norm is conserved. The associated inner product reads

$$\langle \varphi^f | \psi^f \rangle := \sum_\alpha \int \varphi_\alpha^f(\mathbf{r}, t)^* \psi_\alpha^f(\mathbf{r}, t) d^3 r, \tag{6.130}$$

and in symbolic notation it can be written

$$\langle \varphi^f | \psi^f \rangle = \varphi_I^* G_{II'}^{-1} \psi_{I'} = \varphi_I^* G^{II'} \psi_{I'} \tag{6.131}$$

with

$$G_{II'} := \delta_{\alpha\alpha'} \delta(\mathbf{r} - \mathbf{r}') = G^{II'}. \tag{6.132}$$

With respect to the transformation properties of the scalar product (6.130) we insert the spinor transformation

$$\psi_\alpha(x) = S(a)_{\alpha\alpha'}^{-1} \psi_{\alpha'}'(x') \tag{6.133}$$

for ψ and φ into (6.130) and observe that (6.130) is invariant only under the little group of classes (m_+) and (m_-), i.e.,

$$\langle \varphi^f | \psi^f \rangle \stackrel{\text{little group}}{=} \langle \varphi^{f'} | \psi^{f'} \rangle. \tag{6.134}$$

Only for the little group the transformation matrices $S(a)$ are unitary and cancel each other in (6.130) and the volume element is invariant. Altogether this is a remarkable result:

Even in the most elementary case of a relativistically invariant dynamics, norm conservation, algebraic state representation, symmetry transformations and the dynamical law are intimately connected.

Naturally we expect this to hold in more complicated cases, too. Before continuing this discussion we remark that usually with the transition from the quantum mechanical interpretation to the field-theoretic interpretation of the Dirac equation a change of the representation is performed by the introduction of the Dirac vacuum. This vacuum enforces the definition of particles and antiparticles with opposite charges. But with respect to (unobservable) subfermions with a sole spinorial self-interaction there is no reason to introduce charges.

In addition the definition of distinguishable particles and antiparticles breaks the antisymmetry of the original spinor field wave functions which does not fit into our calculational program. So we apply the Dirac vacuum and representation only to the case of the treatment of effective theories, where one intends to describe real physical particles with positive energies.

Next we turn to the subfermions. As a consequence of global gauge invariance of the spinor field Lagrangian (2.52) under the transformations (2.118), (2.120) we have conserved currents, in particular the abelian current

$$j^\mu(x) := \sum_i \lambda_i^{-1} \bar{\psi}_{A\alpha i}(x) \gamma_{\alpha\beta}^\mu \psi_{A\beta i}(x) \tag{6.135}$$

with

$$\partial_\mu j^\mu(x) = 0. \tag{6.136}$$

In quantum mechanical interpretation the density $\rho = j_0$ is not positive definite as the λ_i have alternating signs.

As we are interested in the quantum field theoretic version of (2.52) we study the hardcore subfermion sector. According to (2.68) the spinor field operators ψ_I are defined by $\psi_I := \psi_{A\alpha i\Lambda}(\mathbf{r}, t) \equiv \left(\psi_{A\alpha i}(\mathbf{r}, t), \psi^C_{A\alpha i}(\mathbf{r}, t) \right)$ with $\psi^C_{A\alpha i} := (C\gamma^0)_{\alpha\beta}\psi^+_{A\beta i}$, where with regard to the single time formalism we suppress the time coordinate in the following. According to (4.43) the lowest order of the metric tensor in the fermion sector reads

$$G_{II'} \equiv G_{A\alpha i\Lambda, A'\alpha' i'\Lambda'}(\mathbf{r}, \mathbf{r}') = \langle 0|\psi_{A\alpha i\Lambda}(\mathbf{r})\psi^+_{A'\alpha' i'\Lambda'}(\mathbf{r}')|0\rangle. \tag{6.137}$$

To relate (6.137) to state representations we consider the matrix elements of single dressed subfermion states $\varphi^{(1)}_{A\alpha i\Lambda}(\mathbf{r}|k)$ and classify their quantum numbers. By definition these matrix elements belong to state functionals which are solutions of the full energy equation (4.100). But apart from an unknown normalization constant for single subfermions and the mass μ_S, their quantum numbers and their functional form solely follow from symmetry considerations. In particular the symbolic quantum number k is composed of the three-momentum $\mathbf{k}$, the dressed subfermion mass μ_S, the spin σ, the sign of energy z, the superspinor–isospinor numbers a and the auxiliary field number j. For the following we need a more detailed breakdown of a.

With respect to a we observe that the quantized fields ψ and ψ^c are considered as independent from each other, *i.e.*, that there exist solutions either for ψ or for ψ^c which are compatible with (4.100) and lead to state quantum numbers $\Gamma = \gamma_1, \gamma_2$ or equivalently to fermion quantum numbers $\gamma_1, \gamma_2 = 1, -1$ as a consequence of the global subfermion $U(1)$ gauge group.

Furthermore, the isospin quantum numbers can simply be expressed by $B = 1, 2$ and number the eigenvalues t_1, t_2 of τ_3. For simplicity we suppress μ_S in the following and combine σ and z in $s = (\sigma, z)$. Then the symbolic quantum number k can be represented by $k \equiv (\mathbf{k}, s, \Gamma, B, j)$ and the matrix elements are given by

$$\varphi^{(1)}_{A\alpha i\Lambda}(\mathbf{r}|k) := \langle 0|\psi_{A\alpha i\Lambda}(\mathbf{r})|k\rangle. \tag{6.138}$$

Having unambiguously defined these matrix elements we can take over Proposition 5.6 from [Stu 94]. It reads

- **Proposition 6.7:** The metric tensor (6.137) of the one-fermion sector is uniquely determined by the one-fermion transition matrix elements (6.138) and by the anticommutators (2.72). Apart from a universal normalization constant it is given by

$$G_{II'} = \frac{1}{2}\lambda_i \delta_{ii'} \delta_{AA'} \delta_{\Lambda\Lambda'} \delta_{\alpha\alpha'} \delta(\mathbf{r} - \mathbf{r}'). \tag{6.139}$$

If instead of dressed subfermions free subfermions are considered, the normalization constant is $\mathcal{N} = 1$.

In [Stu 94] free subfermions are treated, but apart from the universal normalization constant for dressed subfermion states the proof is the same. So we refer to [Stu 94], as this proof is rather lengthy.

If we represent the density j_0 of the current (6.135) in the manner of (6.131) we observe that both in quantum mechanics and in quantum field theory the metric tensor $G_{II'}^{-1} \equiv G^{II'}$ is given by

$$G^{II'} = 2\lambda_i^{-1}\delta_{ii'}\delta_{AA'}\delta_{\Gamma\Gamma'}\delta_{\alpha\alpha'}\delta(\mathbf{r} - \mathbf{r}'). \tag{6.140}$$

Hence a strict deduction of the state space metric in quantum field theory leads in the lowest order fermion sector to the same result as in ordinary Dirac theory, in particular with respect to the symmetry properties. As for reasons of consistency the higher sectors of the quantum field-theoretic metric tensor cannot deviate from the transformation properties of the lowest sector, this means that also in the quantum field theory of subfermions the proper state construction is connected with the Hamiltonian formalism and the invariance of the metric and of inner products under the little group. Of course this result holds independently of the auxiliary fields having, as a consequence, indefinite λ_I, and it can be directly read or concluded from (4.43).

In this deduction of the lowest order subfermion metric (6.140) only symmetry properties of the hard core subfermion states were used for the proof. But this is an exception. In the general case of composite particles, symmetries are not enough for deducing norms, scalar products, and norm conservation. Although in Section 4.8 we derived by general arguments the relativistic invariance of scalar products for composite particle states this statement is of no practical help as long as the hypothetical covariant left hand solutions are not explicitly known, and from the above remarks it is obvious that in addition to symmetries a dynamical law is also needed for a successful treatment of these problems.

Therefore we generalize and abstract these considerations by the statement: *In contrast to the interaction representation, in the nonperturbative Algebraic Schrödinger Representation we leave it to the field dynamics to generate configurations which correspond to relativistic particles. This includes that the information about composite particle configurations, i.e., the state space cannot be a priori a representation space of the full inhomogeneous Lorentz group. Rather in order to allow the dynamical law to generate such representations, the state space can only be a representation space of the little group in accordance with the preceding statements.*

Nevertheless: *If, hypothetically, the spinor field self-interaction between the subfermionic constituents of different composite particles is switched off, the evolution of the free composite particle state space must lead to representations of the inhomogeneous Lorentz group together with norm conservation and probability conservation.*

A simple example of this mechanism is the invariance of the scalar product (6.134) under the little group, where it is left to the Dirac equation to generate by time evolution the full Lorentz covariant Dirac spinor.

If one intends to apply this principle to the more complicated composite particle states treated in this chapter, the exact solutions of the single composite particle equations cannot be used for the construction of an appropriate state space in the nonperturbative Algebraic Schrödinger Representation. As these exact solutions are already elements of a representation space of the full Lorentz group they carry too much information and leave no room for the influence of the dynamical law on the state formation. Thus exact states must be decomposed into suitable pieces which are endowed with elements which can be modulated by the dynamical law. This means that for these pieces no genuine dynamical equations exist and neither norms nor norm conservation can be deduced with their help.

In this situation we take up the proposal of Section 2.2: *the only information about the composite particles can be gained from the effective theories.* And in this setting it is also possible to study the properties of free composite particle states.

We demonstrate this principle for the case of vector bosons. Their free motion is determined by the hard core equations (6.1) and (6.2) for single boson states. But instead of the treatment of the exact solutions in Sections 6.2 and 6.3, we now consider the decomposed solutions (7.56), where the quantities A_μ and $F_{\mu\nu}$ are assumed to be independent free variables. This assumption is the very heart of Weak Mapping, because these new quantities allow the formulation and description of dynamical boson reactions and processes, and it leads to the effective boson dynamics in terms of functional equations.

If the full interaction of the spinor field is switched on, the effective boson dynamics is given by Proposition 5.2. Parts of the corresponding functional equation (5.25) represent the effective functional energy operator for free many-particle boson states. In order to study these states we anticipate the results of Chapter 7.

For brevity we refer only to formula numbers without explicitly reproducing them. The effective functional energy operator for free boson states is given by (7.86) and evaluated by Prop. 7.2. If we define the free boson functional state by $|B\rangle^0$ the corresponding equation reads

$$\mathcal{H}_b^0 |B\rangle^0 = \Delta E_0 |B\rangle^0. \tag{6.141}$$

For free many-particle boson states the ansatz (7.114), (7.115) with $t = 0$ is an exact solution of (6.141) as the corresponding boson matrix elements are uncorrelated, *i.e.*, factorize. For convenience in this ansatz the magnetic field was introduced by definition (7.109). If we substitute (7.114) in (6.141), the set of equations

$$\begin{aligned}
-i\Delta E_0 A_a^i(\mathbf{r}, 0) &= -E_a^i(\mathbf{r}, 0), \tag{6.142}\\
-i\Delta E_0 B_a^i(\mathbf{r}, 0) &= \epsilon_{ijk}\partial_j(\mathbf{r})E_a^k(\mathbf{r}, 0),\\
-i\Delta E_0 E_a^i(\mathbf{r}, 0) &= -\epsilon_{ijk}\partial_j(\mathbf{r})B_a^k(\mathbf{r}, 0)
\end{aligned}$$

results, where the quantities A, B, E are assumed to be transition matrix elements, *i.e.*, $X \equiv \langle 0|X|b\rangle$ for the corresponding field operators. Now we leave it to the system dynamics to generate a relativistic motion of the bosons. In this simple case we merely replace $-i\Delta E_0$ by ∂_0 and $X(\mathbf{r}, 0)$ by $e^{i\omega t}X(\mathbf{r}, 0)$ with $\omega \equiv \Delta E_0$ and obtain from (6.142) the set

$$\begin{aligned}
\partial_0 A_a^i(x) &= -E_a^i(x), \tag{6.143}\\
\partial_0 B_a^i(x) &= -\epsilon_{ijk}\partial_j E_a^k,
\end{aligned}$$

$$\partial_0 E_a^i(x) = \epsilon_{ijk}\partial_j B_a^k.$$

Apart from the transversality condition this system represents the Maxwell equations for free abelian and non-abelian bosons, and it admits plane wave solutions which represent dissipation-free relativistic motions of the bosons. From this result we conclude that the free boson states are stable although we have avoided any resort to the microscopic (subfermion) norm and norm conservation in our arguments.

6.8 Regularization and Probability Interpretation

So far we have formally developed the Algebraic Schrödinger Representation, but we have nothing said about regularization and physical interpretation. Owing to the necessity to perform a well defined canonical quantization of the spinor field model, its quantum field theory was formulated by way of auxiliary fields, and this means that in this formulation each auxiliary field can be separately considered and prepared for 'measurement'. The result is that the GNS matrix elements are singular and the state space metric becomes indefinite.

Obviously these difficulties can only be removed if the regularization of the original 'classical' spinor field equation can be transferred into its quantum field theory formulation. Considering perturbative Pauli–Villars regularization we remember that in the process of calculating vacuum expectation values the singular propagators are replaced by additive regularized expressions. Can one find a nonperturbative counterpart in the Algebraic Schrödinger Representation?

This problem and the related problem of probability interpretation was treated by Stumpf [Stu 00]. We give a simplified version and refer to the original paper for details.

In the absence of condensation phenomena we assumed the free auxiliary field propagator to be a reasonable approximation of the self-consistency calculation. We write it in the form

$$F^{\alpha_1\alpha_2}_{\substack{\kappa_1\kappa_2\\i_1i_2}}(x_1,x_2) \tag{6.144}$$

$$= -i\lambda_{i_1}\delta_{i_1i_2}(\gamma^5)_{\kappa_1\kappa_2}\frac{1}{(2\pi)^4}\int d^4p\, e^{-ip(x_1-x_2)}\left(\frac{1}{\gamma^\mu p_\mu - m_{i_1}}C\right)_{\alpha_1\alpha_2}.$$

Summation over i_1, i_2 and application of (2.26) yields

$$\overline{F}^{\alpha_1\alpha_2}_{\kappa_1\kappa_2}(x_1,x_2) := \sum_{i_1,i_2} F^{\alpha_1\alpha_2}_{\substack{\kappa_1\kappa_2\\i_1i_2}}(x_1,x_2) \tag{6.145}$$

$$= -i(\gamma^5)_{\kappa_1\kappa_2}\frac{1}{(2\pi)^4}\int d^4p\, e^{-ip(x_1-x_2)}\left(\prod_{i=1}^{3}\frac{1}{\gamma^\mu p_\mu - m_{i_1}}C\right)_{\alpha_1\alpha_2},$$

i.e., a regularized fermion propagator.

From this result it follows imperatively that the only way to achieve a nonperturbative regularization consists in summing over the auxiliary field indices. Therefore we define the physical, *i.e.*, regularized normal ordered state amplitudes $\hat{\varphi}^{(n)}$ by

$$\hat{\varphi}^{(n)}_{\substack{a_1 a_n \\ \kappa_1 \kappa_n}}(\mathbf{r}_1 \ldots \mathbf{r}_n | a) := \sum_{\substack{i_1 \ldots i_n \\ i_1 i_n}} \varphi^{(n)}_{\substack{a_1 a_n \\ \kappa_1 \kappa_n}}(\mathbf{r}_1 \ldots \mathbf{r}_n | a), \tag{6.146}$$

and, of course, apply this definition also to the covariant amplitudes (3.88). One immediately realizes that $\hat{\varphi}^{(n)}$ has the same transformation properties as the original $\varphi^{(n)} \; \forall \; n$.

Can one derive an associated dynamical law for (6.146), *i.e.*, the single time physical amplitudes? To answer this question we consider the hard core equation (6.1) in the two-fermion sector in order to avoid lengthy calculations and to argue as transparent as possible. Its explicit form is given by (6.32). We decompose the index $Z := (A, \alpha, \Lambda, i)$ into $Z = (z, i)$ with $z := (A, \alpha, \Lambda)$ and sum over i_1, i_2 in (6.32). Observing the definitions (2.70) one obtains

$$
\begin{aligned}
k_0 \hat{\varphi}^{(2)}_{z_1 z_2}(\mathbf{r}_1, \mathbf{r}_2) \;=\;& i D^0_{z_1 x_1} D^k_{x_1 x_2} \partial_k(\mathbf{r}_1) \hat{\varphi}^{(2)}_{x_2 z_2}(\mathbf{r}_1, \mathbf{r}_2) \\
&+ i D^0_{z_2 x_1} D^k_{x_1 x_2} \partial_k(\mathbf{r}_2) \hat{\varphi}^{(2)}_{z_1 x_2}(\mathbf{r}_1, \mathbf{r}_2) \\
&- i \sum_{i_1, i_2} \left[D^0_{Z_1 X_1} m_{X_1 X_2} \varphi^{(2)}_{X_2 Z_2}(\mathbf{r}_1, \mathbf{r}_2) + D^0_{Z_2 X_1} m_{X_1 X_2} \varphi^{(2)}_{Z_1 X_2}(\mathbf{r}_1, \mathbf{r}_2) \right] \\
&- 3i \sum_{i_1, i_2} \left[D^0_{Z_1 X_1} \hat{U}_{X_1 X_2 X_3 X_4} F_{X_4 Z_2}(\mathbf{r}_1 - \mathbf{r}_2) \varphi^{(2)}_{X_2 X_3}(\mathbf{r}_1, \mathbf{r}_1) \right. \\
&\qquad\qquad \left. - D^0_{Z_2 X_1} \hat{U}_{X_1 X_2 X_3 X_4} F_{X_4 Z_1}(\mathbf{r}_2 - \mathbf{r}_1) \varphi^{(2)}_{X_2 X_3}(\mathbf{r}_2, \mathbf{r}_2) \right].
\end{aligned}
\tag{6.147}
$$

Owing to $\sum_{i_1, i_2} D^0_{x_1 x_2} \delta_{i_1 i_2} \lambda_{i_2} = 0$ the last bracket in (6.147) vanishes and the equation

$$
\begin{aligned}
k_0 \hat{\varphi}^{(2)}_{z_1 z_2}(\mathbf{r}_1, \mathbf{r}_2) \;=\;& i D^0_{z_1 x_1} D^k_{x_1 x_2} \partial_k(\mathbf{r}_1) \hat{\varphi}^{(2)}_{x_2 z_2}(\mathbf{r}_1, \mathbf{r}_2) \\
&+ i D^0_{z_2 x_1} D^k_{x_1 x_2} \partial_k(\mathbf{r}_2) \hat{\varphi}^{(2)}_{z_1 x_2}(\mathbf{r}_1, \mathbf{r}_2) \\
&- i \sum_{i_1, i_2} \left[D^0_{Z_1 X_1} m_{X_1 X_2} \varphi^{(2)}_{X_2 Z_2}(\mathbf{r}_1, \mathbf{r}_2) + D^0_{Z_2 X_1} m_{X_1 X_2} \varphi^{(2)}_{Z_1 X_2}(\mathbf{r}_1, \mathbf{r}_2) \right]
\end{aligned}
\tag{6.148}
$$

results. Evidently (6.148) is no self-consistent equation for the calculation of $\hat{\varphi}^{(2)}$. Rather (6.148) brings about a connection between regularized physical amplitudes and auxiliary amplitudes. This holds *a fortiori* for the full equation (5.1) and (5.2) too, as the operator of equation (6.1) is part of the full equation (5.1).

Hence the auxiliary canonical field formulation in Algebraic Schrödinger Representation is indispensible for the calculation of the dynamical evaluation of the spinor field. But how can one work with the latter representation if it produces singular functions?

Without trying to give a full treatment of this subject, we concentrate on the development of composite particle theory. In this respect the following statements can be made:

i) Bound states of two subfermions (subfermion fields) can be exactly calculated in the auxiliary field formalism. These bound states have to transform themselves in a relativistically covariant way. Therefore their defining equations must be covariant, too. They are given by (6.2) with the associated energy equation (6.1). The relation between covariant solutions of (6.2) and noncovariant (single time) solutions of (6.1) is established in Proposition 6.2.

The exact solutions of (6.2) for vector boson states are derived in Proposition 6.4. From this proposition it follows that the associated secular equation for the vector boson mass μ_B can be exclusively formulated by means of fully regularized Green's functions and propagators (6.11) and (6.12), in spite of the use of the auxiliary field formalism. For the explicit numerical evaluation of this secular equation which leads to finite values of μ_B we refer to the thesis of Sand [Sand 91]. The corresponding exact vector boson states depend on the auxiliary field indices, and as a consequence of this they contain singular distributions. In contrast to that, the physical amplitudes $\hat{\varphi}^{(2)}$ are fully regularized as can be verified by summation over i_1, i_2 in (6.42).

ii) In the auxiliary field formalism the single time normal ordered state amplitudes $\hat{\varphi}^{(n)}$ are assumed to be representable by expansions into series of m-particle field states, $m < n$. Such expansions can (to some extent) be justified by insertion of intermediate states. For instance, the corresponding series expansions for $m = 2$ are given by

$$\varphi^{(2n)}(I_1 \dots I_{2n}|a) = \sum_{k_1 \dots k_n} C^{[I_1 I_2}_{k_1} \dots C^{I_{2n-1} I_{2n}]}_{k_n} \rho^{(n)}(k_1 \dots k_n|a), \qquad (6.149)$$

where $\left\{ C^{II'}_k \right\}$ is a complete set of two-particle states derived from the solutions of the hard core equation.

Hence the singularities of $\varphi^{(2n)}$ are expressed by the singularities of their substates, i.e., of the $C^{II'}_k$. The occurrence of such singularities in the set $\left\{ \varphi^{(2n)} \right\}$, however, does not disturb the derivation of an effective dynamics for these composite states. In order to guarantee a finite map in Chapter 5, dual states (test functions) are introduced by (5.5) which completely compensate the singularities of the original states and lead to finite expressions of the effective dynamics. In addition, if $\hat{\varphi}^{(2n)}$ is formed one obtains a regular function, as in (6.149) all $\widehat{C}^{II'}_k$ are regular.

From this discussion it follows that at least in the realm of composite particle theory the classical regularization of the spinor field can be transferred into its quantized version as far as energies and states are concerned. But this rather subtle relation between physical states and its auxiliary field counterparts is not enough to ensure the existence of quantum observables.

What additionally is needed is the specification of an inner product space which allows a probability interpretation of the physical states. *This inner product space cannot be the Krein space of the auxiliary field states, as by definition these states are unphysical states. But in contrast to the map (6.146) between auxiliary field amplitudes and physical amplitudes, no relation between the auxiliary field space metric and the physical metric can be established.* In the inner product space of the auxiliary field states $\langle a|, |b\rangle$ given by

$$\langle a|b \rangle = \sum_{m,n} \tau_m^*(I_1' \ldots I_m'|a) G^{I_1' \ldots I_m', I_1 \ldots I_n} \tau_n(I_1 \ldots I_n|b) \tag{6.150}$$

(compare Section 4.3), no independent summation over $i_1' \ldots i_m'$ and $i_1 \ldots i_n$ is possible which would allow to apply (6.146).

As for a closed system without external forces the inner product structure has to be self-consistently defined in accordance with the system dynamics, the only way to proceed is the use of the dynamical equations. Again we demonstrate this for the two-fermion sector. With

$$\hat{\varphi}_{z_1 z_2}^{(2)}(\mathbf{r}_1, \mathbf{r}_2, t) = e^{-iEt} \hat{\varphi}_{z_1 z_2}^{(2)}(\mathbf{r}_1, \mathbf{r}_2) \tag{6.151}$$

equation (6.148) goes over into

$$\frac{\partial}{\partial t} \hat{\varphi}_{z_1 z_2}^{(2)}(\mathbf{r}_1, \mathbf{r}_2, t) \tag{6.152}$$

$$= -D_{z_1 x_1}^0 D_{x_1 x_2}^k \partial_k(x_1) \hat{\varphi}_{x_2 z_2}^{(2)}(x_1, x_2) - D_{z_2 x_1}^0 D_{x_1 x_2}^k \partial_k(x_2) \hat{\varphi}_{z_1 x_2}^{(2)}(x_1, x_2)$$

$$+ \sum_{i_1, i_2} \left[D_{Z_1 X_1}^0 m_{X_1 X_2} \varphi_{X_2 Z_2}^{(2)}(x_1, x_2) + D_{Z_2 X_1}^0 m_{X_1 X_2} \varphi_{Z_1 X_2}^{(2)}(x_1, x_2) \right].$$

From this equation one can derive a current conservation law. For abbreviation we suppress all indices and coordinates aside from the auxiliary field indices. With $\alpha^k(1) := \alpha^k \otimes 1$ and $\alpha^k(2) := 1 \otimes \alpha^k$ and with $\hat{\varphi} \equiv \hat{\varphi}^{(2)}$ we obtain

$$\frac{\partial}{\partial t}(\hat{\varphi}^+ \hat{\varphi}) + \partial_k^1(\hat{\varphi}^+ \alpha^k(1) \hat{\varphi}) + \partial_k^2(\hat{\varphi}^+ \alpha^k(2) \hat{\varphi}) \tag{6.153}$$

$$+ i\hat{\varphi}^+ \beta(1) \sum_{i_1, i_2} \varphi_{i_1 i_2} m_{i_1} - i \sum_{i_1, i_2} \varphi_{i_1 i_2}^+ m_{i_1} \beta(1) \hat{\varphi}$$

$$+ i\hat{\varphi}^+ \beta(2) \sum_{i_1, i_2} \varphi_{i_1 i_2} m_{i_2} - i \sum_{i_1, i_2} \varphi_{i_1 i_2}^+ m_{i_2} \beta(2) \hat{\varphi} = 0.$$

We remember that (6.153) is no self-consistent equation for $\hat{\varphi}^{(2)}$ but only an identity following from the time-dependent version of (6.32). Therefore the most general functions which identically satisfy (6.153) are derived from linear combinations of time-dependent energy eigensolutions of (6.32), i.e.,

$$\hat{\varphi}_{z_1 z_2}^{(2)}(\mathbf{r}_1, \mathbf{r}_2, t) = \sum_k c_k e^{-iE_k t} \hat{\varphi}_{z_1 z_2}^{(2)}(\mathbf{r}_1, \mathbf{r}_2). \tag{6.154}$$

We substitute the corresponding functions into (6.153) and consider the second line of (6.153). This line then reads

$$i \sum_k c_k^* e^{iE_k t} \hat{\varphi}^+(k) \beta(1) \sum_{k'} \sum_{i_1, i_2} \varphi_{i_1 i_2}(k') m_{i_1} c_{k'} e^{-iE_{k'} t} \tag{6.155}$$

$$- i \sum_k \sum_{i_1, i_2} \varphi_{i_1 i_2}^+(k) m_{i_1} c_k^* e^{iE_k t} \beta(1) \sum_{k'} c_{k'} e^{-iE_{k'} t} \hat{\varphi}(k'),$$

and a corresponding expression is obtained for the third line of (6.153).

We now exclude all energy eigenstates from the expansion (6.154) which refer to the auxiliary field indices i_1 or i_2. This exclusion corresponds to the decoupling of the subfermion sector from the bound state sector. Then (6.155) goes over into

$$\sum_{k,k'} c_k^* c_{k'} e^{i(E_k - E_{k'})t} \left[\hat{\varphi}^+(k)\beta(1) \sum_{i_1,i_2} \varphi_{i_1 i_2}(k') m_{i_1} - \sum_{i_1,i_2} \varphi_{i_1 i_2}^+(k) m_{i_1} \beta(1)\hat{\varphi}(k') \right]. \quad (6.156)$$

Then with $m_i = m + \delta m_i$ the identity (6.153) reads

$$\frac{\partial}{\partial t}(\hat{\varphi}^+ \hat{\varphi}) + \partial_k^1(\hat{\varphi}^+ \alpha^k(1)\hat{\varphi}) + \partial_k^2(\hat{\varphi}^+ \alpha^k(2)\hat{\varphi}) \quad (6.157)$$

$$+ \sum_{k,k'} c_k^* c_{k'} e^{i(E_k - E_{k'})t}$$

$$\times \left\{ \sum_{\rho=1}^{2} \left[\hat{\varphi}^+(k)\beta(\rho) \sum_{i_1,i_2} \varphi_{i_1 i_2}(k')\delta m_{i_\rho} - \sum_{i_1,i_2} \varphi_{i_1 i_2}^+(k)\delta m_{i_\rho}\beta(\rho)\hat{\varphi}(k') \right] \right\} = 0,$$

and for vanishing last term we have current conservation.

Without any assumption about $\varphi^{(2)}$ the last term vanishes in the limit of vanishing mass differences δm_i. This situation resembles Heisenberg's introduction of a dipole ghost with the essential difference that the transition to a dipole ghost and a subfermion for $\delta m_i \to 0$ is performed only in the regularized $\hat{\varphi}^{(n)}$ after all calculations were done.

Then from (6.157) it follows that $\hat{\varphi}^+ \hat{\varphi}$ is the density. This means: the physical state amplitudes $\hat{\varphi}^{(2)}$ describe stable bound states and are elements of a corresponding Hilbert space.

Without explicit proof it is obvious that this statement holds for higher order bound states $\varphi^{(n)}$, $n > 2$, too. Thus within the nonperturbative bound state formalism we have found the corresponding metric which is in accordance with the field dynamics and which makes it possible to calculate physical observables for comparison with experiments.

We can summarize the results of nonperturbative Pauli–Villars regularization of the spinor field model in the following way:

The hard core equations (6.1) and (6.2) possess finite self-energy solutions for composite particle (field) states (if one extrapolates from vector bosons to other species!). The associated 'wave functions' $\varphi^{(m)}$ are singular, but the physical amplitudes $\hat{\varphi}^{(m)}$ are regularized.

By means of a series expansion of $\varphi^{(n)}$ in terms of composite particle bound states and by the introduction of corresponding dual states a singularity-free dynamical law for the effective dynamics of the substates can be derived, and $\hat{\varphi}^{(n)}$ becomes regularized. The single composite particle bound states $\hat{\varphi}^{(m)}$ are elements of a corresponding Hilbert space in accordance with the field dynamics, and every $\hat{\varphi}^{(n)}$ which is composed of these substates is an element of a Hilbert space, too.

According to this summary a probability interpretation of the physical state space should be possible. As we are mainly interested in the derivation of effective dynamics for composite particles and as no sufficient information about the whole state space of the spinor field is available at present, we only make a rough sketch of a probability interpretation of a regularized theory which differs greatly from the other treatments of indefinite metric theories.

In the other attempts undertaken so far emphasis was laid on a physical probability interpretation of the original ghost field theories themselves, see for instance the older book by Nagy [Nag 66] or the recent treatment of Faddeev–Popov ghosts, etc., by Nakanishi and Ojima [Nak 90]. In our approach the auxiliary (ghost) field theory is considered as unphysical and in accordance with our discussion the physical theory is separated from the unphysical theory by a noninvertible map.

On the other hand, the unphysical theory is indispensible, as it provides the basis for that map. To express these facts quantitatively we formulate them by some assumptions.

- **Assumption 6.1:** In the effective theories of composite subfermion bound states owed to extremely large masses the sector of free and scattered subfermions decouples from the bound state sector.

- **Assumption 6.2:** The energies of the composite particles themselves and of the many-composite particle states do not depend on the auxiliary field indices, i.e., on subfermion numbers.

- **Assumption 6.3:** In the bound state sector the set of all possible composite many-particle states is complete.

- **Assumption 6.4:** In the regularized, i.e., physical, amplitudes in the final step of all calculations the transition to a massive subfermion of mass m and a massive dipole ghost at m has to be performed.

These assumptions summarize what more or less tacitly is used for the derivation of effective theories in the preceeding and the following chapters and what leads to current conservation for the physical amplitudes, i.e., to their probability interpretation. As we are mainly interested in the derivation of effective theories we do not continue the probability discussion but we only remark that the proof for $\hat{\varphi}^{(2)}$ can be generalized to $\hat{\varphi}^{(n)}$, $n < 2$.

Finally, in the preceding section it was mentioned that for physical reasons bound state wave functions may be decomposed into suitable pieces for the purpose of Weak Mapping and that these pieces are to be used for the definition of the set of expansion functions $C_k^{II'}$ of (6.149). Can then probability conservation be maintained?

Obviously the above given proof must be modified in order to comprehend the whole functional state (4.98) and the whole functional equation (4.100). This can be done, but for this proof we refer to Stumpf [Stu 00].

7 Effective $SU(2) \times U(1)$ Gauge Theory

7.1 Introduction

In Chapter 1 it was shown that de Broglie used wave functions of single composite photons for which he demonstrated that these functions satisfy the Maxwell equations. So we can consider this approach as a precursor of Weak Mapping. With respect to this fact it would be natural to discuss quantum electrodynamics as an effective dynamics of composite subfermion states first.

We will, however, at once start with the derivation of non-abelian gauge field dynamics as effective dynamics of bound subfermion states. More precisely: our aim is the derivation of a local $SU(2) \times U(1)$ gauge theory as an effective theory for two- and three-particle bound states of elementary subfermions. With such a gauge theory we will be able to derive the essentials of the electro–weak interaction sector of the standard model in the next chapter and to discover quantum electrodynamics as an effective theory, too.

Before beginning our calculations we give a short comment on other attempts at deriving effective field dynamics in high energy physics.

Whilst in numerous papers the formation of composite particles from elementary (relativistic) constituents was described by Bethe–Salpeter equations and their solutions, the search for an appropriate description of the reactions, $i.e.$, of the interactions of composite particles deviates from this concept. In contrast to quantum mechanics and Bethe–Salpeter equations which describe composite objects by wave functions, the reactions of composite particles were theoretically studied by means of field operators or field operator products.

For instance, the early approach to describe the photon as a composite object in quantum field theory was performed by operator products. A photon operator was assumed to be a product of a neutrino and an antineutrino operator, see, $e.g.$, Pryce [Pry 38]. Similar assumptions were made with respect to other composite objects either to derive an S matrix description for them or to generate effective Lagrangians.

In particular the transition from a Lagrangian for elementary constituents to an effective Lagrangian for composite particles by use of field operator products can be considered as a mapping. Because this kind of mapping is concerned with operators, one can denote it as 'strong' mapping, in contrast to 'Weak' Mapping discussed in the preceding chapters. Gradually these techniques were modified and incorporated into the evaluation of path integrals; for a critical survey of path integrals see Section 2.1.

As far as genuine strong mapping is concerned one has to observe that the representation of an operator product, $i.e.$, its various matrix elements between all possible states of the representation space under consideration, describe much more physical situations than only the formation of a composite particle, and only in S matrix theory it is possible to asymptotically project out just the information about such a particle. With respect to the formation of a local effective Lagrangian such an asymptotic projection is not possible and the concept of strong mapping has to be rejected. Indeed, this concept was frequently criticized. Reviews of papers about strong mapping and its criticism can be found in Stumpf [Stu 85a,d], [Stu 86a], [Stu 94]; see also Section 2.1.

At present most of the attempts of explaining the hierarchy of particles and their dynamics are related to the standard model which is the generally accepted high energy theory of matter. The standard model has been extremely successful in the theoretical verification and prediction of experimental data. Nevertheless, it is generally assumed that the standard model is only a preliminary theory of matter and forces because a considerable number of basic problems remains unsolved. The large number of basic particles, their repetition in generations, their mass spectra, the Higgs mechanism, etc., are phenomenologically introduced and need an explanation. For a more detailed discussion see, e.g., Buchmüller [Buch 85], Mohapatra [Moha 91], D'Souza and Kalman [Sou 92], Belotti [Belo 94].

The answer to this theoretical challenge is twofold: either one insists that the basic particles of the standard model are elementary and tries to embed them into higher symmetry groups and corresponding representations which leads to GUTs, supersymmetric theories and finally to strings, or one applies the atomistic idea of substructures in order to solve these problems.

The first way possibly enables one to unify the forces, but at the expense of a rapidly increasing number of basic 'elementary' particles. The strings, with which one tries to avoid this proliferation, need rather artificial constructions. In addition, the inclusion of gravity is still unclear and there is no convincing bridge over the gap between GeV physics and the Planck mass. Hence this way seems to be highly unsatisfactory, although such theories produce stuff for experimental research.

On the other hand, guided by the already known layers in the structure of matter one may expect that the proliferation of leptons and quarks as well as of gauge bosons will find a natural explanation in a simple underlying substructure which avoids the drawbacks of the first way. In the last decades lots of substructure models have indeed been proposed (see, e.g., Lyons [Lyo 83]) and it seems that in substructure model building almost every conceivable combination of substructure constituents (preons) has been employed. Thus the drawback of this way is the arbitrariness of the substructure model construction.

In order to avoid this, all complicated substructure models have to be rejected and one has to postulate that a substructure model has to be as simple as possible. The simplest model is constituted by solely assuming subfermions with two degrees of freedom, i.e., *isospin.*

In Chapters 2–4 we have proposed such a subfermion theory based on models of de Broglie, Heisenberg and Nambu, Jona–Lasinio. In this fusion model of the standard theory the fermions (leptons and quarks) should be composites of three subfermions, whilst the electro–weak gauge bosons have to be considered as composites of two subfermions and the gluons are assumed to be built up by six subfermions.

A short comment should be given concerning this choice for gauge bosons and quarks or leptons, respectively. As gauge bosons are spin 1 particles, it is convenient to consider them as a fusion product of at least two spin $1/2$ subfermions. In the same way an observable spin $1/2$ lepton or baryon leads to a nontrivial fusion product of at least three spin $1/2$ subfermions. Owing to the analogy with chromodynamics these assumptions are widespread, see Lyons [Lyo 83].

The differences and difficulties arise in assigning the other phenomenological quantum numbers of the standard model to these fusion products and to justify their choice, see for instance [Lyo 83], [Sou 92] and the numerous contributions in terms of fields by Dürr and Saller [Dürr 78,79] and Saller [Sal 81,82a,b,c,94].

Our approach is an alternative way to all these attempts. The main problem of justification of the assigned phenomenological quantum numbers will be solved by application of the Weak Mapping to the derivation of the effective dynamics and a thorough group-theoretical analysis of the wave functions themselves.

In previous papers by Stumpf [Stu 86a,b], [Stu 87] a subfermion shell model of electro–weak gauge bosons, leptons, quarks and gluons was proposed and effective theories were derived. These calculations were confirmed and improved in a series of papers by Stumpf and Pfister, see [Stu 92], [Pfi 94], [Pfi 95a,b] and by Grimm [Gri 94a,b,c]. Our representation is based on these papers.

The general task is to derive the standard model by means of the spinor field model of Chapter 2 as an effective theory with the inclusion of all symmetries, symmetry breakings, mass spectra, coupling constants and dynamical equations.

This means in particular that the effective dynamics has to create local gauge symmetries and their partial breakdown which are not contained in the original spinor field equation. In view of the present state of the art this is an ambitious program which cannot be realized in one step. In this chapter we therefore restrict ourselves to the discussion of the effective dynamics of gauge bosons, leptons and quarks and the derivation of their quantum numbers, leaving aside gluon formation and dynamics, symmetry breaking, Higgs bosons and mass spectra as well as the calculation of all coupling constants. But we emphasize that all these topics are within the range of our calculational program.

The chapter is organized as follows: first we give a short introduction into the phenomenological unbroken local $SU(2) \times U(1)$ gauge theory including the quantization of this theory in the framework of the Functional Quantum Theory. The results of this representation are to be used for comparison with the corresponding effective theory later on. After a review of the derivation of the effective bound state dynamics of Section 5.8 we will then treat the pure gauge boson bound states and their effective dynamics in an approximation which leads to the classical effective theory in order to illustrate our proceeding. Then, based on the results of Chapter 6, we will discuss the leptons and quarks as three-particle bound states and derive the full dynamics of the coupled boson–fermion system. The result will be an unbroken local $SU(2) \times U(1)$ gauge theory including leptons and quarks.

With respect to the generations we explicitly treat the simpler one-generation calculation and refer for the more complicated three-generation calculations to the literature.

7.2 Local $SU(2) \times U(1)$ Gauge Theory

It is one of the aims of our subfermion program to reproduce the equations of the standard model for leptons, quarks and gauge bosons as effective equations resulting from an underlying subfermion dynamics. We have already emphasized in the introduction that this program can only be realized in several steps. In a first step we confine ourselves to the study of the effective dynamics of leptons, quarks and gauge bosons without symmetry breaking effects.

Thus we expect to obtain an unbroken $SU(2) \times U(1)$ quantum dynamics for bosons and fermions forming the theoretical basis for a description of electro–weak interactions. In order to verify this conjecture we have to compare the results obtained from the subfermion dynamics by Weak Mapping with those of a quantum field theoretic formulation of a 'phenomenological' $SU(2) \times U(1)$ theory. *A comparison is only possible if the latter theory is formulated in terms of functional energy equations analogously to those of the subfermion theory given in Chapter 4.*

In order to perform this comparison we will first formulate the classical local $SU(2) \times U(1)$ gauge theory and then pass to its quantized version. In this context we are faced with the problem of the quantization of non-abelian gauge theories.

Owing to their applications in quantum chromodynamics, electro–weak theory and in numerous more general elementary particle theories, non-abelian gauge fields are at present one of the centers of interest in theoretical physical research. But in contrast to the abelian gauge group of quantum electrodynamics, the non-abelian gauge groups lead to nonlinear relations between vector potentials and field strength tensors causing severe difficulties with respect to quantization and renormalization.

As one way of circumventing these difficulties the use of path integrals has been proposed. But this again offers conceptual as well as technical difficulties which were gradually recognized and for a discussion of which we again refer to [Stu 94]. Hence to escape this situation, with increasing intensity canonical quantization of non-abelian gauge fields have been studied, see, *e.g.*, Gaigg, Kummer and Schweda [Gai 90]. As quantization is connected with gauge fixing, numerous versions of quantization related to different gauges do exist and are applied.

In a rough classification one can distinguish the quantization prescriptions by the use of covariant or noncovariant gauges. With respect to the treatment of covariant gauges the most prominent one seems to be the BRS formalism which leads to new global symmetries. Because of their relativistic invariance the covariant gauges are of basic theoretical interest. But on the other hand, owing to the inherent redundant degrees of freedom these gauges are burdened with indefinite metric problems which still cannot be satisfactorily treated.

In contrast, noncovariant gauges partially or totally avoid these redundant degrees of freedom, and thus in general do not suffer from indefinite metric problems. However, the price to be paid for these less formal and more physical representations of gauge theories are difficulties with the formulation of propagators. It would exceed the scope of this book to describe the present situation in this field in detail; we refer for instance to Gaigg, Kummer and Schweda [Gai 90].

In this context it is remarkable that by various authors the temporal gauge is considered as the 'physical' gauge which sets the standard for the treatment of non-abelian gauge theories, see Cheng [Che 90], Tsai [Tsai 90], Haller [Hall 90], Huang [Hua 92], Jackiw [Jack 80], Goldstone and Jackiw [Gold 78] and Leibbrand [Leib 87]. This opinion is in accordance with our results which distinguish the temporal gauge, because it is related to the energy representation of quantum fields. *Thus, in order to compare the results of Weak Mapping with phenomenological theories we derive the energy representation of non-abelian gauge fields in temporal gauge.*

In the framework of the standard model the electro–weak interactions are described by a spontaneously broken local $SU(2) \times U(1)$ gauge theory, the so called Glashow–Salam–Weinberg theory. Disregarding the phenomenological breaking of the left–right symmetry of fermions we consider the lepton field combination $\chi = \chi_A = \begin{pmatrix} \nu_e \\ e \end{pmatrix}$ which transforms as a doublet under the weak isospin group $SU(2)_W$:

$$\chi \longrightarrow U(\alpha_k)\chi = e^{-i\alpha_k T^k}\chi = e^{-i\alpha_k \sigma^k/2}\chi \tag{7.1}$$

and under the weak hypercharge group $U(1)_Y$ like

$$\chi \longrightarrow U(\beta)\chi = e^{-i\beta Y}\chi = e^{-i\beta y}\chi . \tag{7.2}$$

For the generators of the weak $SU(2)_W$ we have $T^k = \sigma^k/2$ with the Pauli matrices $\sigma^k, k = 1,2,3$, and the generator of the weak hypercharge group $U(1)_Y$ is given by $Y = y\mathbb{1}$. We consider the doublet χ to be that state configuration from which the phenomenological left handed electrons and neutrinos arise after symmetry breaking; by setting $y = -1/2$ we assign to them the hypercharge $-1/2$. The connection between the electric charge Q, the hypercharge and the isospin is given by the relation $Q = T_3 + Y$.

By the application of the principle of local gauge invariance to the fermion theory, *i.e.*, by assuming the invariance of the fermionic Lagrangian density with respect to gauge transformations with local parameters $\alpha_k(x)$ and $\beta(x)$ we are forced to introduce local gauge fields. For the non-abelian weak isospin group we obtain the gauge potential triplet A_μ^a, $a = 1,2,3$, and for the hypercharge group we have the singlet $B_\mu =: A_\mu^4$ with the Lorentz vector indices $\mu = 0 \ldots 3$. The corresponding field strengths are given by

$$F_{\mu\nu}^a := \partial_\mu A_\nu^a - \partial_\nu A_\mu^a + g_{\mathrm{ph}}\hat{\varepsilon}_{abc}A_\mu^b A_\nu^c \tag{7.3}$$

with the structure constants $\hat{\varepsilon}_{abc}$

$$\hat{\varepsilon}_{abc} := \begin{cases} \varepsilon_{abc} & \text{if } a,b,c \in \{1,2,3\} \\ 0 & \text{else} \end{cases} \tag{7.4}$$

The Lagrangian density of this local $SU(2) \times U(1)$ gauge theory reads

$$\mathcal{L} = -\frac{1}{4} \sum_{a=1}^{4} F_{\mu\nu}^{a} F_{a}^{\mu\nu} + i\bar{\chi}\gamma^{\mu}D_{\mu}\chi \tag{7.5}$$

with the covariant derivative

$$D_{\mu} := \partial_{\mu} - ig_{\text{ph}} \sum_{a=1}^{4} A_{\mu}^{a} L_{a} \equiv \partial_{\mu} - ig_{\text{ph}} \sum_{a=1}^{3} T^{a} A_{\mu}^{a} - ig_{\text{ph}}' Y A_{\mu}^{4}, \tag{7.6}$$

where

$$L_{a} := \begin{cases} \dfrac{1}{2}\sigma^{a} & \text{if } a \in \{1,2,3\} \\ y\dfrac{g_{\text{ph}}'}{g_{\text{ph}}}\mathbb{1} & \text{if } a = 4 \end{cases} \tag{7.7}$$

is the matrix representation of the generators of the group $SU(2)_W \times U(1)_Y$ with the Lie algebra relations $[L_a, L_b]_- = i\hat{\varepsilon}_{abc}L_c$. Here and in the following we adopt the convention $A^a = A_a$ etc..

We emphasize that we have collected the $SU(2)$ and the $U(1)$ field strenghts in the quantity A^a, $a = 1, 2, 3, 4$. Nevertheless, one should keep in mind that we deal with two independent gauge groups; the independent coupling constant g_{ph}' for the hyper charge in (7.6) is hidden in the definition of the generators L_a.

In this stage the Glashow–Salam–Weinberg theory contains no mass terms, neither for the fermions nor for the gauge bosons. This is enforced by the requirement to satisfy the simultaneous postulates of left–right asymmetry, $SU(2)$ gauge invariance and renormalizability and leads to the introduction of the controversial Higgs field concept of spontaneously broken $SU(2)$ symmetry. We will see that in our composite particle formalism the fermion and gauge boson masses occur in a natural manner by the assumption of asymmetrical vacuum expectation values, i.e., by appropriate algebraic representations.

For brevity we consider only one isospin doublet $\chi = \chi_A = \begin{pmatrix} \nu_e \\ e \end{pmatrix}$ which corre-

sponds to $y = -\frac{1}{2}$. The inclusion of further doublets, e.g., of $\begin{pmatrix} u \\ d \end{pmatrix}$ (or corresponding Cabibbo rotated quark states) as well as the inclusion of the right handed isospin singlets is obvious; in phenomenology to each of them a specific hypercharge is attributed in order to obtain the correct electric charge. This rather arbitrary procedure has to be compared with the natural way in which the hypercharge arises in our formalism.

The field equations corresponding to the Lagrangian density (7.5) read

$$\nabla_{\mu}^{ab} F_{b}^{\nu\mu} = j_{a}^{\nu} \tag{7.8}$$

and

$$i\gamma^\mu \partial_\mu \chi + g_{\text{ph}} \gamma^\mu A_\mu^a L_a \chi = 0, \tag{7.9}$$
$$i\gamma^\mu \partial_\mu \chi^c - g_{\text{ph}} \gamma^\mu A_\mu^a L_a^T \chi^c = 0,$$

with

$$\nabla_\mu^{ab} := \delta_{ab} \partial_\mu - g_{\text{ph}} \hat{\varepsilon}_{abc} A_\mu^c, \tag{7.10}$$

$$j_a^\mu := g_{\text{ph}} (: \bar{\chi} \gamma^\mu L_a \chi :) \tag{7.11}$$

$$=: \frac{g_{\text{ph}}}{2} \left[\bar{\chi}_{\alpha A} \gamma_{\alpha\beta}^\mu (L_a)_{AB} \chi_{\beta B} - \chi_{\beta B} \gamma_{\alpha\beta}^\mu (L_a)_{AB} \bar{\chi}_{\alpha A} \right],$$

and $\chi_{\alpha A}^c = C_{\alpha\beta} \bar{\chi}_{\beta A}$ as the 'charge' conjugated spinor field.

We combine χ and χ^c into one superspinor Ψ and introduce the index κ according to (6.7) by writing

$$\Psi_{\alpha\kappa} := \begin{cases} \chi_{\alpha A=1} & \text{for} \quad \kappa = 1 \\ \chi_{\alpha A=2} & \text{for} \quad \kappa = 2 \\ \chi_{\alpha A=1}^c & \text{for} \quad \kappa = 3 \\ \chi_{\alpha A=2}^c & \text{for} \quad \kappa = 4 \end{cases} . \tag{7.12}$$

In this notation the gauge group generators are given by

$$\hat{L}_{\kappa_1 \kappa_2}^a := \begin{cases} \dfrac{1}{2} \begin{pmatrix} \sigma^a & 0 \\ 0 & (-1)^a \sigma^a \end{pmatrix}_{\kappa_1 \kappa_2} & \text{if } a \in \{1,2,3\} \\[4mm] y \dfrac{g_{\text{ph}}'}{g_{\text{ph}}} \begin{pmatrix} \mathbb{I} & 0 \\ 0 & -\mathbb{I} \end{pmatrix}_{\kappa_1 \kappa_2} & \text{if } a = 4 \end{cases} . \tag{7.13}$$

With these definitions the current (7.11) reads

$$j_a^\mu = \frac{g_{\text{ph}}}{2} \Psi_{\alpha_1 \kappa_1} (C\gamma^\mu)_{\alpha_1 \alpha_2} (\gamma^5 \hat{L}^a)_{\kappa_1 \kappa_2} \Psi_{\alpha_2 \kappa_2}, \tag{7.14}$$

and the field equations (7.9) go over into

$$i\gamma_{\alpha_1 \alpha_2}^\mu \partial_\mu \Psi_{\alpha_2 \kappa_1} + g_{\text{ph}} A_\mu^a \gamma_{\alpha_1 \alpha_2}^\mu \hat{L}_{\kappa_1 \kappa_2}^a \Psi_{\alpha_2 \kappa_2} = 0, \tag{7.15}$$

where $C \equiv C_{\alpha\beta}$ is the charge conjugation matrix in spinor space.

If we introduce the generalized 'electric' field strength $E_a^i := F_a^{i0}$, we obtain from (7.8) for $\nu = i = 1, 2, 3$

$$\partial_0 E_a^i = -g_{\text{ph}} \hat{\varepsilon}_{abc} A_0^b E_c^i - \nabla_k^{ab} F_b^{ik} + j_a^i \tag{7.16}$$

and for $\nu = 0$

$$\nabla_k^{ab} E_b^k = -j_a^0, \tag{7.17}$$

whilst with the definition of E_a^k it follows from (7.3)

$$\partial_0 A_a^i = -E_a^i - \partial_i A_a^0 - g_{\text{ph}} \hat{\varepsilon}_{abc} A_b^0 A_c^i. \tag{7.18}$$

We now pass from the general form of equations (7.8), (7.15) into a special gauge, namely the temporal gauge, by the gauge fixing condition $A_a^0 \equiv 0$, which we assume to apply to classical as well as to quantum theory. If we substitute definitions (7.3), (7.10) into (7.16), (7.17), these equations and (7.15), (7.18) read in temporal gauge

$$i\partial_0 \Psi_{\alpha_1\kappa_1} = -i(\gamma^0\gamma^k)_{\alpha_1\alpha_2}\partial_k\Psi_{\alpha_2\kappa_1} + g_{\mathrm{ph}}A_a^k(\gamma^0\gamma^k)_{\alpha_1\alpha_2}\hat{L}_{\kappa_1\kappa_2}^a\Psi_{\alpha_2\kappa_2}, \tag{7.19}$$

$$i\partial_0 A_a^i = -iE_a^i, \tag{7.20}$$

$$i\partial_0 E_a^i = i\Big[-\Delta A_a^i + \partial_i\partial_j A_a^j - g_{\mathrm{ph}}\hat{\varepsilon}_{abc}A_b^i(\partial_j A_c^j) + 2g_{\mathrm{ph}}\hat{\varepsilon}_{abc}A_b^j(\partial_j A_c^i), \tag{7.21}$$

$$-g_{\mathrm{ph}}\hat{\varepsilon}_{abc}A_b^j(\partial_i A_c^j) - g_{\mathrm{ph}}^2\hat{\varepsilon}_{abc}\hat{\varepsilon}_{cde}A_b^j A_d^j A_e^i\Big] + ij_a^i,$$

$$\partial_k E_a^k = g_{\mathrm{ph}}\hat{\varepsilon}_{abc}A_b^k E_c^k - j_a^0. \tag{7.22}$$

We remember that equations (7.20)–(7.22) contain the equations for a non-abelian $SU(2)$ dynamics ($a = 1, 2, 3$) as well as a conventional abelian $U(1)$ dynamics ($a = 4$).

Equations (7.19)–(7.22) describe the local $SU(2) \times U(1)$ gauge theory in terms of the canonical conjugated fields A_a and E_a; this formulation is the starting point for the quantization of this gauge theory. For comparison with our effective dynamical equations on the *classical* level it is, however, convenient to introduce an alternative formulation in terms of A_a, E_a, and B_a fields.

The generalized magnetic fields B_a^i are introduced by

$$B_a^i(x) := -\frac{1}{2}\varepsilon_{ijk}F_a^{jk}(x). \tag{7.23}$$

With this definition we obtain for the field equations of the classical local $SU(2) \times U(1)$ gauge theory in temporal gauge

$$i\partial_0 A_a^i = -iE_a^i, \tag{7.24}$$

$$i\partial_0 B_a^i = -i\varepsilon_{ijk}\partial_j E_a^k + ig_{\mathrm{ph}}\varepsilon_{ijk}\hat{\varepsilon}_{abc}E_b^j A_c^k,$$

$$i\partial_0 E_a^i = i\varepsilon_{ijk}\partial_j B_a^k - ig_{\mathrm{ph}}\varepsilon_{ijk}\hat{\varepsilon}_{abc}A_b^j B_c^k + ij_a^i$$

together with the fermion field equation (7.19) and the Gauss law (7.22).

The quantization of this theory is performed in the framework of the Functional Quantum Theory analogously to the treatment of the spinor field theory in Chapter 4. For that purpose we go back to the formulation of the gauge field theory in terms of the conjugated variables E_a^i and A_a^i, i.e., to equations (7.19)–(7.22). For convenience we introduce the super-index $K := (\mathbf{z}, i, a, \eta)$ and define

$$B_K := \begin{cases} A_a^i(\mathbf{z}) & \text{for } \eta = 1 \\ E_a^i(\mathbf{z}) & \text{for } \eta = 2, \end{cases} \tag{7.25}$$

and $\Psi_I := \Psi_{\alpha A\Lambda}(\mathbf{r})$ with $I := (\mathbf{r}, \alpha, A, \Lambda)$. We remark, that the super-fields B_K are combinations of the conjugated variables A_a^i and E_a^i and should not be confused with the generalized magnetic field B_a^i from (7.23).

In this notation the dynamical equations (7.19), (7.20), (7.21) read

$$i\partial_0 \Psi_{I_1} = D_{I_1 I_2} \Psi_{I_2} + W_{I_1 I_2}^K B_K \Psi_{I_2}, \tag{7.26}$$

$$i\partial_0 B_{K_0} = L_{K_0 K_1} B_{K_1} + M_{K_0}^{K_1 K_2} B_{K_1} B_{K_2} \tag{7.27}$$

$$+ V_{K_0}^{K_1 K_2 K_3} B_{K_1} B_{K_2} B_{K_3} + J_{K_0}^{I_1 I_2} \Psi_{I_1} \Psi_{I_2},$$

whilst the constraint equation (7.22), the 'Gauss law', can be written as

$$N_K^L B_K + S_{K_1 K_2}^L B_{K_1} B_{K_2} + R_{I_1 I_2}^L \Psi_{I_1} \Psi_{I_2} = 0. \tag{7.28}$$

For the formulation of these equations the following definitions are used:

$$D_{I_1 I_2} := -i(\gamma^0 \gamma^k)_{\alpha_1 \alpha_2} \partial_k \delta_{\kappa_1 \kappa_2} \delta(\mathbf{r}_1 - \mathbf{r}_2), \tag{7.29}$$

$$W_{I_1 I_2}^K := g_{\text{ph}}(\gamma^0 \gamma^j)_{\alpha_1 \alpha_2} \widehat{L}_{\kappa_1 \kappa_2}^a \delta_{1\eta} \delta(\mathbf{r}_1 - \mathbf{r}_2) \delta(\mathbf{r}_1 - \mathbf{z}), \tag{7.30}$$

$$L_{K_1 K_2} := -i\delta_{j_1 j_2} \delta_{a_1 a_2} \Big[\delta_{\eta_1 1} \delta_{\eta_2 2} \delta(\mathbf{z}_1 - \mathbf{z}_2) - \delta_{\eta_1 2} \delta_{\eta_2 1} \partial_k^2(\mathbf{z}_1) \delta(\mathbf{z}_1 - \mathbf{z}_2) \Big]$$

$$+ i\delta_{\eta_1 2} \delta_{\eta_2 1} \delta_{a_1 a_2} \partial_{j_1}(\mathbf{z}_1) \partial_{j_2}(\mathbf{z}_1) \delta(\mathbf{z}_1 - \mathbf{z}_2), \tag{7.31}$$

$$M_{K_0}^{K_1 K_2} := ig_{\text{ph}} \Big\{ \delta_{\eta_0 2} \delta_{\eta_1 1} \delta_{\eta_2 1} \widehat{\mathcal{E}}_{a_0 a_1 a_2} \tag{7.32}$$

$$\times \Big[\delta_{j_0 j_2} \delta(\mathbf{z}_0 - \mathbf{z}_2) \partial_{j_1}(\mathbf{z}_0) \delta(\mathbf{z}_0 - \mathbf{z}_1)$$

$$+ 2\delta_{j_0 j_2} \delta(\mathbf{z}_0 - \mathbf{z}_1) \partial_{j_1}(\mathbf{z}_0) \delta(\mathbf{z}_0 - \mathbf{z}_2)$$

$$- \delta_{j_1 j_2} \delta(\mathbf{z}_0 - \mathbf{z}_1) \partial_{j_0}(\mathbf{z}_0) \delta(\mathbf{z}_0 - \mathbf{z}_2) \Big] \Big\}_{\text{sym}(K_1, K_2)},$$

$$V_{K_0}^{K_1 K_2 K_3} := -\Big[ig_{\text{ph}}^2 \delta_{\eta_0 2} \delta_{\eta_1 1} \delta_{\eta_2 1} \delta_{\eta_3 1} \widehat{\mathcal{E}}_{a_0 a_1 b} \widehat{\mathcal{E}}_{b a_2 a_3} \tag{7.33}$$

$$\times \delta_{j_1 j_2} \delta_{j_0 j_3} \delta(\mathbf{z}_0 - \mathbf{z}_1) \delta(\mathbf{z}_0 - \mathbf{z}_2) \delta(\mathbf{z}_0 - \mathbf{z}_3) \Big]_{\text{sym}(K_1, K_2, K_3)},$$

$$J_K^{I_1 I_2} := i\frac{g_{\text{ph}}}{2} \delta_{\eta 2} (C\gamma^j)_{\alpha_1 \alpha_2} (\gamma^5 \widehat{L}^a)_{\kappa_1 \kappa_2} \delta(\mathbf{z} - \mathbf{r}_1) \delta(\mathbf{z} - \mathbf{r}_2), \tag{7.34}$$

$$N_{K_1}^L := \delta_{a a_1} \delta_{\eta_1 2} \partial_{j_1}(\mathbf{z}) \delta(\mathbf{z} - \mathbf{z}_1), \tag{7.35}$$

$$S_{K_1 K_2}^L := -g_{\text{ph}} \Big[\widehat{\mathcal{E}}_{a a_1 a_2} \delta_{j_1 j_2} \delta(\mathbf{z} - \mathbf{z}_1) \delta(\mathbf{z} - \mathbf{z}_2) \delta_{\eta_1 1} \delta_{\eta_2 2} \Big]_{\text{sym}(K_1, K_2)}, \tag{7.36}$$

$$R_{I_1 I_2}^L := \frac{g_{\text{ph}}}{2} (C\gamma^0)_{\alpha_1 \alpha_2} (\gamma^5 \widehat{L}^a)_{\kappa_1 \kappa_2} \delta(\mathbf{z} - \mathbf{r}_1) \delta(\mathbf{z} - \mathbf{r}_2) \tag{7.37}$$

with the reduced index $L = (\mathbf{z}, a)$.

Next we have to quantize the dynamical variables of our coupling theory. As in quantum electrodynamics, also for non-abelian fields it can be shown that the fermion field quantization fixes the boson field quantization, see [Stu 94]. From the canonical fermion field anticommutator (2.72)

$$\big[\Psi_{I_1}, \Psi_{I_2} \big]_+ = A_{I_1 I_2} \tag{7.38}$$

with

$$A_{I_1 I_2} := (C\gamma^0)_{\alpha_1 \alpha_2} \delta_{A_1 A_2} \sigma_{\Lambda_1 \Lambda_2}^1 \delta(\mathbf{r}_1 - \mathbf{r}_2) \mathbf{1}, \tag{7.39}$$

and from vanishing commutators between B and Ψ fields it follows that the boson field commutator

$$\left[B_{K_1}, B_{K_2} \right]_- = S_{K_1 K_2} \qquad (7.40)$$

has to have the form

$$S_{K_1 K_2} = -\delta_{i_1 i_2} \sigma^2_{\eta_1 \eta_2} \delta_{a_1 a_2} \delta(\mathbf{z}_1 - \mathbf{z}_2) \mathbf{1} \qquad (7.41)$$

for the $U(1)$ singlets $a = 4$ and the $SU(2)$ triplets $a = 1, 2, 3$.

With the relations (7.38)–(7.41) the fermion and boson field quantization is defined and we can apply it to the system under consideration. This in particular means that the dynamical equations (7.26) and (7.27) are considered as operator equations in the Heisenberg picture, whilst in order to avoid contradictions the constraint equations (7.28) are assumed to be valid only by acting on physical states $|a\rangle$ of the corresponding state space. This weak implementation is in accordance with our general algebraic treatment of quantum fields, i.e., we can proceed along the same lines as we did for theories without constraint. First we define the 'phenomenological' state functionals for all physical states $|a\rangle$ by

$$|\mathcal{G}(f, b; a)\rangle^{\mathrm{ph}} := \sum_{n,m=0}^{\infty} \frac{i^n}{n! m!} \langle 0| \mathcal{S} \left(B_{K_1} \ldots B_{K_m} \right) \mathcal{A} \left(\Psi_{I_1} \ldots \Psi_{I_n} \right) |a\rangle \qquad (7.42)$$

$$\times f_{I_1} \ldots f_{I_1} b_{K_1} \ldots b_{K_m} |v\rangle$$

with the symmetrization and antisymmetrization operators $\mathcal{S}$ and $\mathcal{A}$.

The dynamical equation for the functional states (7.42) follows by analogous rearrangements to those which we performed in Section 4.6 at great length. For short we do not explicitly repeat such calculations for the present case; rather we give only the results of these extensive calculations. The corresponding functional energy equation was explicitly calculated by Stumpf and Pfister [Stu 97] and reads

$$E_{p0} |\mathcal{G}(f, b; p)\rangle^{\mathrm{ph}} \qquad (7.43)$$

$$= \left[f_{I_1} D_{I_1 I_2} \partial^f_{I_2} + f_{I_1} W^K_{I_1 I_2} \partial^f_{I_2} \partial^b_K + b_{K_1} L_{K_1 K_2} \partial^b_{K_2} \right.$$

$$+ J^K_{I_1 I_2} b_K \left(\partial^f_{I_1} \partial^f_{I_2} + \frac{1}{4} A_{I_1 I_3} A_{I_2 I_4} f_{I_3} f_{I_4} \right)$$

$$+ M^{K_1 K_2}_{K_0} b_{K_0} \left(\partial^b_{K_1} \partial^b_{K_2} + \frac{1}{12} S_{K_1 K_3} S_{K_2 K_4} b_{K_3} b_{K_4} \right)$$

$$\left. + V^{K_1 K_2 K_3}_{K_0} b_{K_0} \left(\partial^b_{K_1} \partial^b_{K_2} \partial^b_{K_3} + \frac{1}{4} S_{K_0 K_4} S_{K_2 K_5} b_{K_4} b_{K_5} \partial^b_{K_3} \right) \right] |\mathcal{G}(f, b; p)\rangle^{\mathrm{ph}}$$

with definitions (7.29)–(7.33).

For our purpose it is not necessary to explicitly treat the constraint equation (7.28); we only remark that it also may be represented as a functional equation in analogy to (7.43). An extensive treatment of phenomenological non-abelian gauge theories in temporal gauge and their constraints in functional space was given in [Stu 97].

As can be seen in (7.43) the functional equation contains the genuine dynamical terms and the quantization terms, *i.e.*, those terms which are composed of commutators S or anticommutators A respectively. Apart from dynamical terms the Weak Mapping of the subfermion theory leads to quantization terms, too. But naturally, owing to the compositeness of their basic objects, the quantization terms are deformed, *i.e.*, modified in comparison with those terms arising by canonical quantization of local objects. A study of such deformed quantization terms can be found in [Stu 94].

For subsequent applications we concentrate on the dynamical part of the functional equation, and for comparison with the subfermion theory we suppress the quantization terms in (7.43), *i.e.*, the terms containing the fermion anticommutator A and the boson commutator S. This procedure will be justified in Section 8.6, where we will examine in detail the question of effective quantization. Then equation (7.43) may be written as

$$E_{p0}|\mathcal{G}\rangle^{\mathrm{ph}} = \left(\mathcal{H}_{\mathrm{f}}^{\mathrm{ph}} + \mathcal{H}_{\mathrm{b}}^{\mathrm{ph}} + \mathcal{H}_{\mathrm{bf}}^{\mathrm{ph}}\right)|\mathcal{G}\rangle^{\mathrm{ph}} \tag{7.44}$$

with

$$\mathcal{H}_{\mathrm{f}}^{\mathrm{ph}} := f_{I_1} D_{I_1 I_2} \partial_{I_2}^f, \tag{7.45}$$

$$\mathcal{H}_{\mathrm{b}}^{\mathrm{ph}} := b_{K_1} L_{K_1 K_2} \partial_{K_2}^b + b_{K_0} M_{K_0}^{K_1 K_2} \partial_{K_1}^b \partial_{K_2}^b + b_{K_0}^b V_{K_0}^{K_1 K_2 K_3} \partial_{K_1}^b \partial_{K_2}^b \partial_{K_3}^b,$$

$$\mathcal{H}_{\mathrm{bf}}^{\mathrm{ph}} := f_{I_1} W_{I_1 I_2}^K \partial_{I_2}^f \partial_K^b + b_K J_{I_1 I_2}^K \partial_{I_1}^f \partial_{I_2}^f,$$

where the superscript 'ph' indicates that the phenomenological theory is treated.

For the explicit evaluation of (7.44) we have to resolve the super-indices. According to (7.25) we have for the bosonic functional generators

$$b_K \equiv b_{i,a,\eta}(\mathbf{z}) = \begin{cases} b_{i,a}^A(\mathbf{z}) & \text{for} \quad \eta = 1 \\ b_{i,a}^E(\mathbf{z}) & \text{for} \quad \eta = 2, \end{cases} \tag{7.46}$$

and corresponding expressions for the functional duals ∂_K^b. We remember that $i = 1, 2, 3$ is the vector index of the corresponding fields A and E in temporal gauge, whilst $a = 4$ describes the $U(1)$ singlets and $a = 1, 2, 3$ the $SU(2)$ triplets.

The resolution of the super-indices in (7.45) by means of (7.29)–(7.34) then yields

$$\mathcal{H}_{\mathrm{f}}^{\mathrm{ph}} = -i \int f_{\alpha\kappa}(\mathbf{r}) (\gamma^0 \gamma^k)_{\alpha\beta} \partial_k \partial_{\beta\kappa}^f(\mathbf{r}) \, d\mathbf{r}, \tag{7.47}$$

$$\mathcal{H}_{\mathrm{b}}^{\mathrm{ph}} = -i \sum_{a=1}^{4} \int b_{j,a}^A(\mathbf{r}) \partial_{j,a}^E(\mathbf{r}) d\mathbf{r} \tag{7.48}$$

$$+ i \sum_{a=1}^{4} \int b_{j,a}^E(\mathbf{r}) \left[\partial_k \partial_k \partial_{j,a}^A(\mathbf{r}) + \partial_j \partial_k \partial_{k,a}^A(\mathbf{r}) \right] d\mathbf{r}$$

$$+ i g_{\mathrm{ph}} \sum_{a,b,c=1}^{3} \int b_{j,a}^E(\mathbf{r}) \varepsilon_{abc} \left[\left(\partial_k \partial_{k,b}^A(\mathbf{r}) \right) \partial_{j,c}^A(\mathbf{r}) + 2\partial_{k,b}^A(\mathbf{r}) \partial_k \partial_{j,c}^A(\mathbf{r}) \right.$$

$$\left. - \partial_{k,b}^A(\mathbf{r}) \partial_j \partial_{k,c}^A(\mathbf{r}) \right] d\mathbf{r}$$

$$-ig_{\mathrm{ph}}^2 \sum_{a\ldots e=1}^{3} \int b_{j,a}^E(\mathbf{r})\varepsilon_{abc}\varepsilon_{cde}\partial_{k,b}^A(\mathbf{r})\partial_{k,d}^A(\mathbf{r})\partial_{j,e}^A(\mathbf{r})d\mathbf{r},$$

$$\mathcal{H}_{\mathrm{bf}}^{\mathrm{ph}} = g_{\mathrm{ph}}\sum_{a=1}^{4}\int f_{\alpha_1\kappa_1}(\mathbf{r})(\gamma^0\gamma^k)_{\alpha_1\alpha_2}\hat{L}_{\kappa_1\kappa_2}^a\partial_{\alpha_2\kappa_2}^f(\mathbf{r})\,\partial_{k,a}^A(\mathbf{r})\,d\mathbf{r} \tag{7.49}$$

$$-\frac{i}{2}g_{\mathrm{ph}}\sum_{a=1}^{4}\int b_{k,a}^E(\mathbf{r})\gamma_{\alpha_1\alpha_2}^k\hat{L}_{\kappa_1\kappa_2}^a\bar{\partial}_{\alpha_1\kappa_1}^f(\mathbf{r})\partial_{\alpha_2\kappa_2}^f(\mathbf{r})\,d\mathbf{r},$$

where the gauge group generators $\hat{L}^a$ are given by (7.13).

In our notation (7.12) it is easy to see that the charge conjugated superspinor $\Psi_{\alpha\kappa}^C := C_{\alpha\beta}\bar{\Psi}_{\beta\kappa}$ is given by

$$\Psi_{\alpha\kappa}^C = \gamma_{\kappa\kappa'}^5\Psi_{\alpha\kappa'}. \tag{7.50}$$

According to this relation, in (7.49) we have introduced the functional dual operators

$$\bar{\partial}_{\alpha\kappa}^f := \left(C^{-1}\right)_{\alpha\alpha'}\gamma_{\kappa\kappa'}^5\partial_{\alpha'\kappa'}^f. \tag{7.51}$$

In order to avoid confusion we remember that ∂_k denotes the partial derivative $\partial_k \equiv \partial_k(\mathbf{r}) := \partial/\partial x^k$. Furthermore, we did not explicitly write the summations over vector indices, $etc.$.

By extracting the corresponding classical equations in analogy to the pure boson case, $cf.$ Section 7.5, one easily convinces oneself that the fermion term (7.47) yields the kinetic part of the classical fermion equation (7.19). The first term of the coupling part (7.49) gives the gauge boson coupling in this equation, whilst the second term of (7.49) describes the fermion current coupling to the E field equation (7.21).

7.3   Effective Boson–Fermion Dynamics

The effective dynamical equations for composite two- and three-particle bound states arising from an elementary subfermion dynamics were already derived in Section 5.8. We summarize the results.

The basic subfermion theory is defined by the spinor field model of Chapter 2. Its quantum field theoretic energy representation is given by equations (4.100) with the subfermion field functional

$$|\mathcal{F}(j^t;a)\rangle := \sum_n \frac{i^n}{n!}\varphi_t^{(n)}(I_1\ldots I_n|a)j_{I_1}^t\ldots j_{I_n}^t|0\rangle_{\mathrm{F}}. \tag{7.52}$$

In the following for brevity we suppress the single time index t.

In a symbolic notation we write for (4.100) or (5.1), respectively,

$$E_{p0}|\mathcal{F}(j;p)\rangle = \mathcal{H}_{\mathrm{F}}(j,\partial)|\mathcal{F}(j;p)\rangle, \tag{7.53}$$

where the functional Hamiltonian operator $\mathcal{H}_{\mathrm{F}}(j,\partial)$ is given by (5.2).

With respect to the boson–fermion coupling theory we define the effective functional states, compare (7.42):

$$|\mathcal{G}(b, f; a)\rangle :=$$ (7.54)

$$\sum_{n,m=0}^{\infty} \frac{i^n}{n!m!} \, \rho^{(n+m)}(k_1 \ldots k_m; I_1 \ldots I_n|a) b_{k_1} \ldots b_{k_m} f_{I_1} \ldots f_{I_n} |0\rangle_B \otimes |0\rangle_F \, ,$$

which may be regarded as a compact notation for the set of effective two- and three-particle correlation functions $\rho^{(n+m)}$. These functions are again single time functions; we interpret them as matrix elements of effective boson and fermion field operators:

$$\rho^{(n+m)}(k_1 \ldots k_m; I_1 \ldots I_n|a) = \langle 0| \mathcal{S} \left(B_{K_1} \ldots B_{K_m} \right) \mathcal{A} \left(\Psi_{I_1} \ldots \Psi_{I_n} \right) |a\rangle \, .$$ (7.55)

The quantities b_k and f_I are the boson or fermion sources, respectively, and $|0\rangle_B$ and $|0\rangle_F$ are the corresponding functional vacua. Here and in the following we use the abbreviations b_k, f_I, etc., for the operators $b_k \otimes 1$, $1 \otimes f_I$, etc., on the direct product space of (7.54). We denote the functional 'duals' of the boson and fermion sources by ∂_k^b and ∂_I^f and assume the following commutation relations:

$$\left[\partial_k^b, b_{k'} \right]_- = \delta_{kk'} \, ,$$ (7.56)

$$\left[\partial_I^f, f_{I'} \right]_+ = \delta_{II'} \, ,$$

$$\left[\partial_k^b, f_{I'} \right]_- = \left[\partial_I^f, b_k \right]_- = 0,$$

whilst all other commutators (anticommutators) vanish.

Furthermore, we assume $\partial_k^b |0\rangle_B = \partial_I^f |0\rangle_F = 0$, which enforces the functional basis states in (7.52) to be Fock states. This functional Fock space is completely independent of the subfermionic functional space used for the definition of $|\mathcal{F}(j; a)\rangle$ as long as one considers the exact Weak Mapping theorems. Approximately by application of the chain rule both spaces become correlated.

In Section 5.8 the Weak Mapping of the functional subfermion dynamics onto the effective dynamics for two- and three-particle bound states was carried out by means of the functional chain rule. For a pure hard core boson dynamics it was shown in Chapter 5 that the application of the chain rule corresponds to an exact mapping with a subsequent omission of exchange forces; if Weak Mapping is enlarged to include polarization clouds we assume that the application of the functional chain rule yields the leading direct forces of the effective dynamics.

The transformation relations for the functional chain rule in the two- and three-particle case read

$$b_{2,k} := \frac{1}{2} C_{2,k}^{I_1 I_2} j_{I_1} j_{I_2} \, ,$$ (7.57)

$$f_{3,l} := C_{3,l}^{I_1 I_2 I_3} j_{I_1} j_{I_2} j_{I_3}$$ (7.58)

with the two- and three-particle hard core bound states $C_{2,k}$ and $C_{3,l}$. The dual relations

$$j_{I_1} \cdots j_{I_{2r+1}} = \sum_l R^{3,l}_{I_1 \ldots I_{2r+1}} f_{3,l} \,, \tag{7.59}$$

$$j_{I_1} \cdots j_{I_{2r}} = 2 \sum_k R^{2,k}_{I_1 \ldots I_{2r}} b_{2,k} \tag{7.60}$$

introduce the duals of the polarization cloud terms of dressed particle states. In (7.57) and (7.60) we have performed a rescaling $C_{2,k} \to 1/2 C_{2,k}$ and $R^{2,k} \to 2R^{2,k}$; for details we refer to Chapter 5.

We formulate the results of the mapping of Section 5.8 as

- **Proposition 7.1:** If exchange forces between subfermions belonging to different composite particles are neglected, then for a combined two- and three-particle effective dynamics the following effective state equation results from (7.53):

$$E_{p0}|\mathcal{G}(b,f;p)\rangle = \mathcal{H}_{\text{eff}}(b, \partial^b, f, \partial^f)|\mathcal{G}(b,f;p)\rangle, \tag{7.61}$$

where $\mathcal{H}_{\text{eff}}$ is given by

$$\mathcal{H}_{\text{eff}} = \mathcal{H}_f + \mathcal{H}_b + \mathcal{H}_{bf} + \mathcal{H}_{\text{corr}} \tag{7.62}$$

with the effective fermion part

$$\mathcal{H}_f := f_\ell \big(\widehat{K}^{\ell\ell'} \partial^f_{\ell'} + \widehat{M}^{\ell\ell'} \partial^f_{\ell'} \big), \tag{7.63}$$

the effective boson part

$$\mathcal{H}_b := b_k \big(K^{kk'} \partial^b_{k'} + M^{kk'} \partial^b_{k'} + N^{kk'_1 k'_2} \partial^b_{k'_1} \partial^b_{k'_2} \big) \tag{7.64}$$

and the effective boson–fermion interaction term

$$\mathcal{H}_{bf} := b_k S^{k\ell'_1\ell'_2} \partial^f_{\ell'_1} \partial^f_{\ell'_2} + f_\ell \widehat{N}^{\ell k\ell'} \partial^b_k \partial^f_{\ell'}. \tag{7.65}$$

The corresponding definitions are

$$K^{kk'} := 2\widehat{K}_{I_1 I_2} C^{I_2 I_3}_{2,k'} R^{2,k}_{I_1 I_3} \,, \tag{7.66}$$

$$M^{kk'} := -6\widehat{W}_{I_4 I_1 I_2 I_3} F_{I_3 K} C^{I_1 I_2}_{2,k'} R^{2,k}_{I_4 K} \,,$$

$$N^{kk'_1 k'_2} := 12\widehat{W}_{I_4 I_1 I_2 I_3} C^{I_1 I_2}_{2,k'_1} C^{I_3 K}_{2,k'_2} R^{2,k}_{I_4 K} \,,$$

$$S^{k\ell'_1 \ell'_2} := -36\widehat{W}_{I_4 I_1 I_2 I_3} C^{I_1 I_3 K_1}_{3,\ell'_1} C^{I_2 K_2 K_3}_{3,\ell'_2} R^{2,k}_{I_4 K_1 K_2 K_3} \,,$$

and

$$\widehat{K}^{\ell\ell'} := 3\widehat{K}_{I_1 I_2} C^{I_2 K_1 K_2}_{3,\ell'} R^{3,\ell}_{I_1 K_1 K_2} \,, \tag{7.67}$$

$$\widehat{M}^{\ell\ell'} := -18\widehat{W}_{I_4 I_1 I_2 I_3} F_{I_3 K_1} C^{I_1 I_2 K_2}_{3,\ell'} R^{3,\ell}_{I_4 K_1 K_2} \,,$$

$$\widehat{N}^{\ell k\ell'} := 18\widehat{W}_{I_4 I_1 I_2 I_3} \big(C^{I_3 K_1 K_2}_{3,\ell'} C^{I_1 I_2}_{2,k} + 2C^{I_2 I_3 K_2}_{3,\ell'} C^{I_1 K_1}_{2,k} \big) R^{3,\ell}_{I_4 K_1 K_2} \,.$$

In these definitions only the lowest order terms of the polarization cloud are included. The other terms are shifted to $\mathcal{H}_{\text{corr}}$, which we do not explicitly write out.

Proof: The complete set of terms can be read off in Section 5.8. It remains the task to show that the terms (7.66) and (7.67) are the leading terms of the effective boson–fermion dynamics, *i.e.*, that $\mathcal{H}_{\text{corr}}$ can be neglected compared with $\mathcal{H}_{\text{f}}$, $\mathcal{H}_{\text{b}}$ and $\mathcal{H}_{\text{bf}}$. This proof is postponed to Section 8.7. $\diamond$

To explicitly evaluate equation (7.61) it is convenient to proceed in two steps: first we discuss a pure boson theory, then we treat the full equation (7.61) including the coupling to the fermion sector.

The pure boson theory is a subsector of the solutions of equation (7.61). This subsector is obtained if (7.61) is projected with the pure boson projectors $_{\text{F}}\langle 0|_{\text{B}}\langle 0|\partial^b_{k_1} \dots \partial^b_{k_l}$. By means of the commutation relations (7.56) one easily convinces oneself that the resulting equation can equivalently be written as the bosonic equation

$$E_{p0}|\mathcal{B}(b;p)\rangle = \mathcal{H}_{\text{b}}|\mathcal{B}(b;p)\rangle, \tag{7.68}$$

where the boson state functional $|\mathcal{B}(b;p)\rangle$ is given by (5.14), and where we have already neglected the terms from $\mathcal{H}_{\text{corr}}$. Equation (7.68) may be obtained from equation (7.61) by setting $f = \partial^f \equiv 0$.

In Chapter 5 we derived an effective two-particle boson dynamics by means of a hard core mapping. The result was (5.46), which has to be compared with the dressed particle effective boson equation (7.68). We see that the first three terms of (5.46) are identical with the first three terms of $\mathcal{H}_{\text{b}}$ given by (7.64). These terms constitute the leading terms of the effective Hamiltonian, see Section 8.6. The other terms of (5.46) go over into the correction terms $\mathcal{H}_{\text{corr}}$ because of the modified substitution (7.60) for a dressed particle mapping.

We expect that equation (7.68) describes the dynamics of the gauge bosons of a local $SU(2) \times U(1)$ gauge theory. To show this we have to evaluate (7.68); the first step for this evaluation is the investigation of the boson hard core states $C^{I_1 I_2}_{2,k}$ and their duals $R^{2,k}_{I_1 I_2}$ and the specification of their quantum numbers k.

7.4 Boson States and Dual States

The basis of Weak Mapping is the definition of a set of appropriate bound state functions. In the case of a pure boson mapping this set is defined by the boson states $C^{I_1 I_2}_{2,k}$ together with their dual states $R^{2,k}_{I_1 I_2}$.

As we shall see the hard core states $C^{I_1 I_2}_{2,k}$ are closely related to the solutions of the generalized de Broglie–Bargmann–Wigner equation (6.1). The set of solutions of these equations contains bound state and scattering state solutions. The energy of the latter at least amounts to $2m_0$, where m_0 is the lowest mass of m_1, m_2 and m_3 in (2.24). In the following we assume m_0 to be very large, *i.e.*, larger than any observable energy limit. If we confine the evaluation of (7.68) to an energy range $E < m_0 c^2$, then owing to the energy conservation of the spinor field processes no break up processes of bound fermion states can occur, *i.e.*, the dynamics of the bound states and the dynamics of the scattering states should decouple, see also Section 5.6.

This decoupling should exactly be derived from (7.68). However, a general investigation of this hypothesis has not yet been performed. But one can apply a simple argument to justify decoupling: the dynamics of scattering states of fermions contained in (7.68) can equivalently be described by the original dynamics of the spinor field (4.100). Thus the evaluation of (7.68) with inclusion of scattering states must result in a coupling theory between composite bosons and subfermions, and if one is exclusively interested in the boson dynamics one can omit the coupling terms to the subfermions. But this is equivalent to calculate (7.68) from the beginning only with bound states.

Furthermore, from the beginning we confine ourselves to the spin 1 vector boson solutions of Section 6.3 and thus omit the treatment of scalar and other fields. This procedure will be justified in Section 8.4; for a combined treatment of all these fields see Grimm [Gri 94a,b].

According to Section 6.3 in s wave approximation the vector boson bound states are given by

$$\varphi^{(2)}_{\substack{i_1 i_2 \\ \kappa_1 \kappa_2 \\ \alpha_1 \alpha_2}} (\mathbf{r}_1, \mathbf{r}_2, t | k) \tag{7.69}$$

$$= N_B T^a_{\kappa_1 \kappa_2} e^{-k_0 t} e^{ik\frac{\mathbf{r}_1 + \mathbf{r}_2}{2}}$$

$$\times \left[A_\mu f^A_{i_1 i_2} (\mathbf{r}_1 - \mathbf{r}_2) (\gamma^\mu C)_{\alpha_1 \alpha_2} + F_{\mu\nu} f^F_{i_1 i_2} (\mathbf{r}_1 - \mathbf{r}_2) (\Sigma^{\mu\nu} C)_{\alpha_1 \alpha_2} \right].$$

The vector k is the four-momentum of the boson; the superspin–isospin matrices $T^a_{\kappa_1 \kappa_2}$ will be discussed later. The four-vector A^μ satisfies the condition $A^\mu k_\mu = 0$.

In this form it is easy to see that our bound states constitute spin 1 states, see Section 6.4.

We now come to an important point of Weak Mapping. In order to introduce the dynamical quantities of the effective theory, we have to observe that the expansion functions of Weak Mapping are not the full boson bound states (7.69), but a modified set of functions. As we have to consider the quantities A_l and $F_{l\nu} := iA_{[l}k_{\nu]}$ as the new dynamical variables of an effective boson theory in temporal gauge, we separate these factors from (7.69) and define the following basis functions for the Weak Mapping procedure:

$$C^{2,k}_{I_1 I_2} \equiv \left\{ C^l_{Z_1 Z_2}(\mathbf{r}_1, \mathbf{r}_2 | k, a), \; C^{l\nu}_{Z_1 Z_2}(\mathbf{r}_1, \mathbf{r}_2 | k, a) \right\} \tag{7.70}$$

with $Z = (i, \kappa, \alpha)$, where $k = (\mathbf{k}, a)$ denotes the full set of quantum numbers of the states, and where C^l and $C^{l\nu}$ are given by

$$C^l_{Z_1 Z_2}(\mathbf{r}_1, \mathbf{r}_2 | k, a) := N_B T^a_{\kappa_1 \kappa_2} e^{ik(\mathbf{r}_1 + \mathbf{r}_2)/2} f^A_{i_1 i_2} (\mathbf{r}_1 - \mathbf{r}_2) (\gamma^l C)_{\alpha_1 \alpha_2} \tag{7.71}$$

and

$$C^{l\nu}_{Z_1 Z_2}(\mathbf{r}_1, \mathbf{r}_2 | k, a) := N_B T^a_{\kappa_1 \kappa_2} e^{ik(\mathbf{r}_1 + \mathbf{r}_2)/2} f^F_{i_1 i_2} (\mathbf{r}_1 - \mathbf{r}_2) (\Sigma^{l\nu} C)_{\alpha_1 \alpha_2}. \tag{7.72}$$

Because $C^{l\nu}$ is antisymmetric in l, ν for $\nu = 1, 2, 3$, we have to restrict the set of basis functions (7.70) by the condition $l > \nu$.

We consider the temporal gauge, *i.e.*, $A_0 = 0$. Then in s wave approximation a general solution $\varphi^{(2)}_{I_1 I_2}$ of equation (6.1) is given by the following linear combination:

$$\varphi^{(2)}_{I_1 I_2} = \sum_{l,\mathbf{k}} \left\{ A_l(\mathbf{k}) C^l_{Z_1 Z_2}(\mathbf{r}_1, \mathbf{r}_2 | \mathbf{k}, a) + \sum_{\nu} i A_l(\mathbf{k}) k_\nu C^{l\nu}_{Z_1 Z_2}(\mathbf{r}_1, \mathbf{r}_2 | \mathbf{k}, a) \right\}. \tag{7.73}$$

Comparison of the solution (7.73) with the vector boson ansatz

$$\Lambda(x) = m \widehat{A}_\mu(x) \gamma^\mu C + \frac{1}{2} \widehat{F}_{\mu\nu}(x) \Sigma^{\mu\nu} C \tag{7.74}$$

for an ordinary de Broglie–Bargmann–Wigner equation

$$(i\gamma^\mu \partial_\mu - m)_{\alpha\alpha'} \Lambda_{\alpha'\beta}(x) = 0 \tag{7.75}$$

with

$$\widehat{F}_{\mu\nu} = \partial_\mu \widehat{A}_\nu - \partial_\nu \widehat{A}_\mu \tag{7.76}$$

suggests itself to interpret A_l, $l = 1, 2, 3$, as vector potential in temporal gauge and $F_{\mu\nu} := \frac{1}{2}(k_\mu A_\nu - k_\nu A_\mu)$ as field strength tensor in momentum space.

This fixed relation between A_l and $F_{l\nu}$ is a consequence of the linearized field dynamics in (6.1). If by the Weak Mapping procedure for non-abelian fields the full field dynamics with nonlinear terms is switched on, we expect that this relation between A_l and $F_{l\nu}$ will be modified. Thus, with respect to the Weak Mapping, we consider the set of boson functions (7.70) as independent states and leave it to the full field dynamics to establish a relation between vector potentials and field strengths.

Furthermore, the representation (7.73) is based on the temporal gauge $A_0 \equiv 0$. At first this choice seems to be arbitrary. But in connection with the Weak Mapping of the renormalized energy representation of quantum fields it will turn out that for gauge theories the use of the temporal gauge is conclusive, *i.e.*, there is a unique connection between the Hamiltonian formalism and temporal gauge. This was already emphasized in the literature, see for instance Christ and Lee [Chri 80]. For a more detailed discussion we refer to Section 7.2.

The basis functions C^l and $C^{l\nu}$ of Weak Mapping are symmetric with respect to a simultaneous interchange of the index sets $(\mathbf{r}_h, \alpha_h, i_h)$, $h = 1, 2$, see (7.71) and (7.72). As the complete functions (7.70) must be antisymmetric with respect to a simultaneous interchange of the index sets $(\mathbf{r}_h, \alpha_h, \kappa_h, i_h)$, $h = 1, 2$, it follows that the superspin–isospin matrices $T^a_{\kappa_1 \kappa_2}$ must be antisymmetric. In accordance with Section 6.3 we choose the antisymmetric matrices

$$T^a := \begin{cases} \begin{pmatrix} 0 & \sigma^a \\ (-1)^a \sigma^a & 0 \end{pmatrix} & \text{if} \quad a \in \{1, 2, 3\} \\[3mm] \begin{pmatrix} 0 & \mathbb{1} \\ -\mathbb{1} & 0 \end{pmatrix} & \text{if} \quad a = 4 \end{cases}, \tag{7.77}$$

where $\{\sigma^a\}$ is the set of Pauli matrices.

Substitution of (7.77) into the wave functions (7.69) or (7.71), (7.72), respectively, and a subsequent calculation of the good quantum numbers of these states according to Section 6.4 shows that these wave functions represent an isospin triplet of vector bosons with isospin 1 for $a = 1, 2, 3$ and an isospin singlet with isospin 0 for $a = 4$. All states $a = 1, 2, 3, 4$ are eigenstates to the fermion quantum number 0, $i.e.$, we consider only vector bosons which contain combinations of subfermion–anti-subfermion pairs or fields and conjugated fields, respectively.

We remark that the quantum numbers a denote eigenstates of the basic global $SU(2)$ and $U(1)$ invariance groups of the underlying subfermion theory. By the evaluation of the Weak Mapping equations it will be shown that these quantum numbers directly lead to the triplet ($a = 1, 2, 3$) and the singlet ($a = 4$) of gauge bosons for a $local$ effective $SU(2) \times U(1)$ theory.

The investigation of the full state spectrum of two-fermion bound states of equation (6.1) and its physical interpretation can be found in the thesis of Grimm [Gri 94a]. Here we will not discuss this topic.

With respect to the calculation of the dual functions $R^{2,k}_{I_1 I_2}$ we observe that in principle we need the complete set of functions $C^{I_1 I_2}_{2,k}$, $i.e.$, bound states and scattering states. Then the duals $R^{2,k}_{I_1 I_2}$ can be explicitly calculated by applying Schmidt's orthonormalization procedure. In order to avoid this laborious procedure we use only a projector on the bound states which guarantees orthonormality. With respect to the algebraic part of the set (7.70) this dual projector is unique. The non-uniqueness is solely concerned with the orbital part of (7.70). But the explicit evaluation of the map will show that the algebraic part essentially determines the resulting terms, and the orbital part only fixes numerical values of constants. In analogy to the decomposition (7.70) we define the following set of dual functions:

$$R^{2,k}_{I_1 I_2} := \left\{ (T^a)^+_{\kappa_1 \kappa_2} (\gamma^l C)^+_{\alpha_1 \alpha_2} e^{-ik(\mathbf{r}_1 + \mathbf{r}_2)/2} \breve{f}^A_{i_1 i_2}(\mathbf{r}_1 - \mathbf{r}_2) , \right. \tag{7.78}$$
$$\left. (T^a)^+_{\kappa_1 \kappa_2} (\Sigma^{l\nu} C)^+_{\alpha_1 \alpha_2} e^{-ik(\mathbf{r}_1 + \mathbf{r}_2)/2} \breve{f}^F_{i_1 i_2}(\mathbf{r}_1 - \mathbf{r}_2) \right\}$$
$$=: \left\{ R^{l,a}_{I_1 I_2}(\mathbf{k}), R^{l\nu,a}_{I_1 I_2}(\mathbf{k}) \right\} .$$

The set of these functions satisfies orthonormality relations and in the algebraic submanifold also completeness relations with respect to the original set (7.70). We will specify the functions $\breve{f}^{A/F}$ in Section 8.5.

Finally, for the evaluation of the map we need the explicit form of the single time limit of the propagator (3.115) of the fermion fields. This limit can be performed by standard methods and is given by

$$F^t_{Z_1 Z_2}(\mathbf{r}_1, \mathbf{r}_2) = \frac{1}{(2\pi)^2} \lambda_{i_1} \delta_{i_1 i_2} \gamma^5_{\kappa_1 \kappa_2} \tag{7.79}$$
$$\times \left\{ -m^2_{i_1} \frac{K_1(m_{i_1} r)}{r} C_{\alpha_1 \alpha_2} + i m_{i_1} (\gamma^k C)_{\alpha_1 \alpha_2} x^k \left[\frac{2K_1(m_{i_1} r)}{r^3} + m_{i_1} \frac{K_0(m_{i_1} r)}{r^2} \right] \right\}$$

with the modified Bessel functions $K_i(r)$.

It should be noticed that after summation over the auxiliary field indices i_1, i_2 in (7.79) the corresponding propagator is linearly divergent at the origin $\mathbf{r}_1 = \mathbf{r}_2$. This does not prevent regularization as in the generalized de Broglie–Bargmann–Wigner equations only convolutions $G \otimes F$ appear which are regular, see for instance equation (6.20).

7.5 Evaluation of the Boson Dynamics

The evaluation of the map in the bosonic sector simply consists in substituting the bound states (7.70) and the dual projectors (7.78) into equation (7.68). The necessary calculations are lengthy, but elementary. We will demonstrate these calculations only for the most important term. The treatment of the other terms goes along similar lines; for these terms we will give only the results. Details of the calculation with simplified functions can be found in [Stu 86a]; the full calculation with an extended version of the functions (7.70) is contained in [Pfi 90].

To express the results of these calculations we have to specify the notation of (7.57). We have defined the expansion functions of (7.57) by the functions (7.71) and (7.72), and accordingly we explain the set of bosonic sources b_k and their corresponding 'duals' ∂_k by

$$\{b_k\} \equiv \left\{b^A_{l,a}(\mathbf{k}), b^F_{l\nu,a}(\mathbf{k})\right\}, \qquad \{\partial_k\} \equiv \left\{\partial^A_{l,a}(\mathbf{k}), \partial^F_{l\nu,a}(\mathbf{k})\right\} \qquad (7.80)$$

with quantum numbers corresponding to (7.70). Furthermore, the Fourier transforms of the quantities (7.80) are defined by

$$b^A_{l,a}(\mathbf{r}) \quad := \quad (2\pi)^3 \int b^A_{l,a}(\mathbf{k}) e^{i\mathbf{k}\mathbf{r}} \, d\mathbf{k}, \qquad (7.81)$$

$$\partial^A_{l,a}(\mathbf{r}) \quad := \quad \int \partial^A_{l,a}(\mathbf{k}) e^{-i\mathbf{k}\mathbf{r}} \, d\mathbf{k}, \qquad (7.82)$$

etc.. With

$$[\partial^A_{l,a}(\mathbf{k}), b^A_{l',a'}(\mathbf{k}')]_- = \delta_{aa'} \delta_{ll'} \delta(\mathbf{k} - \mathbf{k}') \qquad (7.83)$$

one easily shows that

$$[\partial^A_{l,a}(\mathbf{r}), b^A_{l',a'}(\mathbf{r}')]_- = \delta_{aa'} \delta_{ll'} \delta(\mathbf{r} - \mathbf{r}'), \qquad (7.84)$$

etc. holds for the fourier transformed quantities.

We assume $b^F_{l\nu,a}$ and $\partial^F_{l\nu,a}$ to be antisymmetric in l and ν for $\nu = 1, 2, 3$. In order to work with independent quantities we postulate the restriction $l > \nu$ for all b^F and ∂^F.

We first consider the dynamical terms of (7.68) with the bosonic functional Hamiltonian operator (7.64). With the decomposition $\mathcal{H}_b = \mathcal{H}^0_b + \mathcal{H}^1_b$ the term $\mathcal{H}^0_b$ describes the kinetic energy and the masses of the effective boson dynamics. It is given by

$$\mathcal{H}^0_b := 2R^{2,k}_{I_1 I_3} \widehat{K}_{I_1 I_2} C^{I_2 I_3}_{2,k'} b_k \partial^b_{k'} - 6\widehat{W}_{I_1[I_2 I_3 I_4]} F_{I_4 K} R^{2,k}_{I_1 K} C^{I_2 I_3}_{2,k'} b_k \partial^b_{k'} . \qquad (7.85)$$

• **Proposition 7.2:** The evaluation of the effective kinetic energy term $\mathcal{H}_b^0$ yields

$$\mathcal{H}_b^0 = i \int d^3r \sum_{a=1}^4 b_{ij,a}^F(\mathbf{r})_{|i>j} \varepsilon_{ijk} \varepsilon_{lmk} \partial_m \partial_{l0,a}^F(\mathbf{r}) \tag{7.86}$$

$$+ i \int d^3r \sum_{a=1}^4 b_{i0,a}^F(\mathbf{r}) \varepsilon_{ijk} \varepsilon_{lmk} \partial_j \partial_{lm,a}^F(\mathbf{r})_{|l>m}$$

$$- i \int d^3r \sum_{a=1}^4 b_{i,a}^A(\mathbf{r}) \partial_{i0,a}^F(\mathbf{r})$$

$$+ i \mu_B^2 \int d^3r \sum_{a=1}^4 b_{i0,a}^F(\mathbf{r}) \partial_{i,a}^A(\mathbf{r})$$

with the effective boson mass μ_B and the partial derivative with respect to the space coordinate $\partial_k(\mathbf{r})$.

Proof: The expansion functions (7.71) and (7.72) are given by

$$C_{Z_1 Z_2}^{l,a}(\mathbf{z},\mathbf{u}|\mathbf{k}) := N_B T_{\kappa_1 \kappa_2}^a e^{i\mathbf{k}\mathbf{z}} f_{i_1 i_2}^A(\mathbf{u}) (\gamma^l C)_{\alpha_1 \alpha_2} \tag{7.87}$$

and

$$C_{Z_1 Z_2}^{l\nu,a}(\mathbf{z},\mathbf{u}|\mathbf{k}) := N_B T_{\kappa_1 \kappa_2}^a e^{i\mathbf{k}\mathbf{z}} f_{i_1 i_2}^F(\mathbf{u}) (\Sigma^{l\nu} C)_{\alpha_1 \alpha_2} \tag{7.88}$$

with the coordinates $\mathbf{z} := (\mathbf{r}_1 + \mathbf{r}_2)/2$ and $\mathbf{u} := \mathbf{r}_1 - \mathbf{r}_2$.
The dual functions (7.78) read

$$R_{Z_1 Z_2}^{l,a}(\mathbf{z},\mathbf{u}|\mathbf{k}) = (T^a)_{\kappa_1 \kappa_2}^+ (\gamma^l C)_{\alpha_1 \alpha_2}^+ e^{-i\mathbf{k}\mathbf{z}} \breve{f}_{i_1 i_2}^A(\mathbf{u}), \tag{7.89}$$

$$R_{Z_1 Z_2}^{l\nu,a}(\mathbf{z},\mathbf{u}|\mathbf{k}) = (T^a)_{\kappa_1 \kappa_2}^+ (\Sigma^{l\nu} C)_{\alpha_1 \alpha_2}^+ e^{-i\mathbf{k}\mathbf{z}} \breve{f}_{i_1 i_2}^F(\mathbf{u}).$$

Furthermore, we need the propagator (7.79) which can be written in the general form

$$F_{Z_1 Z_2}^t(\mathbf{z},\mathbf{u}) = \lambda_{i_1} \delta_{i_1 i_2} \gamma_{\kappa_1 \kappa_2}^5 \left[g_{i_1}(\mathbf{u}) C_{\alpha_1 \alpha_2} + g_{i_1}^k(\mathbf{u}) (\gamma^k C)_{\alpha_1 \alpha_2} \right]. \tag{7.90}$$

We consider the expression

$$\widehat{H}_{II'}^0 := 2\widehat{K}_{IK} C_{k'}^{KI'} \partial_{k'}^b - 6\widehat{W}_{IK_1 K_2 K_3} F_{K_3 I'} C_{k'}^{K_1 K_2} \partial_{k'}^b. \tag{7.91}$$

By inserting the kinetic operator (2.82), the vertex (2.83), the functions (7.87) and (7.88) and observing $I = (\kappa, \alpha, i, \mathbf{r})$, etc., we obtain

$$\widehat{H}_{II'}^0 = N_B \int d\mathbf{k}' \left\{ 2 \left[-i(\gamma^0 \gamma^k)_{\alpha\beta} \left(\frac{1}{2}\partial_k(\mathbf{z}) + \partial_k(\mathbf{u}) \right) + m_i \gamma_{\alpha\beta}^0 \right] \right. \tag{7.92}$$

$$\times T_{\kappa\kappa'}^{a'} \left[(\gamma^l C)_{\beta\alpha'} e^{i\mathbf{k}'\mathbf{z}} f_{ii'}^A(\mathbf{u}) \partial_l^{a'}(\mathbf{k}') + (\Sigma^{l\nu} C)_{\beta\alpha'} e^{i\mathbf{k}'\mathbf{z}} f_{ii'}^F(\mathbf{u}) \partial_{l\nu}^{a'}(\mathbf{k}') \right]$$

$$- g\lambda_i \left[2\gamma_{\alpha\beta_1}^0 C_{\beta_2 \beta_3} \delta_{\kappa\rho_1} \gamma_{\rho_2 \rho_3}^5 - 2(\gamma^0 \gamma^5)_{\alpha\beta_1} (\gamma^5 C)_{\beta_2 \beta_3} \delta_{\kappa\rho_1} \gamma_{\rho_1 \rho_2}^5 \right.$$

$$\left. + \gamma_{\alpha\beta_3}^0 C_{\beta_1 \beta_2} \delta_{\kappa\rho_3} \gamma_{\rho_1 \rho_2}^5 - (\gamma^0 \gamma^5)_{\alpha\beta_3} (\gamma^5 C)_{\beta_1 \beta_2} \delta_{\kappa\rho_3} \gamma_{\rho_1 \rho_2}^5 \right]$$

$$\times \lambda_{i'} \gamma_{\rho_3 \kappa'}^5 \left[g_{i'}(\mathbf{u}) C_{\beta_3 \alpha'} + g_{i'}^k(\mathbf{u}) (\gamma^k C)_{\beta_3 \alpha'} \right]$$

$$\times T_{\rho_1 \rho_2}^{a'} \left[(\gamma^l C)_{\beta_1 \beta_2} e^{i\mathbf{k}'(\mathbf{z}+\mathbf{u}/2)} \breve{f}^A \partial_l^{a'}(\mathbf{k}') + (\Sigma^{l\nu} C)_{\beta_1 \beta_2} e^{i\mathbf{k}'(\mathbf{z}+\mathbf{u}/2)} \breve{f}^F \partial_{l\nu}^{a'}(\mathbf{k}') \right] \right\}$$

where α, β are the spin indices, κ, ρ the isospin/superspin indices and i the auxiliary field indices of the nonperturbative regularization, see Chapter 2. Furthermore, we have used the abbreviation $\check{f}^{A/F} := \sum_{j,j'=1}^{3} f_{jj'}^{A/F}(0)$.

Next we evaluate the isospin/superspin algebra. Owing to the antisymmetry of (7.77) it holds $T^a \in \{C, \gamma^5 C, \gamma^5 \gamma^\mu C\}$. Thus we have $\mathrm{tr}[\gamma^5 T^a] = 0$ and owing to this vanishing trace an overall T^a matrix can be extracted from (7.92). We obtain

$$
\begin{aligned}
\widehat{H}_{II'}^0 = N_B T_{\kappa\kappa'}^{a'} \int d\mathbf{k'} \Big\{ & 2\Big[-i(\gamma^0\gamma^k\gamma^l C)_{\alpha\alpha'} \Big(\tfrac{1}{2}\partial_k(\mathbf{z}) + \partial_k(\mathbf{u})\Big) e^{i\mathbf{k'z}} f_{ii'}^A(\mathbf{u})\partial_l^{a'}(\mathbf{k'}) \\
& -i(\gamma^0\gamma^k\Sigma^{l\nu}C)_{\alpha\alpha'} \Big(\tfrac{1}{2}\partial_k(\mathbf{z}) + \partial_k(\mathbf{u})\Big) e^{i\mathbf{k'z}} f_{ii'}^F(\mathbf{u})\partial_{l\nu}^{a'}(\mathbf{k'}) \\
& + m_i(\gamma^0\gamma^l C)_{\alpha\alpha'} e^{i\mathbf{k'z}} f_{ii'}^A(\mathbf{u})\partial_l^{a'}(\mathbf{k'}) \\
& + m_i(\gamma^0\Sigma^{l\nu}C)_{\alpha\alpha'} e^{i\mathbf{k}z} f_{ii'}^F(\mathbf{u})\partial_{l\nu}^{a'}(\mathbf{k'})\Big] \\
& + 4g\lambda_i\lambda_{i'} \Big[(\gamma^0\gamma^l\gamma^k C)_{\alpha\alpha'} g_{i'}^{k'}(\mathbf{u}) e^{i\mathbf{k'}(\mathbf{z}+\mathbf{u}/2)} f^A\partial_l^{a'}(\mathbf{k'}) \\
& + (\gamma^0\gamma^l C)_{\alpha\alpha'} g_{i'}(\mathbf{u}) e^{i\mathbf{k'}(\mathbf{z}+\mathbf{u}/2)} f^A\partial_l^{a'}(\mathbf{k'})\Big]\Big\}. \quad (7.93)
\end{aligned}
$$

Now we calculate the term $\mathcal{H}_b^0$ itself by projecting $\mathcal{H}_{II'}^0$ onto $R_{II'}^k, b_k$. We expand the spinor parts on the symmetric Dirac algebra and observe the orthogonality relations between the C functions and their duals. Introducing the fouriertransformed functional operators we finally obtain

$$
\begin{aligned}
\mathcal{H}_b^0 = {} & i\int d^3r \sum_{a=1}^{4} b_{ij,a}^F(\mathbf{r})_{|i>j}\varepsilon_{ijk}\varepsilon_{lmk}\partial_m\partial_{l0,a}^F(\mathbf{r}) && (7.94) \\
& + i\int d^3r \sum_{a=1}^{4} b_{i0,a}^F(\mathbf{r})\varepsilon_{ijk}\varepsilon_{lmk}\partial_j\partial_{lm,a}^F(\mathbf{r})_{|l>m} \\
& - ic_0 \int d^3r \sum_{a=1}^{4} b_{i,a}^A(\mathbf{r})\partial_{i0,a}^F(\mathbf{r}) \\
& + i(c_1 + gc_2)\int d^3r \sum_{a=1}^{4} b_{i0,a}^F(\mathbf{r})\partial_{i,a}^A(\mathbf{r})
\end{aligned}
$$

with

$$
c_0 := -2^9\pi^3 N_B \sum_{i,i'}(m_i + m_{i'}) \int \check{f}_{ii'}^A(\mathbf{r}) f_{ii'}^F(\mathbf{r}) d\mathbf{r}, \quad (7.95)
$$

$$
c_1 := -2^9\pi^3 N_B \sum_{i,i'}(m_i + m_{i'}) \int \check{f}_{ii'}^F(\mathbf{r}) f_{ii'}^A(\mathbf{r}) d\mathbf{r}, \quad (7.96)
$$

$$
c_2 := -2^9\pi^3 N_B \sum_{i,i',j,j'} \lambda_i\lambda_{i'} \int \check{f}_{ii'}^F(\mathbf{r}) g_i(\mathbf{r}) f_{jj'}^A(0) d\mathbf{r}. \quad (7.97)
$$

The kinetic term with $\partial_k(\mathbf{u})$ vanishes because of the integrations over the even functions $f(\mathbf{u}) \equiv f(|\mathbf{u}|)$.

In Section 8.5 we will show that for the dual functions (7.78) it holds $\check{f}^F_{ii'}(\mathbf{u}) = \frac{1}{2}(m_i + m_{i'})\check{f}^A_{ii'}(\mathbf{u})$. From the orthogonality between the C functions (7.87), (7.88) and their duals (7.89) it follows $\sum_{i,i'} \int \check{f}^{A/F}_{ii'}(\mathbf{u}) f^{A/F}_{ii'}(\mathbf{u}) \, du = -2^{-2}(2\pi)^{-3} N_B^{-1}$. Thus we have $c_0 = 1$ and obtain (7.86) with $\mu^2_B := c_1 + g c_2$. $\diamond$

In [Stu 94] and [Pfi 90] the effective boson mass μ_B was calculated with the result

$$\mu^2_B = 10m^2 + \frac{g}{\pi^2}, \tag{7.98}$$

where m is the mean value of the auxiliary field masses m_1, m_2 and m_3. Our calculations differ from those in [Stu 94] and [Pfi 90] by the neglect of nontrivial orbital angular momentum parts, which have no influence on the structure of the resulting equations.

If one insists on massless vector bosons, this can be achieved by a suitable choice of the coupling constant g; with

$$g = -\frac{c_1}{c_2} = -10\pi^2 m^2 \tag{7.99}$$

we obtain $\mu_B = 0$.

As the next term we consider

$$\mathcal{H}^1_b := 6\widehat{W}_{I_1[I_2 I_3 I_4]} R^{2,k}_{I_1 K} C^{I_2 I_3}_{2,k'_1} C^{I_4 K}_{2,k'_2} b_k \partial^b_{k'_1} \partial^b_{k'_2}. \tag{7.100}$$

Its evaluation yields

$$\mathcal{H}^1_b = iG \int d^3r \sum_{a,b,c=1}^4 \hat{\varepsilon}_{abc} \left[\varepsilon_{ijk}\varepsilon_{lmk} b^F_{ij,b}(\mathbf{r})_{|i>j}\partial^F_{l0,a}(\mathbf{r})\partial^A_{m,c}(\mathbf{r}) \right. \tag{7.101}$$

$$\left. + \varepsilon_{ijk}\varepsilon_{lmk} b^F_{l0,b}(\mathbf{r})\partial^F_{ij,a}(\mathbf{r})_{|i>j}\partial^A_{m,c}(\mathbf{r}) \right]$$

with

$$\hat{\varepsilon}_{abc} := \begin{cases} \varepsilon_{abc} & \text{if } a,b,c \in \{1,2,3\} \\ 0 & \text{otherwise} \end{cases} \quad ; \tag{7.102}$$

in [Stu 94] and [Pfi 90] the coupling constant G was calculated and yielded the value

$$G = \frac{32\sqrt{2}\pi^{7/2}}{105} \frac{g^2 N_B}{m^4}. \tag{7.103}$$

According to the definition of $\hat{\varepsilon}_{abc}$ the self-coupling term $\mathcal{H}^1_b$ vanishes for the isospin singlets $a = 0$. The terms $\mathcal{H}^0_b$ and $\mathcal{H}^1_b$ represent the genuine gauge boson dynamics.

According to our decomposition of $\mathcal{H}_b$ we thus write (7.68) as

$$E_{p0}|\mathcal{B}(b;p)\rangle = \left(\mathcal{H}^0_b + \mathcal{H}^1_b\right)|\mathcal{B}(b;p)\rangle. \tag{7.104}$$

For relating this single time energy equation to dynamical boson equations which explicitly contain time derivatives we remember the Heisenberg equation (4.79), which was the starting point for the derivation of our functional eigenvalue equations. The functional version of the Heisenberg equation is given by

$$E_{p0}|\mathcal{B}(b;p)\rangle = i \int d^3r \, b_k(\mathbf{r},t) \left[\frac{\partial}{\partial t}\partial_k^b(\mathbf{r},t)\right] |\mathcal{B}(b;p)\rangle, \tag{7.105}$$

and the Heisenberg equation (4.79) is identical with the projection of equations (7.105) into the configuration space. Thus by shifting the reference time t of the single time formalism from $t = 0$ to an arbitrary value t the dynamical time parameter is restored by equations (7.105).

By substitution of (7.105) into (7.104) and with the independence of b^A and b^F, equation (7.104) can be interpreted as the energy representation of the functional equations

$$i\frac{\partial}{\partial t}\partial_{i,a}^A(\mathbf{r},t)|\mathcal{B}(b;p)\rangle = -i\partial_{i0,a}^F(\mathbf{r},t)|\mathcal{B}(b;p)\rangle, \tag{7.106}$$

$$i\frac{\partial}{\partial t}\partial_{i0,a}^F(\mathbf{r},t)|\mathcal{B}(b;p)\rangle = \left[\frac{i}{2}\varepsilon_{ijk}\varepsilon_{lmk}\partial_j\partial_{lm,a}^F(\mathbf{r},t) \right. \tag{7.107}$$

$$\left. - \frac{i}{2}G\hat{\varepsilon}_{abc}\varepsilon_{ijk}\varepsilon_{lmk}\partial_{lm,b}^F(\mathbf{r},t)\partial_{j,c}^A(\mathbf{r},t)\right]|\mathcal{B}(b;p)\rangle,$$

$$\frac{i}{2}\frac{\partial}{\partial t}\partial_{ij,a}^F(\mathbf{r},t)|\mathcal{B}(b;p)\rangle = \left[i\varepsilon_{ijk}\varepsilon_{lmk}\partial_m\partial_{l0,a}^F(\mathbf{r},t)\right. \tag{7.108}$$

$$\left. - iG\hat{\varepsilon}_{abc}\varepsilon_{ijk}\varepsilon_{lmk}\partial_{l0,b}^F(\mathbf{r},t)\partial_{m,c}^A(\mathbf{r},t)\right]|\mathcal{B}(b;p)\rangle,$$

where the time derivatives only act on the subsequent functional dual ∂, but not on the time parameter of the functionals $|\mathcal{B}\rangle$.

We now subdivide (7.106)–(7.108) into non-abelian electric and magnetic fields by introducing the definitions

$$b_{k,a}^E(\mathbf{r},t) := b_{k0,a}^F(\mathbf{r},t), \tag{7.109}$$

$$b_{k,a}^B(\mathbf{r},t) := -\sum_{l<m}\varepsilon_{klm}b_{lm,a}^F(\mathbf{r},t)$$

$$= -\frac{1}{2}\sum_{l,m}\varepsilon_{klm}b_{lm,a}^F(\mathbf{r},t)$$

for the sources b_k.

With corresponding definitions for the duals ∂_k one can easily derive the commutation relations for these transformed functional operators

$$\left[\partial_{k,a}^X(\mathbf{r},t), b_{k',a'}^X(\mathbf{r}',t)\right]_- = \delta_{kk'}\delta_{aa'}\delta(\mathbf{r}-\mathbf{r}') \tag{7.110}$$

for $X = A, E, B$.

Then equations (7.106)–(7.108) can be equivalently written as

$$i\frac{\partial}{\partial t}\partial_{i,a}^A(\mathbf{r},t)|\mathcal{B}(b;p)\rangle = -i\partial_{i,a}^E(\mathbf{r},t)|\mathcal{B}(b;p)\rangle, \tag{7.111}$$

$$i\frac{\partial}{\partial t}\partial_{i,a}^B(\mathbf{r},t)|\mathcal{B}(b;p)\rangle \tag{7.112}$$
$$= \left[-i\varepsilon_{ijk}\partial_k\partial_{j,a}^E(\mathbf{r},t) + iG\hat{\varepsilon}_{abc}\varepsilon_{ijk}\partial_{j,b}^E(\mathbf{r},t)\partial_{k,c}^A(\mathbf{r},t)\right]|\mathcal{B}(b;p)\rangle$$

and

$$i\frac{\partial}{\partial t}\partial_{i,a}^E(\mathbf{r},t)|\mathcal{B}(b;p)\rangle \tag{7.113}$$
$$= \left[i\varepsilon_{ijk}\partial_k\partial_{j,a}^B(\mathbf{r},t) - iG\hat{\varepsilon}_{abc}\varepsilon_{ijk}\partial_{j,b}^B(\mathbf{r},t)\partial_{k,c}^A(\mathbf{r},t)\right]|\mathcal{B}(b;p)\rangle.$$

Equations (7.111), (7.112), (7.113) describe a functional composite boson dynamics without quantization terms. In Section 8.6 we will perform the comparison with the quantized phenomenological gauge boson dynamics; here for a first outlook we derive the 'classical' equations for classical fields from our equations. The transition from the functional version of the theory to its classical formulation can be performed by means of the ansatz, *cf.* Stumpf [Stu 88]:

$$|\mathcal{B}(b;p)\rangle = \exp\left[Z_0(b;p)\right]|0\rangle_B \tag{7.114}$$

with

$$Z_0(b;p) := \int \sum_X \sum_{k,a} X_a^k(\mathbf{r},t)b_{k,a}^X(\mathbf{r},t)d^3r \tag{7.115}$$

for the classical fields $X = A, E, B$. This ansatz corresponds to an Ursell expansion in statistical mechanics and factorizes the correlation functions of the 'true' functional (5.14) with respect to one-particle correlation functions or wave functions. The latter are then interpreted as classical field functions.

If we project the functional equations (7.111)–(7.113) onto the functional vacuum sector and observe the ansatz (7.114) we obtain the following equations for the covariant classical fields:

$$\partial_0 A_a^i(\mathbf{r},t) = -E_a^i(\mathbf{r},t), \tag{7.116}$$
$$\partial_0 B_a^i(\mathbf{r},t) = \varepsilon_{ijk}\partial_j E_a^k(\mathbf{r},t) + G\hat{\varepsilon}_{abc}\varepsilon_{ijk}E_b^j(\mathbf{r},t)A_c^k(\mathbf{r},t),$$
$$\partial_0 E_a^i(\mathbf{r},t) = -\varepsilon_{ijk}\partial_j B_a^k(\mathbf{r},t) - G\hat{\varepsilon}_{abc}\varepsilon_{ijk}B_b^j(\mathbf{r},t)A_c^k(\mathbf{r},t).$$

For comparison we consider the field equations (7.24) of a classical $SU(2) \times U(1)$ gauge theory which for the pure gauge boson dynamics in temporal gauge read

$$i\partial_0 A_a^i = -iE_a^i, \tag{7.117}$$
$$i\partial_0 B_a^i = -i\varepsilon_{ijk}\partial_j E_a^k + ig_{\text{ph}}\varepsilon_{ijk}\hat{\varepsilon}_{abc}E_b^j A_c^k,$$
$$i\partial_0 E_a^i = i\varepsilon_{ijk}\partial_j B_a^k - ig_{\text{ph}}\varepsilon_{ijk}\hat{\varepsilon}_{abc}A_b^j B_c^k.$$

Shifting now B_a^i to $-B_a^i$ and identifying G with $-g_{\text{ph}}$, we immetiately see the equivalence of the effective equations (7.116) and the phenomenological equations (7.117).

Thus with (7.116) we have derived a classical $SU(2)$ Yang–Mills dynamics as well as the classical $U(1)$ electrodynamics in temporal gauge, except for the generalized Gauss law (7.22). In [Stu 86a] the Gauss law was introduced by the condition of relativistic invariance. In the meanwhile it was shown, *cf.* Magg [Magg 84] and Grimm [Gri 94a], that the Gauss law can be derived from the dynamical equations (7.116) with some suitable initial conditions. From this it follows that with the additional assumption of the Gauss law the Weak Mapping reproduces a complete local abelian and non-abelian gauge field dynamics in the temporal gauge.

We add that by an investigation of the normalization constant N_B of the vector boson functions we can show the coupling constant G to be dimensionless whereas the fields A, B, E obtain the usual dimensions; for a more detailed discussion of the dimensions of the effective fields we refer to [Pfi 90].

7.6   Subfermion Shell Model States

From our previous considerations and calculations it follows that apart from the basic spinor field equation, the theoretical derivation of the phenomenological laws depends on the kind of algebraic representation and on the appropriate choice of the subfermion bound states which are used for Weak Mapping.

At present the number of so called elementary particles in phenomenology and the number of corresponding elementary processes is rather large. Thus one cannot expect to cover this subject completely with one book. We therefore confine ourselves to demonstrate the applicability of our method for some important selected topics.

In phenomenological theory numerous attempts have been made to describe high energy phenomena, see for instance Mohapatra [Moha 86]. But undoubtedly the $SU(2) \times U(1)$ theories are the most elementary and successful access to this field, and thus we shall demonstrate that it is possible to interpret them as effective theories.

As far as the wave functions are concerned, for bosons exact wave functions are available. But we already remarked, that for bound states of three or more subfermions only approximate solutions can be found, see Section 6.5. So the question arises whether one can trust the results of Weak Mapping if exact solutions of the defining equation are not known.

The answer is affirmative because by the Weak Mapping theorems (see Chapter 5) any contribution to bound states, which in the approximate bound states themselves is not contained, is automatically taken into account in the interaction terms of the effective functional equation. Hence, as Weak Mapping is originally formulated as an invertible map between functional state spaces, in principle nothing is lost in the Weak Mapping transformation from the original theory to the effective theory. *This result allows to investigate Weak Mappings with idealized states instead of using exact solutions of the generalized de Broglie–Bargmann–Wigner equations.*

One of the most successful idealizations of states in many-body theory is the use of shell model states for atomic and molecular as well as for nuclear systems. Hence, owing to the abovementioned 'conservation' property of Weak Mapping we are justified to investigate shell model states of subfermions, although this idea is not to be taken too literally.

States for electro–weak gauge bosons, leptons and quarks were introduced and discussed in papers by Stumpf [Stu 86a,b], [Stu 87]. The corresponding map was studied and led to the dynamics of Han–Nambu quarks and leptons coupled to electro–weak gauge bosons. We will verify and improve these results by the application of improved three-particle states including their systematic group-theoretical analysis.

The three-particle states which we will use for the Weak Mapping are based on the functions which were discussed in Section 6.5. However, like in the vector boson case in Section 7.4, these solutions have to be adapted to the Weak Mapping procedure with respect to their degrees of freedom. In this section we will perform this adaption by means of the shell model idea; in order to do this, we shortly discuss the original model proposed in [Stu 86a,b].

The original states were assumed to be of the following form (for $t = 0$)

$$C_{2,\mathbf{k}}^{I_1 I_2} := e^{i\mathbf{k}(\mathbf{r}_1+\mathbf{r}_2)/2}\chi(\mathbf{r}_1 - \mathbf{r}_2|\mathbf{k})S_{\alpha_1\alpha_2}^b T_{\kappa_1\kappa_2}^a \qquad (7.118)$$

for electro–weak gauge bosons with symmetrical spin part

$$\{S^b\} := \{\gamma^\mu C, \Sigma^{\mu\nu} C\} \qquad (7.119)$$

and antisymmetric isospinor–superspinor part

$$T^a := \begin{cases} \begin{pmatrix} 0 & \sigma^a \\ (-1)^a\sigma^a & 0 \end{pmatrix} & \text{if} \quad a \in \{1,2,3\} \\[2ex] \begin{pmatrix} 0 & \mathbb{I} \\ -\mathbb{I} & 0 \end{pmatrix} & \text{if} \quad a = 4 \end{cases} \qquad (7.120)$$

According to the scheme of Section 6.4 these states are characterized by the subfermion quantum number $f = 0$, i.e.,, these bosons are constituted by linear combinations of subfermion/antisubfermion pairs or fields and conjugated fields, respectively.

The functional dependence of the states (7.118) was given by

$$\chi(\mathbf{r}|\mathbf{k}) = 2m^{3/2}e^{-mr}, \qquad (7.121)$$

where these functions were assumed to be independent of the center of mass momentum $\mathbf{k}$. In the shell model picture these functions describe $(1s)$ states; thus the composite bosons (7.118) occupy the $(1s)$ shell.

Comparing these vector boson states with the states (7.71) and (7.72) we immediately see that the latter coincide with the shell model states except for a modified relative coordinate function and except for the nonperturbative auxiliary field regularization.

On the other hand, the lepton and quark states were assumed to be of the form (for $t = 0$):

$$C_{3,k}^{I_1 I_2 I_3} := e^{ik(\mathbf{r}_1 + \mathbf{r}_2 + \mathbf{r}_3)/3} \Theta_{\kappa_1 \kappa_2 \kappa_3}^{l} \chi_{\alpha_1 \alpha_2 \alpha_3}^{\rho,s} \left(\mathbf{r}_2 - \mathbf{r}_3, \mathbf{r}_1 - \frac{1}{2}(\mathbf{r}_2 + \mathbf{r}_3) \right) \qquad (7.122)$$

with the symmetric isospinor–superspinor part Θ^l. The antisymmetric part was given by

$$\chi_{\alpha_1 \alpha_2 \alpha_3}^{\rho,s} := \left[\chi_{\alpha_1 \alpha_2 \alpha_3}^{s} g(\mathbf{r}_2 - \mathbf{r}_3) h^{\rho} \left(\mathbf{r}_1 - \frac{1}{2}(\mathbf{r}_2 + \mathbf{r}_3) \right) \right]_{\text{as}[1,2,3]} \qquad (7.123)$$

with $g \equiv \chi$ of (7.121), whilst h^{ρ} was given by

$$h^{\rho}(\mathbf{r}) := \begin{cases} 2m^{3/2} e^{-mr}(1 - mr) & \text{if} \quad \rho = 0 \\ 2m^{3/2} e^{-mr} r^{\rho} & \text{if} \quad \rho \in \{1, 2, 3\} \end{cases} \qquad (7.124)$$

with $\mathbf{r} = (r^{\rho})$ and $r = |\mathbf{r}|$.

In this ansatz the function h^0 describes another rotational invariant s state, whilst $h^i, i = 1, 2, 3$, are p states. Furthermore, we assume that h^0 describes lepton states and h^i, $i = 1, 2, 3$, quark states. This means that whilst the gauge bosons are constituted by $(1s)$ states, the fermions are described by $(1s, 2s)$ states (leptons) or $(1s, 2p)$ states (quarks).

If one compares these states with configurations in atomic physics, the quark states appear as 'excited' states of leptons. But a comparison with atomic physics is misleading. On the level of subfermion processes the interactions and decays for bound states are governed by the corresponding effective theory and the remaining residual forces and not by the rules of quantum mechanics.

In particular we shall demonstrate that the effective theory which results from the set of bound states (7.118) and (7.122), is an unbroken $SU(2) \times U(1)$ gauge theory coupled to fermions. In such a theory one has strict conservation of baryon number and lepton number, compare the instructive Table 3.2 in [Moha 86], i.e., although on the subfermion level the quarks appear as 'excited' states of leptons, the effective theory admits no decay process of quarks into leptons.

Only by the residual forces it might be possible to violate these conservation laws, but for large subfermion masses the residual forces are extremely weak as we will show in Section 8.7. So one cannot expect that these forces produce observable effects, in particular with respect to the violation of baryon and lepton number conservation.

The spin part $\chi_{\alpha_1 \alpha_2 \alpha_3}^{s}$ was assumed to be constituted either by positive or by negative energy solutions of the Dirac equation for large m with total spin $1/2$. For the isospinor–superspinor part $\Theta_{\kappa_1 \kappa_2 \kappa_3}^{l}$ a total of eight symmetric tensors in correlation with color was used. By means of this arrangement of (shell model) states a unique relation to the first generation of leptons and quarks was established and quantum numbers and effective dynamics were derived, see [Stu 86a,b], [Stu 87].

We now proceed to improve these shell model states in the light of the results of Section 6.5 for the three-particle problem. The results of Section 6.5 justify the form (7.122) for the state functions and motivate the following simple ansatz for the spin–orbit wave functions χ:

$$\chi_{\alpha_1 \alpha_2 \alpha_3}^{\rho,s}(\mathbf{r}_1, \mathbf{r}_2, \mathbf{r}_3) = \left\{ \chi_{\alpha_1 \alpha_2 \alpha_3}^{s} \zeta^{\rho}(\mathbf{r}_1 - \mathbf{r}_2, \mathbf{r}_2 - \mathbf{r}_3) \right\}_{\text{as}[1,2,3]} \qquad (7.125)$$

with the spin tensors

$$\chi^s_{\alpha_1\alpha_2\alpha_3} := (\gamma^5 C)_{\alpha_1\alpha_2} \chi^s_{\alpha_3}. \tag{7.126}$$

The quantum number s describes the set of basis vectors in spin space defined by

$$\chi^s_\alpha \in \left\{ \begin{pmatrix} 1 \\ 0 \\ 0 \\ 0 \end{pmatrix}, \begin{pmatrix} 0 \\ 1 \\ 0 \\ 0 \end{pmatrix}, \begin{pmatrix} 0 \\ 0 \\ 1 \\ 0 \end{pmatrix}, \begin{pmatrix} 0 \\ 0 \\ 0 \\ 1 \end{pmatrix} \right\}. \tag{7.127}$$

We mention that the antisymmetrization in (7.125) has to be performed with respect to the product of the spin tensors χ^s and the orbital functions ζ^ρ. This can be done by means of Young diagrams, where the nontrivial representations of the permutation group for each factor of a product wave function are reduced to one irreducible representation of the whole expression. For brevity we do not explicitly cite the application to the present case but refer to the original paper by Pfister and Stumpf [Pfi 95a].

The real functions $\zeta^\rho(\mathbf{r}_1 - \mathbf{r}_2, \mathbf{r}_2 - \mathbf{r}_3) =: \zeta^\rho(\mathbf{u}_1, \mathbf{u}_2)$ with $\rho = 0, 1, 2, 3$ are required to have the following angular momentum properties:

$$\left(\vec{L}_{\mathbf{u}_1} + \vec{L}_{\mathbf{u}_2} \right)^2 \zeta^0(\mathbf{u}_1, \mathbf{u}_2) = 0 \tag{7.128}$$

and

$$\zeta^l(\mathbf{u}_1, \mathbf{u}_2) = \sum_{i=1}^2 u^l_i \zeta'_i(\mathbf{u}_1, \mathbf{u}_2), \qquad l \in \{1, 2, 3\} \tag{7.129}$$

with

$$\left(\vec{L}_{\mathbf{u}_1} + \vec{L}_{\mathbf{u}_2} \right)^2 \zeta'_i(\mathbf{u}_1, \mathbf{u}_2) = 0, \tag{7.130}$$

from which we obtain

$$\left(\vec{L}_{\mathbf{u}_1} + \vec{L}_{\mathbf{u}_2} \right)^2 \zeta^l(\mathbf{u}_1, \mathbf{u}_2) = 2\zeta^l(\mathbf{u}_1, \mathbf{u}_2). \tag{7.131}$$

According to Section 6.4 these properties ensure that the functions ζ^0 describe s states whilst the functions ζ^l are connected with p states. Thus these functions are in accordance with our shell model ansatz.

Furthermore, we require

$$\lim_{|\mathbf{u}_1|\to\infty} \zeta^\rho(\mathbf{u}_1, \mathbf{u}_2) = \lim_{|\mathbf{u}_2|\to\infty} \zeta^\rho(\mathbf{u}_1, \mathbf{u}_2) = 0 \tag{7.132}$$

and normalizability:

$$\int \zeta^{\rho'}(\mathbf{u}_1, \mathbf{u}_2)\zeta^\rho(\mathbf{u}_1, \mathbf{u}_2)d\mathbf{u}_1\, d\mathbf{u}_2 = \langle \zeta^{\rho'}|\zeta^\rho \rangle \sim \delta_{\rho\rho'} < \infty. \tag{7.133}$$

As for the present investigation we are interested in qualitative results rather than in numerical values of coupling constants, *etc.*, it is not necessary to further specify the functions ζ^ρ.

Associating the s function to the leptons and the p functions to the quarks and simultaneously separating the superspin/isospin parts Θ^l for leptons and quarks we obtain the following arrangement:

$$
C_{3,k}^{I_1 I_2 I_3} := \begin{cases} e^{ik\frac{1}{3}(\mathbf{r}_1+\mathbf{r}_2+\mathbf{r}_3)} \Theta^j_{\kappa_1 \kappa_2 \kappa_3} \chi^{0,s}_{\alpha_1 \alpha_2 \alpha_3}(\mathbf{r}_1, \mathbf{r}_2, \mathbf{r}_3), & j \in M_1 := \{1, \dots 8\} \\ e^{ik\frac{1}{3}(\mathbf{r}_1+\mathbf{r}_2+\mathbf{r}_3)} \Theta^j_{\kappa_1 \kappa_2 \kappa_3} \chi^{i,s}_{\alpha_1 \alpha_2 \alpha_3}(\mathbf{r}_1, \mathbf{r}_2, \mathbf{r}_3), & j \in M_2 := \{9 \dots 16\} \end{cases}
$$

$$(7.134)$$

with $i \in \{1, 2, 3\}$. The set of indices M_1 is assumed to describe lepton states, whilst M_2 is connected with the quark states.

In Section 6.5 this assumption was justified by an analysis of the superspin/isospin tensors Θ^j. By fixing the fermion number of the subfermions to be $f = \frac{1}{3}$ we succeeded in reproducing the phenomenological quantum numbers of charge, hypercharge, and isospin.

On the other hand, the calculations of Section 6.5 show that as long as one does not treat the generation problem, one obtains superflous states which cannot be interpreted. We remarked that for mixed symmetry solutions also the family structure of leptons and quarks can be reproduced. *In accordance with these mixed symmetry solutions we will intentionally exclude the superflous states from our subsequent calculations.* Therefore in the following we will only consider states which are denoted by suitable subsets $\overline{M}_1$ and $\overline{M}_2$ of M_1 and M_2 and which are in accordance with the classification in Section 6.5. These subsets are defined by

$$\overline{M}_1 := \{2, 3, 6, 7\}, \qquad \overline{M}_2 := \{10, 11, 14, 15\}. \tag{7.135}$$

The corresponding superspin/isospin tensors are explicitly given by (6.108).

For convenience we renumber the sets $\overline{M}_1$ and $\overline{M}_2$ by

$$
\begin{aligned}
\overline{M}_1 &:= \{2 \to 1, 3 \to 2, 6 \to 3, 7 \to 4\}, \\
\overline{M}_2 &:= \{10 \to 1, 11 \to 2, 14 \to 3, 15 \to 4\}.
\end{aligned}
\tag{7.136}
$$

As leptons as quarks are distinguished by additional quantum numbers, this does not lead to any confusion.

Thus by (7.134) we have defined the expansion functions for the Weak Mapping which contain in their index $k = (\mathbf{k}, j, l, s)$ all necessary quantum numbers and indices to reproduce the indices of the phenomenological leptons and quarks. With respect to the color degrees of freedom of the quarks we refer to the investigations of Stumpf in [Stu 87] which demonstrated the possibility of interpreting the index l of the shell model states as color.

For the explicit evaluation of the various terms of the effective functional equation (7.61), besides the wave functions the corresponding duals are needed. The hard core duals of (7.134) are defined by

$$
R_{I_1 I_2 I_3}^{3,k} := \frac{N_{\mathrm{L,Q}}}{(2\pi)^3} e^{-ik\frac{1}{3}(\mathbf{r}_1+\mathbf{r}_2+\mathbf{r}_3)} \Theta^j_{\kappa_1 \kappa_2 \kappa_3} \chi^{\rho,s}_{\alpha_1 \alpha_2 \alpha_3}(\mathbf{r}_1, \mathbf{r}_2, \mathbf{r}_3)^+, \tag{7.137}
$$

where N_L or N_Q, respectively, are normalization constants for leptons or quarks. These constants are determined by the orthogonality relations of the fermion functions and their duals, see [Pfi 95a]. For the lepton duals we have $j \in \overline{M}_1$ and for the quark duals $j \in \overline{M}_2$.

As far as the boson wave functions are concerned we take over the ansatz (7.118) and its dual without any change. Dressed duals will be treated in the next section.

7.7   Effective Boson–Fermion Interaction

After the investigation of the pure boson sector of the effective functional equation (7.61) in the preceding sections we now turn to the evaluation of the fermion and the interaction part of this equation.

The evaluation of the boson sector was performed by the application of the non-perturbative subfermion regularization which we presented in Chapter 2. Thus the applicability of this regularization was demonstrated; it enabled us to calculate approximately the parameters of the effective boson theory, i.e., effective boson masses and the coupling constant G.

For the effective fermion and coupling terms, however, we deviate from this procedure, as the calculation of the three-particle states in Section 6.5 was performed by means of an (unspecified) *ad hoc* regularization of the subfermion propagator. Consequently we have to evaluate the mapping terms under these preconditions. This means that we cannot make any predictions for the numerical values of the parameters of the effective fermion and coupling sectors; rather we concentrate on the investigation of the structure of the effective theory.

The basic ingredients of the terms concerned are the shell model functions (7.134) for leptons ($j \in \overline{M}_1$) and quarks ($j \in \overline{M}_2$). According to the chain rule transformation (7.58) the corresponding functional fermion operators read $f_{s,j}(\mathbf{k})$ for $j \in \overline{M}_1$ (leptons) and $f_{s,j,l}(\mathbf{k})$ for $j \in \overline{M}_2$ (quarks). These sources f and their duals ∂_f are required to satisfy the usual anticommutation relations, i.e.,

$$\left[f_{s,j}(\mathbf{k}), \partial^f_{s',j'}(\mathbf{k}') \right]_+ = \delta_{jj'}\delta_{ss'}\delta(\mathbf{k}-\mathbf{k}'), \qquad \text{etc..} \qquad (7.138)$$

For the comparison with the phenomenological equations it is, however, convenient to introduce the transformed fermion sources

$$f_{j,\alpha}(\mathbf{r}) \quad := \quad (2\pi)^{-3/2} \int \exp(-i\mathbf{kr})\, \chi^s_\beta (\gamma^0\gamma^5)_{\beta\alpha} f_{s,j}(\mathbf{k})\, d\mathbf{k}, \qquad (7.139)$$

$$f_{j,l,\alpha}(\mathbf{r}) \quad := \quad (2\pi)^{-3/2} \int \exp(-i\mathbf{kr})\, \chi^s_\beta (\gamma^0\gamma^5)_{\beta\alpha} f_{s,j,l}(\mathbf{k})\, d\mathbf{k} \qquad (7.140)$$

with χ^s_β given by (7.127) and corresponding duals for leptons and quarks, respectively. These transformed functional operators again satisfy the anticommutation relations

$$\left[f_{j,\alpha}(\mathbf{r}), \partial^f_{j',\alpha'}(\mathbf{r}') \right]_+ = \delta_{jj'}\delta_{\alpha\alpha'}\delta(\mathbf{r}-\mathbf{r}'), \qquad \text{etc..} \qquad (7.141)$$

We start with the investigation of the fermion part (7.63) of the effective equation (7.62). The term (7.63) is constituted by two parts: the first gives the kinetic fermion term, the second a mass correction. In (5.74) the fermion part $\mathcal{H}_f$ contains an additional fermion self-coupling term. As this term leads to the formation of gluons, in the following calculations we omit this term and postpone the gluon theory to future investigations.

Thus we are left with the two terms of (7.63). We only cite the result of their evaluation and refer for the detailed calculations to [Pfi 95a]. *The calculations are performed by exact evaluations of the algebraic part of the corresponding terms and with an ad hoc regularization of the subfermion propagator according to Section 6.5.* The result is the effective operator $\mathcal{H}_f^0$ which is the diagonal part of the effective functional equation for leptons and quarks and reads

$$
\mathcal{H}_f^0 = \sum_{j \in \overline{M}_1} \int f_{j,\alpha}(\mathbf{r}) \left[-i(\gamma^0 \gamma^k) \partial_k + \bar{m}_1 \gamma^0 \right]_{\alpha \alpha'} \partial_{j,\alpha'}^f(\mathbf{r}) \, d\mathbf{r} \qquad (7.142)
$$

$$
+ \sum_{j \in \overline{M}_2} \int f_{j,l,\alpha}(\mathbf{r}) \left[-i(\gamma^0 \gamma^k) \partial_k + \bar{m}_2 \gamma^0 \right]_{\alpha \alpha'} \partial_{j,l,\alpha'}^f(\mathbf{r}) \, d\mathbf{r}.
$$

The effective lepton and quark masses $\bar{m}_1$ and $\bar{m}_2$ are given by expressions containing the subfermion coupling constant g, the subfermion mass m and the three-particle functions φ.

We now consider the effective boson–fermion interaction terms (7.65). The term $S^{k \ell'_1 \ell'_2}$ is expected to describe the coupling of the boson fields to the fermionic current in the field equations and the term $\widehat{N}^{\ell k \ell'}$ leads to the covariant coupling of fermions to bosons. The evaluation of the latter yields

$$
f_\ell \widehat{N}^{\ell k \ell'} \partial_k^b \partial_{\ell'}^f = iK_1 \sum_{j,j' \in \overline{M}_1} \sum_{a=1}^{3} \int f_{j,\alpha}(\mathbf{r})(\gamma^0 \gamma^k)_{\alpha \alpha'} G_{jj'}^a \partial_{j',\alpha'}^f(\mathbf{r}) \partial_{k,a}^A(\mathbf{r}) \, d\mathbf{r} \quad (7.143)
$$

$$
+ iK_1' \sum_{j,j' \in \overline{M}_1} \int f_{j,\alpha}(\mathbf{r})(\gamma^0 \gamma^k)_{\alpha \alpha'} G_{jj'}^4 \partial_{j',\alpha'}^f(\mathbf{r}) \partial_{k,4}^A(\mathbf{r}) \, d\mathbf{r}
$$

$$
+ iK_1 \sum_{j,j' \in \overline{M}_2} \sum_{a=1}^{3} \int f_{j,l,\alpha}(\mathbf{r})(\gamma^0 \gamma^k)_{\alpha \alpha'} G_{jj'}^a \partial_{j',l,\alpha'}^f(\mathbf{r}) \partial_{k,a}^A(\mathbf{r}) \, d\mathbf{r}
$$

$$
+ iK_1' \sum_{j,j' \in \overline{M}_2} \int f_{j,l,\alpha}(\mathbf{r})(\gamma^0 \gamma^k)_{\alpha \alpha'} G_{jj'}^4 \partial_{j',l,\alpha'}^f(\mathbf{r}) \partial_{k,4}^A(\mathbf{r}) \, d\mathbf{r}
$$

with the constants K_1, K_1'. We remember that according to Section 6.5, (6.99) and (6.100), the $SU(2)$ and $U(1)$ generators in superspin/isospin space are given by

$$
T_{\kappa_1 \kappa_2}^k = \frac{1}{2} \begin{pmatrix} \sigma^k & 0 \\ 0 & (-1)^k \sigma^k \end{pmatrix}_{\kappa_1 \kappa_2} \qquad (7.144)
$$

$$
F_{\kappa_1 \kappa_2} = \frac{1}{3} \begin{pmatrix} 1 & 0 \\ 0 & -1 \end{pmatrix}_{\kappa_1 \kappa_2} \qquad (7.145)
$$

Introducing the combinations $T^\pm := T^1 \pm iT^2$, the set of generators is to be denoted by $\tilde{G}^a_{\kappa_1\kappa_2}$ with $\tilde{G}^a \in \{T^\pm, T^3, F\}$. With these definitions the quantities $G^a_{jj'}$ which occur in (7.143) are given by

$$G^a_{jj'} := \Theta^j_{\kappa_1\kappa_2\kappa_3}\tilde{G}^a_{\kappa_1\kappa'_1}\Theta^{j'}_{\kappa'_1\kappa_2\kappa_3} + \Theta^j_{\kappa_1\kappa_2\kappa_3}\tilde{G}^a_{\kappa_2\kappa'_2}\Theta^{j'}_{\kappa_1\kappa'_2\kappa_3} + \Theta^j_{\kappa_1\kappa_2\kappa_3}\tilde{G}^a_{\kappa_3\kappa'_3}\Theta^{j'}_{\kappa_1\kappa_2\kappa'_3}. \quad (7.146)$$

The functions Θ^j constitute the superspin/isospin part of our three-particle shell model states; they were chosen in Section 6.5 just to be eigenstates of the generators $(T)^2, T^k, F$. Thus the quantities $G^a_{jj'}$ are the representations of the $SU(2)$ and $U(1)$ generators in j space, $i.e.$, in the space spanned by the sets $\overline{M}_1$ or $\overline{M}_2$ respectively, and can be explicitly calculated. With that the right hand side of (7.143) is well defined.

It remains the evaluation of the S term of the effective boson–fermion interaction (7.65), which according to (7.66) is given by

$$b_k S^{k\ell\ell'}\partial^f_\ell \partial^f_{\ell'} = -108\widehat{W}_{I_4 I_1 I_2 I_3}C^{3,\ell}_{I_1 I_3 K_1}C^{I_2 K_2 K_3}_{3,\ell'}R^{2,k}_{I_4 K_1 K_2 K_3}b_k\partial^f_\ell \partial^f_{\ell'}. \quad (7.147)$$

This term contains the first order boson polarization cloud $R^{2,k}_{I_1...I_4}$. For this polarization cloud we make the ansatz

$$\begin{aligned}
R^{2,k}_{I_1 I_2 I_3 I_4} &\equiv R_{Z_1 Z_2 Z_3 Z_4}(\mathbf{r}_1 \ldots \mathbf{r}_4 | \mathbf{k}, l\nu, a) &(7.148) \\
&= \exp\{-\frac{i}{4}\mathbf{k}(\mathbf{r}_1 + \mathbf{r}_2 + \mathbf{r}_3 + \mathbf{r}_4)\} \\
&\quad \times \omega(\mathbf{r}_1 - \mathbf{r}_2, \mathbf{r}_2 - \mathbf{r}_3, \mathbf{r}_3 - \mathbf{r}_4)(T^a)^+_{\kappa_1\kappa_2}\gamma^5_{\kappa_3\kappa_4}(\Sigma^{l\nu}C)^+_{\alpha_1\alpha_2}C_{\alpha_3\alpha_4},
\end{aligned}$$

which has to be antisymmetrized in all indices.

It can be seen by dimensional arguments (leading term approximation) that the functions which correspond to the gauge field *potentials* give a vanishing contribution in (7.147) for very high subfermion mass m.

In contrast to the hard core two-particle dual function, which is proportional to $(T^a)^+_{\kappa_1\kappa_2}(\Sigma^{i\nu}C)^+_{\alpha_1\alpha_2}$, we have in (7.148) additional matrices $\gamma^5_{\kappa_3\kappa_4}$, $C_{\alpha_3\alpha_4}$. These matrices describe fermion–antifermion pairs with isospin 0 and spin 0 (*cf.* Section 6.4). The symmetric function ω is required to be even and should vanish for $|\mathbf{r}_i - \mathbf{r}_j| \to \infty$. Furthermore, ω is required to be orthogonal to the spatial part of those four-particle hard core states which have the same isospin, spin and fermion quantum numbers as our two-particle states.

The final result of the evaluation of the term (7.147) is:

$$\begin{aligned}
b_k S_1^{k\ell\ell'}\partial^f_\ell \partial^f_{\ell'} &= iK_2 \sum_{j,j'\in\overline{M}_1}\sum_{a=1}^{3}\int b^E_{k,a}(\mathbf{r})\bar{\partial}^f_{j,\alpha}(\mathbf{r})G^a_{jj'}\gamma^k_{\alpha\alpha'}\partial^f_{j',\alpha'}(\mathbf{r})\,d\mathbf{r} &(7.149) \\
&\quad + iK'_2 \sum_{j,j'\in\overline{M}_1}\int b^E_{k,4}(\mathbf{r})\bar{\partial}^f_{j,\alpha}(\mathbf{r})G^4_{jj'}\gamma^k_{\alpha\alpha'}\partial^f_{j',\alpha'}(\mathbf{r})\,d\mathbf{r} \\
&\quad + iK_3 \sum_{j,j'\in\overline{M}_2}\sum_{a=1}^{3}\int b^E_{k,a}(\mathbf{r})\bar{\partial}^f_{j,l,\alpha}(\mathbf{r})G^a_{jj'}\gamma^k_{\alpha\alpha'}\partial^f_{j',l,\alpha'}(\mathbf{r})\,d\mathbf{r} \\
&\quad + iK'_3 \sum_{j,j'\in\overline{M}_2}\int b^E_{k,4}(\mathbf{r})\bar{\partial}^f_{j,l,\alpha}(\mathbf{r})G^4_{jj'}\gamma^k_{\alpha\alpha'}\partial^f_{j',l,\alpha'}(\mathbf{r})\,d\mathbf{r},
\end{aligned}$$

where the quantites G^a are given by (7.146). Analogously to equation (7.50) the Hermitian operator $\gamma^5_{\kappa_1\kappa'_1}\gamma^5_{\kappa_2\kappa'_2}\gamma^5_{\kappa_3\kappa'_3}$ transforms a superspin/isospin state Θ^j into its charge-conjugated state.

Expanding the charge conjugated tensors we may write

$$(\Theta^j)^C = \gamma^5_1\gamma^5_2\gamma^5_3\Theta^j =: L_{jj'}\Theta^{j'}. \tag{7.150}$$

With these coefficients $L_{jj'}$ we introduced the functional fermion duals

$$\begin{aligned}
\bar{\partial}^f_{j,\alpha}(\mathbf{r}) &:= (C^{-1})_{\alpha\alpha'}L_{jj'}\partial^f_{j,\alpha}(\mathbf{r}), \\
\bar{\partial}^f_{j,l,\alpha}(\mathbf{r}) &:= (C^{-1})_{\alpha\alpha'}L_{jj'}\partial^f_{j,l,\alpha}(\mathbf{r})
\end{aligned} \tag{7.151}$$

in (7.149). The constants K_2, K_3 and K'_2, K'_3, respectively, can chosen to be equal if additional requirements on the three-particle functions φ^ρ are imposed.

Summarizing the results we obtain in the low energy limit the effective functional equation

$$E_{p0}|\mathcal{G}(b,f;p)\rangle = (\mathcal{H}_{\mathrm{b}} + \mathcal{H}_{\mathrm{f}} + \mathcal{H}_{\mathrm{bf}})\,|\mathcal{G}(b,f;p)\rangle \tag{7.152}$$

with the boson parts (7.86) and (7.101):

$$\begin{aligned}
\mathcal{H}_{\mathrm{b}} = {}& -i\int \sum_{a=1}^4 b^A_{i,a}(\mathbf{r})\partial^F_{i0,a}(\mathbf{r})\,d\mathbf{r} \\
&+i\int \sum_{a=1}^4 b^F_{ij,a}(\mathbf{r})_{|i>j}\varepsilon_{ijk}\varepsilon_{lmk}\partial_m\partial^F_{l0,a}(\mathbf{r})\,d\mathbf{r} \\
&+i\int \sum_{a=1}^4 b^F_{i0,a}(\mathbf{r})\varepsilon_{ijk}\varepsilon_{lmk}\partial_j\partial^F_{lm,a}(\mathbf{r})_{|l>m}\,d\mathbf{r} \\
&+iG\int \sum_{a,b,c=1}^4 \hat{\varepsilon}_{abc}\Big[\varepsilon_{ijk}\varepsilon_{lmk}b^F_{ij,b}(\mathbf{r})_{|i>j}\partial^F_{l0,a}(\mathbf{r})\partial^A_{m,c}(\mathbf{r}) \\
&\qquad\qquad + \varepsilon_{ijk}\varepsilon_{lmk}b^F_{l0,b}(\mathbf{r})\partial^F_{ij,a}(\mathbf{r})_{|i>j}\partial^A_{m,c}(\mathbf{r})\Big]\,d\mathbf{r},
\end{aligned} \tag{7.153}$$

the fermion part (7.142):

$$\begin{aligned}
\mathcal{H}_{\mathrm{f}} = {}& \sum_{j\in\overline{M}_1}\int f_{j,\alpha}(\mathbf{r})\left[-i(\gamma^0\gamma^k)\partial_k + \bar{m}_1\gamma^0\right]_{\alpha\alpha'}\partial^f_{j,\alpha'}(\mathbf{r})\,d\mathbf{r} \\
&+ \sum_{j\in\overline{M}_2}\int f_{j,l,\alpha}(\mathbf{r})\left[-i(\gamma^0\gamma^k)\partial_k + \bar{m}_2\gamma^0\right]_{\alpha\alpha'}\partial^f_{j,l,\alpha'}(\mathbf{r})\,d\mathbf{r},
\end{aligned} \tag{7.154}$$

and the coupling parts (7.143) and (7.149):

$$\begin{aligned}
\mathcal{H}_{\mathrm{bf}} = {}& iK_1\sum_{j,j'\in\overline{M}_1}\sum_{a=1}^3\int f_{j,\alpha}(\mathbf{r})(\gamma^0\gamma^k)_{\alpha\alpha'}G^a_{jj'}\partial^f_{j',\alpha'}(\mathbf{r})\partial^A_{k,a}(\mathbf{r})\,d\mathbf{r} \\
&+ iK'_1\sum_{j,j'\in\overline{M}_1}\int f_{j,\alpha}(\mathbf{r})(\gamma^0\gamma^k)_{\alpha\alpha'}G^4_{jj'}\partial^f_{j',\alpha'}(\mathbf{r})\partial^A_{k,4}(\mathbf{r})\,d\mathbf{r}
\end{aligned} \tag{7.155}$$

$$+iK_1 \sum_{j,j' \in \overline{M}_2} \sum_{a=1}^{3} \int f_{j,l,\alpha}(\mathbf{r})(\gamma^0 \gamma^k)_{\alpha\alpha'} G_{jj'}^{a} \partial_{j',l,\alpha'}^{f}(\mathbf{r}) \partial_{k,a}^{A}(\mathbf{r}) \, d\mathbf{r}$$

$$+iK_1' \sum_{j,j' \in \overline{M}_2} \int f_{j,l,\alpha}(\mathbf{r})(\gamma^0 \gamma^k)_{\alpha\alpha'} G_{jj'}^{4} \partial_{j',l,\alpha'}^{f}(\mathbf{r}) \partial_{k,4}^{A}(\mathbf{r}) \, d\mathbf{r}$$

$$+iK_2 \sum_{j,j' \in \overline{M}_1} \sum_{a=1}^{3} \int b_{k,a}^{E}(\mathbf{r}) \bar{\partial}_{j,\alpha}^{f}(\mathbf{r}) G_{jj'}^{a} \gamma_{\alpha\alpha'}^{k} \partial_{j',\alpha'}^{f}(\mathbf{r}) \, d\mathbf{r}$$

$$+iK_2' \sum_{j,j' \in \overline{M}_1} \int b_{k,4}^{E}(\mathbf{r}) \bar{\partial}_{j,\alpha}^{f}(\mathbf{r}) G_{jj'}^{4} \gamma_{\alpha\alpha'}^{k} \partial_{j',\alpha'}^{f}(\mathbf{r}) \, d\mathbf{r}$$

$$+iK_3 \sum_{j,j' \in \overline{M}_2} \sum_{a=1}^{3} \int b_{k,a}^{E}(\mathbf{r}) \bar{\partial}_{j,l,\alpha}^{f}(\mathbf{r}) G_{jj'}^{a} \gamma_{\alpha\alpha'}^{k} \partial_{j',l,\alpha'}^{f}(\mathbf{r}) \, d\mathbf{r}$$

$$+iK_3' \sum_{j,j' \in \overline{M}_2} \int b_{k,4}^{E}(\mathbf{r}) \bar{\partial}_{j,l,\alpha}^{f}(\mathbf{r}) G_{jj'}^{4} \gamma_{\alpha\alpha'}^{k} \partial_{j',l,\alpha'}^{f}(\mathbf{r}) \, d\mathbf{r}.$$

These terms have to be compared with the phenomenological equation (7.43). We start with its phenomenological boson part which is given by (7.48) and reads

$$\mathcal{H}_b^{ph} = -i \sum_{a=1}^{4} \int b_{j,a}^{A}(\mathbf{r}) \partial_{j,a}^{E}(\mathbf{r}) \, d\mathbf{r} \tag{7.156}$$

$$+i \sum_{a=1}^{4} \int b_{j,a}^{E}(\mathbf{r}) \left[\partial_k \partial_k \partial_{j,a}^{A}(\mathbf{r}) + \partial_j \partial_k \partial_{k,a}^{A}(\mathbf{r}) \right] d\mathbf{r}$$

$$+i g_{ph} \sum_{a,b,c=1}^{3} \int b_{j,a}^{E}(\mathbf{r}) \varepsilon_{abc} \left[\left(\partial_k \partial_{k,b}^{A}(\mathbf{r}) \right) \partial_{j,c}^{A}(\mathbf{r}) + 2 \partial_{k,b}^{A}(\mathbf{r}) \partial_k \partial_{j,c}^{A}(\mathbf{r}) \right.$$

$$\left. - \partial_{k,b}^{A}(\mathbf{r}) \partial_j \partial_{k,c}^{A}(\mathbf{r}) \right] d\mathbf{r}$$

$$-i g_{ph}^2 \sum_{a \ldots e=1}^{3} \int b_{j,a}^{E}(\mathbf{r}) \varepsilon_{abc} \varepsilon_{cde} \partial_{k,b}^{A}(\mathbf{r}) \partial_{k,d}^{A}(\mathbf{r}) \partial_{j,e}^{A}(\mathbf{r}) \, d\mathbf{r}.$$

In Section 7.5 we already compared the *classical* equations resulting from (7.153) and (7.156) and found that with the identification $G = -g_{ph}$ they are completely identical.

In order to compare the full functional equations we have to observe, that the primary effective variables of the Weak Mapping in the boson sector are the fields A_a^μ and $F_a^{\mu\nu}$. Contrary to this is the formulation of the phenomenological quantized gauge theory, which is done in terms of the canonically conjugated fields A_a^μ and E_a^μ. Thus we have to transform the effective boson term (7.153) onto these variables.

This transformation is performed by eliminating the B field equation from equations (7.111)–(7.113), which in Section 7.5 were shown to be the equivalent formulation of the boson theory in terms of the fields A, E and B. We use (7.111) to rewrite (7.112) as

$$-i\partial_t \partial_{i,a}^{B}(\mathbf{r}, t) | \mathcal{B} \rangle \tag{7.157}$$

$$= \left\{-i\partial_t \varepsilon_{ijk}\partial_k\partial_{j,a}^A(\mathbf{r},t) + \frac{i}{2}G\varepsilon_{abc}\varepsilon_{ijk}\partial_t\left[\partial_{j,b}^A(\mathbf{r},t)\partial_{k,c}^A(\mathbf{r},t)\right]\right\}|\mathcal{B}\rangle.$$

With $\partial_{i,a} \equiv \partial_{i,a}(\mathbf{r},t)$, etc., we can conclude from (7.157)

$$\partial_{i,a}^B|\mathcal{B}\rangle = \left[\varepsilon_{ijk}\partial_k\partial_{j,a}^A - \frac{1}{2}G\varepsilon_{abc}\varepsilon_{ijk}\partial_{j,b}^A\partial_{k,c}^A\right]|\mathcal{B}\rangle \tag{7.158}$$

with a vanishing integration constant. We remark that (7.158) corresponds to the phenomenological equation $\mathbf{B} = \partial \times \mathbf{A} - \frac{1}{2}\hat{g}\mathbf{A} \times \mathbf{A}$.

Substitution of (7.158) into (7.113) and again use of (7.105) finally yields

$$\mathcal{H}_{\mathrm{b}} = -i\sum_{a=1}^{4}\int b_{i,a}^A(\mathbf{r})\partial_{i,a}^E(\mathbf{r})\,d\mathbf{r} \tag{7.159}$$

$$+i\varepsilon_{ijk}\varepsilon_{lmk}\sum_{a=1}^{4}\int b_{i,a}^E(\mathbf{r})\partial_j\partial_l\partial_{m,a}^A(\mathbf{r})\,d\mathbf{r}$$

$$+i\frac{G}{2}\sum_{a\ldots c=1}^{3}\int \varepsilon_{abc}\varepsilon_{ijk}\varepsilon_{lmk}b_{i,a}^E(\mathbf{r})\partial_j\partial_l\partial_{l,b}^A(\mathbf{r})\partial_{m,c}^A(\mathbf{r})\,d\mathbf{r}$$

$$+iG\sum_{a\ldots c=1}^{3}\int \varepsilon_{abc}\varepsilon_{ijk}\varepsilon_{lmk}b_{i,a}^E(\mathbf{r})[\partial_m\partial_{l,b}^A(\mathbf{r})]\partial_{j,c}^A(\mathbf{r})\,d\mathbf{r}$$

$$-i\frac{G^2}{2}\sum_{a\ldots e=1}^{3}\int \varepsilon_{abc}\varepsilon_{cde}\varepsilon_{ijk}\varepsilon_{lmk}b_{i,a}^E(\mathbf{r})\partial_{m,e}^A(\mathbf{r})\partial_{l,d}^A(\mathbf{r})\partial_{j,b}^A(\mathbf{r})\,d\mathbf{r},$$

where in the following we restrict ourselves to the independent boson sources b^A and b^E, i.e., we put $b^B = 0$ in $|\mathcal{B}\rangle$.

Carrying out the differentation in the third term of the right hand side of (7.159) and evaluating the ε tensors, the agreement of (7.159) and (7.156) is easily proved, if we make the identification $G = -g_{\mathrm{ph}}$. We remember, however, that (7.156) is not the full functional equation of a canonically quantized gauge boson theory. This equation is given by the boson sector of (7.43) and contains additional quantization terms. The question of the comparison of the full quantized effective and phenomenological theories will be treated in Section 8.6.

Next we compare the pure fermion term (7.154) with its phenomenological counterpart (7.47):

$$\mathcal{H}_{\mathrm{f}}^{\mathrm{ph}} = \int f_{\alpha\kappa}(\mathbf{r})\left[-i\gamma^0\gamma^k\partial_k\right]_{\alpha\beta}\partial_{\beta\kappa}^f(\mathbf{r})\,d\mathbf{r}. \tag{7.160}$$

First we observe that for brevity we formulated the phenomenological equations only in terms of electron and neutrino fields, i.e., there are no quark states. The restriction of the effective terms, which contain these quark states, is easily done by setting $\overline{M}_2 = 0$.

We have already mentioned that in our presentation of the three-particle problem we obtained isospin quadruplets instead of doublets for the leptons and quarks. Leaving aside this problem for the moment, with (7.160) we have reproduced the phenomenological kinetic fermion terms. However, it is important to remark that the Weak Mapping yields in a natural manner an effective fermion mass term (in particular, a neutrino mass) which in the phenomenological theory has to be introduced in a rather artificial way by the Higgs mechanism. We regard this feature as an important advantage of our fusion concept.

In principle the effective fermion masses can be calculated; this has to be done by means of an improved regularization procedure and the determination of the three-particle functions. As already mentioned in this presentation we restrict ourselves to the reproduction of the structure of the electro–weak theory.

The same holds for the coupling constants of the effective coupling term (7.155). This term has to be compared with (7.49):

$$
\mathcal{H}_{\text{bf}}^{\text{ph}} = g_{\text{ph}} \sum_{a=1}^{4} \int f_{\alpha_1\kappa_1}(\mathbf{r})(\gamma^0\gamma^k)_{\alpha_1\alpha_2} \widehat{L}_{\kappa_1\kappa_2}^a \partial_{\alpha_2\kappa_2}^f(\mathbf{r}) \, \partial_{k,a}^A(\mathbf{r}) \, d\mathbf{r} \qquad (7.161)
$$
$$
- \frac{i}{2} g_{\text{ph}} \sum_{a=1}^{4} \int b_{k,a}^E(\mathbf{r}) \gamma_{\alpha_1\alpha_2}^k \widehat{L}_{\kappa_1\kappa_2}^a \bar{\partial}_{\alpha_1\kappa_1}^f(\mathbf{r}) \partial_{\alpha_2\kappa_2}^f(\mathbf{r}) \, d\mathbf{r} \ .
$$

An inspection of the super-generators $G_{jj'}$ and $T_{jj'}^{\pm}$ in (7.155) shows, that for the electron–neutrino doublet they reproduce the corresponding generators of equation (7.161). If we leave aside the numerical discussion of the coupling constants the agreement of both terms is obvious. Thus we have reproduced the phenomenological unbroken $SU(2) \times U(1)$ gauge theory in temporal gauge.

A comment should be given with respect to the fermion isospin quadruplets, which according to Section 6.5 lead to superflous states.

In this section we already indicated that this problem was solved by the calculation of mixed symmetry three-particle states, cf. [Pfi 95c]. This ansatz also yields a natural explanation of the family structure of the standard model.

The evaluation of the Weak Mapping with mixed symmetry three-particle states was performed by Pfister [Pfi 95b]. The result of these calculations is the justification of the reproduction of the local $SU(2) \times U(1)$ gauge theory, but now with the correct isospin doublets and with the inclusion of the three fermion families.

8 Effective Electro–weak Theory

8.1 Phenomenological Electro–weak Boson Dynamics

The functional equation (7.43) describes the quantum version of an unbroken $SU(2) \times U(1)$ local gauge theory coupled to fermion fields. Although its group structure is basic for the standard model, in the unbroken form the theory cannot be correctly related to experiments. Rather, the symmetry breaking Higgs mechanism is needed to obtain a renormalizable phenomenological field theory which is in accordance with experiments.

One consequence of this Higgs mechanism is the necessity of introducing new fields as linear combinations of the original gauge fields A_μ^a by the so called Glashow–Salam–Weinberg transformation. It is only by this transformation that one can discover the photon in this theory. Thus the derivation of quantum electrodynamics in this wider context is closely related to symmetry breaking and is a rather sophisticated matter. On the other hand, the derivation of electrodynamics was the decisive result of de Broglie's fusion theory of fermions. Hence such a derivation is of fundamental importance, but as long as in the fusion theory of the preceding chapters one does not treat symmetry breaking one cannot arrive at photons and quantum electrodynamics.

We will treat this problem in the present chapter. For comparison we first briefly discuss the phenomenological theory. In concentrating on the $SU(2)$ symmetry breaking we consider a left–right symmetric model for the fermions in order to avoid the complications of the full standard model. The Lagrangian of such a model reads

$$\mathcal{L} = -\frac{1}{4} \sum_{k=1}^{3} F_{\mu\nu}^k F_k^{\mu\nu} - \frac{1}{4} F_{\mu\nu}^4 F_4^{\mu\nu} + (D^\mu \phi_A^*)(D_\mu \phi_A) - V(\phi) + i\bar{\chi}\gamma^\mu D_\mu \chi, \qquad (8.1)$$

where the field strengths $F_{\mu\nu}$ are given by (7.3).

Compared with the Lagrangian (7.5) we have added the Higgs field ϕ, which is responsible for the symmetry breaking and which in the simplest case is a complex scalar doublet under the $SU(2)$ and a scalar singlet under the $U(1)$ transformations. The potential $V(\phi)$ is invariant under these groups.

The covariant derivatives are defined by

$$D_\mu := \partial_\mu - \frac{i}{2} g_V \sum_{a=1}^{3} \sigma^a A_\mu^a - \frac{i}{2} g_S \sigma^0 A_\mu^4 \qquad (8.2)$$

with the real non-abelian gauge field triplet A_μ^a, $a = 1, 2, 3$, the real abelian singlet A_μ^4, and the Pauli matrices σ^a, $a = 1, 2, 3$, and $\sigma^0 = \mathbb{I}_2$. The Higgs field potential $V(\phi)$ is given by

$$V(\phi) := -m_{\mathrm{H}}^2 \phi_A^* \phi_A - \frac{f}{4}(\phi_A^* \phi_A)^2. \tag{8.3}$$

In the standard model one works with a chiral theory and the fermions have to be coupled to the ϕ field, too; after symmetry breaking this coupling induces the fermion masses. But with respect to a comparison with a left–right symmetric map we make no use of this extension.

The complex doublet ϕ_A transforms in the following way under local $SU(2)$ transformations:

$$\phi_A'(x) = \exp\left[-\frac{i}{2}\sigma^k \omega_k(x)\right] \phi_A(x), \tag{8.4}$$

$$(\phi_A^*)'(x) = \exp\left[\frac{i}{2}\sigma^k \omega_k(x)\right] \phi_A^*(x)$$

with real gauge functions $\omega_k(x)$, $k = 1, 2, 3$.

Splitting up ϕ_A into real and imaginary parts one can use this gauge representation for a parametrization of ϕ_A, as an arbitrary ϕ_A can always be written in the form

$$\begin{pmatrix} \phi_1 \\ \phi_2 \end{pmatrix} = \exp\left[-\frac{i}{2}\sigma^k \xi_k(x)\right] \begin{pmatrix} 0 \\ \rho(x) \end{pmatrix} \tag{8.5}$$

with four real functions $\xi_k(x)$, $k = 1, 2, 3$, and $\rho(x)$.

This reparametrization can be subjected to a gauge transformation which leaves the Lagrangian invariant and just compensates the exponential. Hence for the further treatment we break the classical $SU(2)$ covariance of the fields ϕ_A and fix their gauge by assuming them to be of the form

$$\phi_A = \begin{pmatrix} 0 \\ \rho(x) \end{pmatrix}. \tag{8.6}$$

The function $\rho(x)$ is real and thus has no charge, *i.e.*, no electromagnetic coupling.

We now assume that the ϕ_A field in the form (8.6) possesses a nonvanishing vacuum expectation value

$$\langle 0|\phi_A|0 \rangle = \frac{\sqrt{2}m_{\mathrm{H}}}{\sqrt{f}} \begin{pmatrix} 0 \\ 1 \end{pmatrix} = \rho_0 \begin{pmatrix} 0 \\ 1 \end{pmatrix}, \tag{8.7}$$

thereby inducing a spontaneous breaking of the $SU(2)$ symmetry of the quantum ground state.

The field $\rho(x)$ can be expanded at ρ_0 in the form

$$\phi_A = \begin{pmatrix} 0 \\ \rho_0 + \eta(x) \end{pmatrix}. \tag{8.8}$$

As a consequence of this expansion from the covariant derivative of ϕ_A terms are shiftet to the vector fields, giving the following Lagrangian:

$$\mathcal{L} = -\frac{1}{4}\sum_{a=1}^{3} F_{\mu\nu}^a F_a^{\mu\nu} - \frac{1}{4} F_{\mu\nu}^4 F_4^{\mu\nu} \tag{8.9}$$

$$+ \frac{g_V^2 m_H^2}{2f} A_\mu^a A_a^\mu + \frac{g_S^2 m_H^2}{2f} A_\mu^4 A_4^\mu - \frac{g_S g_V m_H^2}{f} A_\mu^4 A_3^\mu$$

$$+ \frac{1}{2}\partial_\mu \eta \partial^\mu \eta - m_H^2 \eta^2 + i\bar{\chi}\gamma^\mu D_\mu \chi + \mathcal{L}_I,$$

where $\mathcal{L}_I$ contains all other remaining terms.

As an effect of this transformation the vector fields have aquired mass terms, but the mixed term $A_\mu^4 A_3^\mu$ does not allow a meaningful interpretation. The latter term, however, can be removed from (8.9) by application of the Glashow–Salam–Weinberg transformation

$$A_\mu^1 = \frac{1}{\sqrt{2}}(W_\mu + W_\mu^*), \tag{8.10}$$

$$A_\mu^2 = -i\frac{1}{\sqrt{2}}(W_\mu - W_\mu^*),$$

$$A_\mu^3 = Z_\mu \cos\Theta_W + A_\mu \sin\Theta_W,$$

$$A_\mu^4 = -Z_\mu \sin\Theta_W + A_\mu \cos\Theta_W.$$

We remark that the new field A_μ, which is to describe the electromagnetic Maxwell field, should not be confused with the original fields A_μ^a, $a = 1, 2, 3, 4$.

With this transformation we obtain

$$\frac{g_V^2 m_H^2}{2f} A_3^\mu A_3^\mu + \frac{g_S^2 m_H^2}{2f} A_\mu^4 A_4^\mu - \frac{g_S g_V m_H^2}{f} A_\mu^4 A_3^\mu \tag{8.11}$$

$$= \frac{m_H^2}{f}\left\{ Z_\mu Z^\mu \left(\frac{1}{2}g_V^2 \cos^2\Theta_W + \frac{1}{2}g_S^2 \sin^2\Theta_W + g_S g_V \sin\Theta_W \cos\Theta_W \right) \right.$$

$$+ A_\mu A^\mu \left(\frac{1}{2}g_V^2 \sin^2\Theta_W + \frac{1}{2}g_S^2 \cos^2\Theta_W - g_S g_V \sin\Theta_W \cos\Theta_W \right)$$

$$\left. + A_\mu Z^\mu \left[\frac{1}{2}(g_V^2 - g_S^2)\sin 2\Theta_W - g_S g_V \cos 2\Theta_W \right] \right\}.$$

For the values

$$\sin\Theta_W = \frac{g_S}{\sqrt{g_S^2 + g_V^2}}, \tag{8.12}$$

$$\cos\Theta_W = \frac{g_V}{\sqrt{g_S^2 + g_V^2}},$$

the coefficients of the terms $A_\mu Z^\mu$ as well as of $A_\mu A^\mu$ vanish and (8.9) goes over into

$$\mathcal{L}' = -\frac{1}{2}(\partial^\mu W^\nu - \partial^\nu W^\mu)(\partial_\mu W_\nu^* - \partial_\nu W_\mu^*) + \frac{g_V^2 m_H^2}{f} W^\mu W_\mu^* \tag{8.13}$$

$$-\frac{1}{4}(\partial_\mu Z_\nu - \partial_\nu Z_\mu)(\partial^\mu Z^\nu - \partial^\nu Z^\mu) + \frac{(g_S^2 + g_V^2)m_H^2}{2f}Z_\mu Z^\mu$$

$$-\frac{1}{4}(\partial_\mu A_\nu - \partial_\nu A_\mu)(\partial^\mu A^\nu - \partial^\nu A^\mu)$$

$$+\frac{1}{2}\partial_\mu\eta\partial^\mu\eta - m_H^2\eta^2 + i\bar\chi\gamma^\mu D'_\mu\chi + \mathcal{L}'_I.$$

This Lagrangian contains a charged complex vector field $W_\mu(x)$ with mass $g_V m_H/\sqrt{f}$, a real vector field $Z_\mu(x)$ with mass $m_H g_S g_V/\sqrt{2f}$, and a massless real vector field $A_\mu(x)$. Furthermore, it contains the Higgs field $\eta(x)$, the fermion fields $\chi(x)$ and $\mathcal{L}'_I$ with $\mathcal{L}'_I$ as well as all interaction terms resulting from the transformation of $F^a_{\mu\nu}F_a^{\mu\nu}$ and $F^4_{\mu\nu}F_4^{\mu\nu}$.

Given the value of Θ_W, the Weinberg angle, the covariant derivative now reads

$$
\begin{aligned}
D'_\mu = {} & \partial_\mu - \frac{i}{2}g_V W_\mu \frac{1}{\sqrt{2}}(\sigma^1 - i\sigma^2) - \frac{i}{2}g_V W^*_\mu \frac{1}{\sqrt{2}}(\sigma^1 + i\sigma^2) \\
& -\frac{i}{2}Z_\mu(g_V\sigma^3 \cos\Theta_W - g_S\sigma^0 \sin\Theta_W) \\
& -\frac{i}{2}A_\mu(g_V\sigma^3 \sin\Theta_W + g_S\sigma^0 \cos\Theta_W).
\end{aligned}
\tag{8.14}
$$

Within this scheme it is obvious to interpret the A_μ field as the electromagnetic Maxwell field. Thus the coefficient of A_μ in (8.14) has to be the charge operator eQ, where Q is defined by $Q := \sigma^3 + \sigma^0$. This leads to the condition

$$g_V\sigma^3 \sin\Theta_W + g_S\sigma^0 \cos\Theta_W \overset{!}{=} e(\sigma^3 + \sigma^0), \tag{8.15}$$

from which the relation

$$e = \frac{g_S g_V}{\sqrt{g_S^2 + g_V^2}} \tag{8.16}$$

results. Hence we can write for D'_μ

$$
\begin{aligned}
D'_\mu = {} & \partial_\mu - \frac{i}{2}g_V W_\mu \frac{1}{\sqrt{2}}(\sigma^1 - i\sigma^2) - \frac{i}{2}g_V W^*_\mu \frac{1}{\sqrt{2}}(\sigma^1 + i\sigma^2) \\
& -\frac{i}{2}Q'Z_\mu - \frac{i}{2}eQA_\mu.
\end{aligned}
\tag{8.17}
$$

Although the transformation (8.10) is needed to discover the electromagnetic field which is hidden in (8.9), it is obvious that once one has obtained (8.9) the transformation (8.10) is only an additional operation which does not influence the effect of symmetry breaking. Hence from a structural point of view the Lagrangian (8.9) is fundamental as it contains all ingredients which are necessary (in this kind of simplified model) for a physical interpretation.

If by Weak Mapping one can derive an equivalent to (8.9) one can of course add the transformation (8.10), but the success of the Weak Mapping procedure is already guaranteed by having obtained (8.9). Therefore in the following we concentrate on (8.9).

At this stage of our presentation of the phenomenological theory we have to realize that we consider the Lagrangian (8.9) as a Lagrangian for an effective field dynamics resulting from a fundamental subfermion theory. In this context we intend to explain the phenomenological symmetry breaking by means of a symmetry breaking on the subfermion level. Thus the further treatment of (8.9) depends upon the mechanism which we imagine to be responsible for symmetry breaking at the subfermion level.

In contrast to the phenomenological treatment we do not try to explain symmetry breaking by the action of Higgs fields. Rather, in accordance with Heisenberg we assume the $SU(2)$ asymmetry of the vacuum to be the cause of the symmetry breaking effect. This assumption is also in accordance with the algebraic formalism, as in this formalism the dynamics is explicitly determined by the special representation of the field operator algebra which in turn depends on the vacuum.

Experimentally so far no Higgs field has been discovered, and in addition the imaginary mass in (8.1) is no recommendation for considering the Higgs field as a physical field. Therefore in (8.9) we omit all terms with the Higgs field $\eta(x)$ and restrict ourselves only to that part of the Lagrangian (8.9) which can be experimentally verified, *i.e.*, which leads to the masses for the W and Z bosons.

In this case the Lagrangian is reduced to the expression

$$\widehat{\mathcal{L}} = -\frac{1}{4}\sum_{a=1}^{3} F_{\mu\nu}^a F_a^{\mu\nu} - \frac{1}{4}F_{\mu\nu}^4 F_4^{\mu\nu} \tag{8.18}$$

$$+\frac{g_{\rm V}^2 m_{\rm H}^2}{2f} A_\mu^a A_a^\mu + \frac{g_{\rm S}^2 m_{\rm H}^2}{2f} A_\mu^4 A_4^\mu - \frac{g_{\rm S} g_{\rm V} m_{\rm H}^2}{f} A_\mu^4 A_3^\mu$$

$$+i\bar{\chi}\gamma^\mu D_\mu\chi,$$

and the constants in (8.18) are to be considered as a suitable parametrization, the physical meaning of which has to be explored by the postulated symmetry breaking mechanism.

In the form (8.18) it is not possible to directly perform a comparison with the results of Weak Mapping. Rather, for that purpose we need the canonical equations of motion corresponding to the Lagrangian (8.18). In the preceding chapter we preferred the temporal gauge in the formulation of Weak Mapping. On the other hand, the temporal components A_0^a, A_0^4 are usually assumed to be unequal zero for a massive vector boson theory, although in the symmetry broken case for the conjugated momenta $(\pi^A)_0^a = (\pi^A)_0^4 = 0$ holds, too, *i.e.*, the A_0 fields have no conjugated momentum variables.

Hence in formulating the canonical equations of motion we do not fix $A_0^a = A_0^4 = 0$ a priori, but leave it to the dynamics to distinguish redundant variables. In the next section we will, however, discuss the compatibility of the temporal gauge with the equations of motion of the symmetry broken theory.

The classical canonical equations of motion are readily deduced from the Hamiltonian formalism. As we will see, for comparison with the results of Weak Mapping we have to derive the canonical equations of motion after the transformation (8.10) has been performed. In the context of Hamiltonian formalism, however, (8.10) has to be completed in order to become a canonical transformation.

The canonical conjugate variables of the gauge vector potentials $A_k^a(x)$ are the 'electric' gauge field strengths $-E_k^a(x)$, $k = 1, 2, 3$, whilst $A_0^a(x)$ has no counterpart for $a = 1, 2, 3, 4$. Therefore we restrict the canonical transformation to the spatial degrees of freedom, $i.e.$, to $A_k^a(x)$, $E_k^a(x)$ etc.. Then from (8.10) it follows the complete set of equations

$$A_k^1 = \frac{1}{\sqrt{2}}(W_k + W_k^*), \tag{8.19}$$

$$A_k^2 = -i\frac{1}{\sqrt{2}}(W_k - W_k^*),$$

$$A_k^3 = Z_k \cos \Theta_W + A_k \sin \Theta_W,$$

$$A_k^4 = -Z_k \sin \Theta_W + A_k \cos \Theta_W,$$

$$E_k^1 = -\frac{1}{\sqrt{2}}(E_k^W + E_k^{W*}),$$

$$E_k^2 = i\frac{1}{\sqrt{2}}(E_k^W - E_k^{W*}),$$

$$E_k^3 = -E_k^Z \cos \Theta_W - E_k^A \sin \Theta_W,$$

$$E_k^4 = E_k^Z \sin \Theta_W - E_k^A \cos \Theta_W,$$

where E_k^X is the canonically conjugated field of X_k ($X \in \{W, Z, A\}$), and for which the conservation of the commutation relations or in the classical case of the Poisson brackets can easily be proved.

We first consider the Hamiltonian which corresponds to the Lagrangian (8.18). With the conjugated field momenta

$$(\pi^A)_k^a = -E_k^a, \qquad (\pi^A)_0^a = 0, \tag{8.20}$$

$$(\pi^A)_k^4 = -E_k^4, \qquad (\pi^A)_0^4 = 0,$$

$$(\pi^\chi)_\alpha = i\chi_\alpha^+,$$

where

$$E_k^a = \dot{A}_k^a + \partial_k A_0^a + g_V \varepsilon_{abc} A_0^b A_k^c, \tag{8.21}$$

$$E_k^4 = \dot{A}_k^4 + \partial_k A_0^4,$$

it reads

$$\widehat{\mathcal{H}} = -E_k^a \dot{A}_k^a - E_k^4 \dot{A}_k^4 + i\chi_\alpha^+ \dot{\chi}_\alpha - \widehat{\mathcal{L}} \tag{8.22}$$

$$= \frac{1}{2}(B_k^a B_k^a + E_k^a E_k^a) + \frac{1}{2}(B_k^4 B_k^4 + E_k^4 E_k^4)$$

$$- \frac{g_V^2 m_H^2}{2f} A_\mu^a A_a^\mu - \frac{g_S^2 m_H^2}{2f} A_\mu^4 A_4^\mu + \frac{g_S g_V m_H^2}{f} A_\mu^4 A_3^\mu$$

$$- i\bar{\chi}\gamma^0 \gamma^k \partial_k \chi - i\bar{\chi}\gamma^0 \gamma^\mu (D_\mu - \partial_\mu)\chi + \widehat{\mathcal{H}}_c$$

with

$$\widehat{\mathcal{H}}_c \quad := \quad \partial_k(E_k^a A_0^a) - (\partial_k E_k^a)A_0^a + g_V \varepsilon_{abc} E_k^a A_k^b A_0^c \tag{8.23}$$
$$+\partial_k(E_k^4 A_0^4) - (\partial_k E_k^4)A_0^4$$

and

$$B_k^a \quad = \quad \varepsilon_{klm}\partial_l A_m^a - \frac{g_V}{2}\varepsilon_{abc}\varepsilon_{klm}A_l^b A_m^c , \tag{8.24}$$
$$B_k^4 \quad = \quad \varepsilon_{klm}\partial_l A_m^4 .$$

In its final form (8.22), $\widehat{\mathcal{H}}$ does not contain any time derivative of the dynamical variables, which are defined to be $A_k^a, A_k^4, E_k^a, E_k^4$ and χ, χ^+. Thus the quantities B_k^a and B_k^4 are only to be considered as constraints introduced for abbreviation. Having obtained (8.22) the canonical transformation (8.19) can be simply performed by substitution of (8.19) into (8.22), whilst for the redundant A_0^a fields one has to substitute the formulas corresponding to (8.10).

If we denote the set of variables in (8.19) by $\Lambda := (A_k^a, A_k^4, E_k^a, E_k^4, A_0^a, A_0^4, \chi, \chi^+)$ then the transformed Hamiltonian is given by

$$\widehat{\mathcal{H}}' = \widehat{\mathcal{H}}(\Lambda)_{|\Lambda=(8.19)} . \tag{8.25}$$

These definitions permit to relate the dynamical equations of the unbroken $SU(2) \times U(1)$ theory to those of the symmetry broken theory: a fact which is helpful for recognizing the modification of the dynamics by symmetry breaking and for comparing this effect with the results of Weak Mapping calculations.

The first step is to decompose the Hamiltonian (8.22) into broken and unbroken parts. We define

$$\widehat{\mathcal{H}} = \widehat{\mathcal{H}}_0 + \widehat{\mathcal{H}}_1 \tag{8.26}$$

with the symmetry unbroken part

$$\widehat{\mathcal{H}}_0 \quad := \quad \frac{1}{2}(B_k^a B_k^a + E_k^a E_k^a) + \frac{1}{2}(B_k^4 B_k^4 + E_k^4 E_k^4) \tag{8.27}$$
$$-i\bar{\chi}\gamma^0\gamma^k\partial_k\chi - i\bar{\chi}\gamma^0\gamma^\mu(D_\mu - \partial_\mu)\chi + \widehat{\mathcal{H}}_c$$

and the symmetry breaking part

$$\widehat{\mathcal{H}}_1 := -\frac{g_V^2 m_H^2}{2f}A_\mu^a A_a^\mu - \frac{g_S^2 m_H^2}{2f}A_\mu^4 A_4^\mu + \frac{g_S g_V m_H^2}{f}A_\mu^4 A_3^\mu . \tag{8.28}$$

According to (8.11) in the new variables the term (8.28) is given by

$$\widehat{\mathcal{H}}_1' := -\frac{g_V^2 m_H^2}{f}W^\mu W_\mu^* - \frac{(g_S^2 + g_V^2)m_H^2}{2f}Z^\mu Z_\mu . \tag{8.29}$$

For evaluating the corresponding canonical equations we use a symbolical notation without referring to space coordinates. Furthermore, we observe that it is convenient for formal calculations to use the form of $\widehat{\mathcal{H}}_0$ in the original variables. The derivatives with respect to the new variables can then be obtained by means of the chain rule according to

$$\dot{E}_k^W = -\frac{\delta \widehat{\mathcal{H}}'}{\delta W_k^*} = -\frac{\delta \widehat{\mathcal{H}}_0}{\delta A_k^1}\frac{\delta A_k^1}{\delta W_k^*} - \frac{\delta \widehat{\mathcal{H}}_0}{\delta A_k^2}\frac{\delta A_k^2}{\delta W_k^*} - \frac{\delta \widehat{\mathcal{H}}_1'}{\delta W_k^*}. \tag{8.30}$$

If we proceed in this manner the following canonical equations result:

$$\dot{W}_k = \frac{\delta \widehat{\mathcal{H}}'}{\delta E_k^{W*}} = -\frac{1}{\sqrt{2}}\frac{\delta \widehat{\mathcal{H}}_0}{\delta E_k^1} - \frac{i}{\sqrt{2}}\frac{\delta \widehat{\mathcal{H}}_0}{\delta E_k^2}, \tag{8.31}$$

$$\dot{W}_k^* = \frac{\delta \widehat{\mathcal{H}}'}{\delta E_k^W} = -\frac{1}{\sqrt{2}}\frac{\delta \widehat{\mathcal{H}}_0}{\delta E_k^1} + \frac{i}{\sqrt{2}}\frac{\delta \widehat{\mathcal{H}}_0}{\delta E_k^2},$$

$$\dot{Z}_k = \frac{\delta \widehat{\mathcal{H}}'}{\delta E_k^Z} = -\cos\Theta_{\mathrm{W}}\frac{\delta \widehat{\mathcal{H}}_0}{\delta E_k^3} + \sin\Theta_{\mathrm{W}}\frac{\delta \widehat{\mathcal{H}}_0}{\delta E_k^4},$$

$$\dot{A}_k = \frac{\delta \widehat{\mathcal{H}}'}{\delta E_k^A} = -\sin\Theta_{\mathrm{W}}\frac{\delta \widehat{\mathcal{H}}_0}{\delta E_k^3} - \cos\Theta_{\mathrm{W}}\frac{\delta \widehat{\mathcal{H}}_0}{\delta E_k^4},$$

$$\dot{E}_k^W = \frac{\delta \widehat{\mathcal{H}}'}{\delta W_k^*} = \frac{1}{\sqrt{2}}\frac{\delta \widehat{\mathcal{H}}_0}{\delta A_k^1} + \frac{i}{\sqrt{2}}\frac{\delta \widehat{\mathcal{H}}_0}{\delta A_k^2} + \frac{g_{\mathrm{V}}^2 m_{\mathrm{H}}^2}{f}W_k,$$

$$\dot{E}_k^{W*} = \frac{\delta \widehat{\mathcal{H}}'}{\delta W_k} = \frac{1}{\sqrt{2}}\frac{\delta \widehat{\mathcal{H}}_0}{\delta A_k^1} - \frac{i}{\sqrt{2}}\frac{\delta \widehat{\mathcal{H}}_0}{\delta A_k^2} + \frac{g_{\mathrm{V}}^2 m_{\mathrm{H}}^2}{f}W_k^*,$$

$$\dot{E}_k^Z = \frac{\delta \widehat{\mathcal{H}}'}{\delta Z_k} = \cos\Theta_{\mathrm{W}}\frac{\delta \widehat{\mathcal{H}}_0}{\delta A_k^3} - \sin\Theta_{\mathrm{W}}\frac{\delta \widehat{\mathcal{H}}_0}{\delta A_k^4} + \frac{m_{\mathrm{H}}^2(g_{\mathrm{S}}^2 + g_{\mathrm{V}}^2)}{f}Z_k,$$

$$\dot{E}_k^A = \frac{\delta \widehat{\mathcal{H}}'}{\delta A_k} = \sin\Theta_{\mathrm{W}}\frac{\delta \widehat{\mathcal{H}}_0}{\delta A_k^3} + \cos\Theta_{\mathrm{W}}\frac{\delta \widehat{\mathcal{H}}_0}{\delta A_k^4},$$

$$\dot{\chi} = i\frac{\delta \widehat{\mathcal{H}}'}{\delta \chi^+}, \qquad \dot{\chi}^+ = -i\frac{\delta \widehat{\mathcal{H}}'}{\delta \chi}.$$

The derivatives of $\widehat{\mathcal{H}}_0$ which appear in these equations of course have to be taken at the values of the new variables in the sense of (8.25). But now it is evident that the transformed canonical equations arise from linear combinations of the old ones apart from the symmetry breaking mass terms. In the following we will use (8.31) for a comparison with the results of Weak Mapping calculations.

8.2 Phenomenological Constraints

The field equations of abelian and non-abelian gauge theories can either be derived by the canonical formalism or by the action principle. In the latter case a set of field equations results which in addition to the canonical equations contains equations with no time derivatives. These equations are considered as constraints of the field dynamics, and by gauge fixing further constraints are imposed. So the question arises of whether or not such constraints can be derived in the canonical formalism and in which way they are influenced by symmetry breaking.

The first part of this question was discussed by Grimm [Gri 94a], so we concentrate on the problem of constraints in the case of symmetry breaking. As in the case of the canonical equations themselves we will also relate the constraints for broken symmetry to those for unbroken symmetry, and in doing so we follow the pattern for the unbroken case treated by Grimm [Gri 94a].

In a first step we consider the divergence of the E field equations in (8.31); this means that we have to calculate the quantities $\partial_k(\delta \widehat{\mathcal{H}}_0/\delta A_k^a)$ and $\partial_k(\delta \widehat{\mathcal{H}}_0/\delta A_k^4)$. These calculations require some rearrangements. With respect to these rearrangements it is essential that they can be performed either in the original fields or equally well in the transformed fields (8.19), as all rearrangements are referred to the algebraic properties of the fields which are invariant under this transformation. So in the original fields we obtain, see [Gri 94]:

$$\partial_k\left(\frac{\delta \widehat{\mathcal{H}}_0}{\delta A_k^a}\right) = \varepsilon_{klm}\left[-\partial_k\partial_m B_l^a + g_V\partial_k\varepsilon_{abc}B_l^b A_m^c\right] \tag{8.32}$$

$$+g_V\partial_k\varepsilon_{abc}A_0^b E_k^c + \frac{1}{2}g_V\partial_k\bar{\chi}\gamma^k\sigma^a\chi$$

$$= -\partial_0\left(g_V\varepsilon_{abc}A_k^b E_k^c - \frac{1}{2}g_V\bar{\chi}\gamma^0\sigma^a\chi\right) + g_V\varepsilon_{abc}A_0^b \mathcal{G}^c$$

with the Gauss law operator

$$\mathcal{G}^a := -\partial_k E_k^a - g_V\varepsilon_{abc}A_k^b E_k^c + \frac{1}{2}g_V\bar{\chi}\gamma^0\sigma^a\chi \tag{8.33}$$

and

$$\partial_k\left(\frac{\delta \widehat{\mathcal{H}}_0}{\delta A_k^4}\right) = -\frac{g_S}{2}\partial_0\bar{\chi}\gamma^0\chi. \tag{8.34}$$

For the derivation of these equations current conservation has to be assumed.

With (8.33) we can represent (8.32) in the following form:

$$\partial_k\left(\frac{\delta \widehat{\mathcal{H}}_0}{\delta A_k^a}\right) = -\partial_0\partial_k E_k^a - \partial_0\mathcal{G}^a - g_V\varepsilon_{abc}A_0^b\mathcal{G}^c, \tag{8.35}$$

and with

$$\mathcal{G}^4 := -\partial_k E_k^4 + \frac{1}{2}g_S\bar{\chi}\gamma^0\chi \tag{8.36}$$

one obtains

$$\partial_k\left(\frac{\delta \widehat{\mathcal{H}}_0}{\delta A_k^4}\right) = -\partial_0\partial_k E_k^4 - \partial_0\mathcal{G}^4. \tag{8.37}$$

We remark that the Gauss law for the unbroken $SU(2)$ or $U(1)$ gauge theory is given by $\mathcal{G}^a = 0$ or $\mathcal{G}^4 = 0$, respectively.

Taking the divergence of the E equations in (8.31), substituting into them (8.35) and (8.37) and observing the inverse transformations (8.19) one obtains the set of Gauss constraints for broken symmetry

$$\frac{1}{\sqrt{2}}(\partial_0 \mathcal{G}^1 + g_V \varepsilon_{1bc} A_0^b \mathcal{G}^c) \qquad (8.38)$$

$$+\frac{i}{\sqrt{2}}(\partial_0 \mathcal{G}^2 + g_V \varepsilon_{2bc} A_0^b \mathcal{G}^c) - \frac{g_V^2 m_H^2}{f}\partial_k W_k = 0,$$

$$\frac{1}{\sqrt{2}}(\partial_0 \mathcal{G}^1 + g_V \varepsilon_{1bc} A_0^b \mathcal{G}^c) \qquad (8.39)$$

$$-\frac{i}{\sqrt{2}}(\partial_0 \mathcal{G}^2 + g_V \varepsilon_{2bc} A_0^b \mathcal{G}^c) - \frac{g_V^2 m_H^2}{f}\partial_k W_k^* = 0,$$

$$\cos \Theta_W (\partial_0 \mathcal{G}^3 + g_V \varepsilon_{3bc} A_0^b \mathcal{G}^c) \qquad (8.40)$$

$$- \sin \Theta_W (\partial_0 \mathcal{G}^4) - \frac{m_H^2(g_S^2 + g_V^2)}{f}\partial_k Z_k = 0,$$

and

$$\sin \Theta_W (\partial_0 \mathcal{G}^3 + g_V \varepsilon_{3bc} A_0^b \mathcal{G}^c) + \cos \Theta_W (\partial_0 \mathcal{G}^4) = 0, \qquad (8.41)$$

where A_0^a and $\mathcal{G}^a, \mathcal{G}^4$ have to be taken in the transformed variables according to (8.19).

To interpret these equations we have to discuss the question of 'gauge fixing' of the symmetry broken theory. We observe that by aquiring mass for the W and Z bosons the original $SU(2)$ invariance is destroyed. Therefore the resulting field theory for W and Z bosons cannot be described in terms of $SU(2)$ covariant vector potentials, field strengths and covariant derivatives. Rather, this system has to be considered as a set of three massive vector fields and one massless field with self-interaction and fermionic sources.

This means that constraints like the Lorentz condition or the temporal gauge condition cannot be immediately interpreted as gauge fixing conditions, but their compatibility with the equations of the broken symmetry theory has to be investigated.

With respect to the covariant Lorentz condition it can be shown that already in the unbroken theory its use leads to complications beyond perturbation treatment, see [Hua 92]. On the other hand, one verifies by direct inspection of the field equations that the temporal gauge $A_0^a = 0$, $A_0^4 = 0$ of the initially unbroken $SU(2) \times U(1)$ gauge theory and subsequent symmetry breaking are compatible.

Hence in the following we start the symmetry breaking with a gauge theory in temporal gauge. Owing to the canonical tranformations (8.10) the conditions $A_0^a = 0$, $A_0^4 = 0$ are immediately transferred to the massive vector fields W, W^* and Z and to the massless field A. Hence we can apply this condition to equations (8.38)–(8.41) and evaluate them in the original or just as well in the set of transformed fields. Using the former we obtain

$$\frac{1}{\sqrt{2}}(\dot{\mathcal{G}}^1 + i\dot{\mathcal{G}}^2) - \frac{g_V^2 m_H^2}{\sqrt{2}f}\partial_k(A_k^1 + iA_k^2) = 0, \qquad (8.42)$$

$$\frac{1}{\sqrt{2}}(\dot{\mathcal{G}}^1 - i\dot{\mathcal{G}}^2) - \frac{g_V^2 m_H^2}{\sqrt{2}f}\partial_k(A_k^1 - iA_k^2) = 0, \qquad (8.43)$$

$$\cos \Theta_W \dot{\mathcal{G}}^3 - \sin \Theta_W \dot{\mathcal{G}}^4 - \frac{m_H^2(g_S^2 + g_V^2)}{f}\partial_k(\cos \Theta_W A_k^3 - \sin \Theta_W A_k^4) = 0, \qquad (8.44)$$

and

$$\sin \Theta_W \dot{\mathcal{G}}^3 + \cos \Theta_W \dot{\mathcal{G}}^4 = 0. \tag{8.45}$$

With respect to the Gauss law we remember that in the case of unbroken symmetry this law is an expression of a residual gauge invariance. If the symmetry is broken one cannot expect to maintain the same Gauss law. *Indeed equations (8.42) are merely consequences of the dynamical equations without a possibility to reformulate it as constraints, i.e., without time derivatives.*

Only (8.45) can be integrated to give

$$\sin \Theta_W \mathcal{G}^3 + \cos \Theta_W \mathcal{G}^4 = h(\mathbf{r}), \tag{8.46}$$

where by suitable initial conditions $h(\mathbf{r})$ can be assumed to vanish. Hence we finally obtain

$$\sin \Theta_W \mathcal{G}^3 + \cos \Theta_W \mathcal{G}^4 = 0. \tag{8.47}$$

This is a constraint which after symmetry breaking describes the remaining $U(1)$ gauge group. In the transformed coordinates this equation reads

$$\partial_k E_k^A - ig_V \sin \Theta_W (W_k^* E_k^W - W_k E_k^{W^*}) \tag{8.48}$$
$$+ \frac{1}{2} \bar{\chi} \gamma^0 (\sin \Theta_W \sigma^3 + \cos \Theta_W \sigma^0) \chi = 0.$$

We summarize our results in:

- **Proposition 8.1:** Apart from initial conditions after symmetry breaking of the $SU(2) \times U(1)$ gauge theory the Gauss law (8.48) of the residual $U(1)$ gauge group is a consequence of the canonical equations (8.31).

8.3 Effective Dynamics with Symmetry Breaking

For the mechanism of symmetry breaking we refer to an idea of Heisenberg: by comparison with ferromagnetism Heisenberg proposed a vacuum state which breaks isospin symmetry [Hei 66]. We apply this idea to the subfermion model of an unbroken $SU(2) \times U(1)$ gauge theory with composite bosons and composite fermions which we treated in the preceding chapter. In accordance with the algebraic formalism the non-invariance of the vacuum under the $SU(2)$ isospin group has to manifest itself in the non-invariance of the vacuum expectation values, in particular in the two-point function.

The invariant propagator was defined in (6.5). In order to study the modifications due to isospin symmetry breaking we consider the superspin/isospin part of (6.5) given by $\gamma_{\kappa_1 \kappa_2}^5$ which by separation of isospin and superspin indices Λ and A, respectively, can equivalently be written as $\gamma_{\kappa_1 \kappa_2}^5 \equiv \sigma_{\Lambda_1 \Lambda_2}^1 \sigma_{A_1 A_2}^0$, where $\sigma_{A_1 A_2}^0 \equiv \mathbb{I}_2$ represents the isospin invariance of the vacuum.

Symmetry breaking takes place if the degeneracy of the masses in (6.5) with respect to isospin is lifted. The propagator then reads

$$F_{I_1 I_2} = -i\lambda_{i_1}\delta_{i_1 i_2}\sigma^1_{\Lambda_1\Lambda_2}\sigma^0_{A_1 A_2}\left[(i\gamma^\mu\partial_\mu(x_1) + m^{A_1}_{i_1})C\right]_{\alpha_1\alpha_2}\Delta\left(x_1 - x_2, m^{A_1}_{i_1}\right) \qquad (8.49)$$

with

$$m^A_i := m_i + \mu \qquad \text{for} \qquad A = 1, \qquad (8.50)$$
$$m^A_i := m_i - \mu \qquad \text{for} \qquad A = 2.$$

For Weak Mapping calculations we need the single time limit of the propagator (8.49) modified by (8.50). The single time limit of the symmetric propagator is given by (7.79). As in combination with boson functions the p wave parts of (7.79) do not contribute to the corresponding integrals, we suppress them from the beginning and consider only the s wave parts of (7.79). Taking into account the symmetry breaking by (8.50) this yields

$$
\begin{aligned}
F^t_{Z_1 Z_2}(\mathbf{r}_1, \mathbf{r}_2) &= -\frac{\lambda_{i_1}}{(2\pi)^2}\delta_{i_1 i_2}\sigma^1_{\Lambda_1\Lambda_2}\sigma^0_{A_1 A_2}\left(m^{A_1}_{i_1}\right)^2\frac{K_1(m^{A_1}_{i_1}r)}{r}C_{\alpha_1\alpha_2} \qquad (8.51)\\
&\approx -\frac{\lambda_{i_1}}{(2\pi)^2}\delta_{i_1 i_2}\sigma^1_{\Lambda_1\Lambda_2}\\
&\quad \times \left\{\sigma^0_{A_1 A_2}m^2_{i_1}\frac{K_1(m_{i_1}r)}{r}C_{\alpha_1\alpha_2}\right.\\
&\quad \left. +\mu\sigma^3_{A_1 A_2}\left[m_{i_1}\frac{\partial}{\partial r}K_1(m_{i_1}r) + 2m_{i_1}\frac{K_1(m_{i_1}r)}{r}\right]C_{\alpha_1\alpha_2}\right\},
\end{aligned}
$$

where we have expanded F with respect to small mass corrections μ; the quantities $K_i(r)$ are the modified Bessel functions.

For later applications we rewrite (8.51) in the form

$$F^t_{I_1 I_2} = F^0_{I_1 I_2} + F^1_{I_1 I_2}, \qquad (8.52)$$

where $F^0_{I_1 I_2}$ is the invariant part and $F^1_{I_1 I_2}$ is the symmetry breaking part of the propagator.

Symmetry breaking by the vacuum means that the Lagrangian and the corresponding dynamical equations remain invariant under the full symmetry group, but admit (vacuum) solutions which break that symmetry.

Thus symmetry breaking manifests itself only in (8.51) whilst the functional equations of the subfermions (4.90) are not changed, and as (8.51) is part of an associated algebraic state this means that symmetry breaking can be expressed by a suitable choice of the representation, which, of course, should be selected by a self-consistent calculation of the vacuum field dynamics.

In accordance with phenomenological theory we start with the unbroken $SU(2) \times U(1)$ theory of the preceding chapter. By Weak Mapping this theory was derived as an effective theory in temporal gauge. In Section 8.2 it was demonstrated that the temporal gauge fixing can be applied to the symmetry broken theory although the corresponding gauge group is lost.

The temporal gauge expresses the redundancy of the A_0 fields which also in massive vector meson theories have no conjugated momentum variables. In addition the phenomenological theory shows that in the transition from the unbroken to the broken case the set of associated physical variables remains unchanged. Owing to these arguments we are able to take over the calculations of the preceding chapter with the only modification of replacing $F_{I_1 I_2}$ by the symmetry broken expressions discussed above.

In order to clearly illustrate our proceeding we repeat the essential formulas of Chapter 7. The functional Hamiltonian (7.62) is given by

$$\mathcal{H}_{\text{eff}} = \mathcal{H}_{\text{f}} + \mathcal{H}_{\text{b}} + \mathcal{H}_{\text{bf}} + \mathcal{H}_{\text{corr}} \tag{8.53}$$

with the effective fermion part

$$\mathcal{H}_{\text{f}} := f_\ell \big(\widehat{K}^{\ell \ell'} \partial^f_{\ell'} + \widehat{M}^{\ell \ell'} \partial^f_{\ell'} \big) , \tag{8.54}$$

the effective boson part

$$\mathcal{H}_{\text{b}} := b_k \big(K^{kk'} \partial^b_{k'} + M^{kk'} \partial^b_{k'} + N^{kk'_1 k'_2} \partial^b_{k'_1} \partial^b_{k'_2} \big), \tag{8.55}$$

and the effective boson–fermion interaction term

$$\mathcal{H}_{\text{bf}} := b_k S^{k \ell'_1 \ell'_2} \partial^f_{\ell'_1} \partial^f_{\ell'_2} + f_\ell \widehat{N}^{\ell k \ell'} \partial^b_k \partial^f_{\ell'} . \tag{8.56}$$

In this representation of $\mathcal{H}_{\text{eff}}$ all exchange forces are neglected and we suppress the quantization terms and the residual interactions as they are not relevant for our further proceeding. This will be justified in Section 8.7 in detail.

The quantities appearing in (8.54)–(8.56) are defined as follows:

$$
\begin{aligned}
K^{kk'} &:= 2K_{I_1 I_2} C^{I_2 I_3}_{2,k'} R^{2,k}_{I_1 I_3} , \\
M^{kk'} &:= -6W_{I_4 I_1 I_2 I_3} F_{I_3 K} C^{I_1 I_2}_{2,k'} R^{2,k}_{I_4 K} , \\
N^{kk'_1 k'_2} &:= 6W_{I_4 I_1 I_2 I_3} C^{I_1 I_2}_{2,k'_1} C^{I_3 K}_{2,k'_2} R^{2,k}_{I_4 K} , \\
S^{k \ell'_1 \ell'_2} &:= -108 W_{I_4 I_1 I_2 I_3} C^{I_1 I_3 K_1}_{3,\ell'_1} C^{I_2 K_2 K_3}_{3,\ell'_2} R^{2,k}_{I_4 K_1 K_2 K_3} ,
\end{aligned}
\tag{8.57}
$$

and

$$
\begin{aligned}
\widehat{K}^{\ell \ell'} &:= 3K_{I_1 I_2} C^{I_2 K_1 K_2}_{3,\ell'} R^{3,\ell}_{I_1 K_1 K_2} , \\
\widehat{M}^{\ell \ell'} &:= -18 W_{I_4 I_1 I_2 I_3} F_{I_3 K_1} C^{I_1 I_2 K_2}_{3,\ell'} R^{3,\ell}_{I_4 K_1 K_2} , \\
\widehat{N}^{\ell k \ell'} &:= 9 W_{I_4 I_1 I_2 I_3} \big(C^{I_3 K_1 K_2}_{3,\ell'} C^{I_1 I_2}_{2,k} + 2 C^{I_2 I_3 K_2}_{3,\ell'} C^{I_1 K_1}_{2,k} \big) R^{3,\ell}_{I_4 K_1 K_2} .
\end{aligned}
\tag{8.58}
$$

From (8.57) and (8.58) we see that the symmetry breaking propagators are (only) involved in the mass terms $M^{kk'}$ and $\widehat{M}^{\ell \ell'}$ and thus are closely connected with the mass generation by Weak Mapping. In addition in order to study this mass generating mechanism we do no longer insist on the vanishing of the last term of (7.86), i.e., the vanishing of the boson mass, which we assumed in the context of the unbroken effective theory. Hence we have to include the term

$$b_k \widetilde{M}^{kk'} \partial_{k'}^b := i\mu_B^2 \int dr \sum_{a=1}^{4} \sum_{i=1}^{3} b_{i0,a}^F(\mathbf{r}) \partial_{i,a}^A(\mathbf{r}), \tag{8.59}$$

and instead of relation (7.99) for $\mu_B = 0$ we apply the general formula (7.98)

$$\mu_B^2 = 10m^2 + g/\pi^2 \tag{8.60}$$

in our calculations. Substitution of (8.52) into (8.57) and (8.58) and addition of (8.59) to (8.55) yields the mass terms

$$M^{kk'} = M_0^{kk'} + M_1^{kk'} + \widetilde{M}^{kk'} \tag{8.61}$$

for bosons and

$$\widehat{M}^{\ell\ell'} = \widehat{M}_0^{\ell\ell'} + \widehat{M}_1^{\ell\ell'} \tag{8.62}$$

for fermions. Thus we obtain for the various parts of the functional Hamiltonian operator $\mathcal{H}_{\text{eff}}$

$$\begin{aligned} \mathcal{H}_f &= \mathcal{H}_f^0 + \mathcal{H}_f^1 \\ &:= f_\ell\big(\widehat{K}^{\ell\ell'} \partial_{\ell'}^f + \widehat{M}_0^{\ell\ell'} \partial_{\ell'}^f\big) + f_\ell \widehat{M}_1^{\ell\ell'} \partial_{\ell'}^f \end{aligned} \tag{8.63}$$

and

$$\begin{aligned} \mathcal{H}_b &= \mathcal{H}_b^0 + \mathcal{H}_b^1 \\ &:= b_k\big(K^{kk'} \partial_{k'}^b + M_0^{kk'} \partial_{k'}^b + N^{kk_1'k_2'} \partial_{k_1'}^b \partial_{k_2'}^b\big) + b_k\big(M_1^{kk'} + \widetilde{M}^{kk'}\big) \partial_{k'}^b \end{aligned} \tag{8.64}$$

whilst $\mathcal{H}_{bf} =: \mathcal{H}_{bf}^0$ remains unchanged, if the influence of the symmetry breaking propagator term on the polarization cloud in $S^{k\ell_1'\ell_2'}$ is neglected. This is in accordance with our decision not to include the symmetry breaking effects on state calculations in a first step.

For a first inspection of the effects of symmetry breaking we are mainly interested in the modifications of the boson parts as these modifications lead to the formation of electro–weak boson and photon fields which we intended to derive. Therefore we suppress $\mathcal{H}_f^1$ in the following.

Then the functional Hamiltonian operator reads

$$\mathcal{H}_{\text{eff}} = \mathcal{H}_{\text{eff}}^0 + \mathcal{H}_b^1 := \mathcal{H}_f^0 + \mathcal{H}_b^0 + \mathcal{H}_{bf}^0 + \mathcal{H}_b^1. \tag{8.65}$$

The operator $\mathcal{H}_{\text{eff}}^0$ was discussed at full length in the preceding chapter and we can take over the results obtained there.

The boson part $\mathcal{H}_b^0$ is identical with (7.159), the fermion part $\mathcal{H}_f^0$ with (7.154) and the boson–fermion interaction $\mathcal{H}_{bf}^0$ with (7.155).

Explicitly we have according to (7.154)

$$\begin{aligned} \mathcal{H}_f^0 &= \sum_{j \in \overline{M}_1} \int f_{j,\alpha}(\mathbf{r}) \big[-i(\gamma^0 \gamma^k)\partial_k + \overline{m}_1 \gamma^0\big]_{\alpha\alpha'} \partial_{j,\alpha'}^f(\mathbf{r}) \, dr \\ &\quad + \sum_{j \in \overline{M}_2} \int f_{j,l,\alpha}(\mathbf{r}) \big[-i(\gamma^0 \gamma^k)\partial_k + \overline{m}_2 \gamma^0\big]_{\alpha\alpha'} \partial_{j,l,\alpha'}^f(\mathbf{r}) \, dr, \end{aligned} \tag{8.66}$$

where $\overline{M}_1$ and $\overline{M}_2$ distinguish between lepton and quark states, see Section 7.6. Furthermore, we have according to (7.159)

$$
\mathcal{H}_b^0 \;=\; -i \sum_{a=1}^{4} \int b_{i,a}^A(\mathbf{r}) \partial_{i,a}^E(\mathbf{r})\, d\mathbf{r} \tag{8.67}
$$

$$
+i\varepsilon_{ijk}\varepsilon_{lmk} \sum_{a=1}^{4} \int b_{i,a}^E(\mathbf{r}) \partial_j \partial_l \partial_{m,a}^A(\mathbf{r})\, d\mathbf{r}
$$

$$
+i\frac{G}{2} \sum_{a\ldots c=1}^{3} \int \varepsilon_{abc}\varepsilon_{ijk}\varepsilon_{lmk} b_{i,a}^E(\mathbf{r}) \partial_j \partial_{l,b}^A(\mathbf{r}) \partial_{m,c}^A(\mathbf{r})\, d\mathbf{r}
$$

$$
+iG \sum_{a\ldots c=1}^{3} \int \varepsilon_{abc}\varepsilon_{ijk}\varepsilon_{lmk} b_{i,a}^E(\mathbf{r}) [\partial_m \partial_{l,b}^A(\mathbf{r})] \partial_{j,c}^A(\mathbf{r})\, d\mathbf{r}
$$

$$
-i\frac{G^2}{2} \sum_{a\ldots e=1}^{3} \int \varepsilon_{abc}\varepsilon_{cde}\varepsilon_{ijk}\varepsilon_{lmk} b_{i,a}^E(\mathbf{r}) \partial_{m,e}^A(\mathbf{r}) \partial_{l,d}^A(\mathbf{r}) \partial_{j,b}^A(\mathbf{r})\, d\mathbf{r},
$$

and finally according to (7.155)

$$
\mathcal{H}_{bf}^0 \;=\; iK_1 \sum_{j,j'\in \overline{M}_1} \sum_{a=1}^{3} \int f_{j,\alpha}(\mathbf{r})(\gamma^0\gamma^k)_{\alpha\alpha'} G_{jj'}^a \partial_{j',\alpha'}^f(\mathbf{r}) \partial_{k,a}^A(\mathbf{r})\, d\mathbf{r} \tag{8.68}
$$

$$
+iK_1' \sum_{j,j'\in \overline{M}_1} \int f_{j,\alpha}(\mathbf{r})(\gamma^0\gamma^k)_{\alpha\alpha'} G_{jj'}^4 \partial_{j',\alpha'}^f(\mathbf{r}) \partial_{k,4}^A(\mathbf{r})\, d\mathbf{r}
$$

$$
+iK_1 \sum_{j,j'\in \overline{M}_2} \sum_{a=1}^{3} \int f_{j,l,\alpha}(\mathbf{r})(\gamma^0\gamma^k)_{\alpha\alpha'} G_{jj'}^a \partial_{j',l,\alpha'}^f(\mathbf{r}) \partial_{k,a}^A(\mathbf{r})\, d\mathbf{r}
$$

$$
+iK_1' \sum_{j,j'\in \overline{M}_2} \int f_{j,l,\alpha}(\mathbf{r})(\gamma^0\gamma^k)_{\alpha\alpha'} G_{jj'}^4 \partial_{j',l,\alpha'}^f(\mathbf{r}) \partial_{k,4}^A(\mathbf{r})\, d\mathbf{r}
$$

$$
+iK_2 \sum_{j,j'\in \overline{M}_1} \sum_{a=1}^{3} \int b_{k,a}^E(\mathbf{r}) \bar{\partial}_{j,\alpha}^f(\mathbf{r}) G_{jj'}^a \gamma_{\alpha\alpha'}^k \partial_{j',\alpha'}^f(\mathbf{r})\, d\mathbf{r}
$$

$$
+iK_2' \sum_{j,j'\in \overline{M}_1} \int b_{k,4}^E(\mathbf{r}) \bar{\partial}_{j,\alpha}^f(\mathbf{r}) G_{jj'}^4 \gamma_{\alpha\alpha'}^k \partial_{j',\alpha'}^f(\mathbf{r})\, d\mathbf{r}
$$

$$
+iK_3 \sum_{j,j'\in \overline{M}_2} \sum_{a=1}^{3} \int b_{k,a}^E(\mathbf{r}) \bar{\partial}_{j,l,\alpha}^f(\mathbf{r}) G_{jj'}^a \gamma_{\alpha\alpha'}^k \partial_{j',l,\alpha'}^f(\mathbf{r})\, d\mathbf{r}
$$

$$
+iK_3' \sum_{j,j'\in \overline{M}_2} \int b_{k,4}^E(\mathbf{r}) \bar{\partial}_{j,l,\alpha}^f(\mathbf{r}) G_{jj'}^4 \gamma_{\alpha\alpha'}^k \partial_{j',l,\alpha'}^f(\mathbf{r})\, d\mathbf{r}.
$$

Concerning the origin of the constants G, K_i, K_i' etc. we refer to Sections 7.5 and 7.7. It remains the task to calculate $\mathcal{H}_b^1$ in order to obtain a completely explicit expression for $\mathcal{H}_{\mathrm{eff}}$ of (8.65).

Before continuing our discussion it should be pointed out that our procedures are adapted to a low energy approximation of the effective theories. In this case the Lorentz kinematic does not influence the form of the corresponding hard core wave functions, and, in particular, with respect to the wave functions there is no essential difference between massless and massive (particle) states.

But as soon as one wishes to go beyond this rough approximation one cannot further apply the shell model functions to the symmetry broken case. Rather has one self-consistently to calculate the hard core states *and* the effective dynamics. This means that already the hard core bound states spoil the original symmetry and that the effective dynamics adds only renormalization corrections to the symmetry breaking scheme.

In order to avoid extensive calculations of relativistic wave functions in the symmetry breaking case, in our treatment we shift the whole effect of symmetry breaking to the effective dynamics. However, we emphasize that all investigations can be done in more detail and more consistently.

To justify these approximations we additionally remember that with respect to the general structure of the effective theory the exact evaluation of the algebra is essential and that the algebraic structure is not altered by more sophisticated wave functions; we have already emphasized that throughout this book all calculations are exactly evaluated in the algebraic parts.

In the special case of $\mathcal{H}_b^1$ the exact algebraic evaluation yields (8.59), *i.e.*, an already calculated term, and the new term

$$b_k M_1^{kk'} \partial_{k'}^b = -i\mu K_4 \int d\mathbf{r} \left[b_{k,3}^E(\mathbf{r}) \partial_{k,4}^A(\mathbf{r}) + b_{k,4}^E(\mathbf{r}) \partial_{k,3}^A(\mathbf{r}) \right] \tag{8.69}$$

with the constant K_4.

We now proceed along the lines of Section 7.5. Denoting the functional state of the combined boson–fermion system by $|\mathcal{G}(b, f; a)\rangle$ the functional energy equation reads

$$E_{p0}|\mathcal{G}(b, f; p)\rangle = (\mathcal{H}_f^0 + \mathcal{H}_b^0 + \mathcal{H}_{bf}^0 + \mathcal{H}_b^1)|\mathcal{G}(b, f; p)\rangle \tag{8.70}$$

with omission of all terms which are not essential for our investigations, *i.e.*, quantization terms, residual interactions, exchange forces, and symmetry breaking terms for fermions. All these terms can be included in the calculations if a more detailed insight into the effective dynamics of the system is intended.

For a first inspection of the symmetry breaking effect of the vacuum, equation (8.70) is thus sufficient, and although for a successful treatment of Weak Mapping the functional formulation of the Algebraic Schrödinger Representation expressed in (8.70) is indispensible, in a first step its evaluation can be performed on the classical level.

This procedure is justified by the knowledge that in the low energy limit and by omission of the above mentioned terms a close connection between (8.70) and the classical equations exists. This was demonstrated in Section 7.5 for the effective boson dynamics and we now extend this formalism to the combined boson–fermion system.

The corresponding generalization of equation (7.105) reads

$$E_{p0}|\mathcal{G}(b,f;p)\rangle = i\int d\mathbf{r}\left[b_k(\mathbf{r},t)\frac{\partial}{\partial t}\partial_k^b(\mathbf{r},t) + f_l(\mathbf{r},t)\frac{\partial}{\partial t}\partial_l^f(\mathbf{r},t)\right]|\mathcal{G}(b,f;p)\rangle. \quad (8.71)$$

With the ansatz

$$|\mathcal{G}(b,f;p)\rangle = \exp\left[Z_0(b,f;p)\right]|0\rangle_{\rm F}, \quad (8.72)$$

where Z_0 is given by

$$Z_0(b,f;p) := \int\left[\sum_X\sum_{k,a}X_k^a(\mathbf{r},t)b_{k,a}^X(\mathbf{r},t) + \sum_{l,a}i\chi_{l,a}(\mathbf{r},t)f_{l,a}(\mathbf{r},t)\right]d\mathbf{r}, \quad (8.73)$$

and where $X = A, E$ and χ are the corresponding classical boson and fermion fields, one obtains by substitution of (8.71) and (8.72) into (8.70) the classical effective dynamics.

With (8.66)–(8.68) and (8.64), (8.59) and (8.69) the classical effective equations resulting from (8.70) read for bosons and for $a = 1,2,3$:

$$\dot{A}_i^a = -E_i^a, \quad (8.74)$$

$$\begin{aligned}
\dot{E}_i^a = &\ \varepsilon_{ijk}\varepsilon_{lmk}\partial_j\partial_l A_m^a + \mu_{\rm B}^2 A_i^a - \mu K_4 A_i^4\delta_{a3}\\
&+ \frac{G}{2}\varepsilon_{abc}\varepsilon_{ijk}\varepsilon_{lmk}\partial_j A_l^b A_m^c\\
&+ G\varepsilon_{abc}\varepsilon_{ijk}\varepsilon_{lmk}(\partial_m A_l^b)A_j^c\\
&- \frac{G^2}{2}\varepsilon_{abc}\varepsilon_{cde}\varepsilon_{ijk}\varepsilon_{lmk}A_m^e A_l^d A_j^b\\
&+ K_2\,\bar{\chi}_{j,\alpha}G_{jj'}^a\gamma_{\alpha\alpha'}^i\chi_{j',\alpha'} + K_3\,\bar{\chi}_{j,l,\alpha}G_{jj'}^a\gamma_{\alpha\alpha'}^i\chi_{j',l,\alpha'}
\end{aligned}$$

and for $a = 4$:

$$\dot{A}_i^4 = -E_i^4, \quad (8.75)$$

$$\begin{aligned}
\dot{E}_i^4 = &\ \varepsilon_{ijk}\varepsilon_{lmk}\partial_j\partial_l A_m^4 + \mu_{\rm B}^2 A_i^4 - \mu K_4 A_i^3\\
&+ K_2'\,\bar{\chi}_{j,\alpha}G_{jj'}^4\gamma_{\alpha\alpha'}^i\chi_{j',\alpha'} + K_3'\,\bar{\chi}_{j,l,\alpha}G_{jj'}^4\gamma_{\alpha\alpha'}^i\chi_{j',l,\alpha'},
\end{aligned}$$

whilst the fermion equations are given by

$$\begin{aligned}
\dot{\chi}_{j,\alpha} = &\ \left[-(\gamma^0\gamma^k)_{\alpha\alpha'}\partial_k - i\overline{m}_1\gamma_{\alpha\alpha'}^0\right]\chi_{j,\alpha'} \quad (8.76)\\
&+ K_1(\gamma^0\gamma^k)_{\alpha\alpha'}\sum_{a=1}^3 G_{jj'}^a\chi_{j',\alpha'}A_k^a + K_1'(\gamma^0\gamma^k)_{\alpha\alpha'}G_{jj'}^4\chi_{j',\alpha'}A_k^4,
\end{aligned}$$

$$\begin{aligned}
\dot{\chi}_{j,l,\alpha} = &\ \left[-(\gamma^0\gamma^k)_{\alpha\alpha'}\partial_k - i\overline{m}_2\gamma_{\alpha\alpha'}^0\right]\chi_{j,l,\alpha'} \quad (8.77)\\
&+ K_1(\gamma^0\gamma^k)_{\alpha\alpha'}\sum_{a=1}^3 G_{jj'}^a\chi_{j',l,\alpha'}A_k^a + K_1'(\gamma^0\gamma^k)_{\alpha\alpha'}G_{jj'}^4\chi_{j',l,\alpha'}A_k^4.
\end{aligned}$$

Obviously equations (8.74) and (8.75) for $a = 3, 4$ are coupled in their symmetry breaking terms. In order to achieve decoupling we transform the fields by going over to new fields A_k, E_k^A and Z_k, E_k^Z, which according to (8.19) are generally defined by

$$\begin{pmatrix} Z_k \\ A_k \end{pmatrix} = \begin{pmatrix} \cos\theta & -\sin\theta \\ \sin\theta & \cos\theta \end{pmatrix} \begin{pmatrix} A_k^3 \\ A_k^4 \end{pmatrix}, \tag{8.78}$$

$$\begin{pmatrix} E_k^Z \\ E_k^A \end{pmatrix} = \begin{pmatrix} -\cos\theta & \sin\theta \\ -\sin\theta & -\cos\theta \end{pmatrix} \begin{pmatrix} E_k^3 \\ E_k^4 \end{pmatrix}.$$

Furthermore, we observe that

$$\frac{\delta\widehat{\mathcal{H}}_0}{\delta A_i^4} = -\varepsilon_{ijk}\varepsilon_{lmk}\partial_j\partial_l A_m^4 \tag{8.79}$$
$$- K_2'\bar{\chi}_{j,a}G_{jj'}^4\gamma_{\alpha\alpha'}^i\chi_{j',\alpha'} - K_3'\bar{\chi}_{j,l,a}G_{jj'}^4\gamma_{\alpha\alpha'}^i\chi_{j',l,\alpha'}$$

and

$$\frac{\delta\widehat{\mathcal{H}}_0}{\delta A_i^3} = -\varepsilon_{ijk}\varepsilon_{lmk}\partial_j\partial_l A_m^3 \tag{8.80}$$
$$- \frac{G}{2}\varepsilon_{3bc}\varepsilon_{ijk}\varepsilon_{lmk}\partial_j A_l^b A_m^c$$
$$- G\varepsilon_{3bc}\varepsilon_{ijk}\varepsilon_{lmk}(\partial_m A_l^b)A_j^c$$
$$+ \frac{G^2}{2}\varepsilon_{3bc}\varepsilon_{cde}\varepsilon_{ijk}\varepsilon_{lmk}A_m^e A_l^d A_j^b$$
$$- K_2\bar{\chi}_{j,a}G_{jj'}^3\gamma_{\alpha\alpha'}^i\chi_{j',\alpha'} - K_3\bar{\chi}_{j,l,a}G_{jj'}^3\gamma_{\alpha\alpha'}^i\chi_{j',l,\alpha'}$$

hold, where $\widehat{\mathcal{H}}_0$ is the classical Hamiltonian (8.27) corresponding to the effective functional Hamiltonian $\mathcal{H}^0$ without symmetry breaking. Equations (8.74) and (8.75) for $a = 3, 4$ then read for the E fields

$$\dot{E}_i^3 = -\frac{\delta\widehat{\mathcal{H}}_0}{\delta A_i^3} + \mu_B^2 A_i^3 - \mu K_4 A_i^4, \tag{8.81}$$

$$\dot{E}_i^4 = -\frac{\delta\widehat{\mathcal{H}}_0}{\delta A_i^4} + \mu_B^2 A_i^4 - \mu K_4 A_i^3,$$

and if we apply the transformation (8.78) to the A fields as well as to the E fields the following equations result:

$$\dot{Z}_i = -E_i^Z, \tag{8.82}$$
$$\dot{A}_i = -E_i^A,$$
$$\dot{E}_i^Z = \cos\theta\frac{\delta\widehat{\mathcal{H}}_0}{\delta A_i^3} - \sin\theta\frac{\delta\widehat{\mathcal{H}}_0}{\delta A_i^4} + (\mu_B^2 + \mu K_4)Z_i,$$
$$\dot{E}_i^A = \sin\theta\frac{\delta\widehat{\mathcal{H}}_0}{\delta A_i^3} + \cos\theta\frac{\delta\widehat{\mathcal{H}}_0}{\delta A_i^4} + (\mu_B^2 - \mu K_4)A_i.$$

Assuming $\mu_B^2 = \mu K_4$ the last equation reads

$$\dot{E}_i^A = \sin\theta \frac{\delta\widehat{\mathcal{H}}_0}{\delta A_i^3} + \cos\theta \frac{\delta\widehat{\mathcal{H}}_0}{\delta A_i^4}, \qquad (8.83)$$

i.e., we have obtained the Maxwell equations of (8.31) for the massless photon, supplied by electro–weak interactions, as well as the equations for the Z boson.

Applying the transformation (8.78) to the fields A_k^1, A_k^2 and E_k^1, E_k^2 we introduce the fields W_k, W_k^* and $E_k^W, E_k^{W^*}$; in the same manner as for the A and Z fields we reproduce the W field equations of (8.31), provided we assume for the transformation angle θ the value $\theta = 45°$.

With $\mu_B^2 = \mu K_4$ the Z boson aquires the mass 2μ, whilst the W boson mass is μ. The experimental values are $\Theta_W \approx 30°$ for the Weinberg angle, whilst the mass ratio is $m_W/m_Z \approx 77.7/88.6$. Compared with $\theta = 45°$ and $m_W/m_Z = \frac{1}{2}$ our rough approximations lead to a qualitative agreement with the experiment, but, of course, *the most remarkable achievement is the complete agreement of the effective dynamics in the low energy limit with the phenomenological theory (8.31)*. The numerical values can be calculated with greater precision if the deformation of the states by symmetry breaking is included.

Finally we consider the fermion equations. In agreement with the results of Section 7.7, equations (8.76) and (8.77) reproduce the initial equations (7.9) of the phenomenological theory, but now with fermion masses $\overline{m}_1$ and $\overline{m}_2$.

The further treatment of these equations is standard and should not be reproduced here. We only remark that the constants K_2, K_2' and K_3, K_3' in (8.68) can be removed by a scale transformation which does not affect the other terms. In this way the complete agreement with the phenomenological expressions is secured.

8.4   Spectrum of Physical Particles

Apart from left–right symmetry breaking, in the preceding sections we have demonstrated that the electro–weak standard model can be interpreted as an effective dynamics of composite bosons and fermions which result as bound states from a subfermion model, and the interactions of which in the low energy limit just coincide with those of the standard model.

However, this successful treatment leaves the question undecided by which principle the electro–weak bosons and fermions are distinguished compared with the set of other possible states. Or in other words: *a complete subfermion theory of the standard model has to give an answer to the question why some distinguished particles appear in nature (on the level of the standard model) whilst other possible ones seem to be excluded.* In this section we will outline an answer to this question.

We start our discussion with the boson states. These states are derived from the hard core solutions of equation (6.2) for two subfermions. In this case the exact solutions are given by Proposition 6.4 for vector bosons, and in a similar manner the solutions for scalar bosons can be derived, see [Gri 94a,d]. Compared with ordinary quantum mechanical bound state calculations, Proposition 6.4 shows that (6.1) allows no orbital excitations, in contrast for instance to the hydrogen atom. Rather the spectrum is only constituted by the algebraic properties of the state functions.

The algebraic properties themselves are determined by the direct products of spin and superspin/isospin algebra elements. Because of the exact separation of the superspin/isospin algebra from the other degrees of freedom in equation (6.2), one can distinguish two classes of solutions: symmetric and antisymmetric representations of the superspin/isospin algebra elements. The symmetric superspin–isospin class is given by the set of matrices $\{\gamma^\mu C, \Sigma^{0k} C, \Sigma^k C\}$ and the antisymmetric class by $\{C, \gamma^5 C, \gamma^5 \gamma^\mu C\}$. The former algebra elements are connected with scalar bosons, the latter class is connected with vector bosons.

The bosons attached to the standard model are characterized by

$$T_{U(1)} := -i\gamma^5\gamma^2 C = \begin{pmatrix} 0 & \mathbb{1} \\ -\mathbb{1} & 0 \end{pmatrix} \tag{8.84}$$

which constitutes a $U(1)$ singlet state and by the $SU(2)$ triplet

$$T^1_{SU(2)} := \gamma^5\gamma^3 C = \begin{pmatrix} 0 & \sigma^1 \\ -\sigma^1 & 0 \end{pmatrix}, \tag{8.85}$$

$$T^2_{SU(2)} := -iC = \begin{pmatrix} 0 & \sigma^2 \\ -\sigma^2 & 0 \end{pmatrix},$$

$$T^3_{SU(2)} := -\gamma^5\gamma^1 C = \begin{pmatrix} 0 & \sigma^3 \\ -\sigma^3 & 0 \end{pmatrix},$$

where these matrices satisfy the corresponding Lie algebra relations. According to Section 6.4 all these states are eigenstates with the fermion number $f = 0$.

Therefore it remains to explain why all other algebraic combinations, i.e., all other scalar and vector bosons obviously do not appear in physical reality. One could think of the following simple argument: for very heavy subfermions these bosons are very heavy, too, and therefore they completely decouple from a low energy limit effective dynamics.

In order to reconsider this hypothesis we have to investigate the effective masses of the corresponding bosons. We remember that the mass eigenvalues (energies in the rest frame) of the hard core equation (6.1) are not the physical relevant masses. This is because for Weak Mapping the center of mass amplitudes of the bound state wave functions are assumed to be effective dynamical variables, whilst in the hard core equation these quantities are correlated in order to obtain an energy eigenvalue. Consequently the physical relevant mass values have to be studied by means of the effective dynamics itself.

Effective masses have to be studied by means of the effective equations of motion. The relevant term of the effective functional equation containing the effective boson masses is the term $\mathcal{H}^0_b$ which is given by (7.85). The evaluation of this term for the vector bosons of the set (8.84) and (8.85) was performed in Section 7.5 and leads to (7.86):

$$\mathcal{H}_b^0 = i \int d^3r \sum_{a=1}^{4} b_{ij,a}^F(\mathbf{r})_{|i>j} \varepsilon_{ijk} \varepsilon_{lmk} \partial_m \partial_{l0,a}^F(\mathbf{r}) \tag{8.86}$$

$$+ i \int d^3r \sum_{a=1}^{4} b_{i0,a}^F(\mathbf{r}) \varepsilon_{ijk} \varepsilon_{lmk} \partial_j \partial_{lm,a}^F(\mathbf{r})_{|l>m}$$

$$- i \int d^3r \sum_{a=1}^{4} b_{i,a}^A(\mathbf{r}) \partial_{i0,a}^F(\mathbf{r})$$

$$+ i \mu_B^2 \int d^3r \sum_{a=1}^{4} b_{i0,a}^F(\mathbf{r}) \partial_{i,a}^A(\mathbf{r})$$

with

$$\mu_B^2 := c_1 + g c_2 , \tag{8.87}$$

see Proposition 7.2.

The classical field equations for the fields A_a^i, B_a^i and E_a^i which result from $\mathcal{H}_b^0$ are given by the harmonic part of (7.116); for $\mu_B \neq 0$ they read

$$\partial_0 A_a^i(\mathbf{r}, t) = -E_a^i(\mathbf{r}, t), \tag{8.88}$$
$$\partial_0 B_a^i(\mathbf{r}, t) = \varepsilon_{ijk} \partial_j E_a^k(\mathbf{r}, t),$$
$$\partial_0 E_a^i(\mathbf{r}, t) = -\varepsilon_{ijk} \partial_j B_a^k(\mathbf{r}, t) + \mu_B^2 A_a^i(\mathbf{r}, t).$$

So far we have repeated the discussion of Section 7.5. We now turn to the set of vector bosons with the isospin/superspin matrices $T^a \in \{\gamma^5 C, \gamma^5 \gamma^0 C\}$, which are not contained in the set (8.84) and (8.85) and which according to Section 6.4 are states with the fermion number $f = \pm 2$.

An inspection of the calculations of Proposition 7.2 shows that because $\mathrm{tr}[\gamma^5 T^a] = 0$, which holds for *all* vector bosons, *i.e.*, for all $T^a \in \{C, \gamma^5 C, \gamma^5 \gamma^\mu C\}$, the remaining evaluation of $\mathcal{H}_b^0$ for the $f = 2$ vector bosons is identical with that of the 'electro–weak' bosons. Hence *all* vector bosons have the same mass, provided we are concerned with the unbroken $SU(2) \times U(1)$ symmetry. Thus we will have to find another argument why these $f = \pm 2$ vector bosons do not appear in the physical reality.

The reason for their dropping out is the superselection rule for fermion numbers. As the original subfermion equation underlies fermion number conservation, any correct Weak Mapping is not allowed to violate this conservation law for the phenomenological fields. For instance, it is not possible to have covariant derivatives of fermion fields with respect to $f = 2$ vector boson fields or covariant derivatives of charge conjugated fermion fields to $f = -2$ vector bosons, as terms of the kind $\Phi_{f=2} \bar{\psi} \psi$ in the Lagrangian are not fermion number conserving. Only if these vector boson fields were purely imaginary a symmetric coupling to fermions would be possible. But then the fields are non-Hermitian, and thus not observable. Thus the $f = 2$ vector bosons do not play any physical role.

We now turn to the scalar bosons of the symmetric isospin/superspin class. We assume the corresponding wave functions to be given by

$$\varphi^{(2)}_{Z_1 Z_2}(\mathbf{z}, \mathbf{u}) = \int \Phi(\mathbf{k}) T^s_{\kappa_1 \kappa_2} e^{i\mathbf{k}\mathbf{z}} f^{\Phi}_{i_1 i_2}(\mathbf{u})(\gamma^5 C)_{\alpha_1 \alpha_2} d\mathbf{k} \tag{8.89}$$

$$+ \int G_\mu(\mathbf{k}) T^s_{\kappa_1 \kappa_2} e^{i\mathbf{k}\mathbf{z}} f^G_{i_1 i_2}(\mathbf{u})(\gamma^\mu \gamma^5 C)_{\alpha_1 \alpha_2} d\mathbf{k}$$

with the superspin/isospin matrices T^s from the symmetric class. From the wave functions (8.89) the corresponding expansion functions for the Weak Mapping result by separation of the center of mass amplitudes Φ and G_μ.

In order to calculate the effective masses of the scalar bosons we have to determine the functions f^Φ and f^G. For brevity we do not perform this calculation in detail but refer to the corresponding vector boson calculations Proposition 6.4 and Section 8.5. The result of the evaluation of the expressions for the scalar boson case is that for symmetric s states we have

$$f^G_{i_1 i_2}(\mathbf{u}) = f^A_{i_1 i_2}(\mathbf{u}), \tag{8.90}$$
$$f^\Phi_{i_1 i_2}(\mathbf{u}) = f^F_{i_1 i_2}(\mathbf{u}),$$

i.e., for s states we have the identity of the scalar boson functions f^Φ and f^G with the vector boson functions f^F and f^A.

The calculation of $\mathcal{H}^0_b$ for scalar bosons runs along the same lines as that for vector bosons, provided $\text{tr}[\gamma^5 T^s] = 0$ holds. This is the case for all T^s of the symmetric class with exception of $\Sigma^2 C$. For this special element a special treatment of $\mathcal{H}^0_b$ is needed which we do not explicitly perform.

With the dual functions

$$R^{\prime s}_{Z_1 Z_2}(\mathbf{z}, \mathbf{u}|\mathbf{k}) = (T^s)^+_{\kappa_1 \kappa_2}(\gamma^5 C)^+_{\alpha_1 \alpha_2} e^{-i\mathbf{k}\mathbf{z}} \breve{f}^\Phi_{i_1 i_2}(\mathbf{u}), \tag{8.91}$$
$$R^{\prime \mu, s}_{Z_1 Z_2}(\mathbf{z}, \mathbf{u}|\mathbf{k}) = (T^s)^+_{\kappa_1 \kappa_2}(\gamma^\mu \gamma^5 C)^+_{\alpha_1 \alpha_2} e^{-i\mathbf{k}\mathbf{z}} \breve{f}^G_{i_1 i_2}(\mathbf{u}),$$

the straightforward calculation yields for the harmonic part of the scalar boson dynamics

$$\mathcal{H}^{\prime 0}_b = i \int d^3 r \sum_{s=1}^{2} b^G_{0,s}(\mathbf{r}) \partial_k \partial^G_{k,s}(\mathbf{r}) \tag{8.92}$$

$$+ i \int d^3 r \sum_{s=1}^{2} b^G_{k,s}(\mathbf{r}) \partial_k \partial^G_{0,s}(\mathbf{r})$$

$$+ i c'_0 \int d^3 r \sum_{s=1}^{2} b^G_{0,s}(\mathbf{r}) \partial^\Phi_s(\mathbf{r})$$

$$+ i(c'_1 + g c'_2) \int d^3 r \sum_{s=1}^{2} b^\Phi_s(\mathbf{r}) \partial^G_{0,s}(\mathbf{r})$$

with

$$c'_0 := i 2^9 \pi^3 \sum_{i,i'} (m_i + m_{i'}) \int \breve{f}^G_{ii'}(\mathbf{r}) f^\Phi_{ii'}(\mathbf{r}) d\mathbf{r}, \tag{8.93}$$

$$c_1' := i2^9\pi^3 \sum_{i,i'}(m_i + m_{i'}) \int \check{f}_{ii'}^\Phi(\mathbf{r}) f_{ii'}^G(\mathbf{r}) d\mathbf{r},$$

$$c_2' := i2^9\pi^3 \sum_{i,i',j,j'} \lambda_i \lambda_{i'} \int \check{f}_{ii'}^\Phi(\mathbf{r}) g_i(\mathbf{r}) f_{jj'}^G(0) d\mathbf{r}.$$

The corresponding classical field equations read

$$\dot{\Phi}_s = (c_1' + gc_2')G_0^s, \tag{8.94}$$
$$\dot{G}_0^s = \partial_k G_k^s + c_0'\Phi_s,$$
$$\dot{G}_k^s = \partial_k G_0^s.$$

Rescaling G_μ^s by $G_\mu^s \to (c_1' + gc_2')^{-1}G_\mu^s$ we can transform (8.94) into the system

$$\dot{\Phi}_s = G_0^s, \tag{8.95}$$
$$\dot{G}_0^s = \partial_k G_k^s + c_0'(c_1' + gc_2')\Phi_s,$$
$$\dot{G}_k^s = \partial_k G_0^s.$$

Apart from a suitable initial condition this system is equivalent to the Klein–Gordon equation

$$\partial^\mu \partial_\mu \Phi_s - \mu_S^2 \Phi_s = 0, \tag{8.96}$$

see for instance [Gri 94], with

$$\mu_S^2 := -c_0'(c_1' + gc_2'). \tag{8.97}$$

This effective scalar boson mass μ_S has to be compared with the effective mass $\mu_B^2 = (c_1 + gc_2)$ for the vector bosons, see Proposition 7.2.

Observing (8.90) we have $c_i' = -ic_i$ for $i = 0, 1, 2$, and analogously to the vector boson case (see Section 7.5) one can show that $c_0 = -i$. Thus we have the result that *for $s = 0$ states the effective vector boson and scalar boson masses are identical.* The consequence of this result is that we cannot argue with distinct masses in order to suppress the appearance of scalar bosons.

For clarifying the role of the scalar bosons we remember that π mesons usually are assumed to be built up by a quark/antiquark pair, *i.e.*, $\pi \hat{=} q\bar{q}$. In our fusion model quarks are generated by three elementary subfermions, *i.e.*, we assume $\pi \hat{=} fff\bar{f}\bar{f}\bar{f}$. The quantum numbers of this combination, however, are already generated by the combination $f\bar{f}$, which just correspond to our scalar bosons. *Thus we expect these scalar bosons to describe phenomenological π mesons.*

With respect to higher order bound states for which the number of subfermion constituents is $n \geq 3$ a direct classification is not possible. But once the vector bosons are generated (and symmetry is broken) another mechanism comes into operation: calculations by Pfister [Pfi 93] show that these bosons already act on the subfermion level. Therefore if bound state calculations for $n \geq 3$ are performed, vector bosons with $n = 2$ are involved in the subfermion interactions; *i.e.*, for $n \geq 3$ bound states are no longer exclusively constituted by spinorial contact interactions. Hence the calculations of $n = 3$ and $n = 4$ bound states of Chapters 6 and 9 have to be considered as preliminary.

Although the quantum numbers of such state calculations remain exactly valid, the interaction energy has to be supplied by long range forces between subfermions resulting from their interaction with vector bosons. In comparison with the rather weak spinorial interaction (not even a δ distribution!) of the original fields, the switching on of long range forces qualitatively changes the energy spectrum, so that only those states remain where the internal long range forces are attractive, whilst for repulsive Coulomb forces no binding occurs and the states disappear from observable physics by decoupling.

Nevertheless, in the context of Weak Mapping, spinorial contact interactions play a fundamental role with respect to the further formation of matter. In (5.74) the term $\mathcal{H}_{\mathrm{ff}} := WCCCR_p f \partial^f \partial^f \partial^f$ (in symbolic notation) occurs which corresponds to a four-fermion contact interaction for bound subfermion states (5.68) and their dressed duals R_p.

In (7.122) these states are assumed to describe leptons and quarks, and in [Stu 87] it was shown that the evaluation of this term in combination with the kinetic terms leads to separate NJL equations for quarks and leptons, where the quark vertex contains an additional matrix $\delta_{\rho\rho'}$ resulting from the index ρ in (7.122) with the corresponding sources f_ρ (other indices omitted!).

Such terms are not in contradiction with phenomenological theories. They are possible extensions of the Standard model, *etc.* which cannot be excluded, see Weinberg [Wein 96]. But the essential point is that apart from the occurrence of $\delta_{\rho\rho'}$ in the vertex and of ρ in the source numeration nothing can be learned from the effective theory about the origin of these new degrees of freedom as long as form factors and exchange forces can be neglected.

In this case also, by symmetry transformations at the subfermion level nothing of the substructure can be detected in the effective theory. Both the functional energy equation of the subfermions and the effective functional equation for quarks are form invariant under little group transformations of the subfermions.

On the other hand, similarly to the dynamical symmetry generation in the effective boson theory, the effective quark theory allows additional $SU(3)$ symmetry transformations which are not present in the original subfermion theory. Therefore it suggests itself to forget the subfermion background and to interpret these new degrees of freedom as colour indices. Then it is easy to generate gluons. They follow in the same way from the effective NJL equation for quarks as the electro–weak bosons follow from the subfermion equation (2.19).

Although the latter considerations need further elaboration they clearly show the pattern for the generation of the most important elementary constituents of phenomenological theories of matter from a subfermion model. But of course the classification of physical states resulting from subfermion physics is a formidable task, as a comparison with atomic physics shows, where only the hydrogen atom can be exactly calculated.

8.5   Observables of Boson Substructures

Can one observe boson substructures? Is the photon composite? These seem to be simple questions. However, the answer is by no means simple. And this, of course, holds for all other comparable cases of presumed substructures in subnuclear physics.

The formalism which we developed in the preceding chapters allows us to systematically treat such questions. Up to now we have applied this formalism in order to derive effective dynamics of composite particles and to show their agreement with phenomenological theories. In order to detect substructures (at least theoretically) we obviously have to exceed the limits of the derivation of such effective theories and to proceed to corrections which interfere with the complete agreement with phenomenological theories, but, at the same time, improve our ideas about subnuclear processes.

Apart from renormalization and dressed particle corrections, it is easy to see where the corrections have to be performed. One can:

i) improve the hard core state description and investigate form factors
ii) include residual interactions
iii) include exchange forces
vi) study modified quantization terms.

The latter subjects will be treated in the subsequent sections; in this section we concentrate on i). In particular, as an example, we will study the hard core states of electro–weak bosons in more detail. The primary observables of their substructure are the corresponding spatial probability densities and of course the nontrivial form factors in the effective dynamics of these bosons.

For a first inspection we concentrate on the derivation of spatial probability densities, because apart from an insight into the substructure of electro–weak bosons themselves, these informations are needed for the treatment of the above mentioned topics ii)–iv). This is, of course, a minimal investigation program, but the treatment of form factors and corresponding processes would exceed the scope of this book and should be the object of forthcoming research.

By (6.41) and (6.42) we already know the exact many-time boson state functions. What has to be done is to evaluate the single time limit of (6.41) and (6.42) as exactly as possible and to draw further conclusions about the substructure.

We will perform our calculation with the assumption of small differences $m_{i_1} - m_{i_2}$ of the subfermion masses m_i, $i = 1, 2, 3$. Furthermore, we remember that our model is based on very high subfermion masses in order to guarantee the unobservability of the basic subfermions.

As we already emphasized a very detailed evaluation of (6.41) and (6.42) in the restsystem (under the above mentioned assumptions) was performed by Pfister [Pfi 90] and used in full length for the derivation of the effective $SU(2) \times U(1)$ dynamics in the low energy limit. However, of special interest are the wave functions for moving bosons, because by such states scattering processes are (theoretically) governed. Thus we will include these moving states into our considerations.

With respect to such states the relativistic covariance is guaranteed by the use of the relativistic covariant state description (6.41) and (6.42) with the transformation law (6.52). Then in any frame of reference one can perform the single time limit, so that any postulated observer will see 'his' relativistically transformed states.

The exact vector boson functions (6.41) and (6.42) may be written as

$$\varphi^{(2)}_{Z_1 Z_2}(x_1, x_2) = N_B T^a_{\kappa_1 \kappa_2} e^{-\frac{i}{2}k(x_1 + x_2)} \chi_{\substack{i_1 i_2 \\ \alpha_1 \alpha_2}}(x_1 - x_2 | k). \tag{8.98}$$

For the relative function we write

$$\chi_{\substack{i_1 i_2 \\ \alpha_1 \alpha_2}}(x_1 - x_2 | k) = \lambda_{i_1} \lambda_{i_2} \int d^4 p\, e^{-ip(x_1 - x_2)} \frac{A_\mu Z^\mu(p, k)_{\substack{i_1 i_2 \\ \alpha_1 \alpha_2}}}{N(p, k)_{i_1 i_2}} \tag{8.99}$$

(no summation over i_1, i_2) with the numerator

$$Z^\mu := [(p_\nu + k_\nu/2)\gamma^\nu + m_{i_1}]\gamma^\mu[(p_\rho - k_\rho/2)\gamma^\rho + m_{i_2}]C$$

and the denominator

$$N := \left[(p + k/2)^2 - m_{i_1}^2 + i\epsilon\right]\left[(p - k/2)^2 - m_{i_2}^2 + i\epsilon\right]. \tag{8.100}$$

The numerator Z^μ is evaluated by expanding the matrix products on a basis of the Dirac algebra. As we have already emphasized we restrict ourselves to the s wave part of the vector boson functions by neglecting the terms of Z^μ which are proportional to p^μ. In doing so we obtain

$$Z^\mu = \left(m_{i_1} m_{i_2} - p^2 + \frac{k^2}{4}\right)\gamma^\mu C + \frac{i}{2} k_\nu (m_{i_1} + m_{i_2})\Sigma^{\mu\nu} C, \tag{8.101}$$

where, in addition, we used the transversality condition $k_\mu A^\mu = 0$.

The relative function (8.99) contains the vector potential A_μ. To introduce field strengths, too, we collect all terms in $A_\mu Z^\mu$ which after evaluation yield the antisymmetric combination $ik_{[\nu} A_{\mu]}$ and identify the latter with the field strength tensor $F_{\nu\mu}$. The corresponding terms are assumed to constitute the field strength part of the wave function.

The division of (8.98) into a vector potential part and a field strength part is only motivated by the intention to work with independent hard core expansion functions for A_μ and $F_{\mu\nu}$ separately. In particular, these quantities are considered as the dynamical variables of the effective theory, and hence the functional relation between A_μ and $F_{\mu\nu}$ which holds in the original hard core equation is given up in favour of the independent variability of these quantities.

This means that the identification $F_{\nu\mu} = ik_{[\nu} A_{\mu]}$ is only used for a classification of the various terms, but is not maintained in Weak Mapping as in the latter case the exact hard core states are subdivided into independent parts and undergo a mass renormalization.

Dividing up the relative function (8.99) into vector potential and field strength parts we obtain with $F_{\mu\nu} = ik_{[\mu} A_{\nu]}$ and (8.101)

$$\chi_{\substack{i_1 i_2 \\ \alpha_1 \alpha_2}} (x_1 - x_2 | k) \tag{8.102}$$

$$= \int d^4 p \, e^{-ip(x_1 - x_2)} \left[A_\mu f^A_{i_1 i_2}(p, k)(\gamma^\mu C)_{\alpha_1 \alpha_2} + F_{\mu\nu} f^F_{i_1 i_2}(p, k)(\Sigma^{\mu\nu} C)_{\alpha_1 \alpha_2} \right],$$

where

$$f^A_{i_1 i_2}(p, k) \quad := \quad -\lambda_{i_1} \lambda_{i_2} \frac{p^2 - m_{i_1} m_{i_2} - \frac{k^2}{4}}{N_{i_1 i_2}(p, k)}, \tag{8.103}$$

$$f^F_{i_1 i_2}(p, k) \quad := \quad \frac{1}{2} \lambda_{i_1} \lambda_{i_2} \frac{m_{i_1} + m_{i_2}}{N_{i_1 i_2}(p, k)}. \tag{8.104}$$

As far as observables are concerned we remember that vector potentials are in general unobservable. Hence in this respect only the $F_{\mu\nu}$ terms of (8.102) are of interest. In the following we thus concentrate on the evaluation of the corresponding wave functions (8.104). By the Feynman formula [Feyn 49] we represent N^{-1} by

$$N^{-1} \quad = \quad \int_0^1 \frac{dz}{\left[\frac{1}{4} k^2 + p^2 - 2(z-1)kp - m_{i_1}^2 + z(m_{i_1}^2 - m_{i_2}^2) + i\epsilon \right]^2} \tag{8.105}$$

$$= \quad \int_0^1 \frac{dz}{\left\{ [p - (z-1)k]^2 + [\frac{1}{4} - (z-1)^2]k^2 - m_{i_1}^2 + z(m_{i_1}^2 - m_{i_2}^2) + i\epsilon \right\}^2}$$

and with $u = p - (z-1)k$ we obtain for the Fourier transform of (8.104)

$$f^F_{i_1 i_2}(x, k) \quad = \quad \frac{1}{2} \lambda_{i_1} \lambda_{i_2}(m_{i_1} + m_{i_2}) \tag{8.106}$$

$$\times \int_0^1 \frac{e^{-iux} e^{-i(z-1)kx}}{\left\{ u^2 + [\frac{1}{4} - (z-1)^2]\mu^2 - m_{i_1}^2 + z(m_{i_1}^2 - m_{i_2}^2) + i\epsilon \right\}^2} d^4u \, dz,$$

where we have introduced the eigenvalue condition $k^2 = \mu^2$ for the hard core boson mass. This mass is assumed to be very small compared with the subfermion masses m_i. Hence we have two inequalities for an estimate of the integral in (8.106): $\mu^2 \ll m_i^2$ and $|m_{i_1}^2 - m_{i_2}^2| \ll m_i^2$ for all $i = 1, 2, 3$. We use these inequalities to neglect μ^2 completely and to replace $z(m_{i_1}^2 - m_{i_2}^2)$ by $\bar{z}(m_{i_1}^2 - m_{i_2}^2)$, where $\bar{z}$ is a suitable meanvalue in the interval $0 < \bar{z} < 1$. With

$$\overline{M}^2_{i_1 i_2} := m_{i_1}^2 - \bar{z}(m_{i_1}^2 - m_{i_2}^2) \tag{8.107}$$

we can represent f^F in a (very) good approximation by

$$f^F_{i_1 i_2}(x, k) \quad = \quad \frac{1}{2} \lambda_{i_1} \lambda_{i_2}(m_{i_1} + m_{i_2}) \tag{8.108}$$

$$\times \int_0^1 e^{-i(z-1)kx} dz \int \frac{e^{-iux}}{\left[u^2 - M^2_{i_1 i_2} + i\epsilon \right]^2} d^4u,$$

and with

$$\int d^4u\, e^{-iux}(u^2 - m^2 + i\epsilon)^{-2} = \frac{i}{2}\frac{(2\pi)^2}{\Gamma(2)}K_0[m(-x^2 + i\epsilon)^{\frac{1}{2}}] \tag{8.109}$$

with the modified Bessel function $K_0(r)$, see [Gel 60], one obtains for (8.108)

$$f_{i_1 i_2}^F(x,k) = \frac{1 - e^{ikx}}{kx}\pi^2 \lambda_{i_1}\lambda_{i_2}(m_{i_1} + m_{i_2})K_0[\overline{M}_{i_1 i_2}(-x^2 + i\epsilon)^{\frac{1}{2}}]. \tag{8.110}$$

According to (6.33) $\varphi^{(2)}$ is identical with $\tau^{(2)}$, i.e., a time ordered matrix element. Therefore we have to investigate (8.110) in this respect.

• **Proposition 8.2:** For spacelike $x^2 = x_0^2 - \mathbf{r}^2$ the function (8.110) is a causal function.

Proof: We consider the causal Green's function $D^C(x)$, see [Bogo 59]. For $u = -k$ we have

$$\int d^4u\, e^{-iux}(u^2 - m^2 + i\epsilon)^{-1} = -\int d^4k\, e^{ikx}(m^2 - k^2 - i\epsilon)^{-1} = -D^C(x)(2\pi)^4. \tag{8.111}$$

For spacelike x^2 we obtain

$$D^C(x) = \frac{im}{4\pi^2\sqrt{-x^2}}K_1(m\sqrt{-x^2}) \equiv D^{(-)}(x). \tag{8.112}$$

Then it follows for spacelike x^2

$$\begin{aligned}
\frac{\partial}{\partial m}D^C(x) &= \frac{\partial}{\partial m}D^{(-)}(x) \\
&= \frac{\partial}{\partial m}\frac{im}{4\pi^2\sqrt{-x^2}}K_1(m\sqrt{-x^2}) \\
&= \frac{2m}{(2\pi)^4}\int e^{-iux}(u^2 - m^2 + i\epsilon)^{-2}d^4u.
\end{aligned}$$

With the properties of Bessel functions, see [Rade 97], the derivative of K_1 can be evaluated and yields (8.109). ◇

The analytical behaviour of the Bessel function K_0 is given by

$$K_0(mr) \approx -\ln\left(\frac{mr}{2}\right) + 0.57 \quad \text{for} \quad mr \ll 1, \tag{8.113}$$

$$K_0(mr) \approx \left(\frac{\pi}{2mr}\right)^{1/2}e^{-mr} \quad \text{for} \quad mr \gg 1. \tag{8.114}$$

As is to be expected the unregularized functions $f_{i_1 i_2}^F(r)$ (as well as $f_{i_1 i_2}^A$) are divergent at the origin $r = 0$. But it is remarkable that for spacelike x the singularities of (8.109) are weaker than the singularities for time-like x. Hence the use of the single time (Hamiltonian) formalism softens the singularity structure of the theory.

The single time limit of (8.110) simply consists in setting $x_0 \equiv t_1 - t_2 = 0$; in this way we obtain the single time relative functions

$$f^F_{i_1 i_2}(\mathbf{r}, \mathbf{k}) = \frac{1 - e^{-i\mathbf{k}\mathbf{r}}}{\mathbf{k}\mathbf{r}} \pi^2 \lambda_{i_1} \lambda_{i_2} (m_{i_1} + m_{i_2}) K_0[\overline{M}_{i_1 i_2} r] \qquad (8.115)$$

with $r = \sqrt{\mathbf{r}^2}$, where it can be shown that the imaginary part $i\epsilon$ is negligible.

With this simple behaviour we can immediately form the spatial densities of the states, the general form of which follows by (4.24) if one does not integrate over the space coordinates. This expression contains the metric tensor G; for the metric tensor $G^{II', JJ'}$ of the two-coordinate space we use the direct product of the metric tensors $G^{II'}$ of the one-coordinate space. If condensation phenomena on the subfermion level are excluded, $G_{II'}$ was calculated in [Stu 94], Proposition 5.6. Then with $G^{II'} = G_{II'}^{-1}$ one obtains the corresponding contravariant tensor, which reads $G^{II'} = 2\lambda_i^{-1} \delta_{ii'} \delta_{\kappa\kappa'} \delta_{\alpha\alpha'} \delta(\mathbf{r} - \mathbf{r}')$.

The spatial density then follows from (4.24) if $\tau_{II'}^{(2)} \equiv \varphi_{II'}^{(2)}$ is observed and the sum is performed over all algebraic indices, but the integral only over $\mathbf{r}'_1, \mathbf{r}'_2$. This gives with $G^{I_1 I_2 J_1 J_2} = G^{I_1 J_1} G^{I_2 J_2}$ the formula

$$\rho(\mathbf{r}_1, \mathbf{r}_2) = 4 \sum_{\substack{i_1, i_2 \\ \alpha_1, \alpha_2 \\ \kappa_1, \kappa_2}} \varphi^{(2)}_{\substack{i_1 i_2 \\ \alpha_1 \alpha_2 \\ \kappa_1 \kappa_2}} (\mathbf{r}_1, \mathbf{r}_2 | \mathbf{k})^+ \lambda_{i_1}^{-1} \lambda_{i_2}^{-1} \varphi^{(2)}_{\substack{i_1 i_2 \\ \alpha_1 \alpha_2 \\ \kappa_1 \kappa_2}} (\mathbf{r}_1, \mathbf{r}_2 | \mathbf{k}). \qquad (8.116)$$

The complete vector boson function is defined by (8.98). Its observable part results if in (8.102) $A_\mu = 0$ is assumed. If in (8.98) the single time limit is performed and (8.115) is substituted, the following density function for the relative coordinate $\mathbf{r} \equiv \mathbf{r}_1 - \mathbf{r}_2$ results from (8.116):

$$\rho(\mathbf{r}) = \sum_{i_1, i_2} 2^7 \pi^4 |N_{\mathrm{B}}|^2 \frac{1 - \cos(\mathbf{k}\mathbf{r})}{(\mathbf{k}\mathbf{r})^2} |F_{\mu\nu}|^2 \lambda_{i_1} \lambda_{i_2} (m_{i_1} + m_{i_2})^2 |K_0[\overline{M}_{i_1 i_2} r]|^2, \qquad (8.117)$$

whilst for the plane wave of (8.98) the center of mass coordinate $(\mathbf{r}_1 + \mathbf{r}_2)/2$ drops out. In (8.117) one can perform the limit $\mathbf{k} \to 0$ and obtains a finite value. On the other hand, (8.117) shows a relativistic contraction of the wave function or the density function, respectively, in the direction of the wave vector $\mathbf{k}$. This behaviour is to be expected for a Lorentz covariant wave function.

Another problem is the positivity of (8.117) as the summation has to be extended over the auxiliary indices i_1, i_2. As the λ_i are indefinite, by this prescription the positivity of (8.117) cannot be guaranteed. This fact gives rise to the reflection whether the scalar product is correctly defined in the case of nonperturbative Pauli–Villars regularization.

Owing to the infinite number of inequivalent representations of the field operator algebra the metric of a quantum field theory becomes a new, additional variable in the algebraic formulation of the theory. This freedom in the choice of the representation is the essential difference between quantum field theory and quantum mechanics, and although the metric, i.e., the representation has to be compatible with the system dynamics, it offers the possibility for suitable modifications of the scalar product in the present case. Therefore in accordance with Pauli–Villars regularization in perturbation theory we postulate: *The physical states are the regularized states*, see Section 6.8.

This postulate means that with respect to the formation of scalar products the quantities

$$\varphi^{(n)}_{\substack{\kappa_1\dots\kappa_n \\ \alpha_1\dots\alpha_n}} (\mathbf{r}_1,\dots\mathbf{r}_n|a)^{\text{phys}} := \sum_{\substack{i_1\dots i_n \\ \kappa_1\dots\kappa_n \\ \alpha_1\dots\alpha_n}} \varphi^{(n)}_{i_1\dots i_n} (\mathbf{r}_1,\dots\mathbf{r}_n|a) =: \hat{\varphi}^{(n)} \tag{8.118}$$

are the physically relevant amplitudes, and the same definition has to be applied to the metric tensor.

One immediately realizes that if this definition is used to calculate G^{IJ}_{phys} from (8.116) one obtains $G^{IJ}_{\text{phys}} \equiv 0$. But this is no misfortune as we assume that the subfermions are unobservable. It remains the task to define an appropriate metric tensor instead of (8.116).

So far a study of a metric (or equivalently of a representation) beyond perturbation theory has only been started for the BCS model of superconductivity, see Stumpf and Borne [Stu 94], Grebe [Gre 98]. Therefore without a systematic study about inequivalent representations for the model under consideration we have to guess at least one suitable representation by physical reasoning.

This discussion was performed in Section 6.8, and the physical reasoning is concerned with the treatment of composite particle processes. In this case, for the state amplitudes composite particle representations are introduced which, for instance for boson processes, are given by (6.149).

The corresponding state space then solely depends on the properties of these single composite particle functions, which under suitable preconditions satisfy current conservation laws, see Section 6.8.

If we assume that these preconditions are satisfied, we can directly integrate (8.118) with its Hermitian conjugate in order to obtain the norm. In particular for the boson wave functions we obtain the modified density

$$\rho(\mathbf{r})^{\text{phys}} = \tag{8.119}$$

$$2^7 \pi^4 |N_{\text{B}}|^2 \frac{1-\cos(\mathbf{kr})}{(\mathbf{kr})^2} |F_{\mu\nu}|^2 \left| \sum_{i_1,i_2} \lambda_{i_1} \lambda_{i_2} (m_{i_1}+m_{i_2})^2 K_0[\overline{M}_{i_1 i_2} r] \right|^2$$

which obviously is positive definite.

Naturally positivity is only a necessary but not a sufficient condition for the physical interpretation of a theory. But in accordance with Section 6.8 it holds:

- **Proposition 8.3:** The physical amplitudes (8.118) belong to a representation of the original spinor field $\Psi(x)$ defined by (2.37).

Proof: We consider the definition of normal ordering (3.85) and project it into configuration space. From this projection it is obvious that the transition to physical normal ordered state amplitudes (8.118) is transferred to the τ_n functions, *i.e.*, we obtain

$$\tau^{(n)}_{\substack{\kappa_1\dots\kappa_n \\ \alpha_1\dots\alpha_n}} (\mathbf{r}_1,\dots\mathbf{r}_n|a)^{\text{phys}} = \sum_{\substack{i_1\dots i_n \\ \kappa_1\dots\kappa_n \\ \alpha_1\dots\alpha_n}} \tau^{(n)}_{i_1\dots i_n} (\mathbf{r}_1,\dots\mathbf{r}_n|a), \tag{8.120}$$

and by (4.41) and (2.37) this yields

$$\tau^{(n)}_{\substack{\kappa_1\ldots\kappa_n \\ \alpha_1\ldots\alpha_n}} (\mathbf{r}_1,\ldots\mathbf{r}_n|a)^{\text{phys}} = \langle 0|\mathcal{A}(\Psi_{\alpha_1\kappa_1}\ldots\Psi_{\alpha_n\kappa_n})|a\rangle. \tag{8.121}$$

The corresponding propagator F^{phys} is defined by equation (6.11). $\Diamond$

From this proposition it follows immediately that an explicit state representation (4.37) exists for the Ψ fields. This in turn allows to apply the formalism of Section 6.8 to the physical states, i.e., the corresponding dynamics is norm conserving. Together with the positive definiteness we have thus derived a unitary representation.

For further applications in Section 8.7 we compile and recapitulate the vector boson formulas required for Weak Mapping. These formulas include the propagators, the hard core states, their duals and the corresponding constants. With $\mathbf{z} = \frac{1}{2}(\mathbf{r}_1 + \mathbf{r}_2)$ and $\mathbf{r} = \mathbf{r}_1 - \mathbf{r}_2$ the boson functions read in s state approximation and without Lorentz contraction

$$\begin{aligned}
C^A_{I_1I_2} &:= N_A T^a_{\kappa_1\kappa_2}(\gamma^\ell C)_{\alpha_1\alpha_2} e^{ikz} f^A_{i_1i_2}(\mathbf{r}), & (8.122)\\
R^A_{I_1I_2} &:= (T^a)^+_{\kappa_1\kappa_2}(\gamma^\ell C)^+_{\alpha_1\alpha_2} e^{-ikz} \breve{f}^A_{i_1i_2}(\mathbf{r}), \\
f^A &:= \lambda_{i_1}\lambda_{I_2} M_{i_1i_2} r^{-1} K_1(M_{i_1i_2}r), \\
\breve{f}^A &:= N_A^{-1}\kappa_A\lambda_{i_1}^{-1}\lambda_{I_2}^{-1} M_{i_1i_2}\left(e^{-m_{i_1}r} + e^{-m_{i_2}r}\right), \\
C^F_{I_1I_2} &:= N_F T^a_{\kappa_1\kappa_2}(\Sigma^{\mu\nu}C)_{\alpha_1\alpha_2} e^{ikz} f^F_{i_1i_2}(\mathbf{r}), \\
R^F_{I_1I_2} &:= (T^a)^+_{\kappa_1\kappa_2}(\Sigma^{\mu\nu}C)^+_{\alpha_1\alpha_2} e^{-ikz} \breve{f}^F_{i_1i_2}(\mathbf{r}), \\
f^F &:= \lambda_{i_1}\lambda_{i_2} M_{i_1i_2} K_0(M_{i_1i_2}r), \\
\breve{f}^F &:= N_F^{-1}\kappa_F\lambda_{i_1}^{-1}\lambda_{i_2}^{-1} M^2_{i_1i_2}\left(e^{-m_{i_1}r} + e^{-m_{i_2}r}\right).
\end{aligned}$$

Taking for the estimates only the s part of the propagator it reads

$$F_{I_1I_2} = -(2\pi)^{-2}\lambda_{i_1}\delta_{i_1i_2}m^2_{i_1}r^{-1} K_1(m_{i_1}r)\gamma^5_{\kappa_1\kappa_2} C_{\alpha_1\alpha_2}. \tag{8.123}$$

The value of the propagator is absolutely fixed as in principle F stems from the solution of an inhomogeneous equation, whereas the C functions can be multiplied by suitable constants as they are solutions of homogeneous eigenvalue equations or, more precisely: result from such solutions. The physical amplitudes are defined by

$$\widehat{C}_{\substack{\kappa_1\kappa_2 \\ \alpha_1\alpha_2}}(\mathbf{r}_1,\mathbf{r}_2) := \sum_{i_1,i_2} C_{\substack{i_1i_2 \\ \kappa_1\kappa_2 \\ \alpha_1\alpha_2}}(\mathbf{r}_1,\mathbf{r}_2), \tag{8.124}$$

and according to the discussion in Section 6.8 the norm is defined by

$$\| C^{\text{phys}} \| := \int \widehat{C}_{\substack{\kappa_1\kappa_2 \\ \alpha_1\alpha_2}}(\mathbf{r}_1,\mathbf{r}_2)^+ \widehat{C}_{\substack{\kappa_1\kappa_2 \\ \alpha_1\alpha_2}}(\mathbf{r}_1,\mathbf{r}_2) d^3r_1 d^3r_2. \tag{8.125}$$

This expression contains the factor $\delta(0)$ from the plane wave normalization of the center of mass motion. Apart from this conventional factor the normalization condition reads

$$|N_X|^2 \int |\hat{f}^X(\mathbf{r})|^2 \, d^3r \overset{!}{=} 1 \tag{8.126}$$

if one applies it to the independent states C^A and C^F of (8.122), and with the Taylor formula [Rade 97] one obtains

$$\hat{f}^A = \sum_{i_1,i_2} f^A_{i_1 i_2} = n^A_R K_1'''(mr) r^2, \tag{8.127}$$

$$\hat{f}^F = \sum_{i_1,i_2} f^F_{i_1 i_2} = n^F_R K_0'''(mr) r^3,$$

and

$$n^A_R := \frac{1}{96} \sum_{i_1,i_2} \lambda_{i_1} \lambda_{i_2} (m_{i_1} + m_{i_2})^4, \tag{8.128}$$

$$n^F_R := \frac{i}{96} \pi^2 \sum_{i_1,i_2} \lambda_{i_1} \lambda_{i_2} (m_{i_1} + m_{i_2})^4.$$

The value of m is a mean value of $M_{i_1 i_2}$. Owing to the regularization the norm integrals are finite, i.e., $\hat{f}^A, \hat{f}^F \in L^2$. Approximate calculation of the integrals leads to

$$N_A = (4\pi\Gamma(7))^{-\frac{1}{2}} \left(n^A_R\right)^{-1} (2m)^4, \tag{8.129}$$

$$N_F = \left(9\pi^2\Gamma(9)\right)^{-\frac{1}{2}} \left(n^F_R\right)^{-1} (2m)^5.$$

As was discussed in Section 6.8, from this physical state normalization the normalization of duals has to be sharply distinguished as the latter takes place in the auxiliary field space. So in the boson case the normalization constants κ_A, κ_F of the duals are defined by

$$\sum_{I_1,I_2} R^X_{I_1 I_2}(k) C^X_{I_1 I_2}(k') \overset{!}{=} \delta(k - k'). \tag{8.130}$$

Approximate calculation yields

$$\kappa_A = \left(2^4 3^2 \pi^2\right)^{-1}, \tag{8.131}$$

$$\kappa_F = \left(2^3 3^3 \pi^2\right)^{-1}.$$

In Weak Mapping expressions partial regularizations occur. For instance, we have

$$\sum_{i_1} f^A_{i_1 i_2} = n^1_R \lambda_{i_2} K_1'(mr) \tag{8.132}$$

with

$$n^1_R := \sum_{i_1} \frac{1}{4} \lambda_{i_1} m^2_{i_1}. \tag{8.133}$$

Furthermore, it is convenient to approximate the propagator. Its space integral leads to the value

$$\int F_{Z_1 Z_2}(\mathbf{r}) d^3 r = -\frac{1}{2} \lambda_{i_1} \delta_{i_1 i_2} \gamma^5_{\kappa_1 \kappa_2} C_{\alpha_1 \alpha_2} , \tag{8.134}$$

and owing to the concentration of the orbital part $K_1(mr)/r$ around the origin, this part can be considered as the element of a sequence which for $m_1 \to \infty$ converges towards a δ distribution. Thus in our estimates we approximate the propagator F under the integral sign by

$$F_{Z_1 Z_2}(\mathbf{r}) \approx -\lambda_{i_1} \delta_{i_1 i_2} \delta(\mathbf{r}) \gamma^5_{\kappa_1 \kappa_2} C_{\alpha_1 \alpha_2} . \tag{8.135}$$

Finally, we notice that amongst the Weak Mapping expressions some local terms with $C(\mathbf{r}, \mathbf{r})^{\mathrm{phys}}$ occur. Such terms are fully regularized and have to be evaluated without any approximation in order to obtain correct results. We remember that the C amplitudes are derived from (6.41) by identifying C^A with the coefficient functions of A^μ and C^F with the coefficient functions of $F^{\mu\nu}$, where $F^{\mu\nu} = k^{[\mu} A^{\nu]}$, compare (7.69)–(7.73). Then for $\mathbf{r}_1 = \mathbf{r}_2 = \mathbf{r}$ from (7.73) it follows

$$\hat{\varphi}^{(2)}_{\substack{\kappa_1 \kappa_2 \\ \alpha_1 \alpha_2}}(\mathbf{r}, \mathbf{r}) = N_A^{-1} \hat{C}^A_\mu(\mathbf{r}, \mathbf{r}) A^\mu + N_F^{-1} \hat{C}^F_{\mu\nu}(\mathbf{r}, \mathbf{r}) F^{\mu\nu} . \tag{8.136}$$

This formula can be compared with the regularized $\varphi^{(2)}$ of (6.41) and yields the corresponding values of $\hat{C}^A_\mu(\mathbf{r}, \mathbf{r})$ and $\hat{C}^F_{\mu\nu}(\mathbf{r}, \mathbf{r})$. In particular one obtains for temporal gauge and transversality condition with definitions (6.48) and (6.49)

$$\hat{C}^A_{\substack{\ell, \kappa_1 \kappa_2 \\ \alpha_1 \alpha_2}}(\mathbf{r}, \mathbf{r}) = \frac{2ig}{(2\pi)^4} N_A T^a_{\kappa_1 \kappa_2} (\gamma^\ell C)_{\alpha_2 \alpha_2} e^{i\mathbf{k}\mathbf{r}} \tag{8.137}$$

$$\times \int d^4 q \, [S(k-q)S(k) + R_\mu(k-q)R^\nu(q)]$$

and

$$\hat{C}^F_{\substack{\mu\nu, \kappa_1 \kappa_2 \\ \alpha_1 \alpha_2}}(\mathbf{r}, \mathbf{r}) = \frac{2ig}{(2\pi)^4} N_F T^a_{\kappa_1 \kappa_2} e^{i\mathbf{k}\mathbf{r}} (\Sigma^{\mu\nu} C)_{\alpha_2 \alpha_2} \int d^4 q \, S(q) R^{[\nu}(k-q) A^{\mu]} . \tag{8.138}$$

The corresponding integrals were evaluated in [Pfi 89] and numerically calculated by Sand [Sand 91]. They are finite and unequal zero. For details we refer to these papers, because for our estimates we need only the fact that such expressions have non-vanishing finite values. We show this for C^A. Comparing (8.137) with (8.122) we obtain

$$\hat{C}^A_{\substack{\ell \kappa_1 \kappa_2 \\ \alpha_1 \alpha_2}}(\mathbf{r}, \mathbf{r}) = N_A T^a_{\kappa_1 \kappa_2} (\gamma^\ell C)_{\alpha_1 \alpha_2} e^{i\mathbf{k}\mathbf{r}} \hat{f}^A(0) \tag{8.139}$$

with

$$\hat{f}^A(0) := \int d^4 q \, [S(k-q)S(q) + R_\nu(k-q)R^\nu(q)] . \tag{8.140}$$

In the latter definition we left our the factor $(2ig)(2\pi)^{-4}$, as this factor only leads to a renormalization of N_A which cancels out in the final result. The leading term in (8.140) is the S term. With $J_0 := \int d^4 q S(k-q) S(q)$ one obtains [Sand 91]

$$J_0 = -i\pi \sum_{i,j} \lambda_i \lambda_j m_i m_j \int_0^1 dx \, \ln|P_{ji}(x)| \qquad (8.141)$$

with

$$P_{ji}(x) := k^2 x^2 - (m_j^2 - m_i^2 + k^2)x + m_j^2 \,. \qquad (8.142)$$

The boson masses $k^2 = \mu_{\mathrm{B}}^2$ and the auxiliary mass differences are very small compared with the auxiliary masses m_i themselves. Therefore we can expand the ln term around $m_j \approx m$. By means of the Taylor formula [Rade 97] and of the regularization properties of the λ_i the evaluation of J_0 then yields

$$J_0 \propto \sum_{i,j} \lambda_i \lambda_j m_i^3 m_j^3 m^{-2}, \qquad (8.143)$$

i.e., a finite value after regularization.

8.6 Effective Quantum Hamiltonians

With the state functionals (7.114) and (7.115) or (8.72) and (8.73) respectively, we derived classical equations of motion from the effective functional equations gained by Weak Mapping and compared these equations with those of classical phenomenological dynamics. The resulting agreement between effective and phenomenological theory on the classical level is an impressive success of fusion theory, but it is not the whole truth. As we are dealing with microphysical processes, quantum effects play an essential role and thus we have to pose the question of the agreement between effective and phenomenological theory also on the quantum level.

In our fusion program we perform a quantization only on the subfermion level, *i.e.*, on the level of the assumed fundamental constituent elements of matter. Any quantum effect on the level of corresponding bound states is to be explained as an effect of the underlying subfermion theory. Thus we cannot *a priori* assume the resulting effective theories to be (canonically) quantized versions of corresponding classical Hamiltonian theories; from the fusion point of view we rather consider the quantization of *phenomenological* theories as heuristic procedures which with more or less success allow the prediction of experimental results. In this section we investigate the relation between such an heuristically quantized phenomenological theory and the corresponding effective bound state theory.

Surprisingly enough this is not an easy task on the side of the phenomenological theory. Although, for instance, in quantum electrodynamics experiments and theory are considered to be of the highest precision so far obtained in physics, the theoretical basis of phenomenological quantum electrodynamics is uncertain. It cannot be a canonically quantized version of the classical theory because from a mathematical point of view such a version does not exist owing to the well known infinities occurring in the corresponding calculations.

What in conventional theory is calculated and compared with high precision experiments stems from a regularized and renormalized modification of the canonical formalism, and it is adapted to covariant perturbation theory. In particular, in the latter case propagators are manipulated in order to obtain finite results, and if formulated in terms of Lagrangians and quantization rules this leads to consequences which have little similarity with the formalism of (nonexisting) canonical quantum electrodynamics. Compare for instance Itzykson and Zuber [Itz 80], who formulated a Lagrangian of quantum electrodynamics with inclusion of Pauli–Villars regularization to all orders of perturbation theory.

But still more disturbing is that such regularizations are not unique and the formalism depends on the way one prefers to proceed; see, for instance, the contribution of Scharf [Schar 89] to the problem of finite quantum electrodynamics. So for comparison between effective theory and phenomenological theory it is by no means clear what is a suitable counterpart on the latter side.

From an algebraic point of view the manipulations of the propagators correspond to changes in the representation of the field operator algebra, and one is reminded that quantum field theory possesses infinitely many inequivalent representations. Independently of whether one considers composite particle dynamics or not, the dependence of the dynamics on a specific representation has to be observed in any reliable attempt to evaluate this dynamics. In contrast to the obscure situation in conventional theory, in the algebraic formalism presented in this book a clear answer can be given.

In the general case the effective dynamics of a system is described by functional equations in which the corresponding representations are incorporated and by which the symmetries of the system are represented. Therefore if time translation symmetry is not broken, generators of time translation which depend on the representation exist in functional space. Whether such generators have images in Hilbert space which can be formulated by means of a (representation dependent) effective field operator algebra has to be studied separately in each case.

In the algebraic formalism effective Hamiltonians are derived as functional energy operators which can be obtained by canonical transformations. Such transformations have already been introduced in Section 4.6 and extensively studied for quantum electrodynamics and quantum chromodynamics in [Stu 96] and [Stu 97]. We apply this method to the present case of composite particle theory.

In order to demonstrate the essentials of our method by a simple (nontrivial) example and to avoid lengthy calculations, we confine ourselves to the discussion of an effective non-abelian $SU(2)$ boson dynamics which was initially derived by Weak Mapping in Section 7.4. Although we already indicated that from a rigorous point of view the phenomenological canonical theory of a $SU(2)$ boson theory does not exist, we apply the corresponding formalism as an intermediate step in the presentation of our method and use it to show in which way we have to manipulate it to obtain physical meaningful results.

The Hamiltonian of a phenomenological $SU(2)$ gauge field theory reads

$$\mathbf{H} = \frac{1}{2} \int d\mathbf{r} \left[E_i^a(\mathbf{r}) E_i^a(\mathbf{r}) + B_i^a(\mathbf{r}) B_i^a(\mathbf{r}) \right] \tag{8.144}$$

with

$$B_i^a := \varepsilon_{ijk}\partial_j A_k^a - \frac{1}{2}g_{\text{ph}}f_{abc}\varepsilon_{ijk}A_j^b A_k^c, \tag{8.145}$$

where $f_{abc} = \varepsilon_{abc}$ are the structure constants of the gauge group $SU(2)$. The canonical commutation relations are defined by

$$\left[A_i^a(\mathbf{r},t), E_{i'}^{a'}(\mathbf{r}',t)\right]_- = -i\delta_{aa'}\delta_{ii'}\delta(\mathbf{r}-\mathbf{r}'). \tag{8.146}$$

The mapping of this theory into functional space is performed by the definition of the functional states

$$|\mathcal{B}(b;p)\rangle = \sum_{n=0}^{\infty}\frac{1}{n!}\rho^{(n)}(K_1\ldots K_n|p)b_{K_1}\ldots b_{K_n}|0\rangle_{\text{B}}, \tag{8.147}$$

where $\rho^{(n)}(K_1\ldots K_n|p)$ is assumed to be the symmetric matrix element of corresponding boson field operators:

$$\rho^{(n)}(K_1\ldots K_n|p) = \langle 0|\mathcal{S}(B_{K_1}\ldots B_{K_m})_t|p\rangle \tag{8.148}$$

for equal times $t_1 = \ldots = t_n = t$.

The quantities b_K are the functional source operators which correspond to the field operators $B_K \in \{A_k^a(\mathbf{r},t), E_k^a(\mathbf{r},t)\}$ with the super-index K. For more details with respect to the functional state space construction we refer to Chapters 3 and 4.

Analogously to Proposition 4.12 for $|\mathcal{B}\rangle$ a functional eigenvalue equation can be derived, which describes the full quantum dynamics in terms of functionals or matrix elements, respectively, and which is in accordance with the introduction of inequivalent representations. This equation reads

$$\Delta E|\mathcal{B}\rangle = \mathcal{H}(b,\partial^b)|\mathcal{B}\rangle \tag{8.149}$$

with the functional energy operator

$$\mathcal{H}(b,\partial^b) = \dot{\mathbf{H}}\left(\partial^b - \frac{1}{2}Sb\right) - \mathbf{H}\left(\partial^b + \frac{1}{2}Sb\right), \tag{8.150}$$

where $\mathbf{H}$ is the original Hamiltonian (8.144) and $S \equiv S_{KK'}$ is the right hand side of the commutator (8.146) written in the super-indices K, K'.

According to (8.150) and (8.146) we obtain the functional energy operator by means of the following substitutions performed in the original Hamiltonian:

$$A_i^a \longrightarrow \partial_{i,a}^A \pm \frac{i}{2}b_{i,a}^E, \tag{8.151}$$

$$E_i^a \longrightarrow \partial_{i,a}^E \mp \frac{i}{2}b_{i,a}^A,$$

where the upper sign belongs to the first term of the right hand side of (8.150).

In order to pursue the action of the commutator in the subsequent evaluation of (8.150) we write for the commutator (8.146)

$$\left[A_i^a(\mathbf{r}, t), E_{i'}^{a'}(\mathbf{r}', t)\right]_- = -iS_{\underset{aa'}{ii'}}(\mathbf{r}, \mathbf{r}') \tag{8.152}$$

and express the presence of the commutator in (8.151) symbolically by the operator S. For canonical quantization we just have

$$S_{\underset{aa'}{ii'}}(\mathbf{r}, \mathbf{r}') = \delta_{aa'}\delta_{ii'}\delta(\mathbf{r} - \mathbf{r}'), \tag{8.153}$$

but in the course of regularization S has to be modified. Thus we write instead of (8.151)

$$A_i^a \longrightarrow \partial_{i,a}^A \pm \frac{i}{2} S b_{i,a}^E, \tag{8.154}$$

$$E_i^a \longrightarrow \partial_{i,a}^E \mp \frac{i}{2} S b_{i,a}^A.$$

If for brevity we suppress the space coordinates in the following formula, the exact evaluation of (8.150) yields the following expression for $\mathcal{H}$:

$$\begin{aligned}
\mathcal{H}(b, \partial^b) &= -iS b_{i,a}^A \partial_{i,a}^E + i\varepsilon_{ijk}\varepsilon_{lmk} S b_{i,a}^E \partial_j \partial_l \partial_{m,a}^A \tag{8.155}\\
&\quad -\frac{i}{2} g_{\text{ph}}\varepsilon_{ijk}\varepsilon_{lmk} f_{abc} \left[2S b_{i,a}^E \partial_{j,c}^A (\partial_m \partial_{l,b}^A) + S b_{i,a}^E \partial_j (\partial_{l,b}^A \partial_{m,c}^A)\right.\\
&\quad \left. -\frac{1}{4} S b_{l,b}^E \partial_i (S b_{m,c}^E S b_{j,a}^E)\right]\\
&\quad -\frac{i}{2} g_{\text{ph}}^2 \varepsilon_{ijk}\varepsilon_{lmk} f_{abc} f_{cde}\\
&\quad \times \left[S b_{i,a}^E \partial_{m,e}^A \partial_{j,b}^A \partial_{l,d}^A - \frac{1}{4} S b_{i,a}^E S b_{j,b}^E S b_{k,e}^E \partial_{m,d}^A\right],
\end{aligned}$$

where we have to keep in mind the summations over $a, \ldots$ and $k, \ldots$ and the integration over the space coordinates $\mathbf{r}$; furthermore, we remember that $\partial_i \equiv \partial/\partial r^i$ denotes a partial derivative. The expression (8.155) corresponds exactly to the bosonic part of the functional equation (7.43), which we already derived in Section 7.2. We will soon recognize the advantages of the derivation of (8.155) with the Hamiltonian method (8.150).

The transition from (8.150) to (8.155) can be interpreted as a transformation defined by

$$X_{i,a}^1(\mathbf{r}) = \partial_{i,a}^A(\mathbf{r}) + \frac{i}{2} S b_{i,a}^E(\mathbf{r}), \tag{8.156}$$

$$X_{i,a}^2(\mathbf{r}) = \partial_{i,a}^A(\mathbf{r}) - \frac{i}{2} S b_{i,a}^E(\mathbf{r}),$$

$$Y_{i,a}^1(\mathbf{r}) = \partial_{i,a}^E(\mathbf{r}) - \frac{i}{2} S b_{i,a}^A(\mathbf{r}),$$

$$Y_{i,a}^2(\mathbf{r}) = \partial_{i,a}^E(\mathbf{r}) + \frac{i}{2} S b_{i,a}^A(\mathbf{r}).$$

Then the functional energy operator is given by

$$\mathcal{H} = \mathbf{H}\left(X^1, Y^1\right) - \mathbf{H}\left(X^2, Y^2\right) \tag{8.157}$$

with the original Hamiltonian $\mathbf{H} \equiv \mathbf{H}(A, E)$. Substitution of (8.156) into (8.157) yields $\mathcal{H}$ in the form (8.155).

Provided that S is nondegenerate we can invert (8.156) and obtain

$$\begin{aligned}
Sb_{i,a}^A &= i(Y_{i,a}^1 - Y_{i,a}^2), \\
Sb_{i,a}^E &= -i(X_{i,a}^1 - X_{i,a}^2), \\
\partial_{i,a}^A &= \frac{1}{2}(X_{i,a}^1 + X_{i,a}^2), \\
\partial_{i,a}^E &= \frac{1}{2}(Y_{i,a}^1 + Y_{i,a}^2).
\end{aligned} \tag{8.158}$$

Of course, as (8.158) is nonsingular we can substitute (8.158) into (8.155) and obtain (8.157). In this case and by this transformation from a given functional equation (8.155) its corresponding quantum Hamiltonian form (8.157) can be derived (or rederived).

This, on the other, hand is of special interest for our initially stated problem: if by Weak Mapping an effective functional equation in terms of b sources has been obtained and if one wants to explicitly know the corresponding effective quantum Hamiltonian, one can apply the reverse transformation (8.158) (in the case of $SU(2)$ vector bosons) or its equivalent in other cases, in order to try to derive the quantum Hamiltonian form of the functional operator $\mathcal{H}$.

In the course of this derivation the quantities S which describe the quantization of the theory have to be determined; thus we have obtained a general method to compare effective functional equations with corresponding quantized Hamiltonian theories.

We can, however, not exclude that there exists no corresponding exact Hamiltonian form of the effective functional theory; in this case our method allows the identification of 'disturbing' terms and thus again a comparison with a corresponding phenomenological quantum theory, including the possibility of predicting measurable corrections of the phenomenological theory.

However, in the case under consideration we can proceed in a much simpler way. As we already mentioned in its canonical form an $SU(2)$ vector boson quantum theory does not exist. Only by regularization finite values in perturbation theory and other calculational schemes can be obtained, and one method of regularization is the Pauli–Villars regularization. As we have been using a nonperturbative Pauli–Villars regularization for the spinor field, see Chapter 2, it is natural to apply the ordinary Pauli–Villars method to a vector boson gauge theory.

In perturbation theory the propagators are the principal objects of regularization. For instance in quantum electrodynamics a single subtraction in the photon propagator

$$G_{\sigma\rho}^{\text{reg}}(k) = \left[-i\left(\frac{\eta_{\sigma\rho} - k_\sigma k_\rho/\mu^2}{k^2 - \mu^2} - \frac{k_\sigma k_\rho/\mu^2}{k^2 - \mu^2/\lambda}\right)\right] - \left[\mu^2 \to \mu_1^2\right] \tag{8.159}$$

makes it smooth enough to render all diagrams convergent. *But this regularization has further consequences if one insists on a minimum of consistency.* For instance, the causal Green's function has the same spectral mass function as the commutator. This means that simultaneously with the regularization of the causal Green's function the commutator vanishes. This fact can also be verified for the spinor field (see Section 2.5), if one observes that the regularized (*i.e.*, the physical) propagator corresponds to the summation over the auxiliary field indices, see for instance (6.10)–(6.12). Hence if we consider $\mathcal{H}^{\text{reg}} := S^{-1}\mathcal{H}_{|S=0}$ we obtain for the phenomenological energy operator

$$
\begin{aligned}
\mathcal{H}^{\text{reg}}(b, \partial^b) \;=\; & -ib_{i,a}^A \partial_{i,a}^E \\
& +i\varepsilon_{ijk}\varepsilon_{lmk} b_{i,a}^E \partial_j \partial_l \partial_{m,a}^A \\
& -ig_{\text{ph}}\varepsilon_{ijk}\varepsilon_{lmk} f_{abc} \left[\frac{1}{2} b_{i,a}^E \partial_j \partial_{l,b}^A \partial_{m,c}^A + b_{i,a}^E (\partial_m \partial_{l,b}^A)\partial_{j,c}^A \right] \\
& -\frac{i}{2} g_{\text{ph}}^2 \varepsilon_{ijk}\varepsilon_{lmk} f_{abc} f_{cde} b_{i,a}^E \partial_{j,b}^A \partial_{l,d}^A \partial_{m,e}^A .
\end{aligned}
\tag{8.160}
$$

This expression exactly corresponds to (7.48) which was obtained from a functional equation of a canonically quantized phenomenological gauge boson theory by neglection of quantization terms, see Sections 7.3 and 7.7. Thus this procedure has been justified by a more detailed investigation of the quantization procedure of the phenomenological theory.

On the other hand, (7.48) was shown (see Section 7.7) to yield the effective functional energy operator (7.159) if we identify $g_{\text{ph}} = -G$. *Thus we have proved the equivalence of the effective gauge boson theory with the corresponding phenomenological theory also on the quantum level, provided the quantization of the latter is performed as we discussed above.*

Up to now we have discussed only the quantization of the phenomenological theories; there are, however, also quantization terms in the corresponding effective theories which stem from the subfermion quantization and which may give rise to phenomenological quantization effects. This question cannot be discussed in general but only with respect to a specific mapping. In our case of a combined two- and three-subfermion bound state theory there was a single quantization term $b_k Q^k$ in the bosonic sector of the functional equation (7.61) and (7.62). As this term does not reflect the field dynamics and eventually leads to a shift of the renormalized boson energy, we are justified to suppress it also in the context of the question of the effective quantization.

Returning to the phenomenological theory we can say the following: *The Hamiltonian form (8.157) holds for any kind of commutator, hence in the regularized case, too. Therefore the Hamiltonian form (8.157) also holds for (8.160) and one clearly recognizes that in the regularized case the vanishing commutator has to be replaced by the regularized propagator.*

This can be achieved by the application of perturbation theory. In that case perturbation theory is performed by the formal application of the Wick rule, but with a regularized propagator. The same recipe has to be used for effective theories: either one works in covariant perturbation theory with the regularized propagator, or one tries to directly integrate the functional equation after a formal normal ordering has been introduced.

The only difference in this procedure between phenomenological and effective theories concerns the way in which propagators are gained. Whilst in phenomenological theory the propagator is regularized *ad hoc*, in effective theory the propagator is a consequence of Weak Mapping. So in the present case the first signals that vector bosons might be composite can be seen in the propagators. Other possible sources of information about compositeness will be discussed in the next sections.

It has already been mentioned that the method just described needs not to work in any case. There might be situations where the effective dynamics does not admit a formulation by field operators at all. The conditions under which such situations occur can not be definitely fixed at present. But from our discussion it can be concluded that in order to succeed the following conditions should be satisfied by the effective theory:

i) The effective theory should lead to the corresponding classical field equations of the phenomenological theory with or without form factors.

ii) The effective propagator should be sufficiently regular, *i.e.*, the effective commutators should vanish.

iii) The remaining correction terms of the Weak Mapping should be negligible.

With respect to the influence of the exchange forces on the formalism one should expect that experimentally testable effects can be detected, but the introduction of exchange forces into the formalism of quantum field theory is a completely new territory, and one cannot decide without detailed studies whether the Hamiltonian formulation or the functional version are to be preferred for practical calculations.

8.7 Estimate of Residual Interactions

In the preceding chapters we studied effective theories under the assumption that some of the interaction terms produced by Weak Mapping are negligible compared with the leading terms for which a physical interpretation can be given. In this section we will justify this assumption, which is essential for the success of our fusion procedure, *i.e.*, its physical interpretation.

In order to avoid extensive calculations we again confine ourselves to the discussion of the effective $SU(2)$ vector boson dynamics as in all other cases the estimates run along the same lines. To be on firm mathematical ground we take the (exact) Weak Mapping theorem for bosons, Proposition 5.2, as the basis of our considerations. The leading terms of (5.25) were defined by the equation

$$
\begin{aligned}
E_{p0}|\tilde{B}(b;p)\rangle = \ & 2R^k_{I_1K}\widehat{K}_{I_1I_2}C^{I_2K}_k b_k \partial_{k'}|\tilde{B}(b;p)\rangle \\
& -\big\{ 6\widehat{W}_{I_1I_2I_3I_4}F_{I_4K}R^k_{I_1K}C^{I_2I_3}_{k'}b_k\partial_{k'} \\
& -6\,\widehat{W}_{I_1I_2I_3I_4}R^k_{I_1K}C^{[I_2I_3}_{k'_1}C^{I_4K]}_{k'_2}b_k\partial_{k'_1}\partial_{k'_2}\big\}|\tilde{B}(b;p)\rangle,
\end{aligned}
$$

$$(8.161)$$

whilst the residual interactions (including some quantization terms) are given by the remaining terms of (5.25). *From the preceding investigations we know that the use of dressed particle states (at least in lowest order) is unavoidable if one tries to come to an agreement with phenomenological theories for combined boson–fermion systems.* The effect of a mapping with dressed particle states is a modified representation of products of subfermion sources $j_{I_1} \ldots j_{I_l}$ and effective boson sources b_k, see Sections 5.6 and 5.7, including now dressed particle functions instead of products of hard core states.

From this point of view the residual interactions of (5.25) can be partly interpreted as contributions of dressing terms to effective dynamics. In particular they are correction terms to

$$6\widehat{W}_{I_1I_2I_3I_4}F_{I_4K}R_{I_1K}^k C_{k'}^{I_2I_3} b_k \partial_{k'} \;\; =: \;\; \mathcal{S}_1 , \tag{8.162}$$

$$6\widehat{W}_{I_1I_2I_3I_4}R_{I_1K}^k C_{k_1'}^{[I_2I_3} C_{k_2'}^{I_4K]} b_k \partial_{k_1'} \partial_{k_2'} \;\; =: \;\; \mathcal{I}_1 . $$

We have for instance

$$\widehat{W}_{I_1I_2I_3I_4}F_{I_4K_1}R_{I_1K_2}^{k_1} R_{K_1K_3}^{k_2} C_{k_1'}^{[I_2K_2} C_{k_2'}^{I_3K_3]} b_k b_{k_2} \partial_{k_1'} \partial_{k_2'} \tag{8.163}$$

$$\hat{=} \;\; \widehat{W}_{I_1I_2I_3I_4}F_{I_4K_1}R_{I_1K_2K_1K_3}^k C_{k_1'}^{[I_2K_2} C_{k_2'}^{I_3K_3]} b_k \partial_{k_1'} \partial_{k_2'} =: \mathcal{I}_2 , $$

where $R_{K_1K_2K_3K_4}^k \equiv R_{K_1K_2K_3K_4}^{2,k}$ is the first polarization cloud part of a dressed boson state. For the other terms we have in an abbreviated notation with the dual R_p^k of the bosonic polarization cloud

$$\widehat{W}\left(3FF + \frac{1}{4}AA\right) R^{k_1} R^{k_2} C_{k'} b_{k_1} b_{k_2} \partial_{k'} \tag{8.164}$$

$$\hat{=} \;\; \widehat{W}\left(3FF + \frac{1}{4}AA\right) R_p^k C_{k'} b_k \partial_{k'} =: \mathcal{S}_2 , $$

$$\widehat{W}\left(3FF + \frac{1}{4}AA\right) R^{k_1} R^{k_2} R^{k_3} C_{k_1'} C_{k_2'} b_{k_1} b_{k_2} b_{k_3} \partial_{k_1'} \partial_{k_2'} \tag{8.165}$$

$$\hat{=} \;\; \widehat{W}\left(3FF + \frac{1}{4}AA\right) R_p^k C_{k_1'} C_{k_2'} b_k \partial_{k_1'} \partial_{k_2'} =: \mathcal{I}_3 , $$

$$\widehat{W}\left(FF + \frac{1}{4}AA\right) FR^{k_1} R^{k_2} R^{k_3} C_{k'} b_{k_1} b_{k_2} b_{k_3} \partial_{k'} \tag{8.166}$$

$$\hat{=} \;\; \widehat{W}\left(FF + \frac{1}{4}AA\right) FR_p^k C_{k'} b_k \partial_{k'} =: \mathcal{S}_3 , $$

$$\widehat{W}\left(FF + \frac{1}{4}AA\right) FR^{k_1} \ldots R^{k_4} C_{k_1'} C_{k_2'} b_{k_1} b_{k_2} b_{k_3} b_{k_4} \partial_{k_1'} \partial_{k_2'} \tag{8.167}$$

$$\hat{=} \;\; \widehat{W}\left(FF + \frac{1}{4}AA\right) FR_p^k C_{k_1'} C_{k_2'} b_k \partial_{k_1'} \partial_{k_2'} =: \mathcal{I}_4 . $$

The essential point of this representation by means of R_p is the physical interpretation which will be given below except for

$$\widehat{W}_{I_1I_2I_3I_4}\left(F_{I_4K_2}F_{I_3K_3} + \frac{1}{4}A_{I_4K_2}A_{I_3K_3}\right) F_{I_2K_4}R_{[K_1K_2}^{k_1} R_{K_3K_4]}^{k_2} b_{k_1} b_{k_2} \tag{8.168}$$

which admits no dressing cloud interpretation.

In the dressing cloud interpretation (8.163), (8.165), and (8.167) are a first, a second, and a third order correction to the boson–boson interaction, and (8.164) and (8.166) are a first and a second order correction to the self-energy of the bosons.

In this interpretation of the effective theory, apart from (8.168) all other residual interactions are absorbed as a kind of renormalization in the physical part (8.161). For a verification of this statement one would have to directly calculate the relevant terms, *i.e.*, to show that the latter fit into the result of the 'zero order' calculation of (8.161) which was explicitly performed in the preceding chapter.

However, owing to the rather complicated structure of these residual terms such a detailed evaluation has not been set about so far. In particular, at present no dressing state calculation is available which would allow one to perform an estimate with sufficient accuracy and confidence. On the other hand, the hard core vector boson states are well known and thoroughly studied, see for instance Section 8.5. So we confine ourselves to give estimates of the left hand sides of (8.163)–(8.167), keeping in mind their physical meaning.

In this way we dispense with the attempt to show that these terms are suitable supplements to (8.161) as mentioned above. Rather, in the following we will prove that these terms are so tiny that they can be neglected whatever their algebraic structure may be.

The ingredients on the left hand sides of (8.163)–(8.168) are the propagators, the hard core states and the duals of the latter. For an estimate of the mass scale dimension it is sufficient to use rough approximations of these functions which are only assumed to correctly reproduce the mean spatial extension of the original boson wave functions. This for instance means amongst other things that we do without angular momentum terms and consider only s states and suppress the Lorentz contraction factors. All these ingredients are compiled at the end of Section 8.5.

Estimates of residual interactions were already performed in [Stu 85a] and [Pfi 90], but in view of improved wave function calculation, the introduction of physical scalar products, the derivation of exact mapping theorems and the importance of the subject, we once more perform and try to secure such estimates.

In so doing we notice and emphasize that the regularization plays an important role for the success of our approach, whereas the algebraic contributions are not essential. Whether C^A or C^F functions are used makes no great difference in the general tendency of our estimates as these functions are qualitatively not different. Hence we perform all calculations only with C^A and R^A functions and consequently omit the superspin–isospin indices as well as the spin indices, whilst the auxiliary field indices are explicitly included in the calculation.

We demonstrate our calculation techniques by two examples. For the other terms we only give the results.

In order to compare the estimates of the residual interactions with those of the leading terms we first calculate the leading boson–boson interaction term given by $\mathcal{I}_1$ which generates the effective interaction. This term is defined in (8.162). Of course, this term has already been calculated in Section 7.5, see formula (7.101). But we formulate the result of the estimate in a slightly different way as we do not explicitly calculate $\hat{C}(\mathbf{r}, \mathbf{r})$ and in addition drop the algebraic evaluation; in particular we omit the corresponding functional operators b_k and ∂_b^k as they are of no significance for the estimates. In this way we write

$$\mathcal{I}_1 \; \hat{=} \sum_{j_1, j_2, i_2, i_3, i_4} g\lambda_{j_1} \int \delta(\mathbf{u}_1 - \mathbf{r}_2)\delta(\mathbf{u}_1 - \mathbf{r}_3)\delta(\mathbf{u}_1 - \mathbf{r}_4) \tag{8.169}$$

$$\times e^{-i\mathbf{k}(\mathbf{u}_1 + \mathbf{u}_2)/2} \breve{f}_{j_1 j_2}^A (\mathbf{u}_1 - \mathbf{u}_2) N_A e^{i\mathbf{k}_1'(\mathbf{r}_2 + \mathbf{r}_3)/2} f_{i_2 i_3}^A (\mathbf{r}_2 - \mathbf{r}_3)$$

$$\times N_A e^{i\mathbf{k}_2'(\mathbf{r}_4 + \mathbf{u}_2)/2} f_{i_4 j_2}^A (\mathbf{r}_4 - \mathbf{u}_2) d^3 r_2 d^3 r_3 d^3 r_4 d^3 u_1 d^3 u_2.$$

Integrating over $\mathbf{r}_2, \mathbf{r}_3, \mathbf{r}_4$ und introducing $\mathbf{z} = \frac{1}{2}(\mathbf{u}_1 + \mathbf{u}_2)$, $\mathbf{v} = (\mathbf{u}_1 - \mathbf{u}_2)$ we obtain

$$\mathcal{I}_1 \; \hat{=} \; g\sum_{j_1, j_2} \lambda_{j_1} \int e^{-i\mathbf{k}\mathbf{z}} \breve{f}_{j_1 j_2}^A(\mathbf{v}) N_A e^{i\mathbf{k}_1'(\mathbf{z} + \frac{1}{2}\mathbf{v})} \hat{f}^A(0) \tag{8.170}$$

$$\times N_A e^{i\mathbf{k}_2'\mathbf{z}} \left(\sum_{i_4} f_{i_4 j_2}^A(\mathbf{v}) \right) d^3 z d^3 v.$$

The functions $\breve{f}^A$ and f^A are strongly concentrated around the origin $\mathbf{v} = 0$. Therefore we apply the mean value theorem for integrals and assume the mean value $\frac{1}{2}\bar{\mathbf{v}} = 0$ in the plane wave exponential. This gives

$$\mathcal{I}_1 \; \hat{=} \; g\sum_{j_1, j_2} \lambda_{j_1} \int e^{-i\mathbf{k}\mathbf{z}} N_A e^{i\mathbf{k}_1'\mathbf{z}} N_A e^{i\mathbf{k}_2'\mathbf{z}} \hat{f}^A(0) d^3 z \tag{8.171}$$

$$\times \int \breve{f}_{j_1 j_2}^A(\mathbf{v}) \left(\sum_{i_4} f_{i_4 j_2}^A(\mathbf{v}) \right) d^3 v.$$

Now we substitute $\breve{f}^A$ and f^A from (8.122) into (8.171). The remaining λ coefficients and one N_A cancel out. Thus the further calculation can be done by replacing the various m_i by an average value m as a suitable approximation. Then (8.171) goes over into

$$\mathcal{I}_1 \hat{=} gN_A \hat{f}^A(0) n_R^1 \kappa_A \delta(\mathbf{k} - \mathbf{k}_1' - \mathbf{k}_2') \int e^{-mv} K_1'(mv) d^3 v, \tag{8.172}$$

and using the derivative and the asymptotic formula for Bessel functions [Rade 97] we obtain

$$\mathcal{I}_1 \propto gN_A \hat{f}^A(0) n_R^1 m^{-2} \delta(\mathbf{k} - \mathbf{k}_1' - \mathbf{k}_2'). \tag{8.173}$$

This estimate of $\mathcal{I}_1$ is based on the introduction of center of mass and of relative coordinates, the application of the mean value theorem for strongly localized functions and the mutual compensation of λ factors and of N_A factors. Just these elements are used for the evaluation of more complicated expressions. We demonstrate this for $\mathcal{I}_2$.

Among the $\mathcal{I}_2$ terms which arise from index permutations we have the term

$$\mathcal{I}_2 = W_{K_1 I_2 I_3 I_4} F_{I_4 K_2} R_{K_1 K_3}^{k_1} R_{K_2 K_4}^{k_2} C_{k_1'}^{I_2 I_3} C_{K_2'}^{K_3 K_4} b_{k_1} b_{k_2} \partial_{k_1'} \partial_{k_2'}. \tag{8.174}$$

We calculate this term. The relevant part reads

$$\mathcal{I}_2 \;\hat{=}\; \sum_{\substack{j_1\ldots j_4 \\ i_2,i_3,i_4}} g\lambda_{j_1} \int \delta(\mathbf{u}_1-\mathbf{r}_2)\delta(\mathbf{u}_1-\mathbf{r}_3)\delta(\mathbf{u}_1-\mathbf{r}_4)F_{i_4 j_2}(\mathbf{r}_4-\mathbf{u}_2) \tag{8.175}$$

$$\times e^{-i\mathbf{k}_1(\mathbf{u}_1+\mathbf{u}_3)/2}\breve{f}^A_{j_1 j_3}(\mathbf{u}_1-\mathbf{u}_3)e^{-i\mathbf{k}_2(\mathbf{u}_2+\mathbf{u}_4)/2}\breve{f}^A_{j_2 j_4}(\mathbf{u}_2-\mathbf{u}_4)$$
$$\times N_A e^{i\mathbf{k}'_1(\mathbf{r}_2+\mathbf{r}_3)/2}f^A_{i_2 i_3}(\mathbf{r}_2-\mathbf{r}_3)N_A e^{i\mathbf{k}'_2(\mathbf{u}_3+\mathbf{u}_4)/2}f^A_{j_3 j_4}(\mathbf{u}_3-\mathbf{u}_4)$$
$$\times d^3 r_2 d^3 r_3 d^3 r_4 d^3 u_1 \ldots d^3 u_4 .$$

Substitution of (8.135) and integration over $\mathbf{r}_2, \mathbf{r}_3, \mathbf{r}_4$ und $\mathbf{u}_1$ yields

$$\mathcal{I}_2 \;\hat{=}\; g\sum_{j_1\ldots j_4}\lambda_{j_1}\lambda_{j_2}\int e^{-i\mathbf{k}_1(\mathbf{u}_2+\mathbf{u}_3)/2}\breve{f}^A_{j_1 j_3}(\mathbf{u}_2-\mathbf{u}_3)e^{-i\mathbf{k}_2(\mathbf{u}_2+\mathbf{u}_4)/2} \tag{8.176}$$

$$\times \breve{f}^A_{j_2 j_4}(\mathbf{u}_2-\mathbf{u}_4)N_A e^{i\mathbf{k}'_1 \mathbf{u}_2}\hat{f}^A(0)N_A e^{i\mathbf{k}'_2(\mathbf{u}_3+\mathbf{u}_4)/2}$$
$$\times f^A_{j_3 j_4}(\mathbf{u}_3-\mathbf{u}_4)d^3 u_2 d^3 u_3 d^3 u_4 .$$

Substitution of $\breve{f}^A$ and f^A from (8.122) into (8.176) shows that all remaining λ coefficients and all N_A factors cancel out. Hence we can approximate the various m_i values by an average value m. This gives

$$\mathcal{I}_2 \;\hat{=}\; g\int e^{-i\mathbf{k}_1(\mathbf{u}_2+\mathbf{u}_3)/2}\kappa_A m\, e^{-m|\mathbf{u}_2-\mathbf{u}_3|}e^{-i\mathbf{k}_2(\mathbf{u}_2+\mathbf{u}_4)/2} \tag{8.177}$$

$$\times \kappa_A m\, e^{-m|\mathbf{u}_2-\mathbf{u}_4|}e^{i\mathbf{k}'_1 \mathbf{u}_2}\hat{f}^A(0)e^{i\mathbf{k}'_2(\mathbf{u}_3+\mathbf{u}_4)/2}$$
$$\times m|\mathbf{u}_3-\mathbf{u}_4|^{-1}K_1(m|\mathbf{u}_3-\mathbf{u}_4|)d^3 u_2 d^3 u_3 d^3 u_4 .$$

We introduce $\mathbf{z} = \frac{1}{2}(\mathbf{u}_3+\mathbf{u}_4)$ and $\mathbf{v} = (\mathbf{u}_3-\mathbf{u}_4)$ and apply the mean value theorem with respect to the variable $\mathbf{v}$. Then (8.177) goes over into

$$\mathcal{I}_2 \;\hat{=}\; g\int e^{-i\mathbf{k}_1(\mathbf{u}_2+\mathbf{z})/2}\kappa_A m\, e^{-m|\mathbf{u}_2-\mathbf{z}|}e^{-i\mathbf{k}_2(\mathbf{u}_2+\mathbf{z})/2} \tag{8.178}$$

$$\times \kappa_A m\, e^{-m|\mathbf{u}_2-\mathbf{z}|}e^{i\mathbf{k}'_1 \mathbf{u}_2}\hat{f}^A(0)e^{i\mathbf{k}'_2 \mathbf{z}}$$
$$\times m v^{-1}K_1(m v)d^3 u_2 d^3 z d^3 v.$$

Using the asymptotic formula for K_1 we can integrate over $\mathbf{v}$, which results in

$$\mathcal{I}_2 \;\hat{=}\; gm\kappa_A^2 \int e^{-i\mathbf{k}_1(\mathbf{u}_2+\mathbf{z})/2}e^{-m|\mathbf{u}_2-\mathbf{z}|}e^{-i\mathbf{k}_2(\mathbf{u}_2+\mathbf{z})/2} \tag{8.179}$$

$$\times e^{-m|\mathbf{u}_2-\mathbf{z}|}e^{i\mathbf{k}'_1 \mathbf{u}_2}\hat{f}^A(0)e^{i\mathbf{k}'_2 \mathbf{z}}d^3 u_2 d^3 z.$$

With $\mathbf{y} = \frac{1}{2}(\mathbf{u}_2+\mathbf{z})$ and $\mathbf{w} = \mathbf{u}_2-\mathbf{z}$ we transform $\mathcal{I}_2$ and apply the mean value theorem again by assuming $\mathbf{w} = 0$ in the plane wave exponentials. Subsequent integration over $\mathbf{z}$ as well as over $\mathbf{w}$ finally gives

$$\mathcal{I}_2 \propto g\hat{f}^A(0)m^{-2}\delta(\mathbf{k}_1+\mathbf{k}_2-\mathbf{k}'_1-\mathbf{k}'_2). \tag{8.180}$$

Along these lines one can calculate the remaining terms equally well. We summarize the results for the expressions (8.162)–(8.167):

$$\mathcal{I}_1 \propto gn_R^1 m^{-2} \hat{f}^A(0) N_A \delta(\mathbf{k} - \mathbf{k}_1' - \mathbf{k}_2'), \tag{8.181}$$

$$\mathcal{I}_2 \propto gm^{-2} \hat{f}^A(0) \delta(\mathbf{k}_1 + \mathbf{k}_2 - \mathbf{k}_1' - \mathbf{k}_2'), \tag{8.182}$$

$$\mathcal{I}_3 \propto gm^{-4} n_R^1 N_A^{-1} \delta(\mathbf{k}_1 + \mathbf{k}_2 + \mathbf{k}_3 - \mathbf{k}_1' - \mathbf{k}_2'), \tag{8.183}$$

$$\mathcal{I}_4 \propto gm^{-4} \left(N_A^{-1}\right)^2 \delta(\mathbf{k}_1 + \mathbf{k}_2 + \mathbf{k}_3 + \mathbf{k}_4 - \mathbf{k}_1' - \mathbf{k}_2'), \tag{8.184}$$

and

$$\mathcal{S}_1 \propto gm \hat{f}^A(0) \delta(\mathbf{k} - \mathbf{k}'), \tag{8.185}$$

$$\mathcal{S}_2 \propto gm^{-1} N_A^{-1} n_R^1 \delta(\mathbf{k}_1 + \mathbf{k}_2 - \mathbf{k}_1'), \tag{8.186}$$

$$\mathcal{S}_3 \propto gm^{-1} \left(N_A^{-1}\right)^2 \delta(\mathbf{k}_1 + \mathbf{k}_2 + \mathbf{k}_3 - \mathbf{k}'). \tag{8.187}$$

We express these results in terms of the phenomenological coupling constant G, see (7.103). By definition we have

$$G \propto gn_R^1 m^{-2} N_A \hat{f}^A(0). \tag{8.188}$$

From (8.140) and (8.143) it follows

$$\hat{f}^A(0) \approx \mathcal{J}_0 \propto \sum_{i,j} \lambda_i \lambda_j m_i^3 m_j^3 m^{-2}, \tag{8.189}$$

and with (8.133) we have

$$n_R^1 := \sum_{i_1} \frac{1}{4} \lambda_{i_1} m_{i_1}^2. \tag{8.190}$$

Both quantities are dimensionless and thus we leave them out in the following estimates. This leads to the relations

$$\mathcal{I}_1 \propto G, \tag{8.191}$$
$$\mathcal{I}_2 \propto GN_A^{-1},$$
$$\mathcal{I}_3 \propto Gm^{-2} N_A^{-2},$$
$$\mathcal{I}_4 \propto Gm^{-2} N_A^{-3}.$$

Now according to (8.129) we have $N_A \propto m^4$, so we obtain

$$\mathcal{I}_1 \propto G, \quad \mathcal{I}_2 \propto Gm^{-4}, \quad \mathcal{I}_3 \propto Gm^{-10}, \quad \mathcal{I}_4 \propto Gm^{-14}. \tag{8.192}$$

Because of these estimates the residual interactions have exceedingly small coupling constants compared with that of the leading term and thus play no role in the sub-fermion physics of bound states. Similar considerations apply to the residual $\mathcal{S}$ terms which consequently can be neglected, too.

8.8   Estimate of Exchange Forces

In the preceding section we showed by direct calculation that the residual interactions are small in comparison with the 'physical' terms of the effective vector boson dynamics (5.25) which was obtained by an exact mapping procedure, and that for large subfermion masses the former terms can be neglected. With respect to the interpretation of the remaining terms a discussion of the state functional $|\tilde{B}\rangle$ itself is needed in order to obtain an appropriate effective theory, which can be compared with phenomenological theories.

According to (5.26) the state functional $|\tilde{B}\rangle$ contains the operator $\mathcal{U}$ which will be the subject of our discussion. By its definition (5.27), $\mathcal{U}$ is a functional metric tensor collecting the set of 'scalar products' (summation over $I_1, \ldots I_{2n}$) of the generalized determinants

$$D_{2n} \begin{pmatrix} I_1 \ldots I_{2n} \\ k_1 \ldots k_n \end{pmatrix} := C_{k_1}^{[I_1 I_2} \ldots C_{k_n}^{I_{2n-1} I_{2n}]} \tag{8.193}$$

and their dual counterparts

$$D^{2n} \begin{pmatrix} k'_1 \ldots k'_n \\ I_1 \ldots I_{2n} \end{pmatrix} := R_{[I_1 I_2}^{k'_1} \ldots R_{I_{2n-1} I_{2n}]}^{k'_n}. \tag{8.194}$$

In contrast to the case $n = 1$ which is characterized by the orthogonality condition (5.5), the scalar products of the elements (8.193) and (8.194) are *not* orthogonal in general. This non-orthogonality stems from the antisymmetrization in $I_1 \ldots I_{2n}$ and gives rise to exchange terms which need a special treatment, a fact which is well known from atomic, molecular or nuclear quantum mechanics. The effect of such exchange terms on the system dynamics are exchange forces playing an essential role in quantum mechanics. Our aim is to give an estimate of such exchange forces in comparison with the direct forces.

In order to proceed in this way we repeat the result of Proposition 5.3, which allowed the separation of the direct and of the exchange forces. We have

$$\langle D^{2n}, D_{2n} \rangle := R_{K_1 K_2}^{k'_1} \ldots R_{K_{2n-1} K_{2n}}^{k'_n} C_{k_1}^{[K_1 K_2} \ldots C_{k_n}^{K_{2n-1} K_{2n}]} \tag{8.195}$$

$$= \frac{(n!)^2}{(2n)!} \sum_{\text{part } \lambda} (-1)^j \frac{2^{2n-j}}{s_1! \ldots s_l! \, \lambda_1 \ldots \lambda_j} \begin{Bmatrix} k'_1 & \ldots & k'_n \\ k_1 & \ldots & k_n \\ \lambda_1 & \ldots & \lambda_j \end{Bmatrix}_{\text{sym}},$$

where

$$\begin{Bmatrix} k'_1 & \ldots & k'_n \\ k_1 & \ldots & k_n \\ \lambda_1 & \ldots & \lambda_j \end{Bmatrix} \tag{8.196}$$

$$:= \text{tr}\left[R^{k'_1} C_{k_1} \ldots R^{k'_{\lambda_1}} C_{k_{\lambda_1}}\right] \ldots \text{tr}\left[R^{k'_{n-\lambda_j+1}} C_{k_{n-\lambda_j+1}} \ldots R^{k'_n} C_{k_n}\right]$$

with the traces, or 'm-particle correlations',

$$\mathrm{tr}\left[R^{k'_1}C_{k_1}\dots R^{k'_m}C_{k_m}\right] := R^{k'_1}_{I_1 I_2}C^{I_2 I_3}_{k_1}R^{k'_2}_{I_3 I_4}C^{I_4 I_5}_{k_2}\dots R^{k'_m}_{I_{2m-1}I_{2m}}C^{I_{2m}I_1}_{k_m}. \qquad (8.197)$$

The numbers $s_1,\dots s_j$ are the multiplicities of the various partitions $\lambda_1,\dots\lambda_j$, and the direct force terms are characterized by $s_1 = \dots = s_j = 1$, i.e., by $\lambda_1 = \dots = \lambda_j = 1$, $j = n$; for more details we refer to Proposition 5.3.

From (8.195) it follows that the scalar product $\langle D^{2n}, D_{2n}\rangle$ generates a hierarchy of correlations between the various two-particle states $C_{k_1},\dots C_{k_n}$ and their duals $R^{k'_1},\dots R^{k'_n}$, which in turn lead to corresponding exchange forces. Common experience with such hierarchies suggests to arrange them in an ascending series of n-particle correlations and forces, starting with the two-particle correlations. Indeed we will demonstrate that this is a successful strategy for an estimate of (8.195).

Therefore in the first step we consider the n-particle correlations defined by (8.197). In addition we consider these correlations only for bound state functions. This means that the boson states $C^{I_1 I_2}_k$ have to be subdivided into a subset of bound states and a subset of scattering states. The latter subset can be neglected in all calculations if the masses m_1, m_2, m_3 of the original spinor fields are very large. In this case one can apply the decoupling theorem, see for instance Symanzik [Sym 73], Appelquist and Carrazone [App 75], Collins [Coll 87], which justifies the neglect of the contribution of the scattering states to the low energy dynamics.

Without further comment we will make use of this theorem in all further considerations. Thus, as far as the simplification of (5.25) is concerned we can omit the contribution of scattering states and for an estimate of exchange forces we will have to deal only with bound states in the low energy limit. Under these conditions such an estimate was given by Pfister and Stumpf in [Pfi 91]. In this section we will follow the expositions given in that paper.

For a first draft we will not use the realistic (i.e., complicated) boson bound state functions $C^{I_1 I_2}_k$. Rather, with respect to exchange forces we assume that the essential property of wave functions is their spatial extension. In discussing only this aspect we neglect the algebraic parts of such functions and restrict ourselves to a simple class of wave functions which are used in nuclear physics, namely oscillator functions. We thus consider the set of functions

$$\widetilde{C}^{r_1 r_2}_k := \frac{m^{3/2}}{(2\pi)^{3/2}\pi^{3/4}}\exp\left(i\mathbf{k}\mathbf{z} - \frac{m^2}{2}\mathbf{u}^2\right),\ \mathbf{k}\in\mathbb{R}^3 \qquad (8.198)$$

with $\mathbf{z} = \frac{1}{2}(\mathbf{r}_1 + \mathbf{r}_2)$, $\mathbf{u} = (\mathbf{r}_1 - \mathbf{r}_2)$ and $m = \sigma^{-1}$ for $\hbar = c = 1$, where σ is the length scale expressed by the mass scale m. The dual $\widetilde{R}^k_{r_1 r_2}$ of $\widetilde{C}^{r_1 r_2}_k$ is given by its conjugate complex function.

With the set (8.198) for any m any trace expression can be exactly calculated. However, further evaluation of the exchange terms can be considerably simplified if we consider large m. In this case we have

$$\lim_{m\to\infty} m^3 (2\pi)^{-\frac{3}{2}}\exp\left(-\frac{m^2}{2}\mathbf{u}^2\right) = \delta(\mathbf{u}), \qquad (8.199)$$

and thus we can approximately replace the oscillator functions by δ distributions, if they appear under the integral sign.

To carry out the integrations in (8.197) by means of (8.198) and (8.199) we replace the normalization factors of (8.198) and its duals by the identity $m^{3/2} = m^{-3/2}m^3$ and substitute (8.199) under the integral, with the exception of the last functions in (8.197). This yields, see [Pfi 91]

$$\text{tr}\left[\tilde{R}^{k'_1}\tilde{C}_{k_1}\ldots\tilde{R}^{k'_l}\tilde{C}_{k_l}\right] \tag{8.200}$$

$$\approx 2^{\frac{3}{2}}\left(\pi^{-\frac{3}{2}}\right)^{l-1}\delta\left(\sum_{i=1}^{l}(\mathbf{k}_i - \mathbf{k}'_i)\right)\prod_{i=1}^{l} m^3(\mathbf{k}_l)m^{-\frac{3}{2}}(\mathbf{k}_i)m^{-\frac{3}{2}}(\mathbf{k}'_i),$$

where in the original paper the masses were assumed to depend on $\mathbf{k}$. Formula (8.200) shows that the n-th order correlation function appears with the factor $m^{-3(n-1)}$ in the ascending correlation series $n = 2, 3 \ldots$ of (8.195).

To illustrate our discussion we consider the case $n = 2$. In this case our scalar product (8.195) is given by

$$\left\langle D^2, D_2\right\rangle = \frac{1}{6}\mathbf{1} + m^{-3}A(\mathbf{k}_1, \mathbf{k}_2, \mathbf{k}'_1, \mathbf{k}'_2) \tag{8.201}$$

with the direct term

$$\mathbf{1} := \sum_{p\in S_2}\delta(\mathbf{k}_1 - \mathbf{k}'_{p_1})\delta(\mathbf{k}_2 - \mathbf{k}'_{p_2}) \tag{8.202}$$

and the exchange term

$$m^{-3}A(\mathbf{k}_1, \mathbf{k}_2, \mathbf{k}'_1, \mathbf{k}'_2) := 2\sum_{p\in S_2} R^{k_1}_{I_1 I_2}C^{I_2 I_3}_{k'_{p_1}}R^{k_2}_{I_3 I_4}C^{I_4 I_1}_{k'_{p_2}}. \tag{8.203}$$

The latter term can be directly calculated. For brevity we do not explicitly record this result but only remark that it is even smaller than the estimate (8.200).

With increasing n the combinatorics in (8.195) competes with the estimates (8.200) and one has to take into account this effect on the estimates. Without going into details we refer to [Pfi 91], where this extended estimate was extensively treated.

In order to obtain an idea of the order of magnitude of such terms we rewrite the factors of (8.200) in the c.g.s.-system. In this system the quantities m in the exponentials of (8.198) have to be written as $(c/\hbar)m$, and assuming the subfermion mass m to be of the order of the Planck mass one obtains $(c/\hbar)m \approx 10^{32}\text{cm}^{-1}$. In (8.200) there appear the negative powers of this quantities, i.e., the exchange terms in (8.201) are of the order of magnitude $[(c/\hbar)m]^{-3}$, i.e., an enormous small quantity.

According to [Pfi 91] we assume these estimates to be valid also for our vector boson correlations with the corresponding Weak Mapping expansion functions C and their duals R. Generalizing the result (8.201) to arbitrary n we obtain with the neglect of exchange forces

$$\left\langle D^{2n}, D_{2n}\right\rangle \approx \mathbf{1}_n \tag{8.204}$$

with

$$\mathbf{1}_n := \frac{2^n}{(2n)!} \sum_{p \in S_n} \delta_{k'_1 k_{p_1}} \dots \delta_{k'_n k_{p_n}} .$$ (8.205)

In this approximation we have, according to (5.26) and (5.27),

$$|\tilde{\mathcal{B}}(b; a)\rangle \equiv \mathcal{U}|\mathcal{B}(b; a)\rangle \approx |\hat{\mathcal{B}}(b; a)\rangle$$ (8.206)

with the new state functional

$$|\hat{\mathcal{B}}(b; a)\rangle := \sum_{n=0}^{\infty} \frac{1}{n!} \frac{2^n n!}{(2n)!} \rho^{(n)}(k_1 \dots k_n | a) b_{k_1} \dots b_{k_n} |0\rangle_{\mathrm{B}}$$ (8.207)

$$=: \sum_{n=0}^{\infty} \frac{1}{n!} \hat{\rho}^{(n)}(k_1 \dots k_n | a) b_{k_1} \dots b_{k_n} |0\rangle_{\mathrm{B}},$$

where $\rho^{(n)}(k_1 \dots k_n | a)$ are the coefficient functions of the functional state $|\mathcal{B}(b; a)\rangle$ from (5.14); i.e., $|\hat{\mathcal{B}}(b; a)\rangle$ is modified by some statistical factors. Replacing $|\tilde{\mathcal{B}}(b; a)\rangle$ in (5.25) by $|\hat{\mathcal{B}}(b; a)\rangle$ we obtain the effective boson functional equation with the neglect of exchange forces.

Returning to the general problem of the *inclusion* of exchange terms and forces we observe that we can write (5.25) in the form

$$E_{p0}\mathcal{U}|\mathcal{B}(b; p)\rangle = \mathcal{H}_b\mathcal{U}|\mathcal{B}(b; p)\rangle.$$ (8.208)

In accordance with quantum mechanical calculations of exchange forces one is tempted to calculate $\mathcal{U}^{-1}$ in order to obtain the effective Hamiltonian form

$$E_{p0}|\mathcal{B}(b; p)\rangle = \mathcal{U}^{-1}\mathcal{H}_b\mathcal{U}|\mathcal{B}(b; p)\rangle.$$ (8.209)

But the calculation of $\mathcal{U}^{-1}$ would be a formidable task and we do not feel that this is very instructive. Rather we remember that we have arranged (8.195) in terms of increasing correlation functions which are connected with an increasing negative power of $(c/\hbar)m$. The smallness of these powers suggests to proceed in the following way: we project (8.208) into configuration space and obtain

$$E_{p0}U_{k_1 \dots k_n, k'_1 \dots k'_n} \rho^{(n)}_{k'_1 \dots k'_n} = \sum_{m=1}^{n} H^b_{k_1 \dots k_n, k'_1 \dots k'_m} U_{k'_1 \dots k'_m, k''_1 \dots k''_m} \rho^{(n)}_{k''_1 \dots k''_m} ,$$ (8.210)

where U is just the product $\langle D^n, D_n \rangle$. For sufficiently small (and even) n we consider only two-boson correlations and neglect the combinatorics of the higher order contributions. Then for these n the exchange matrix U can be written in the (abbreviated) form

$$U = \mathbf{1}_n + m^{-3}A,$$ (8.211)

where $A \equiv A(I_1 \dots I_n, I'_1 \dots I'_n)$ are the two-body correlation operators. Then we define

$$U^{-1} := \mathbf{1}_n - m^{-3}A$$ (8.212)

and obtain

$$U^{-1}U \; = \; 1_n - m^{-6}A^2 \tag{8.213}$$
$$\approx \; 1_n \,.$$

Thus the definition of U^{-1} is justified in an approximation where we neglect terms of the order m^{-6}. Therefore the effective Hamiltonian of (8.209) with inclusion of exchange terms can be calculated in configuration space up to the order n. This in general suffices for performing low order perturbation calculations, being one of the important applications of effective Hamiltonians.

9 Fermions and Gravitation

9.1 Introduction

In our atomistic program, which is based on de Broglie's fusion theory and Heisenberg's unified spinor field project, we intend to derive the interactions between observable physical particles as effects of an underlying subfermion theory. In the preceding chapters we applied this idea to the electro–weak interactions, and we succeeded in reproducing the main structure of the electro–weak standard model of elementary particles.

In addition, first attempts were made to include gluons into this scheme, see Stumpf [Stu 87]. Hence one can expect to obtain all forces (and particles) of the standard model in this way, and apart from the problem of studying the corrections of the standard model which are induced by the subfermion structure, it is, of course, of interest to know whether or not gravitational forces can also be explained by fusion theory. This application was already proposed and performed by de Broglie (see Chapter 1) and by Heisenberg.

However, in some respect the application of fusion theory to gravitation seems to be fundamentally different from the discussion of the effective standard model. Since Einstein gravitation theory has been regarded as the prototype of a geometrical theory. On the other hand, the fundamental interactions which govern the physics of microscopic systems are rather successfully described in terms of special relativistic fields. Thus there seems to be a conceptual difference between the large scale gravitation theory and microscopic physics which one is faced with in any attempt of a unification of gravitation and the other fundamental interactions.

Following and generalizing the fusion concept of de Broglie we intend to derive a gravitation theory as an effective theory for bound states of an underlying nonlinear spinor field theory in *flat* space. But owing to the universal coupling of the gravitational force to matter any theory of gravitation is strongly related to geometry: the gravitational force acts on all material standards of length and time. This must be reflected in the geometrical structure. In order to incorporate this effect, the consequence of our field-theoretic ansatz is that the observable geometry turns out to be an effective geometry with respect to a fixed flat (and unobservable) background metric (see also Sexl and Urbantke [Sexl 87]).

In addition, this field-theoretic approach to gravitation theory is motivated by several well known problems of conventional gravitation theory. In the Einstein theory of gravitation the energy–momentum of the gravitational field is described only by a pseudo-tensor, which prevents the formulation of proper conservation laws. In contrast, such conservation laws are considered to be important properties of conventional field theories in flat Minkowski space.

Further fundamental difficulties arise if one tries to apply the microscopic concept of quantization to gravitation theories. The Einstein–Hilbert Lagrangian induces a nonrenormalizable quantum theory, and taking seriously the geometrical interpretation of gravitation in the microscopic domain leads to contradictions: the quantization of a metric would lead to fluctuations of this metric and thus would destroy the possibility of conservation laws which are based on a rigid metric structure and which are necessary for the formulation of a quantum theory. In addition, the dependence of the metric on the amount of matter also destroys this possibility.

On the other hand, the quantum concept is well established and tested in microscopic physics; thus in the abovementioned sense we assume the primacy of quantum theory over geometrization; compare also Zel'dovich and Grishchuk [Zel 86], Stumpf and Borne [Stu 94], Wiesendanger [Wie 96].

There have been several attempts to formulate a gravitation theory in a field-theoretic flat space framework, see for instance Gupta [Gup 57], Thirring [Thir 59], [Thir 61], Weinberg [Wein 65], Wyss [Wyss 65], Mittelstaedt and Barbour [Mit 67], Deser and Laurent [Des 68], Dehnen and Ghaboussi [Dehn 85], Logunov [Log 93]. We take over the field-theoretic ansatz of these authors according to which gravitation is described by fields in an unobservable flat pseudo-Euclidean space; if the equations of motion of some 'matter' coupled to the gravitational fields are to be interpreted as equations in a curved (Riemann) geometry, this observable geometry is induced by the gravitational fields.

In our context the derivation of effective gravitational dynamics is governed by the idea of bound states of subfermions and is again performed by means of Weak Mapping. First applications to the problem of gravitation were given in [Stu 88], [Stu 93b]. In our presentation we follow Borne [Bor 97] and derive the effective functional equations for a coupled system of gravitons and elementary fermions. As we restrict ourselves to the discussion of a classical gravitation theory, we extract the classical part of these equations.

For the evaluation of the resulting equations we have to carefully discuss the bound 'graviton' states and their relation to the effective gravitational quantities. The evaluation of the linear part of the bosonic equations will lead to the linear Ricci and Bianchi identities of a curved Riemann (or Riemann–Cartan) space, whilst from the fermion equations we obtain a general covariant Dirac equation, i.e., the minimal coupling of the gravitational field to a 'matter' field according to the equivalence principle.

The effective equations turn out to be equations for geometrical quantities in an *anholonomic* coordinate system which allows the treatment of spinors in curved spaces; this is a result of the flat Minkowski structure of the underlying subfermion theory.

9.2 Anholonomic Gravitation Theory

The connection between our field-theoretic framework in flat Minkowski space and the curved geometries of gravitation theory is established by the introduction of anholonomic coordinates or tetrad systems in the latter. Thus we give a short introduction into the 'phenomenological' gravitation theory in anholonomic coordinates, see for instance Schouten [Schou 54], and in this context we restrict ourselves to the case of the Einstein theory.

The Einstein theory is formulated in a Riemann space V_4 with metric tensor $g_{ij}(x)$, where small Latin indices $i, j \ldots = 0, 1, 2, 3$ denote holonomic quantities. This tensor satisfies the metricity condition

$$\overset{\Gamma}{\nabla}_k g_{ij}(x) = 0 \tag{9.1}$$

with the general covariant derivative $\overset{\Gamma}{\nabla} = \partial + \Gamma$. For a Riemann space one has the metric affine connection

$$\Gamma^k_{ij} = \left\{ \begin{matrix} k \\ ij \end{matrix} \right\} \tag{9.2}$$

where the Christoffel symbol is connected with the metric tensor by

$$\left\{ \begin{matrix} k \\ ij \end{matrix} \right\} := \frac{1}{2} g^{kl} \left(\partial_i g_{jl} + \partial_j g_{il} - \partial_l g_{ij} \right), \tag{9.3}$$

and one has a symmetric connection $\Gamma^k_{[ij]} = 0$. The Riemann–Christoffel, or curvature, tensor can be defined by the so called Ricci identities

$$R_{ijk}{}^l(x) := 2\partial_{[i}\Gamma^l_{j]k}(x) + 2\Gamma^l_{[i|m|}(x)\Gamma^m_{j]k}(x). \tag{9.4}$$

In a Riemann space V_4 one has the well known symmetry and antisymmetry properties

$$R_{(ij)kl} = 0, \tag{9.5}$$
$$R_{ij(kl)} = 0,$$
$$R_{[ijk]l} = 0,$$
$$R_{ijkl} = R_{klij}.$$

The integrability conditions for the curvature tensor are the Bianchi identities

$$\overset{\Gamma}{\nabla}_{[i} R_{jk]l}{}^m = 0. \tag{9.6}$$

The Einstein field equations are given by

$$R_{ij} - \frac{1}{2} g_{ij} R = \kappa T_{ij} \tag{9.7}$$

with the matter energy–momentum tensor T_{ij}, the Ricci tensor $R_{ij} := R_{ikj}{}^k$, and the curvature scalar $R := R_i{}^i$.

The Weyl tensor C_{ijkl} is given by the trace-free part of the curvature tensor:

$$C_{ijkl} := R_{ijkl} - g_{i[k}R_{l]j} - g_{j[l}R_{k]i} + \frac{1}{3}R\, g_{i[k}g_{l]j} \tag{9.8}$$

with $C_{ij}{}^i{}_l = 0$.

We introduce an anholonomic coordinate system on the manifold by means of the tetrads $e_\mu{}^i$ with the anholonomic indices $\mu = 0,1,2,3$. Together with their duals $e^\mu{}_i$ they satisfy the following set of equations:

$$
\begin{aligned}
e_\mu{}^i(x)e^\mu{}_j(x) &= \delta^i_j, \\
e_\mu{}^i(x)e^\nu{}_i(x) &= \delta^\mu_\nu, \\
e_\mu{}^i(x)e_\nu{}^j(x)g_{ij}(x) &= \eta_{\mu\nu}, \\
e^\mu{}_i(x)e^\nu{}_j(x)\eta_{\mu\nu} &= g_{ij}(x),
\end{aligned}
\tag{9.9}
$$

with $\eta_{\mu\nu} = \text{diag}(1,-1,-1,-1)$.

With respect to their anholonomic indices tetrads transform as vectors under Lorentz transformations and thus allow the representation of spinors.

The anholonomic affine connection is given by resolving the anholonomic metricity condition

$$\overset{\Gamma}{\nabla}_\mu \eta_{\nu\rho} = 0 \tag{9.10}$$

as

$$\Gamma_{\mu\nu\rho} = -\Omega_{\mu\nu\rho} + \Omega_{\nu\rho\mu} - \Omega_{\rho\mu\nu} \tag{9.11}$$

with the object of anholonomity

$$\Omega^\rho_{\mu\nu} := e_\mu{}^i e_\nu{}^j \partial_{[i}e^\rho{}_{j]} \tag{9.12}$$

and $\Omega_{\mu\nu\rho} := \eta_{\rho\sigma}\Omega^\sigma_{\mu\nu}$. Some important properties of the anholonomic connection are

$$
\begin{aligned}
\Gamma_{\mu(\nu\rho)} &= 0, \\
\Gamma_{[\mu\nu]\rho} &= -\Omega_{\mu\nu\rho}, \\
\Gamma_{[\mu\nu\rho]} &= -\Omega_{[\mu\nu\rho]}, \\
\eta^{\mu\nu}\Gamma_{\mu\nu\rho} &= -2\Omega^\mu_{\rho\mu}.
\end{aligned}
\tag{9.13}
$$

The curvature tensor $R_{\mu\nu\rho\sigma}$ in anholonomic coordinates is given by the anholonomic Ricci identities

$$R_{\mu\nu\rho\sigma} = 2\partial_{[\mu}\Gamma_{\nu]\rho\sigma} + 2\Gamma_{[\mu|\lambda\rho|}\Gamma^\lambda_{\nu]\sigma} - 2\Gamma^\lambda_{[\mu\nu]}\Gamma_{\lambda\rho\sigma} \tag{9.14}$$

and the anholonomic Bianchi identities read

$$
\begin{aligned}
\overset{\Gamma}{\nabla}_{[\alpha} R_{\mu\nu]\rho\sigma} &\equiv \partial_{[\alpha}R_{\mu\nu]\rho\sigma} + \Gamma_{[\alpha|\lambda\rho}R^\lambda{}_{\sigma|\mu\nu]} - \Gamma_{[\alpha|\lambda\sigma}R^\lambda{}_{\rho|\mu\nu]} + 2R_{\rho\sigma}{}^\lambda{}_{[\nu}\Gamma_{\alpha|\lambda|\mu]} \\
&= 0.
\end{aligned}
\tag{9.15}
$$

With respect to our Weak Mapping procedure we prefer a formulation of Einstein's gravitation theory in terms of the Weyl tensor and the affine connection in anholonomic coordinates. This can be achieved by inserting the anholonomic versions of (9.8) and of the field equations (9.7) into (9.14) and (9.15); one obtains the equations

$$C_{\mu\nu\rho\sigma} - 2\partial_{[\mu}\Gamma_{\nu]\rho\sigma} - 2\Gamma_{[\mu|\lambda\rho}\Gamma^{\lambda}_{\nu]\sigma} + 2\Gamma^{\lambda}_{[\mu\nu]}\Gamma_{\lambda\rho\sigma} \tag{9.16}$$
$$= \kappa\left(-\eta_{\mu[\rho}T_{\sigma]\nu} - \eta_{\nu[\sigma}T_{\rho]\mu} + \frac{2}{3}T\eta_{\mu[\rho}\eta_{\sigma]\nu}\right)$$

and

$$\partial_{\sigma}C_{\mu\nu\rho}{}^{\sigma} + 2\Gamma_{\sigma\lambda[\mu}C^{\lambda}{}_{\nu]\rho}{}^{\sigma} + \Gamma_{\sigma\lambda\rho}C_{\mu\nu}{}^{\lambda\sigma} + \Gamma^{\sigma}_{\lambda\sigma}C_{\mu\nu\rho}{}^{\lambda} \tag{9.17}$$
$$= \kappa\left(2\partial_{[\mu}T_{\nu]\rho} - 2\Gamma^{\lambda}_{[\mu|\rho}T_{\nu]\lambda} - 2\Gamma^{\lambda}_{[\mu\nu]}T_{\lambda\rho} + \frac{1}{3}\partial_{[\mu}T\eta_{\nu]\rho}\right).$$

According to [Edg 80] together with (9.11) these equations are a redundant but complete set of equations for the formulation of the Einstein theory of gravitation. It is this formulation which is suitable for the comparison with our effective gravitation equations.

According to the equivalence principle field equations of matter in a gravitational field are obtained by the transition from Lorentz covariant equations to equations, which are covariant with respect to general coordinate transformations. The general covariance is achieved by minimal coupling of the matter fields to the affine connection of the Riemann–(Cartan) space. In the case of spinorial matter the general covariant derivative is given by

$$D_{\mu}\Psi(x) = \partial_{\mu}\Psi(x) + \frac{i}{4}\Gamma_{\mu\nu\sigma}(x)\Sigma^{\nu\sigma}\Psi(x) \tag{9.18}$$

with the anholonomic affine connection $\Gamma_{\mu\nu\sigma}$ and the generators of the Lorentz group $\Sigma^{\mu\nu} = \frac{i}{2}[\gamma^{\mu}, \gamma^{\nu}]_{-}$, see Brill and Wheeler [Brill 57]. Thus the free Dirac equation in a Riemann space V_4 (and in a Riemann–Cartan space) reads in anholonomic coordinates, i.e., in tangent space (see e.g. Hehl and Datta [Hehl 71], [Dat 71]):

$$\left[i\gamma^{\mu}\partial_{\mu} - m - \frac{1}{4}\Gamma_{\mu\rho\sigma}(x)\gamma^{\mu}\Sigma^{\rho\sigma}\right]\Psi(x) = 0. \tag{9.19}$$

Furthermore, the parity transformed equation (9.19) can be shown to read

$$\left[i\gamma^{\mu}\partial_{\mu}(x') - m - \frac{1}{4}\Gamma_{\mu\rho\sigma}(x')\Sigma^{\rho\sigma}\gamma^{\mu}\right]\Psi'(x') = 0, \tag{9.20}$$

where Ψ is the parity transformed spinor $\Psi'(x') = \eta_p\gamma^0\Psi(x)$ with the phase η_p and $x' = (x^0, -x^1, -x^2, -x^3)$.

It is this equation which we will reproduce as an effective equation for the coupling of elementary fermions to four-subfermion bound states.

9.3 Weak Mapping with Gravitons

We consider the occurrence of the gravitational force as a composite particle effect resulting from the subfermion dynamics (5.1). The gravitational force is assumed to be mediated by gravitons, which are generated by bound states of four elementary subfermion fields in accordance with the spin fusion theory of de Broglie (see Chapter 1). In order to perform our program we have to define four-particle bound states ('graviton' states) and to discuss the relation between these states and conventional graviton states with zero mass and spin 2, and we have to derive dynamical equations for these bound states from our basic spinor field dynamics.

This subfermion dynamics is governed by equation (5.1), which describes all possible reactions and processes between the elementary fermions. Among these reactions there are bound state processes. The suitable means for the extraction of such processes, *i.e.*, for the derivation of an effective dynamics of composite particles, is Weak Mapping.

Weak Mapping was defined in Chapter 5 as a rearrangement of the subfermion dynamics with respect to certain bound states. This rearrangement can be achieved by a mapping of the subfermion functional equation onto an effective functional equation for the corresponding bound states. As we intend to explain gravitation as a four-particle bound state effect, we choose these bound states to be just the four-subfermion bound states of Section 6.6 which we couple to fermion 'matter' fields. For simplicity we represent this fermion matter by elementary subfermions.

Owing to our fusion program any kind of matter (or energy) is represented by subfermion complexes and owing to the universal coupling of these subfermions to gravitation this suffices to guarantee the correct coupling of any matter to gravitation.

Exact weak mapping theorems were derived for the two-fermion hard core bound state dynamics in Chapter 5 and for the combined two-fermion and three-fermion hard core bound state dynamics by Kerschner [Ker 94]. These theorems lead to rather complicated effective theories. If exchange effects between the bound states can be approximately neglected the effective dynamics is essentially simplified and can be derived by a short cut functional calculation technique, see Section 5.6. By this chain rule method one can treat any kind of composite particle configurations. Here we apply it to gravitational mapping; the derivation of the effective mapping equations is in full analogy to the Weak Mapping of Chapter 7.

In accordance with previous investigations by Stumpf [Stu 88] for a successful derivation of an effective gravitation theory we have to deal with dressed bound states. The formal theory of Weak Mapping with dressed particle states was developed in [Stu 87] and [Stu 94], see Section 5.7. As solutions of the full equation (5.1), dressed particle states contain an infinite number of polarization cloud parts, which are induced by a hard core part. For our application we consider the polarization cloud formalism only in the lowest order. This can be justified by an estimate of higher polarization cloud terms [Stu 94] and is in full analogy to the Weak Mapping of Chapter 7.

The effective coupled graviton–fermion system is again described in a functional space. The corresponding functional states are given by

$$|\mathcal{G}(b,f;a)\rangle := \sum_{m,n} \frac{i^n}{n!}\frac{1}{m!}\Theta^{(m,n)}(s_1\ldots s_m; l_1\ldots l_n, t|a)b_{s_1}\ldots b_{s_m}f_{l_1}\ldots f_{l_n}|v\rangle \quad (9.21)$$

with bosonic 'graviton' functional operators b_s and fermion operators f_l. The functions $\Theta^{(m,n)}$ are the correlation functions of the coupled boson–fermion system; they have to be interpreted as matrix elements of corresponding phenomenological field operators describing these particles. The indices s and l are induced by the indices and the quantum numbers of corresponding bosonic and fermionic bound states.

Together with corresponding functional annihilation operators ∂_s^b and ∂_l^f, respectively, the bosonic functional operators b_s generate a CCR algebra, whereas the effective fermion functional operators f_l generate a CAR algebra. Both algebras are represented on the direct product of two Fock spaces with the cyclic vacuum state $|v\rangle$. For a more detailed discussion of the functional space construction we refer to Chapter 3.

We remark that a mapping for four-particle states should start from a modified normal ordered subfermion functional equation, in which besides the two-particle correlations also the four-particle correlations are extracted. This 'normal ordering' can be performed without further difficulties analogously to the normal ordering of Section 3.5. However, it can be shown that this additional normal ordering does not effect the section of the effective equations which we will consider in the following; thus for brevity in the following we do not take into account this extended normal ordering but start from the subfermion equation (5.1).

The transformation of the subfermion functional equation (5.1) into a functional equation for the effective states (9.21) is performed by means of the relations

$$b_s = C_{4,s}^{I_1 I_2 I_3 I_4} j_{I_1} j_{I_2} j_{I_3} j_{I_4} \quad (9.22)$$

and

$$f_l = C_{1,l}^{I_1} j_{I_1}. \quad (9.23)$$

The functions $C_{4,s}$ and $C_{1,l}$ are four-particle and 'one-particle' hard core bound state functions, which were introduced in Chapter 6 and which will be further discussed in Section 9.4; in particular the characterizing indices s and l have to be specified. Apart from bound states the solutions of the diagonal part of (5.1) contain also scattering states. However, owing to the high subfermion masses and to decoupling theorems these scattering parts will be suppressed in the final evaluation.

The polarization cloud terms appear in the inverse relations

$$j_{I_1}\ldots j_{I_{2n-1}} = \sum_l R_{I_1\ldots I_{2n-1}}^{1,l} f_l, \quad (9.24)$$

$$j_{I_1}\ldots j_{I_{4n}} = \sum_s R_{I_1\ldots I_{4n}}^{4,s} b_s \quad (9.25)$$

for $n \in \mathbb{N}$.

The functions $R^{1,l}_{I_1\dots I_{2n-1}}$ are duals corresponding to the fermion polarization cloud parts of the order $(2n-1)$, whereas the functions $R^{4,s}_{I_1\dots I_{4n}}$ are duals for the bosonic polarization clouds. Together with the transformations (9.22) and (9.23) these relations define a consistent lowest order approximation of the full dressed particle formalism in the present case of a coupled system of four-particle bound states and elementary fermions.

For the hard core functions we assume the orthogonality relations

$$R^{1,l}_{I_1} C^{I_1}_{1,l'} = \delta_{ll'} \tag{9.26}$$

and

$$R^{4,s}_{I_1 I_2 I_3 I_4} C^{I_1 I_2 I_3 I_4}_{4,s'} = \delta_{ss'} , \tag{9.27}$$

which have to be considered as approximations of the corresponding dressed particle expressions (5.48) in lowest order.

The mapping of the subfermion functional equation (5.1) onto an effective functional equation

$$\frac{\partial}{\partial t}|\mathcal{G}(b,f;a)\rangle = \mathcal{H}_{\mathrm{GF}}\big(b,f,\partial^b,\partial^f\big)|\mathcal{G}(b,f;a)\rangle \tag{9.28}$$

is performed by means of the invariance relation

$$|\mathcal{F}(j;a)\rangle = |\mathcal{G}(b,f;a)\rangle \tag{9.29}$$

with respect to transformations defined by (9.22) and (9.23). In a second step the functional subfermion Hamiltonian $\mathcal{H}_{\mathrm{F}}$ is transformed into the effective Hamiltonian $\mathcal{H}_{\mathrm{GF}}$ by means of (9.24), (9.25) and the functional chain rule.

As we already mentioned this chain rule is a short cut method of Weak Mapping which already includes the neglect of exchange forces in a low energy limit. In our case the functional chain rule reads

$$\partial_I|\mathcal{G}(b,f;a)\rangle = \big[(\partial_I b_s)\partial^b_s + (\partial_I f_l)\partial^f_l\big]|\mathcal{G}(b,f;a)\rangle \tag{9.30}$$
$$= \big[4C^{IK_1K_2K_3}_{4,s} j_{K_1} j_{K_2} j_{K_3}\partial^b_s + C^I_{1,l}\partial^f_l\big]|\mathcal{G}(b,f;a)\rangle.$$

Repeated application of the chain rule yields $\partial_{I_1}\partial_{I_2}|F\rangle$ and $\partial_{I_1}\partial_{I_2}\partial_{I_3}|F\rangle$ in terms of effective boson and fermion operators acting on the functional state $|\mathcal{G}\rangle$.

By means of these transformations we obtain the following effective functional equation:

$$E_{p0}|\mathcal{G}(b,f;p)\rangle \tag{9.31}$$
$$= \Big\{\widehat{K}_{I_1 I_2}\big[4C^{I_2 I_3 I_4 I_5}_{4,s'} R^{4,s}_{I_1 I_3 I_4 I_5} b_s\partial^b_{s'} + C^{I_2}_{1,l'} R^{1,l}_{I_1} f_l\partial^f_{l'}\big]$$
$$+\widehat{W}_{I_1 I_2 I_3 I_4}\big[24C^{I_2 I_3 I_4 K}_{4,s} R^{1,l_1}_{I_1} R^{1,l_2}_{K} f_{l_1} f_{l_2}\partial^b_s$$
$$+144 C^{I_3 I_4 K_1 K_2}_{4,s'_1} C^{I_2 K_3 K_4 K_5}_{4,s'_2} R^{4,s}_{I_1 K_1 K_2 K_3 K_4 K_5} b_s\partial^b_{s'_1}\partial^b_{s'_2}$$

$$+36C_{4,s}^{I_3I_4K_1K_2}C_{1,l'}^{I_2}R_{I_1K_1K_2}^{1,l}f_l\partial_{l'}^f\partial_s^b$$

$$-12C_{4,s'}^{I_4K_1K_2K_3}C_{1,l_1}^{I_3}C_{1,l_2}^{I_2}R_{I_1K_1K_2K_3}^{4,s}\partial_{l_2}^f\partial_{l_1}^fb_s\partial_{s'}^b$$

$$+C_{1,l_1}^{I_4}C_{1,l_2}^{I_3}C_{1,l_3}^{I_2}R_{I_1}^{1,l''}f_{l'}\partial_{l_1}^f\partial_{l_2}^f\partial_{l_3}^f$$

$$-3F_{I_4K_1}\left(12C_{4,s}^{I_2I_3K_2K_3}R_{I_1K_1K_2K_3}^{4,s'}b_{s'}\partial_s^b+C_{1,l_1}^{I_2}C_{1,l_2}^{I_3}R_{I_1}^{1,l_1''}R_{K_1}^{1,l_2''}f_{l_1'}f_{l_2'}\partial_{l_2}^f\partial_{l_1}^f\right)$$

$$+\left(3F_{I_4K_1}F_{I_3K_2}+\frac{1}{4}A_{I_4K_1}A_{I_3K_2}\right)$$

$$\times\left(C_{1,l'}^{I_2}R_{I_1K_1K_2}^{1,l}f_l\partial_{l'}^f+4C_{4,s}^{I_2K_3K_4K_5}R_{I_1K_1K_2K_3K_4K_5}^{4,s'}b_{s'}\partial_s^b\right)\qquad.$$

$$-(F_{I_4K_1}F_{I_3K_2}+\frac{1}{4}A_{I_4K_1}A_{I_3K_2})F_{I_2K_3}R_{I_1K_1K_2K_3}^{4,s}b_s\bigg]\bigg\}|\mathcal{G}(b,f;p)\rangle,$$

where with the exception of first order polarization cloud terms the higher order polarization cloud terms were neglected. The quantities $\widehat{K}$ and $\widehat{W}$ are defined by (2.82) and (2.83), F and A are given by (7.79) and (2.75). We again mention that we have omitted the single time index t.

We consider equation (9.31) as a formulation of the complete dynamics of the system of four-particle bound states and elementary fermions in a low energy limit, including also quantization effects. For a first evaluation we compare this dynamics with a phenomenological classical dynamics of gravitons and fermions; thus we have to extract a classical part from the functional equation (9.31).

This can be achieved in analogy to the vector boson case in Section 7.5 by means of the ansatz

$$|\mathcal{G}(b,f;a)\rangle=\exp\left(Z_0[b,f;a]\right)|v\rangle \tag{9.32}$$

with

$$Z_0[b,f;a]:=if_l\Theta_l^f(t)+b_s\Theta_s^b(t), \tag{9.33}$$

where Θ^f and Θ^b are classical field variables of the fermion and the boson dynamics. By comparing the ansatz (9.32) with the original functional states (9.21) one verifies that the effect of this ansatz is the transition from one-particle matrix elements $\langle0|\chi_s|a\rangle$ to classical functions χ_s and a factorization of higher order correlation functions $\Theta^{(m,n)}$ with respect to these functions.

Furthermore, in analogy to equation (7.105) we apply the equation

$$E_{po}|\mathcal{G}(b,f;p)\rangle=i\left[b_s\frac{\partial}{\partial t}\partial_s^b+f_\ell\frac{\partial}{\partial t}\partial_\ell^f\right]|\mathcal{G}(b,f;p)\rangle \tag{9.34}$$

to the right hand side of equation (9.31). Afterwards we collect all terms linear in b_s or f_ℓ, respectively, and satisfy the corresponding expressions separately.

Substitution of (9.32), (9.33) into these expressions and projection onto one-particle states then yields the classical boson equation

$$i\frac{\partial}{\partial t}\Theta_s^b=4\widehat{K}_{I_1I_2}C_{4,s'}^{I_2I_3I_4I_5}R_{I_1I_3I_4I_5}^{4,s}\Theta_{s'}^b \tag{9.35}$$

$$+\widehat{W}_{I_1I_2I_3I_4}\Big\{-12C_{4,s'}^{I_4K_1K_2K_3}C_{1,l_1}^{I_3}C_{1,l_2}^{I_2}R_{I_1K_1K_2K_3}^{4,s}\Theta_{l_2}^f\Theta_{l_1}^f\Theta_{s'}^b$$

$$-36F_{I_4K_1}C_{4,s'}^{I_2I_3K_2K_3}R_{I_1K_1K_2K_3}^{4,s}\Theta_{s'}^b$$

$$+144C_{4,s_1'}^{I_3I_4K_1K_2}C_{4,s_2'}^{I_2K_3K_4K_5}R_{I_1K_1K_2K_3K_4K_5}^{4,s}\Theta_{(s_1'}^b\Theta_{s_2')}^b$$

$$+4(3F_{I_4K_1}F_{I_3K_2}+\frac{1}{4}A_{I_4K_1}A_{I_3K_2})$$

$$\times C_{4,s'}^{I_2K_3K_4K_5}R_{I_1K_1K_2K_3K_4K_5}^{4,s}\Theta_{s'}^b\Big\},$$

and the classical fermion equation

$$i\frac{\partial}{\partial t}\Theta_l^f = \widehat{K}_{I_1I_2}C_{1,l'}^{I_2}R_{I_1}^{1,l}\Theta_{l'}^f \tag{9.36}$$

$$+\widehat{W}_{I_1I_2I_3I_4}\Big\{36C_{4,s}^{I_3I_4K_1K_2}C_{1,l'}^{I_2}R_{I_1K_1K_2}^{1,l}\Theta_{l'}^f\Theta_s^b$$

$$-C_{1,l_1'}^{I_4}C_{1,l_1'}^{I_3}C_{1,l_3'}^{I_2}R_{I_1}^{1,l}\Theta_{l_3'}^f\Theta_{l_2'}^f\Theta_{l_1'}^f$$

$$+(3F_{I_4K_1}F_{I_3K_2}+\frac{1}{4}A_{I_4K_1}A_{I_3K_2})R_{I_1K_1K_2}^{1,l}C_{1,l'}^{I_2}\Theta_{l'}^f\Big\}.$$

One can easily understand the meaning of the various terms: the first term of (9.35) is the kinetic boson term, the third a linear and the fourth a quadratic boson term, whilst the second term describes the coupling of a fermion current to the boson equation and the last term is an additional linear term stemming from the subfermion quantization.

Analogously the first term of (9.36) is the kinetic fermion term and the second describes the coupling of bosons to the fermion equations. The last terms are the residual fermion interaction and a linear correction term, which leads to a fermion mass correction. The latter terms give no essential contribution to the investigation of gravitational interactions and will be omitted in the following.

9.4 Graviton States

For the evaluation of the various terms of the effective equations (9.35) and (9.36) we need the explicit form of the hard core states and the first order polarization cloud parts. However, as we restrict ourselves to the linear part of the boson equation, we have to discuss only four-particle hard core states with their duals as well as fermion hard core states and the first fermion polarization cloud part with their corresponding duals, but no boson dressing.

We start with the discussion of 'graviton' states, which according to the fusion idea of de Broglie are assumed to be bound states of four spin $\frac{1}{2}$ subfermions. These 'graviton' bound states were calculated in Section 6.6. We already indicated that we will not *a priori* restrict our graviton states to spin 2 but leave it to the full effective dynamics to select proper eigenstates. Thus we have to discuss the relation of our generalized 'graviton' states to the phenomenological gravitons with spin 2.

The four-particle bound states of Section 6.6 were derived as solutions of the generalized de Broglie–Bargmann–Wigner equation (6.109). For the solution of the integral equation (6.109) we assumed that the four-particle bound states are built up by the fusion of two two-particle bound states with spin 1 and spin 0 parts. As a result we obtained the functions

$$\varphi_{Z_1 Z_2 Z_3 Z_4}^{(4,g)}(\mathbf{r}_1, \mathbf{r}_2, \mathbf{r}_3, \mathbf{r}_4, t|\mathbf{k}) \tag{9.37}$$
$$= N^{(4)} T_{\kappa_1 \kappa_2}^{(1} T_{\kappa_3 \kappa_4}^{2)} e^{iE_k t} e^{-i\mathbf{k}(\mathbf{r}_1+\mathbf{r}_2+\mathbf{r}_3+\mathbf{r}_4)/4}$$
$$\times \Big[f_{i_1 i_2}^A(\mathbf{r}_1 - \mathbf{r}_2) f_{i_3 i_4}^A(\mathbf{r}_3 - \mathbf{r}_4) \theta^X [(\mathbf{r}_1 + \mathbf{r}_2) - (\mathbf{r}_3 + \mathbf{r}_4)]$$
$$\times (\gamma^\mu C)_{\alpha_1 \alpha_2} (\gamma^\nu C)_{\alpha_3 \alpha_4} X_{\mu\nu}(\mathbf{k})$$
$$+ f_{i_1 i_2}^A(\mathbf{r}_1 - \mathbf{r}_2) f_{i_3 i_4}^F(\mathbf{r}_3 - \mathbf{r}_4) \theta^Y [(\mathbf{r}_1 + \mathbf{r}_2) - (\mathbf{r}_3 + \mathbf{r}_4)]$$
$$\times (\gamma^\mu C)_{\alpha_1 \alpha_2} (\Sigma^{\rho\sigma} C)_{\alpha_3 \alpha_4} Y_{\mu\rho\sigma}(\mathbf{k})$$
$$+ f_{i_1 i_2}^F(\mathbf{r}_1 - \mathbf{r}_2) f_{i_3 i_4}^A(\mathbf{r}_3 - \mathbf{r}_4) \theta^{\overline{Y}} [(\mathbf{r}_1 + \mathbf{r}_2) - (\mathbf{r}_3 + \mathbf{r}_4)]$$
$$\times (\Sigma^{\rho\sigma} C)_{\alpha_1 \alpha_2} (\gamma^\mu C)_{\alpha_3 \alpha_4} \overline{Y}_{\rho\sigma\mu}(\mathbf{k})$$
$$+ f_{i_1 i_2}^F(\mathbf{r}_1 - \mathbf{r}_2) f_{i_3 i_4}^F(\mathbf{r}_3 - \mathbf{r}_4) \theta^Z [(\mathbf{r}_1 + \mathbf{r}_2) - (\mathbf{r}_3 + \mathbf{r}_4)]$$
$$\times (\Sigma^{\mu\nu} C)_{\alpha_1 \alpha_2} (\Sigma^{\rho\sigma} C)_{\alpha_3 \alpha_4} Z_{\mu\nu\rho\sigma}(\mathbf{k}) \Big]_{\text{as}[I_1, I_2, I_3, I_4]}$$

which are eigenstates of the three-momentum $\mathbf{k}$. The superspin/isospin matrices $T^{1/2}$ are given by (6.123) and correspond to isospin and subfermion quantum numbers $t^3 = f = 0$. The functions f^A and f^F are given by (8.122), and θ^Y and θ^Z were approximately calculated to yield (6.120) and (6.121).

These bound states are the starting point for the derivation of an effective gravitation theory, which according to Section 9.2 can be formulated in terms of the (classical) field variables $g_{\mu\nu}$ (metric tensor), $\Gamma_{\mu\rho\sigma}$ (affine connection), and $R_{\mu\nu\rho\sigma}$ (curvature tensor). Thus we have to discuss the relation between these gravitational variables and our bound states.

In Section 7.4 we demonstrated for the vector boson case that the center of mass amplitudes of the bound states were in close connection to the dynamical variables of the effective vector boson theory. We transfer this argument to the graviton case.

From their symmetry and antisymmetry properties we expect that the center of mass amplitudes $X_{\mu\nu}, \ldots Z_{\mu\nu\rho\sigma}$ are in close relation to the phenomenological field variables $g_{\mu\nu}, \Gamma_{\mu\rho\sigma}$ and $R_{\mu\nu\rho\sigma}$. Thus we assume that the expansion functions in (9.22) and (9.23) differ from the bound state functions (6.124) just by these center of mass amplitudes, and if Weak Mapping induces a nontrivial field dynamics, these quantities can not be fixed by (6.109). Rather, like in the vector boson case we have to give up the couplings between these quantities following from (6.109) and regard them as independent quantities.

In the graviton case these amplitudes are given by the set

$$\overset{\circ}{X}_s(\mathbf{k}) \in \left\{ X_{\mu\nu}(\mathbf{k}), \ Y_{\mu\rho\sigma}(\mathbf{k}), \ \overline{Y}_{\rho\sigma\mu}(\mathbf{k}), \ Z_{\mu\nu\rho\sigma}(\mathbf{k}) \right\} \tag{9.38}$$

with

$$s = (s_1, s_2, s_3, s_4) = \big((\mu\nu), (\mu\rho\sigma), (\rho\sigma\mu), (\mu\nu\rho\sigma)\big). \tag{9.39}$$

The separation of these quantities from the bound state functions (6.124) yields the four sets of expansion functions

$$C_{4,s,k}^{I_1 I_2 I_3 I_4} \tag{9.40}$$

$$:= N_G T_{\kappa_1 \kappa_2}^{(1} T_{\kappa_3 \kappa_4}^{2)} e^{-ik(r_1+r_2+r_3+r_4)/4}$$

$$\times \left\{ \begin{array}{l} f_{i_1 i_2}^A(\mathbf{r}_1 - \mathbf{r}_2) f_{i_3 i_4}^A(\mathbf{r}_3 - \mathbf{r}_4) \theta^X \Big(\dfrac{\mathbf{r}_1 + \mathbf{r}_2}{4} - \dfrac{\mathbf{r}_3 + \mathbf{r}_4}{4}\Big)(\gamma^\mu C)_{\alpha_1 \alpha_2}(\gamma^\nu C)_{\alpha_3 \alpha_4} \\[2mm] f_{i_1 i_2}^A(\mathbf{r}_1 - \mathbf{r}_2) f_{i_3 i_4}^F(\mathbf{r}_3 - \mathbf{r}_4) \theta^Y \Big(\dfrac{\mathbf{r}_1 + \mathbf{r}_2}{4} - \dfrac{\mathbf{r}_3 + \mathbf{r}_4}{4}\Big)(\gamma^\mu C)_{\alpha_1 \alpha_2}(\Sigma^{\rho\sigma} C)_{\alpha_3 \alpha_4} \\[2mm] f_{i_1 i_2}^F(\mathbf{r}_1 - \mathbf{r}_2) f_{i_3 i_4}^A(\mathbf{r}_3 - \mathbf{r}_4) \theta^{\overline{Y}} \Big(\dfrac{\mathbf{r}_1 + \mathbf{r}_2}{4} - \dfrac{\mathbf{r}_3 + \mathbf{r}_4}{4}\Big)(\Sigma^{\rho\sigma} C)_{\alpha_1 \alpha_2}(\gamma^\mu C)_{\alpha_3 \alpha_4} \\[2mm] f_{i_1 i_2}^F(\mathbf{r}_1 - \mathbf{r}_2) f_{i_3 i_4}^F(\mathbf{r}_3 - \mathbf{r}_4) \theta^Z \Big(\dfrac{\mathbf{r}_1 + \mathbf{r}_2}{4} - \dfrac{\mathbf{r}_3 + \mathbf{r}_4}{4}\Big)(\Sigma^{\mu\nu} C)_{\alpha_1 \alpha_2}(\Sigma^{\rho\sigma} C)_{\alpha_3 \alpha_4} \end{array} \right\}$$

Consequently the boson source operators of (9.22) are given by $b_{\mu\nu}^X(\mathbf{k}), \ldots$, and following the general rules of Weak Mapping we have to interpret the coefficient functions Θ of the functional (9.21) as matrix elements of the phenomenological field operators.

To show this we consider a special case of equation (9.29), namely

$$\frac{1}{4!}\varphi^{(4)}(I_1 \ldots I_4 | \mathbf{k}) j_{I_1} \ldots j_{I_4} |0\rangle_F = \int \Theta^{(1,0)}(\mathbf{k}'|\mathbf{k})_{\mu,\rho\sigma} b_{\mu,\rho\sigma}^{4,k'} |v\rangle \, d^3 k' \tag{9.41}$$

with

$$\Theta^{(1,0)}(\mathbf{k}'|\mathbf{k})_{\mu,\rho\sigma} \equiv \langle 0|\Gamma_{\mu\rho\sigma}^{op}(\mathbf{k}')|\mathbf{k}\rangle,$$

where we suppressed the quantum numbers with exception of the momentum $\mathbf{k}$, and where for simplicity all other contributions to $\phi^{(4)}$ are omitted.

Then we obtain with (9.37) and (9.22)

$$\frac{1}{4!}\varphi^{(4)}(I_1 \ldots I_4 | \mathbf{k}) j_{I_1} \ldots j_{I_4} |0\rangle_F \tag{9.42}$$

$$= \frac{1}{4!} C^{\mu,\rho\sigma}(I_1 \ldots I_4 | \mathbf{k}) Y_{\mu\rho\sigma}(0|\mathbf{k}) j_{I_1} \ldots j_{I_4} |0\rangle_F$$

$$= \frac{1}{4!} b_{\mu,\rho\sigma}^{4,k} Y_{\mu\rho\sigma}(0|\mathbf{k}) |v\rangle.$$

Comparison of (9.41) and (9.42) leads to

$$\Theta^{(1,0)}(\mathbf{k}'|\mathbf{k})_{\mu,\rho\sigma} = \frac{1}{4!} Y_{\mu\rho\sigma}(0|\mathbf{k})\delta(\mathbf{k}' - \mathbf{k}),$$

i.e., it is demonstrated that if the expansion functions of (9.22) are given by the functions (9.40), the matrix elements of the effective theory are represented by $Y_{\mu\rho\sigma}(0|\mathbf{k})$. Of course this result does not mean that all higher effective correlation functions $\Theta^{(n,0)}(\mathbf{k}'_1 \ldots \mathbf{k}'_n | \mathbf{k})$ have to be represented by products of $Y_{\mu\rho\sigma}(0|\mathbf{k})$. It simply expresses the fact that (9.40) is the appropriate transformation function and not (9.37).

According to (6.124) the effective gravitation quantities have the symmetry properties

$$Y_{\mu\rho\sigma} = Y_{\mu[\rho\sigma]}, \tag{9.43}$$
$$\overline{Y}_{\rho\sigma\mu} = \overline{Y}_{[\rho\sigma]\mu},$$
$$Z_{\mu\nu\rho\sigma} = Z_{[\mu\nu][\rho\sigma]}.$$

The complete antisymmetrization of the bound state functions (6.124) would induce the further conditions

$$X_{\mu\nu} = X_{\nu\mu}, \tag{9.44}$$
$$Y_{\mu\rho\sigma} = \overline{Y}_{\rho\sigma\mu},$$
$$Z_{\mu\nu\rho\sigma} = Z_{\rho\sigma\mu\nu}.$$

By Weak Mapping these symmetry properties induce the symmetry properties of the effective gravitation quantities. These conditions are just the symmetry properties for the metric tensor, the affine connection, and the curvature tensor of a Riemann space V_4 in anholonomic coordinates [Schou 54]. However, for the derivation of an effective gravitation theory coupled to elementary fermions the concept of the Riemann space has to be enlarged; we assume a Riemann–Cartan geometry to be a suitable framework for the formulation of a generalized gravitation theory, see for instance Hehl [Hehl 95].

In the latter case the symmetry conditions (9.44) do not hold. Thus in order to obtain a generalized geometry we have to break this antisymmetrization. This can be achieved by a suitable breaking of the isospin symmetry of the fundamental subspinors, which lifts the degeneracy of particles in multiplets and makes them distinguishable, i.e., lifts antisymmetrization.

We do not perform this symmetry breaking consequently, but give up only the second of the conditions (9.44), i.e., we consider the quantities Y and $\overline{Y}$ as independent. Concerning X and Z this means that for a first evaluation we restrict ourselves to a Riemann space V_4. This is no contradiction of our treatment of a combined graviton–fermion system as we derive only the linear gravitation equations which are formally identical for Riemann and for Riemann–Cartan space. We stress, however, that breaking the isospin symmetry of the subfermions induces a natural mechanism for the derivation of an effective Riemann–Cartan geometry.

For the evaluation of the linear part of the effective graviton equations (9.35) we need the dual functions $R^{4,s,k}_{I_1I_2I_3I_4}$, which are defined by the orthogonality relations

$$R^{4,s,k}_{I_1I_2I_3I_4} C^{I_1I_2I_3I_4}_{4,s',k'} = \delta_{kk'}\delta_{ss'}. \tag{9.45}$$

It can be shown that the following set of functions satisfies the conditions (9.45):

$$R^{4,s,k}_{I_1I_2I_3I_4} := \frac{1}{4(2\pi)^3}\frac{1}{N_G} T^{(1}_{\kappa_1\kappa_2} T^{2)}_{\kappa_3\kappa_4} e^{ik(r_1+r_2+r_3+r_4)/4} \tag{9.46}$$

$$\times \left\{ \begin{array}{l} N_1 \check{f}^A_{i_1 i_2}(\mathbf{r}_1 - \mathbf{r}_2) \check{f}^A_{i_3 i_4}(\mathbf{r}_3 - \mathbf{r}_4) \hat{\theta}^X \left(\dfrac{\mathbf{r}_1 + \mathbf{r}_2 - \mathbf{r}_3 - \mathbf{r}_4}{4} \right) (\gamma^\mu C)^+_{\alpha_1 \alpha_2} (\gamma^\nu C)^+_{\alpha_3 \alpha_4} \\[2mm] N_2 \check{f}^A_{i_1 i_2}(\mathbf{r}_1 - \mathbf{r}_2) \check{f}^F_{i_3 i_4}(\mathbf{r}_3 - \mathbf{r}_4) \hat{\theta}^Y \left(\dfrac{\mathbf{r}_1 + \mathbf{r}_2 - \mathbf{r}_3 - \mathbf{r}_4}{4} \right) (\gamma^\mu C)^+_{\alpha_1 \alpha_2} (\Sigma^{\rho\sigma} C)^+_{\alpha_3 \alpha_4} \\[2mm] N_3 \check{f}^F_{i_1 i_2}(\mathbf{r}_1 - \mathbf{r}_2) \check{f}^A_{i_3 i_4}(\mathbf{r}_3 - \mathbf{r}_4) \hat{\theta}^{\overline{Y}} \left(\dfrac{\mathbf{r}_1 + \mathbf{r}_2 - \mathbf{r}_3 - \mathbf{r}_4}{4} \right) (\Sigma^{\rho\sigma} C)^+_{\alpha_1 \alpha_2} (\gamma^\mu C)^+_{\alpha_3 \alpha_4} \\[2mm] N_4 \check{f}^F_{i_1 i_2}(\mathbf{r}_1 - \mathbf{r}_2) \check{f}^F_{i_3 i_4}(\mathbf{r}_3 - \mathbf{r}_4) \hat{\theta}^Z \left(\dfrac{\mathbf{r}_1 + \mathbf{r}_2 - \mathbf{r}_3 - \mathbf{r}_4}{4} \right) (\Sigma^{\mu\nu} C)^+_{\alpha_1 \alpha_2} (\Sigma^{\rho\sigma} C)^+_{\alpha_3 \alpha_4} \end{array} \right\}$$

with

$$\hat{\theta}^X(\mathbf{r}) \;\; = \;\; \hat{\theta}^Y(\mathbf{r}) = \hat{\theta}^Z(\mathbf{r}) = 1 \,, \tag{9.47}$$

$$\check{f}^A_{i_1 i_2}(\mathbf{r}) \;\; := \;\; (\lambda_{i_1} \lambda_{i_2})^{-1} M_{i_1 i_2} \,, \tag{9.48}$$

$$\check{f}^F_{i_1 i_2}(\mathbf{r}) \;\; := \;\; (\lambda_{i_1} \lambda_{i_2})^{-1} M^2_{i_1 i_2} \,, \tag{9.49}$$

where $M_{i_1 i_2} = (m_{i_1} + m_{i_2})/2$.

The normalization constants N_k, $k = 1, 2, 3$, are uniquely fixed by (9.45) and can be approximately calculated.

With the functions (9.40) and their duals (9.46) we are able to explicitly evaluate the Weak Mapping equations (9.35) and (9.36).

9.5 Linear Graviton Equations

We start the evaluation of the effective graviton–fermion dynamics with the pure boson part. The classical part of the effective dynamical boson equation is given by (9.35). As we mentioned above, for a first examination we omit the nonlinear parts and the quantization term of this equation and restrict ourselves to the evaluation of the linear part (without coupling to the fermion matter), which we expect to give a linearized vacuum gravitation theory. This linear vacuum part is given by

$$i \frac{\partial}{\partial t} \Theta^b_s \;\; = \;\; 4 K_{I_1 I_2} C^{I_2 I_3 I_4 I_5}_{4, s'} R^{4, s}_{I_1 I_3 I_4 I_5} \Theta^b_{s'} \tag{9.50}$$

$$- 36 W_{I_1 I_2 I_3 I_4} F_{I_4 K_1} C^{I_2 I_3 K_2 K_3}_{4, s'} R^{4, s}_{I_1 K_1 K_2 K_3} \Theta^b_{s'} \,.$$

The effective boson field variables of equation (9.35) were denoted by Θ^b_s with indices s. These indices are specified by the definition of the Weak Mapping expansion functions (9.40), which are characterized by the indices

$$s = (s_1, s_2, s_3, s_4) = \left((\mu\nu), (\mu\rho\sigma), (\rho\sigma\mu), (\mu\nu\rho\sigma) \right) \tag{9.51}$$

and by the three-momentum $\mathbf{k}$. By the definition of relations (9.22) these indices are transferred to the boson operators b_s and by the classical ansatz (9.32), (9.33) the effective boson field variables are specified in the form

$$\Theta_s^b(\mathbf{k}, t) \in \{\Theta_{\mu\nu}^X(\mathbf{k}, t), \ \Theta_{\mu\rho\sigma}^Y(\mathbf{k}, t), \ \Theta_{\rho\sigma\mu}^{\overline{Y}}(\mathbf{k}, t), \ \Theta_{\mu\nu\rho\sigma}^Z(\mathbf{k}, t)\} \qquad (9.52)$$

with the symmetries (see Section 9.4)

$$
\begin{aligned}
\Theta_{\mu\nu}^X &= \Theta_{\nu\mu}^X , \\
\Theta_{\mu\rho\sigma}^Y &= \Theta_{\mu[\rho\sigma]}^Y , \\
\Theta_{\rho\sigma\mu}^{\overline{Y}} &= \Theta_{[\rho\sigma]\mu}^{\overline{Y}} , \\
\Theta_{\mu\nu\rho\sigma}^Z &= \Theta_{[\mu\nu][\rho\sigma]}^Z , \\
\Theta_{\mu\nu\rho\sigma}^Z &= \Theta_{\rho\sigma\mu\nu}^Z .
\end{aligned}
\qquad (9.53)
$$

Thus we are prepared to evaluate the various terms of equation (9.35) by inserting the kinetic operator (2.82), the vertex (2.83), the propagator (7.79), the expansion functions (9.40), and their duals (9.46). For this evaluation one has to take into account the antisymmetrization of the expansion functions according to Section 9.4. The integrations in coordinate space are performed in a strong coupling limit for the Bessel functions; the nonperturbative subfermion regularization, *i.e.*, the summation over the auxiliary field indices with (2.51) and the subsequent transition to the mean subfermion mass m, guarantees the occurence of finite constants only. *The algebraic calculations, which determine the structure of the resulting effective equations, are performed exactly.*

From these calculations we obtain for the Fourier transformed quantities $\Theta_s^b(\mathbf{r}, t)$ of $\Theta_s^b(\mathbf{k}, t)$ the equations

$$
\begin{aligned}
\partial_0 \Theta_{\mu\nu}^X(\mathbf{r}, t) &= \partial_k \left[\delta_{0\mu} \Theta_{k\nu}^X(\mathbf{r}, t) + \delta_{k\mu} \Theta_{0\nu}^X(\mathbf{r}, t) \right] \\
&\quad + (ma_1 + gc_1) \Theta_{0\mu\nu}^Y(\mathbf{r}, t),
\end{aligned}
\qquad (9.54)
$$

$$
\begin{aligned}
\partial_0 \Theta_{\mu\rho\sigma}^Y(\mathbf{r}, t) &= \partial_k \left[\delta_{0\mu} \Theta_{k\rho\sigma}^Y(\mathbf{r}, t) + \delta_{k\mu} \Theta_{0\rho\sigma}^Y(\mathbf{r}, t) \right] \\
&\quad + (ma_2 + gc_2) \Theta_{0\mu\rho\sigma}^Z(\mathbf{r}, t) \\
&\quad + \frac{1}{4} gc_2 \, \varepsilon^{0\mu\mu'\nu'} \varepsilon_\rho{}^{\sigma\rho'\sigma'} \Theta_{\mu'\nu'\rho'\sigma'}^Z(\mathbf{r}, t),
\end{aligned}
\qquad (9.55)
$$

$$
\begin{aligned}
\partial_0 \Theta_{\rho\sigma\mu}^{\overline{Y}}(\mathbf{r}, t) &= 2\partial_k \left[\delta_{0\rho} \Theta_{k\sigma\mu}^{\overline{Y}}(\mathbf{r}, t) + \delta_{k\rho} \Theta_{0\sigma\mu}^{\overline{Y}}(\mathbf{r}, t) \right] \\
&\quad + (ma_3 + gb_1 + gc_3) \, \delta_{0\rho} \Theta_{\mu\sigma}^X(\mathbf{r}, t),
\end{aligned}
\qquad (9.56)
$$

$$
\begin{aligned}
\partial_0 \Theta_{\mu\nu\rho\sigma}^Z(\mathbf{r}, t) &= 2\partial_k \left[\delta_{0\mu} \Theta_{k\nu\rho\sigma}^Z(\mathbf{r}, t) + \delta_{k\mu} \Theta_{0\nu\rho\sigma}^Z(\mathbf{r}, t) \right] \\
&\quad + (ma_4 + gb_2 + gc_4) \, \delta_{0\mu} \Theta_{\nu\rho\sigma}^Y(\mathbf{r}, t) \\
&\quad + \frac{1}{4} gc_4 \, \varepsilon^{\mu\nu0\mu'} \varepsilon_\rho{}^{\sigma\rho'\sigma'} \Theta_{\mu'\rho'\sigma'}^Y(\mathbf{r}, t),
\end{aligned}
\qquad (9.57)
$$

with the mean subfermion mass m; the right hand sides have to be symmetrized or antisymmetrized according to (9.53).

The constants $a_1 \ldots c_4$ can be approximately calculated; for those which we use in the following we obtain

$$a_2 = m\, 3^2 \cdot 5^2, \tag{9.58}$$

$$a_4 = -m^{-1} \frac{2^2}{3^2 \cdot 5^2},$$

$$b_2 = \frac{1}{m^2} \frac{1}{\pi^2 2 \cdot 3^2 \cdot 5^2},$$

$$c_2 = \frac{2^4 \cdot 5}{\pi^3},$$

$$c_4 = -\frac{1}{m^2} \frac{2^7}{\pi^3 3^4 \cdot 5^3}.$$

We emphasize that the applied approximations effect only the values of these constants but not the structure of equations (9.54)–(9.57).

As a consequence of breaking the symmetry $Y = \overline{Y}$ (Section 9.4) we obtain two sets of coupled linear equations for Θ^X and $\Theta^{\overline{Y}}$ and for Θ^Y and Θ^Z. We start with the discussion of the coupled equations (9.55) and (9.57).

The terms containing b_i and some of the terms containing c_i are corrections to the mass terms with a_i. However, there are two terms, which seem not to fit into this scheme; they have to be discussed separately.

Concerning the last term of (9.55) an explicit evaluation shows, that this term is equal to $-gc_2\Theta^Z_{0\mu\rho\sigma}$, provided the quantities Θ^Z satisfy the symmetry conditions (9.53) and the additional constraint

$$\Theta^Z_{\mu\nu}{}^\mu{}_\sigma = 0. \tag{9.59}$$

These are just the symmetry conditions of the Weyl tensor in a Riemann space V_4, see for instance Weinberg [Wein 72]. With (9.59) and the rescaling $ma_2\Theta^Z_{\mu\nu\rho\sigma} \to \Theta^Z_{\mu\nu\rho\sigma}$ we obtain for (9.55)

$$\Theta^Z_{0\mu\rho\sigma}(\mathbf{r},t) = \partial_0\Theta^Y_{\mu\rho\sigma}(\mathbf{r},t) - \partial_k\left[\delta_{0\mu}\Theta^Y_{k\rho\sigma}(\mathbf{r},t) + \delta_{k\mu}\Theta^Y_{0\rho\sigma}(\mathbf{r},t)\right]. \tag{9.60}$$

By explicitly resolving the indices $\mu,\ldots = 0,1,2,3$ it is easy to show that (9.60) is equivalent to the equations

$$\Theta^Z_{\mu\nu\rho\sigma} = 2\partial_{[\mu}\Theta^Y_{\nu]\rho\sigma} \tag{9.61}$$

and

$$\partial^\mu\Theta^Y_{\mu\rho\sigma}(\mathbf{r},t) = 0, \tag{9.62}$$

where again the Weyl tensor condition (9.59) has to observed. *Equation (9.61) is equivalent to the linearized Ricci identities (9.14) for the Weyl tensor in anholonomic coordinates in a Riemann space V_4, provided Θ^Y is interpreted as the corresponding anholonomic affine connection.*

Equation (9.62) can be interpreted as the linear part of a condition which expresses the redundancy of the C–Γ formulation of the Einstein theory, see Edgar [Edg 80].

We now turn to equation (9.57). Concerning the last term of (9.57) an explicit evaluation shows that all mass terms of (9.57) can be removed by the condition

$$ma_4 + g\left(b_2 + \frac{3}{2}c_4\right) = 0 \tag{9.63}$$

provided again that Θ^Z has the symmetry properties (9.53) and (9.59) of a Weyl tensor in a V_4.

By means of this mass 0 condition we obtain for (9.57)

$$\partial_0\Theta^Z_{\mu\nu\rho\sigma}(\mathbf{r},t) = 2\partial_k\left[\delta_{0\mu}\Theta^Z_{k\nu\rho\sigma}(\mathbf{r},t) + \delta_{k\mu}\Theta^Z_{0\nu\rho\sigma}(\mathbf{r},t)\right], \tag{9.64}$$

where the symmetries of Θ^Z have to be observed. By direct calculation it can be shown that (9.64) is equivalent to

$$\partial^\mu\Theta^Z_{\mu\nu\rho\sigma}(\mathbf{r},t) = 0 \tag{9.65}$$

if the Weyl tensor symmetries (9.53) and (9.59) are satisfied. *Equation (9.65) is just the linearized Bianchi identity for the curvature tensor (9.6), (9.15) (holonomic or anholonomic) in a Riemann space.*

There is, however, still another symmetry for the determination of the 10 independent components of a Weyl tensor in a V_4, which reads

$$\varepsilon^{\mu\nu\rho\sigma}\Theta^Z_{\mu\nu\rho\sigma}(\mathbf{r},t) = 0. \tag{9.66}$$

In contrast to the Weyl tensor condition (9.53) this condition holds only in a V_4 and is not induced by our calculations nor necessary for the results which we obtained above. However, it can be shown to be compatible with (9.65) and (9.59). We take this as a further hint that our formalism induces a generalized geometry (a Riemann–Cartan geometry) in a natural way.

Summarizing our results the effective equations for the bound state quantities Θ^Z and Θ^Y may be interpreted as linear Bianchi and Ricci identities for the anholonomic Weyl tensor and the anholonomic affine connection in a Riemann space V_4.

Comparing these results with the formulation of the Einstein gravitation theory in terms of the Weyl tensor and the affine connection, expressed by equations (9.16) and (9.17), we can claim to have derived the linearized Einstein theory in the vacuum, provided we postulate the equations (9.11) for the tetrads, which in principle may be resolved with respect to the tetrads and which in turn induce the metric tensor by means of (9.9).

An analogous evaluation of the effective equations (9.54) and (9.56) for the quantities Θ^X and $\Theta^{\overline{Y}}$ yields equations which have a structure similar to (9.65) and (9.59). One is tempted to interpret these equations as equations for the Ricci tensor or the metric tensor and the affine connection. In spite of these hints it seems not to be possible to give those equations a consistent meaning in terms of gravitation quantities. As with Θ^Z and Θ^Y we have already obtained a set of quantities which suffices for the derivation of an effective gravitation theory, we may set $\Theta^X = \Theta^{\overline{Y}} = 0$ without contradictions.

With respect to these calculations we remark, that the approximations for the evaluation of the effective equations (9.54)–(9.57) do only concern some constants, which are collected in the mass condition (9.63), but do not affect the general form of the resulting equations. In particular the covariant form of the resulting equations is an exact result of Weak Mapping in the single time formalism.

The mass condition (9.63) fixes the coupling constant g, one of the parameters of our subfermion theory; with (9.58) one obtains

$$g = -8m^2\pi^2\left(1 - \frac{2^7}{\pi 3 \cdot 5}\right) \approx -46,0m^2. \tag{9.67}$$

The condition (9.63) was necessary for the interpretation of (9.57) as a linearized Bianchi identity. By an analogous condition resulting from the derivation of the effective vector boson dynamics in Section 7.5 the value $g = -10\pi^2m^2$ was obtained, see (7.99). In view of the rough approximations of our coordinate functions this seems to be a good qualitative agreement with our value of g. This result may be regarded as an additional hint for the success of our atomistic program.

9.6 Dressed Fermion States

The calculation of the (first order) dressed fermion states was done in [Stu 93b]. In contrast to hard core calculations for dressed particle calculations we need the full covariant functional equation (3.92). In compact notation it reads

$$K_{I_1I_2}\partial_{I_2}|\mathcal{F}(j;l)\rangle = -W_{I_1I_2I_3I_4} : d_{I_4}d_{I_3}d_{I_2} : |\mathcal{F}(j;l)\rangle \tag{9.68}$$

with

$$d_{I_1} = \partial_{I_1} - F_{I_1I_2}j_{I_2} \tag{9.69}$$

and definitions (6.3)–(6.5).

In the lowest order of the covariant polarization cloud we represent the dressed fermion state by the functional

$$|\mathcal{F}(j;l)\rangle = i\varphi^{(1)}(I_1|l)j_{I_1}|0\rangle_F - \frac{i}{3!}\varphi^{(3)}(I_1,I_2,I_3|l)j_{I_1}j_{I_2}j_{I_3}|0\rangle_F. \tag{9.70}$$

If we substitute (9.70) into (9.68) and project (9.68) into configuration space we obtain

$$(D^\mu_{I_1I_2}\partial_\mu - m_{I_1I_2})\varphi^{(1)}(I_2|l) = W_{I_1I_2I_3I_4}\varphi^{(3)}(I_2,I_3,I_4|l) \tag{9.71}$$

and

$$(D^\mu_{I_1I_2}\partial_\mu - m_{I_1I_2})\varphi^{(3)}(K_1,K_2,I_2|l) \tag{9.72}$$
$$= 3W_{I_1I_2I_3I_4}\left[F_{I_4K_1}F_{I_3K_2}\varphi^{(1)}(I_2|l) - F_{I_4K_2}\varphi^{(3)}(K_1,I_2,I_3|l)\right]_{\text{as}[K_1,K_2]},$$

where owing to our three-source approximation in (9.70) we have suppressed the $\varphi^{(5)}$ term in (9.72). In the same way, because of this approximation we also suppress all higher order equations for $\varphi^{(r)}$ with $r > 3$ which result from equation (9.68).

For a straightforward integration even the remaining equations (9.71) and (9.72) are too complicated. For a first inspection we apply an iteration procedure which we break off in the lowest order. We denote the lowest order functions by $\varphi_0^{(1)}$ and $\varphi_0^{(3)}$ and define the lowest order equations by

$$(D_{I_1 I_2}^\mu \partial_\mu - m_{I_1 I_2})\varphi_0^{(1)}(I_2|l) = 0 \tag{9.73}$$

and

$$(D_{I_1 I_2}^\mu \partial_\mu - m_{I_1 I_2})\varphi_0^{(3)}(K_1, K_2, I_2|l) \tag{9.74}$$
$$= 3W_{I_1 I_2 I_3 I_4}\left(F_{I_4 K_1}F_{I_3 K_2} - F_{I_4 K_2}F_{I_3 K_1}\right)\varphi_0^{(1)}(I_2|l).$$

The other terms of equations (9.71) and (9.72) which do not appear in equations (9.73) and (9.74) are included into the first iteration step.

Equations (9.73) and (9.74) can be directly integrated. Whilst $\varphi_0^{(1)}$ is a Dirac superspinor, $\varphi_0^{(3)}$ is given by

$$\varphi_0^{(3)}(K_1, K_2, K_3) = N_3\Big[3G_{K_3 I_1}W_{I_1 I_2 I_3 I_4}(F_{I_4 K_1}F_{I_3 K_2} - F_{I_4 K_2}F_{I_3 K_1})\varphi_0^{(1)}(I_2)\Big]_{\mathrm{as}[K_1 K_2 K_3]} \tag{9.75}$$

with a normalization factor N_3.

For the further evaluation we consider the Green's function $G_{I_1 I_2}$ in (9.75), which reads

$$G_{I_1 I_2} = \frac{1}{(2\pi)^4}\delta_{\kappa_1 \kappa_2}\delta_{i_1 i_2}\int(q_\mu \gamma^\mu - m_i)_{\alpha_1 \alpha_2}^{-1} e^{iq(x_1-x_2)}dq. \tag{9.76}$$

Because of the necessity of decoupling, we have assumed very large subfermion masses in (9.76). This limit of large m_i justifies the strong coupling limit in which the integral in (9.76) can be approximately represented by

$$\frac{1}{(2\pi)^4}\lim_{m\to\infty}\int(q_\mu \gamma^\mu - m)_{\alpha_1 \alpha_2}^{-1} e^{iq(x_1-x_2)}dq \approx -\lim_{m\to\infty}\frac{1}{m}\delta_{\alpha_1 \alpha_2}\delta(x_1 - x_2). \tag{9.77}$$

Hence we obtain

$$\lim_{m_i\to\infty} G_{I_1 I_2} \approx -\lim_{m_i\to\infty}\frac{1}{m_i}\delta_{\kappa_1 \kappa_2}\delta_{\alpha_1 \alpha_2}\delta_{i_1 i_2}\delta(x_1 - x_2). \tag{9.78}$$

If we substitute (9.78) into (9.75) we can perform the integration and afterwards proceed to equal times $t_1 = t_2 = t_3 = 0$. For brevity we do not reproduce the explicit calculations. The result reads in the low energy limit:

$$C_{1,l}^{I_1 I_2 I_3} = N_P A_{[I_1 I_2 I_3]I_4} C_{1,l}^{I_4} \tag{9.79}$$

with

$$A_{I_1 I_2 I_3 I_4} = g \left[C_{\alpha_1 \alpha_2} \delta_{\alpha_3 \alpha_4} - (\gamma^5 C)_{\alpha_1 \alpha_2} \gamma^5_{\alpha_3 \alpha_4} \right] \gamma^5_{\kappa_1 \kappa_2} \delta_{\kappa_3 \kappa_4} \tag{9.80}$$

$$\times \lambda_{i_1} \lambda_{i_2} \lambda_{i_3} \frac{m^2_{i_1} m^2_{i_2}}{m_{i_3}} \delta_{i_3 i_4} \frac{K_1(m_{i_1} |\mathbf{r}_3 - \mathbf{r}_1|)}{|\mathbf{r}_3 - \mathbf{r}_1|} \frac{K_1(m_{i_2} |\mathbf{r}_3 - \mathbf{r}_2|)}{|\mathbf{r}_3 - \mathbf{r}_2|} \delta(\mathbf{r}_3 - \mathbf{r}_4) \ .$$

In (9.79) we have already identified the functions $\varphi_0^{(3)}(I_1, I_2, I_3 | l)$ and $\varphi_0^{(1)}(I | l)$ with the Weak Mapping expansion functions $C_{1,l}^{I_1 I_2 I_3}$ and $C_{1,l}^I$. Later on we will comment upon this identification. The quantum numbers l of the function $C_{1,l}^I$ can be explicitly written as the momentum $\mathbf{k}$, the spin quantum number s, the isospin–superspin a, and the auxiliary field index j. From the above iteration procedure, in particular from (9.79), it follows that the quantum numbers l of the polarization cloud indeed are those of the corresponding basic hard core function, in our case of a free Dirac super–isospinor.

These free Dirac fields (the fermion hard core states) $C_{1,l}^I$ explicitly read in our notation

$$C_{1,l}^I \equiv C(\alpha, \kappa, i, \mathbf{r} | s, \rho, j, \mathbf{k}) \tag{9.81}$$

$$= \lambda_j \delta_{ij} e^{-i\mathbf{r}\mathbf{k}} \delta_{\kappa\rho} \chi^s_\alpha(\mathbf{k})$$

with the ordinary Dirac spinors $\chi^s_\alpha(\mathbf{k})$ for spin $s = \pm\frac{1}{2}$. Compared with [Stu 93b] the functions (9.79), (9.80) contain a modified coordinate dependence because of an improved evaluation of the regularization.

The calculation of the corresponding duals $R_{I_1 I_2 I_3}^{1,l}$ was performed in [Stu 93b] by means of the procedure which was sketched in Section 5.7. In the graviton case, however, we do not apply the full formalism of Section 5.7. Rather, we make the ansatz

$$R_{I_1 I_2 I_3}^{1,l} = N_P^{-1} g^{-2} \frac{A_{[I_1 I_2 I_3] I_4}}{\lambda_{i_1} \lambda_{i_2} \lambda_{i_3}} \frac{R_{I_4}^{1,l}}{\lambda_j} \tag{9.82}$$

with the fermion hard core dual

$$R_I^{1,l} := (\lambda_j)^{-2} C_{1,l}^I , \tag{9.83}$$

where we remember $l = (s, \rho, j, \mathbf{k})$. It can be shown that (9.82) exactly satisfies the orthogonality conditions

$$R_{I_1 I_2 I_3}^{1,l} C_{1,l'}^{I_1 I_2 I_3} = N \delta_{ll'} . \tag{9.84}$$

This relation justifies formula (9.82) which we will use for the evaluation of the effective graviton–fermion sector.

Finally it should be mentioned that in contrast to Section 9.4 in $\varphi_0^{(3)}$ no center of mass amplitude is separated. This corresponds to the choice of the dynamical fermionic quantities in our effective theory. As we already pointed out one of the rules of Weak Mapping is to identify the coefficient functions of the corresponding effective functionals, in the present case of (9.21), with matrix elements of the dynamical quantities of the theory. In the fermionic sector these dynamical quantities are the Dirac super–isospinors $\varphi(I | l)$ or their projections onto another spinor $\varphi(I | l')$, respectively.

We illustrate the situation in analogy to (9.41), (9.42) by considering a special case in the pure fermion sector of (9.29), namely

$$i\varphi_t^{(1)}(I|k)j_I - \frac{i}{3!}\varphi_t^{(3)}(I_1, I_2, I_3|k)j_{I_1}j_{I_2}j_{I_3} = i\Theta^{(0,1)}(l|k)f_{1,l}, \tag{9.85}$$

where $\Theta^{(0,1)}$ is the first pure fermion correlation function of the effective state functional (9.21). We introduce a complete set of one-particle states $C_{1,l}^I$ with $C_{1,l}^I R_I^{1,l'} = \delta_{ll'}$ and obtain

$$i\varphi_t^{(1)}(I|k)j_I - \frac{i}{3!}\varphi_t^{(3)}(I_1, I_2, I_3|k)j_{I_1}j_{I_2}j_{I_3} = iC_{1,l}^I\Theta^{(0,1)}(l|k)R_I^{1,l'}f_{1,l'} \tag{9.86}$$

$$=: i\tilde\Theta^{(0,1)}(I|k)\tilde f_{1,I}$$

with the 'Fourier transforms' $\tilde\Theta$ and $\tilde f$, which are now related to the coordinate space, and we interpret $\tilde\Theta$ as expectation values of Dirac super–isospinors in coordinate space, i.e.,

$$\tilde\Theta^{(0,1)}(I|k) \equiv \langle 0|\psi_I|k\rangle. \tag{9.87}$$

On the other hand, according to (9.23) we have in lowest order

$$i\varphi_t^{(1)}(I|k)j_I - \frac{i}{3!}\varphi_t^{(3)}(I_1, I_2, I_3|k)j_{I_1}j_{I_2}j_{I_3} \tag{9.88}$$

$$= iC_k^I j_I + iC_k^{I_1,I_2,I_3}j_{I_1}j_{I_2}j_{I_3} \tag{9.89}$$

$$= if_{1,k}$$

$$= iC_{1,k}^I \tilde f_{1,I}.$$

Comparing this with (9.86) and (9.87) we have

$$\tilde\Theta^{(0,1)}(I|k) \equiv C_{1,k}^I \equiv \langle 0|\psi_I|k\rangle. \tag{9.90}$$

Thus the choice of $\varphi_t^{(3)}$ for the Weak Mapping functions C is consistent with the interpretation (9.87).

As the superspinor mass m_i also occurs amongst the superspinor quantum numbers, the above argument is only correct if it is accompanied by a mass renormalization. However, in the present case mass renormalization can be neglected, as the superspinor field is super-renormalizable with finite mass corrections which are absorbed into the practically infinite original masses in the limit of very large m_i.

9.7   Fermion–Graviton Coupling

According to the equivalence principle, field equations of matter in a gravitational field are obtained by the transition from Lorentz covariant equations to equations which are covariant with respect to general coordinate transformations. The general covariance is achieved by minimal coupling of the matter fields to the affine connection of the Riemann–(Cartan) space. In the case of spinorial matter the general covariant derivative is given by

$$D_\mu \Psi(x) = \partial_\mu \Psi(x) + \frac{i}{4}\Gamma_{\mu\nu\sigma}(x)\Sigma^{\nu\sigma}\Psi(x) \tag{9.91}$$

with the anholonomic affine connection $\Gamma_{\mu\nu\sigma}$ and the generators of the Lorentz group $\Sigma^{\mu\nu} = \frac{i}{2}[\gamma^\mu, \gamma^\nu]_-$. Thus the free Dirac equation in a Riemann space V_4 (and in a Riemann–Cartan space) reads in anholonomic coordinates, i.e., in tangent space

$$\left[i\gamma^\mu \partial_\mu - m - \frac{1}{4}\Gamma_{\mu\rho\sigma}(x)\gamma^\mu\Sigma^{\rho\sigma}\right]\Psi(x) = 0. \tag{9.92}$$

For a confirmation of the interpretation of our bound state quantity Θ^Y as the anholonomic affine connection of a Riemann space V_4 we have to show that it couples to matter fields according to the equivalence principle. As in the preceeding sections we derived an effective theory for a coupled graviton–fermion system, the evaluation of the linear part of our effective classical fermion equation (9.36) should yield equation (9.92).

We are only interested in the covariant derivative terms of (9.36). Thus we omit the residual fermion self-interaction and quantization terms and are left with

$$
\begin{aligned}
i\frac{\partial}{\partial t}\Theta_l^f(t) &= K_{I_1 I_2}C_{1,l'}^{I_2}R_{I_1}^{1,l}\Theta_{l'}^f(t) \\
&\quad + 36 W_{I_1 I_2 I_3 I_4}C_{4,r}^{I_3 I_4 K_1 K_2}C_{1,l'}^{I_2}R_{I_1 K_1 K_2}^{1,l}\Theta_{l'}^f(t)\Theta_r^b(t).
\end{aligned}
\tag{9.93}
$$

The evaluation of this equation is performed by inserting the kinetic operator (2.82), the vertex (2.83), the graviton expansion function (9.40), the dual of the fermion polarization cloud term (9.82), (9.80), and the fermion hard core state (9.81). We remember the abbreviations $I = (\alpha, \kappa, i, \mathbf{r})$, $l = (s, \rho, j, \mathbf{k})$ as well as the bosonic indices (9.51).

Multiplying (9.93) with the Dirac spinors $C_{1,l}^I$ from (9.81), summarizing over l, and taking into account the completeness relations for elementary Dirac spinors we obtain equations for the fields $\Theta_I^f := C_{1,l}^I\Theta_l^f$:

$$
\begin{aligned}
i\frac{\partial}{\partial t}\Theta_{I_1}^f &= K_{I_1 I_2}\Theta_{I_2}^f \\
&\quad + 36 W_{I_2 I_3 I_4 I_5}C_{4,r}^{I_4 I_5 K_1 K_2}C_{1,l}^{I_1}R_{I_2 K_1 K_2}^{1,l}\Theta_{I_3}^f\Theta_r^b.
\end{aligned}
\tag{9.94}
$$

For the calculations we have to take into account the antisymmetrization of the vertex and the expansion functions according to Section 9.4. Owing to the partial subsymmetries we eventually have to calculate $18 \cdot 3 \cdot 3 = 162$ terms for the interaction part. Analogously with the bosonic equation in Section 9.5 the algebraic parts are exactly calculated.

Owing to the structure of the involved functions the auxiliary field indices are connected with the space coordinates. Regularization is again performed by a systematic expansion of the various terms with respect to the deviations of the auxiliary masses from a mean value m, by application of the regularization relations (2.51) and a subsequent approximate evaluation of the remaining integrals. In this manner we obtain for (9.94):

$$i\frac{\partial}{\partial t}\Theta^f_{\alpha\kappa i}(\mathbf{r},t) = \left[-i(\gamma^0\gamma^k)_{\alpha\beta}\partial_k(\mathbf{r}) + \gamma^0_{\alpha\beta}m_i\right]\Theta^f_{\beta\kappa i}(\mathbf{r},t) \tag{9.95}$$

$$+ \delta_{ii_1}\left[d_1\eta^{\mu\nu}\Theta^X_{\mu\nu}(\mathbf{r},t)\gamma^0_{\alpha\beta}\sum_{i_2}\Theta^f_{\beta\kappa i_2}(\mathbf{r},t)\right.$$

$$+ d_2\Theta^Y_{\mu\rho\sigma}(\mathbf{r},t)\left(\gamma^0\Sigma^{\rho\sigma}\gamma^\mu\right)_{\alpha\beta}\sum_{i_2}\Theta^f_{\beta\kappa i_2}(\mathbf{r},t)$$

$$\left.+ d_2\Theta^{\overline{Y}}_{\rho\sigma\mu}(\mathbf{r},t)\left(\gamma^0\Sigma^{\rho\sigma}\gamma^\mu\right)_{\alpha\beta}\sum_{i_2}\Theta^f_{\beta\kappa i_2}(\mathbf{r},t)\right]$$

with some constants d_1, d_2.

We stress that the quantities Θ^Z, which we interpreted as Weyl tensor components in a Riemann space, do not couple to the fermion fields; *this is a result of the exact algebraic evaluation of the effective equations.* According to Section 9.5 we set $\Theta^X = \Theta^{\overline{Y}} = 0$ and obtain for the regularized spinor fields $\Theta^f_{\alpha\kappa}(\mathbf{r},t) = \sum_i \Theta^f_{\alpha\kappa i}(\mathbf{r},t)$:

$$i\frac{\partial}{\partial t}\Theta^f_{\alpha\kappa}(\mathbf{r},t) = \left[-i(\gamma^0\gamma^k)_{\alpha\beta}\partial_k(\mathbf{r}) + \gamma^0_{\alpha\beta}m\right]\Theta^f_{\beta\kappa}(\mathbf{r},t) \tag{9.96}$$

$$+ d_2\Theta^Y_{\mu\rho\sigma}(\mathbf{r},t)\left(\gamma^0\Sigma^{\rho\sigma}\gamma^\mu\right)_{\alpha\beta}\Theta^f_{\beta\kappa}(\mathbf{r},t).$$

After transforming $d_2\Theta^Y \to \frac{1}{4}\Theta^Y$, which is consistent with the results from Section 9.5, this equation is equivalent to the parity transformed equation (9.92), *cf.* [Stu 93b], [Stu 94]. Thus the effective fermion–graviton equation (9.36) may be interpreted as the general covariant Dirac equation for elementary fermions in a Riemann space V_4.

We remark that we have not used the restricted V_4 symmetries (9.43) for the derivation of this result. Thus it holds also for the extended Riemann–Cartan space.

We now summarize our results: we have derived effective equations for the dynamics of a coupled system of composite four-fermion bound states and elementary fermions from an underlying nonlinear subfermion equation. We have shown that the linear and classical parts of the resulting boson equation can be interpreted as the linear Bianchi and Ricci identities of a Riemann geometry and as well as the linearized (classical) Einstein gravitation theory in vacuum. This interpretation was confirmed by the derivation of the correct coupling of the gravitation quantities to elementary fermions.

We started with a Lorentz covariant subfermion field theory in Minkowski space, and thus our effective graviton–fermion theory is also referred to a Minkowski space. In the framework of this effective field theory we derived effective gravitation quantities which may be interpreted as geometrical quantities: the anholonomic Weyl tensor and the anholonomic affine connection of a Riemann space.

By means of the last equation of (9.9), the relation

$$\Gamma^\rho_{\mu\nu} = -e_\mu{}^i e_\nu{}^j \overset{\Gamma}{\nabla}_i e^\rho{}_j \tag{9.97}$$

and the assumption of a suitable set of tetrads one can perform the transition to holonomic coordinates. In this sense we have derived an effective curved geometry as the framework for a gravitation theory resulting from a field theory in flat Minkowski space.

In the course of our investigations the Weyl tensor condition (9.53) turned out to be necessary in order to give an appropriate interpretation to our effective graviton equations. Thus we have obtained a formulation of a gravitation theory in terms of the Weyl tensor and the affine connection. According to Niederer [Nie 75] the Weyl tensor carries a unitary representation of the Poincaré group for spin 2 and mass 0, where the linearized Bianchi identities are just the unitarity conditions. Thus although we started with massive generalized 'graviton' states (6.124) in the linear approximation we arrived at conventional massless gravitons with spin $s = 2$.

In the complete effective functional equations our formalism of Weak Mapping yields nonlinear terms for graviton self-interactions together with the coupling terms to other composite particles. Thus *in principle* we should be able to derive a full nonlinear gravitation theory.

As a first investigation we have evaluated only the linear and the vacuum part of the effective graviton equations. There are, however, arguments that the full nonlinear Einstein theory can already be induced from the linearized Einstein vacuum equations by means of a consistent coupling to their sources, *cf.* for instance Wyss [Wyss 65], Mittelstaedt and Barbour [Mit 67], Deser and Halpern [Des 70a], [Des 70b]. This may be considered as a hint that de Broglie's original idea and our extended approach are physically meaningful.

We have restricted ourselves to the case of a Riemann space V_4 by a certain choice of the symmetry conditions of our effective gravitation quantities. This restriction turned out to be an artificial one; our formalism seems to induce a Riemann–Cartan space as an effective geometry in a natural way. This is in accordance with the generalization of Einstein gravitation to microscopic domains according to Poincaré gauge theories, *cf.* Hehl *et al.* [Hehl 76], [Hehl 95]. The consequences of our effective theory with respect to a Riemann–Cartan geometry as well as the coupling of the fermion 'matter' to the gravitation equations have to be investigated in the future.

Bibliography

[Ald 80] Aldaya, V. and de Azcarraga, J.A.
Higher-order Hamiltonian formalism in field theory, *J. Phys. A* **13**
(1980) 2545–2551

[Alk 95] Alkofer, R. and Reinhard, H.
Chiral quark dynamics, Springer, Berlin 1995

[App 75] Appelquist, F. and Carrazone, J.
Infrared singularities and massive fields, *Phys. Rev D* **11** (1975)
2856–2861

[Barg 48] Bargmann, V. and Wigner, E.P.
Group theoretical discussion of relativistic wave equations, *Proc. Nat.
Acad. Sci. (USA)* **34** (1948) 211–223

[Bass 55] Bass L. and Schrödinger, E.
Must the photon mass be zero? *Proc. Roy. Soc., Series A,* **232** (1955) N
1182, 1–6

[Belo 94] Bellotti, E.
Searching for darkness and decay, *Phys. World* **7** (1994) 37–42

[Bigi 74] Bigi, I.I., Dürr, H.P. and Winter, N.J.
Lagrange theory for spinor fields of subcanonical dimension 1/2, *Nuovo
Cim.* **22A** (1974) 420–447

[Bogn 74] Bognar, J.
Indefinite inner product spaces, Springer, Berlin 1974

[Bogo 75] Bogoliubov, N.N., Logunov, A.A. and Todorov, I.T.
Introduction to axiomatic quantum field theory, W.A. Benjamin Inc.,
Reading, Mass., 1975

[Bogo 59] Bogoliubov, N.N. and Shirkov, D.V.
Introduction to the theory of quantized fields, Interscience Publ., New
York 1959

[Bopp 40] Bopp, F.
Eine lineare Theorie des Elektrons, *Ann. Phys. (Germ.)* **38** (1940)
345–384

[Bopp 58] Bopp, F.
Lorentzinvariante Wellengleichungen für Mehrbahnsysteme, *Bayr. Akad. Wissensch. Math. Naturw. Klasse Sitzungsber.* (1958) 167–225

[Bor 91] Borne, Th.
Quantenfeldtheoretische Energiedarstellung und schwache Abbildung (Diplomarbeit), University of Tübingen, 1991

[Bor 98] Borne, Th. and Stumpf, H.
Weak field gravitation as a composite particle effect, *Ann. Fond. L. de Broglie* **23** (1998) 38–53

[Bor 69] Borneas, M.
Principle of action with higher derivatives, *Phys. Rev.* **186** (1969) 1299–1303

[Born 34a] Born, M.
On the quantum theory of the electromagnetic field, *Proc. Roy. Soc A* **143** (1934) 410–437

[Born 34b] Born, M. and Infeld, L.
Foundations of the new field theory, *Proc. Roy. Soc A* **144** (1934) 425–451

[Brac 76] Bracci, L., Morchio, G. and Strocchi, F.
Description of symmetries in indefinite metric spaces, in Jammer, A., Jansen, T., Boon, M. (eds): *Proceed. 4-th Int. Coll. of grouptheor. Methods in Physics 1975*, Springer, Berlin 1976, 551–556

[Brat 79] Bratteli, O. and Robinson, D.W.
Operator algebras and quantum statistical mechanics I, Springer, Berlin 1979

[Brill 57] Brill, D.R. and Wheeler, J.A.
Interaction of neutrinos and gravitational fields, *Rev. Mod. Phys.* **29** (1957) 465–479

[Brog 22] de Broglie, L.
Rayonnement noir et quanta de lumière, *J. de Physique, Série VI, T. III* (1922) 422–428

[Brog 25] de Broglie, L.
Recherches sur la théorie des quanta (Thesis), Paris, 1924
Ann. de Physique, 10-série, III (1925) 22–128

[Brog 32a] de Broglie, L.
Sur une analogie entre l'électron de Dirac et l'onde électromagnétique, *C. R. Acad. Sci.*, **195** (1932) 536–537

[Brog 32b] de Broglie, L.
Sur le champ électromagnétique de l'onde lumineuse, *C. R. Acad. Sci.*,
195 (1932) 862–864

[Brog 33] de Broglie, L.
Sur la densité de l'énergie dans la théorie de la lumière,
C. R. Acad. Sci., **197** (1933) 1377–1380

[Brog 34a] de Broglie, L.
L'électron magnétique, Hermann, Paris, 1934

[Brog 34b] de Broglie, L.
Une nouvelle conception de la lumière, Hermann (Exposés de physique
théorique), Paris, 1934

[Brog 34c] de Broglie, L.
Sur la nature du photon, *C. R. Acad. Sci.* **198** (1934) 135–138

[Brog 34d] de Broglie, L.
L'équation d'ondes du photon, *C. R. Acad. Sci.* **198** (1934) 445–448

[Brog 34e] de Broglie, L. and Winter, J.
Sur le spin du photon, *C. R. Acad. Sci.* **199** (1934) 813–816

[Brog 36] de Broglie, L.
Nouvelles recherches sur la lumière, Hermann (Exposés de physique
théorique), Paris 1936

[Brog 40] de Broglie, L.
Une nouvelle théorie de la lumière, la mécanique ondulatoire du photon,
Tome I: *La lumière dans le vide*, Hermann, Paris 1940;
Tome II: *L'interaction entre les photons et la matière*, Hermann, Paris
1942

[Brog 41] de Broglie, L.
Sur l'interprétation de certaines équations dans la théorie des particules
de spin 2, *Compt. rend.* **212** (1941) 657–659

[Brog 43] de Broglie, L.
Théorie générale des particules à spin, Gauthier–Villars, Paris 1943

[Brog 49] de Broglie, L.
Mécanique ondulatoire du photon et théorie quantique des champs,
Gauthier–Villars, Paris 1949

[Brog 50] de Broglie, L.
Optique ondulatoire et corpusculaire, Hermann, Paris 1950

[Buch 85] Buchmüller, W.
Composite quarks and leptons, *Acta Phys. Austr. Suppl.* **27** (1985)
517–595

[Cab 62] Cabibbo, N. and Ferrari, G.
Quantum electrodynamics with Dirac monopoles, *Nuovo Cim.* **23** (1962)
1147–1154

[Cha 48] Chang, T.S.
Field theories with high derivatives, *Proc. Cambridge Phil. Soc.* **44**
(1948) 76–86

[Che 90] Cheng, H.
Why canonical quantization is the only known way to quantize correctly
(Lecture notes in physics **361**), 3–14, Springer, Heidelberg 1990

[Chri 80] Christ, N.H. and Lee. T.D.
Operator ordering and Feynman rules in gauge theories, *Phys. Rev. D*
22 (1980) 939–958

[Cole 74] Coleman, S., Jackiw, R. and Politzer, H.D.
Spontaneous symmetry breaking in the $O(N)$ model for large N, *Phys.
Rev. D* **10** (1974) 2491–2499

[Coll 87] Collins, J.
Renormalization, Cambridge University Press, Cambridge 1987

[Cos 43] Costa de Beauregard, O.
Contribution à l'étude de la théorie de l'électron de Dirac (Thesis),
Paris 1943;
J. Math. Pures et Appliquées **22** *(1943) 85–176*

[Cos 95] Costa de Beauregard, O.
Electromagnetic gauge as integration condition: Einstein's mass–energy
equivalence law and action–reaction opposition, in Barrett, T.W. and
Grimes, D.M. (eds): *Advanced Electromagnetism*, World Scientific,
Singapore 1995, 77–104

[Cos 97a] Costa de Beauregard, O.
A new c-2 effect: hidden angular momentum in magnet, *Phys. Essays* **10**
(1997) N. 3, 492–497

[Cos 97b] Costa de Beauregard, O.
Induced electromagnetic inertia and physicality of the 4-vector potential,
Phys. Essays **10** (1997) N. 4, 646–650

[Dat 71] Datta, B.K.
Spinor fields in General Relativity, *Nuovo Cim.* **6B** (1971) 16–28

[Dehn 85] Dehnen, H. and Ghaboussi, F.
Gravity as Yang–Mills gauge theory, *Nucl. Phys.* B **262** (1985) 144–158

[Des 68] Deser, S. and Laurent, B.E .
Gravitation without self-interaction, *Ann. Phys. (USA)* **50** (1968)
76–101

[Des 70a] Deser, S.
Self-interaction and gauge invariance, *Gen. Rel. Grav.* **1** (1970) 9–18

[Des 70b] Deser, S. and Halpern, L.
Self-coupled scalar gravitation, *Gen. Rel. Grav.* **1** (1970) 131–136

[Dir 36] Dirac, P.A.M.
Relativistic wave equations, *Proc. Roy. Soc.* A **155** (1936) 447–459

[Dob 97] Dobado, A., Gomez Nicola, A., Maroto, A.L., Pelaez, I.R.
Effective Lagrangians for the standard model, Springer, Berlin 1997

[Dre 95] Drees, M. and Godbole, M.R.
Resolved photon processes, *J. Phys. G: Nucl. Part. Phys.* **21** (1995)
1559–1642

[Dürr 59] Dürr, H.P., Heisenberg, W., Mitter, H., Schlieder, S. and
Yamazaki, K.
Zur Theorie der Elementarteilchen, *Z. Naturforsch.* **14a** (1959) 441–485

[Dürr 61] Dürr, H.P.
Isospin and parity in nonlinear spinor theory, *Z. Naturforsch.* **16a**
(1961) 327–345

[Dürr 78] Dürr, H.P. and Saller, H.
Too many flavours? Dynamical generation of effective degrees of
freedom, *Nuovo Cim.* **48** A (1978) 505–560

[Dürr 79] Dürr, H.P. and Saller, H.
Unification of isospin and hypercharge in one basic $SU(2)$, *Nuovo Cim.*
53 A (1979) 469–511

[Dürr 80] Dürr, H.P. and Saller, H.
Sketch of a composite quark–lepton model based merely on $SU(2)$
symmetry, *Phys. Rev.* D **22** (1980) 1176–1183

[Dys 49a] Dyson, F.J.
The radiation theories of Tomonaga, Schwinger and Feynman, *Phys.
Rev.* **75** (1949) 486–502

[Dys 49b] Dyson, F.J.
The S matrix in quantum electrodynamics, *Phys. Rev.* **75** (1949)
1736–1755

[Dys 51a] Dyson, F.J.
Heisenberg operators in quantum electrodynamics I, *Phys. Rev.* **82**
(1951) 428–439

[Dys 51b] Dyson, F.J.
Heisenberg operators in quantum electrodynamics II, *Phys. Rev.* **83**
(1951) 608–627

[Dys 51c] Dyson, F.J.
The Schrödinger equation in quantum electrodynamics, *Phys. Rev.* **83**
(1951) 1207–1216

[Dys 51d] Dyson, F.J.
The renormalization method in quantum electrodynamics, *Proc. Roy.
Soc. (London)* **A207** (1951) 395–401

[Dys 53a] Dyson, J.F.
The use of the Tamm–Dancoff method in field theory, *Phys. Rev.* **90**
(1953) 994

[Dys 53b] Dyson, J.F.
The wave function of a relativistic system, *Phys. Rev.* **91** (1953)
1543–1550

[Edg 80] Edgar, S.B.
The structure of the tetrad formalism in General Relativity: The general
case, *Gen. Rel. Grav.* **12** (1980) 347–362

[Ein 16] Einstein, A.
Näherungsweise Integration der Feldgleichungen der Gravitation,
Sitzungsber. Preuss. Akad. Wiss. **1** 1916, 688–696

[Ein 18] Einstein, A.
Über Gravitationswellen, *Sitzungsber. Preuss. Akad. Wiss.* **1** 1918,
154–167

[Emch 72] Emch, G.
Algebraic methods in statistical mechanics and quantum field theory,
Wiley Interscience, New York 1972

[Erd 97] Erdmann, M.
The partonic structure of the photon, Springer, Berlin 1997

[Fadd 80] Faddeev, L.D. and Slavnov, A.A.
Gauge fields — Introduction to quantum theory, Benjamin–Cummings
Publ. Co., 1980

[Fau 97] Fauser, B. and Stumpf, H.
Positronium as an example of algebraic composite calculations, *Adv.
Cliff. Alg.* **7** (S) (1997) 399–418

[Fau 98] Fauser, B.
On an easy transition from operator dynamics to generating functionals
by Clifford algebras, *J. Math. Phys.* **39** (1998) 4928–4947

[Fett 71] Fetter, A.L. and Walecka, J.D.
Quantum theory of many particle systems, McGraw–Hill, New York 1971

[Feyn 49a] Feynman, R.P.
The theory of positrons, *Phys. Rev.* **76** (1949) 749–759

[Feyn 49b] Feynman, R.P.
Space–time approach to quantum electrodynamics, *Phys. Rev.* **76** (1949)
769–789

[Feyn 87] Feynman, R.P.
Negative probability, in Hiley, B.J., Peat, F.D. (eds): *Quantum
Implications*, Routledge & Kegan, London, 1987, 235–248

[Fie 39a] Fierz, M.
Über die relativistische Theorie kräftefreier Teilchen mit beliebigem
Spin, *Helv. Acta* **12** (1939) 3–37

[Fie 39b] Fierz, M. and Pauli, W.
Über relativistische Feldgleichungen von Teilchen mit beliebigem Spin
im elektromagnetischen Feld, *Helv. Acta* **12** (1939) 297–300

[Fie 39c] Fierz, M. and Pauli, W.
On relativistic wave equations for particles of arbitrary spin in an
electromagnetic field, *Proc. Roy. Soc.* **A 173** (1939) 211–232

[Fon 70] Fonda, L. and Ghirardi, G.C.
Symmetry principles in quantum physics, Marcel Dekker Inc., New York
1970

[Free 51] Freese, E.
Gebundene Teilchen und Streuprobleme in der Quantenfeldtheorie, *Z.
Naturforsch.* **8a** (1951) 776–790

[Frie 72] Fried, H.M.
Functional methods and models in quantum field theory, MIT Press,
Cambridge (Mass.) 1972

[Frie 75] Friedrich, A., Gerling, W. and Bleuler, K.
Erweiterung der neuen Tamm–Dancoff-Methode auf Phasenübergänge in Vielteilchensystemen, *Z. Naturforsch.* **30a** (1975) 142–157

[Frie 76] Friedrich, A. and Gerling, W.
Wick-Regeln mit Zwischenzuständen und die neue Tamm–Dancoff-Methode mit Zwischenzuständen, *Z. Naturforsch.* **31a** (1976) 872–886

[Frie 53] Friedrichs, K.O.
Mathematical aspects of the quantum theory of fields, Interscience Publ., New York 1953

[Gai 90] Gaigg, P., Kummer, W. and Schweda, M. (eds.)
Physical and nonstandard gauges (Lecture Notes in Physics **361**), Springer, Heidelberg 1990

[Gel 43] Gelfand, I. and Naimark, M.A.
On the imbedding of normed rings into the ring of operators in Hilbert space, *Mat. Sborn. NS* **12** [54] (1943) 197–213

[Gel 60] Gelfand, I.M. and Schilow, G.E.
Verallgemeinerte Funktionen I, Deutscher Verlag der Wissenschaften, Berlin, 1960

[Ger 85] Gerjuoy, E. and Adhikari, S.K.
Nonuniqueness of solutions to the Lippmann–Schwinger equation in a soluble three-body model, *Phys. Rev.* **A31** (1985) 2005–2019

[Gli 87] Glimm, Y. and Jaffe, A.
Quantum physics — A functional integral point of view, Springer, Berlin 1987 (2nd ed.)

[Gold 78] Goldstone, J. and Jackiw, R.
Unconstrained temporal gauge for Yang–Mills theory, *Phys. Lett.* **74 B** (1978), 81–84

[Gre 98] Grebe, J.
Effektive Hamilton-Operatoren des BCS-Modells (Diplomarbeit), University of Tübingen, 1998

[Gri 91] Grimm, G., Hailer, B. and Stumpf, H.
State representations for renormalized energy equations in quantum field theory, *Ann. Phys. (Germ.)* **48** (1991) 327–336

[Gri 94a] Grimm, G.
Elektro-schwache Standardtheorie als effektive Dynamik von Bindungszuständen im Rahmen einer nichtlinearen Spinortheorie (Thesis), University of Tübingen, 1994

[Gri 94b] Grimm, G.
Effective boson–fermion dynamics for subfermion models, *Z. Naturforsch.* **49a** (1994) 649–662

[Gri 94c] Grimm, G.
Effective gauge theories with symmetry breaking, *Z. Naturforsch.* **49a** (1994) 1093–1101

[Gri 94d] Grimm, G.
The two-particle sector of generalized Bargmann–Wigner equations, *Nuovo Cim.* **107 A** (1994) 1647–1665

[Gri 95] Grimm, G.
Composite electro–weak gauge bosons from subfermion models, *Nuovo Cim.* **10 A** (1995) 857–866

[Gro 74] Gross, D.J. and Neveu, A.
Dynamical symmetry breaking in asymptotically free field theories, *Phys. Rev. D* **10** (1974) 3235–3253

[Gro 83] Grosser, D.
On the decomposition of spinor fields which satisfy a nonlinear higher order equation, *Z. Naturforsch.* **38a** (1983) 1293–1295

[Gro 87] Grosser, D.
Nearly massless vector bosons as preon–antipreon bound states in a spinorfield model, *Nuovo Cim.* **97 A** (1987) 439–450

[Gup 57] Gupta, S.N.
Einstein's and other theories of gravitation, *Rev. Mod. Phys.* **29** (1957) 334–336

[Gup 77] Gupta, S.N.
Quantum electrodynamics, Gordon & Breach, New York 1977

[Haag 92] Haag, R.
Local quantum physics, Springer, Berlin 1992

[Hail 85] Hailer, B.
Norm– und Massenberechnungen skalarer Bosonen als Bindungszustände in einheitlichen nichtlinearen Spinorfeldmodellen (Thesis), University of Tübingen, 1985

[Hall 90] Haller, K.
Canonical quantization of gauge theories in axial gauges (Lecture notes in physics **361**), Springer, Heidelberg 1990, 33–40

[Har 79] Harari, H.
A schematic model of quarks and leptons, *Phys. Lett.* **86 B** (1979) 83–86

[Hehl 71] Hehl, F.W. and Datta, B.K.
Nonlinear spinor equation and asymmetric connection in general
relativity, *J. Math. Phys.* **12** (1971) 1334–1339

[Hehl 76] Hehl, F.W., von der Heyde, P., Kerlick, G.D. and Nester, J.M.
General relativity with spin and torsion: Foundations and prospects,
Rev. Mod. Phys. **48** (1976) 393–416

[Hehl 95] Hehl, F.W., McCrea, J.D., Mielke, E.W., Ne'eman, Y.
Metric-affine gauge theory of gravity: Field equations, Noether
identities, world spinors and breaking of dilation invariance, *Phys. Rep.*
258 (1995) 1–236

[Heis 54] Heisenberg, W.
Zur Quantenfeldtheorie nichtrenormierbarer Wellengleichungen, *Z.
Naturforschung* **9a** (1954) 292–303

[Heis 66] Heisenberg, W.
Introduction to the unified field theory of elementary particles,
Interscience Publ., London 1966

[Hor 86] Hornung, L.
*Spin-0 und Spin-1-Bosonen als Preon-Antipreonbindungszustände in
einheitlichen nichtlinearen Spinorfeldmodellen* (Thesis), University of
Tübingen, 1986

[Hua 75] Huang, K. and Weldon, H.A.
Bound state wave functions and bound state scattering in relativistic
field theory, *Phys. Rev. D* **11** (1975) 257–278

[Hua 92] Huang, K.
Quarks, leptons and gauge fields, World Scientific, Singapore 1992

[Itz 80] Itzykson, C. and Zuber, J.B.
Quantum field theory, McGraw–Hill Inc., New York 1980

[Iwa 53] Iwanenko, D., Sokolow, A.
Klassische Feldtheorie, Akademie Verlag, Berlin 1953

[Jack 80] Jackiw, R.
Introduction to Yang–Mills quantum theory, *Rev. Mod. Physics* **52**
(1980), 661–673

[Jak 84] Jakobczyk, L.
Borchers algebra formulation of an indefinite inner product quantum
field theory, *J. Math. Phys.* **25** (1984) 617–622

[Jak 88] Jakobczyk, L. and Strocchi, F.
Euclidean formulation of quantum field theory without positivity,
Comm. Math. Phys. **119** (1988) 529–541

[Jel 68] Jelinek, F.
BCS-spin-model, its thermodynamic representations and
automorphisms, *Comm. Math. Phys.* **9** (1968) 169–175

[Jor 28] Jordan, P.
Die Lichtquantenhypothese: Entwicklung und gegenwärtiger Stand,
Ergebn. Ex. Naturwiss. **7** (1928) 158–208

[Kaus 94] Kaus, H.-J.
Gravitative Kopplung als Spin-Fusion (Thesis), University of Tübingen,
1994

[Ker 91] Kerschner, R. and Stumpf, H.
Derivation of composite particle dynamics by Weak Mapping of
quantum fields, *Ann. Phys. (Germ.)* **48** (1991) 56–68

[Ker 94] Kerschner, R.
Effektive Dynamik zusammengesetzter Quantenfelder (Thesis),
University of Tübingen, 1994

[Lag 64] Lagally, M. and Franz, W.
Vorlesungen über Vektorrechnung, Akad. Verlagsgesellschaft Geest &
Portig, Leipzig 1964

[Lat 97] Latal, H., Schweiger, W.
Perturbative and nonperturbative aspects of quantum field theory
(Lecture Notes in Physics **479**), Springer, Berlin 1997

[Lau 86] Lauxmann, Th.
*Fermionen als Dreipreonen-Bindungszustände in einheitlichen
nichtlinearen Spinorfeldtheorien* (Thesis), University of Tübingen, 1986

[Laue 22] von Laue, M.
Die Relativitätstheorie **I**, **II**, Vieweg, Braunschweig, 2. Auflage 1922.
French transl.: *La théorie de la relativité*, Gauthier–Villars, Paris, **I**,
1922; **II**, 1924.

[Leh 55] Lehmann, H., Symanzik, K. and Zimmermann, W.
On the formulation of quantized field theories, *Nuovo Cim.* **1** (1955)
205–225

[Leh 90] Lehner, G.
Elektromagnetische Feldtheorie, Springer, Berlin 1990

[Leib 87] Leibbrandt, G.
Introduction to noncovariant gauges, *Rev. Mod. Phys.* **59** (1987),
1067–1119

[Loch 85] Lochak, G.
A magnetic monopole in the Dirac equation, *Int. J. Theor. Phys.* **24**
(1985) 1019

[Loch 95a] Lochak, G.
Sur la présence d'une second photon dans la théorie de la lumière de
Louis de Broglie, *Ann. Fond. L. de Broglie* **20** (1995) 111–114

[Loch 95b] Lochak, G.
The symmetry between electricity and magnetism and the problem of
the existence of a magnetic monopole, in T.W. Barrett and D.M. Grimes
(eds.): *Advanced electromagnetism,* World Scientific, Singapore 1995,
105–147

[Logu 93] Logunov, A.A.
Theory of the gravitational field, *Phys. Part. Nucl.* **24** (1993) 545–554

[Luri 68] Lurié, D.
Particles and fields, Interscience Publ., New York 1968

[Lyo 83] Lyons, L.
An introduction to the possible substructure of quarks and leptons,
Progr. Part. Nucl. Phys. **10** (1983) 227–304

[Magg 84] Magg, M.
Dynamics of classical nonabelian gauge fields, *Fortschr. Phys.* **32** (1984)
353–393

[Mai 66] Maison, D. and Stumpf, H.
On the field theoretic functional calculus for the anharmonic oscillator I,
Z. Naturforsch. **21a** (1966) 1829–1841

[Mie 12a] Mie, G.
Grundlagen einer Theorie der Materie, *Ann. Phys. (Germ.)* **37** (1912)
511–534

[Mie 12b] Mie, G.
Grundlagen einer Theorie der Materie: Das Problem des Elektrons, *Ann.
Phys. (Germ.)* **39** (1912) 1–40

[Mie 13] Mie, G.
Grundlagen einer Theorie der Materie: Kraft und träge Masse. Das
Problem des Wirkungsquantums. Die Gravitation., *Ann. Phys. (Germ.)*
40 (1913) 1–66

[Mit 67] Mittelstaedt, P. and Barbour, J.B.
On the geometrical interpretation of the theory of gravitation in flat
space, *Z. Physik* **203** (1967) 82–90

[Moha 86] Mohapatra, R.N.
Unification and supersymmetry, Springer, Berlin 1986

[Moha 91] Mohapatra, R.N.
Beyond the standard model, *Prog. Nucl. Part. Phys.* **26** (1991) 1–89

[Moll 72] Møller, C.
The theory of relativity, Clarendon Press, Oxford 1972 (2nd ed.)

[Mor 80] Morchio, G. and Strocchi, F.
Infrared singularities, vacuum structure and pure phases in local
quantum field theory, *Ann. Inst. H. Poincaré* **33** (1980) 251–282

[Mos 90] Moschella, U. and Strocchi, F.
The dipole field model, *Lett. Math. Phys. (NL)* **19** (1990) 143–149

[Nach 86] Nachtmann, O.
Phänomene und Konzepte der Elementarteilchenphysik, Vieweg,
Braunschweig 1986

[Nag 66] Nagy, K.L.
State vector spaces with indefinite metric in quantum field theory,
Noordhoff LTD, Groningen 1966

[Naka 90] Nakanishi, N. and Ojima, I.
Covariant operator formalism of gauge theories and quantum gravity,
World Scientific, Singapore 1990

[Nam 61a] Nambu, Y. and Jona–Lasinio, G.
Dynamical model of elementary particles based on an analogy with
superconductivity I, *Phys. Rev.* **122** (1961) 345–358

[Nam 61b] Nambu, Y. and Jona–Lasinio, G.
Dynamical model of elementary particles based on an analogy with
superconductivity II, *Phys. Rev.* **124** (1961) 246–254

[Nie 75] Niederer, U.
Group theory of the massless spin 2 field and gravitation, *Gen. Rel.
Grav.* **6** (1975) 433–437

[Nish 53] Nishijima, K.
Many-body problem in quantum field theory, *Prog. Theor. Phys.* **10**
(1953) 549–574

[Pau 36] Pauli, W.
Contributions mathématiques à la théorie des matrices de Dirac, *Ann. Inst. H. Poincaré* **16** (1936) 109–136

[Pau 49] Pauli, W. and Villars, F.
On the invariant regularization in relativistic quantum theory, *Rev. Mod. Phys.* **21** (1949) 434–444

[Nous 83] Nous, M.H.
The field with spin 2, *Lett. Nuovo Cim.* **38** (1983) 65–76

[Pfi 87] Pfister, W.
Der einzeitige Formalismus in der funktionalen Quantentheorie (Diplomarbeit), University of Tübingen, 1987

[Pfi 89] Pfister, W., Rosa, M. and Stumpf, H.
Vector boson states as solutions of generalized Bargmann–Wigner equations, *Nuovo Cim.* **102A** (1989) 1449–1467

[Pfi 90] Pfister, W.
Yang–Mills Dynamik als effektive Theorie von vektoriellen Spinor–Isospinor Bindungszuständen in einem Preonfeldmodell (Thesis), University of Tübingen, 1990

[Pfi 91] Pfister, W. and Stumpf, H.
Exchange forces of composite particles in quantum field theory, *Z. Naturforsch.* **46a** (1991) 389–400

[Pfi 93] Pfister, W.
On the theory of composite gauge bosons: Calculation of coupling constants and covariant derivatives, *Nuovo Cim.* **106A** (1993) 363–377

[Pfi 94] Pfister, W.
Exact solutions of generalized three-particle Bargmann–Wigner equations in the strong coupling limit, *Nuovo Cim.* **107A** (1994) 1523–1541

[Pfi 95a] Pfister, W. and Stumpf, H.
Effective dynamics in a subfermion shell model of leptons, quarks and electroweak gauge bosons, *Nuovo Cim.* **108A** (1995) 947–973

[Pfi 95b] Pfister, W.
Hidden quantum numbers of the group $S(3)$ as a classification scheme of the generations of leptons and quarks in a subfermion model, *Nuovo Cim.* **108A** (1995) 1365–1389

[Pfi 95c] Pfister, W.
Mixed symmetry solutions of generalized three-particle
Bargmann–Wigner equations in the strong coupling limit, *Nuovo Cim.*
108A (1995) 1427–1444

[Pla 59] Planck, M.
The theory of heat radiation, Dover, New York 1959

[Pod 41] Podolski, B.
A generalized electrodynamics, *Phys. Rev.* **62** (1941) 68–71

[Prim 83] Primas, H.
Chemistry, quantum mechanics and reductionism, Springer, Berlin 1983

[Pro 36] Proca, A.
Sur la théorie ondulatoire des électrons positifs et negatifs, *J. de
Physique* **7** (1936) 347–353

[Pru 94] Prugovecki E.
On foundational and geometric critical aspects of quantum
electrodynamics, *Found. Phys.* **24** (1994) 335–362

[Pry 38] Pryce, M.H.L.
On the neutrino theory of light, *Proc. Roy. Soc. London A* **165** (1938)
247–271

[Rade 97] Rade, L., Westergren, B.
Springers mathematische Formeln, Springer, Berlin 1997

[Raja 78] Rajaraman, R.
On the alleged equivalence of certain field theories, *Pramana J. Phys.*
11 (1978) 491–506

[Ram 65] Rampacher, H., Stumpf, H. and Wagner, F.
Grundlagen der Dynamik in der nichtlinearen Spinortheorie der
Elementarteilchen, *Fortschr. Phys.* **13** (1965) 385–480

[Ray 48] Rayski, J.
On simultaneous interaction of several fields and the self-energy
problem, *Acta Phys. Pol.* **9** (1948) 129–140

[Riec 91] Rieckers, A.
Condensed Cooper pairs and macroscopic quantum phenomena, in
Gans, W. et al. (eds.): *Large-scale molecular systems,* Nato ASI Series
in Physics, Plenum Press, New York 1991, 533–576

[Riew 72a] Riewe, F. and Green, A.E.S.
Quantum dynamics of higher-derivative fields, *J. Math. Phys.* **13** (1972)
1368–1374

[Riew 72b] Riewe, F. and Green, A.E.S.
Physical interpretation of higher-derivative fields, *J. Math. Phys.* **13**
(1972) 1374–1380

[Riv 88] Rivers, R.J.
Path integral methods in quantum field theory, Cambridge University
Press 1988

[Rod 77] Rodrigues, P.R.
On generating forms of K-generalized Lagragian and Hamiltonian
systems, *J. Math. Phys.* **18** (1977) 1720–1723

[Rod 81] Rodrigues, M.A. and Lorente, M.
Bargmann–Wigner equations: Symmetries of multispinors and equations
of motion, *J. Math. Phys.* **22** (1981) 1283–1288

[Rod 84] Rodrigues, M.A. and Lorente, M.
Group-theoretical analysis of Bargmann–Wigner equations for massless
particles, *Nuovo Cim.* **83A** (1984) 249–262

[Roep 91] Roepstorff, G.
Pfadintegrale in der Quantenphysik, Vieweg, Wiesbaden 1991

[Rze 69] Rzewuski, J.
Field theory, Vol. 2, Iliffe Books, London 1969

[Sal 51] Salpeter, E.E. and Bethe, H.E.
A relativistic equation for bound state problems, *Phys. Rev.* **84** (1951)
1232–1242

[Sall 81] Saller, H.
Colour and hypercharge — Are they structural symmetries? *Nuovo Cim.*
64 A (1981) 141–162

[Sall 82a] Saller, H.
Dynamical unification of fermions, gauge and Higgs fields, *Nuovo Cim.*
67 A (1982) 70–98

[Sall 82b] Saller, H.
Patterns of fermion condensation and composite SU_5 gauge bosons,
Nuovo Cim. **68 A** (1982) 324–347

[Sall 82c] Saller, H.
Dynamical unification of fermions and gauge bosons for internal
symmetry and gravity, *Nuovo Cim.* **71 A** (1982) 17–42

[Sall 89] Saller, H.
On the non-decomposable time representations in quantum theories,
Nuovo Cim. **104 B** (1989) 291–337

[Sall 91] Saller, H.
Indefinite metric and asymptotic probabilities, *Nuovo Cim.* **106 B**
(1991) 1319–1356

[Sall 92] Saller, H.
Canonical and noncanonical quantization on space–time, *Nuovo Cim.*
107 B (1992) 1355–1373

[Sall 93] Saller, H.
The probability structure and BRS invariance of noncompact
Hamiltonians, *Nuovo Cim.* **106 A** (1993) 469–479

[Sall 94] Saller, H.
Nonabelian internal symmetries as conditioned by fractional charge
numbers, *Nuovo Cim.* **107 A** (1994) 1679–1692

[Sand 91] Sand, J.
Zur Pauli–Villars–Regularisierung in der Preontheorie (Diplomarbeit),
University of Tübingen, 1991

[Schar 89] Scharf, G.
Finite quantum electrodynamics, Springer, Berlin 1989

[Schli 60a] Schlieder, S.
Indefinite Metrik im Zustandsraum und
Wahrscheinlichkeitsinterpretation, *Z. Naturforsch.* **15 a** (1960) 448–460

[Schli 60b] Schlieder, S.
Indefinite Metrik im Zustandsraum und
Wahrscheinlichkeitsinterpretation II, *Z. Naturforsch.* **15 a** (1960)
460–467

[Schli 60c] Schlieder, S.
Indefinite Metrik im Zustandsraum und
Wahrscheinlichkeitsinterpretation III, *Z. Naturforsch.* **15 a** (1960)
555–565

[Schou 54] Schouten, J.A.
Ricci-calculus, Springer, Berlin, 1954

[Schu 67] Schuler, W. and Stumpf, H.
On the fieldtheoretic functional calculus for the anharmonic oscillator II,
Z. Naturforsch. **22a** (1967) 1842–1865

[Schwe 55] Schweber, S.S., Bethe, H.A. and de Hoffmann, F.
Mesons and fields, Vol. 1, Row, Peterson and Comp., New York 1955

[Schwi 51a] Schwinger, J.
On the Green's functions of quantized fields I, *Proc. Nat. Acad. Sc.* **37** (1951) 452–455

[Schwi 51b] Schwinger, J.
On the Green's functions of quantized fields II, *Proc. Nat. Acad. Sc.* **37** (1951) 455–459

[Schwi 51c] Schwinger, J.
The theory of quantized fields I, *Phys. Rev.* **82** (1951) 914–927

[Schwi 53a] Schwinger, J.
The theory of quantized fields II, *Phys. Rev.* **91** (1953) 713–728

[Schwi 53b] Schwinger, J.
The theory of quantized fields III, *Phys. Rev.* **91** (1953) 728–740

[Schwi 53c] Schwinger, J.
The theory of quantized fields IV, *Phys. Rev.* **92** (1953) 1283–1299

[Schwi 54a] Schwinger, J.
The theory of quantized fields V, *Phys. Rev.* **93** (1954) 615–628

[Schwi 54b] Schwinger, J.
The theory of quantized fields VI, *Phys. Rev.* **94** (1954) 1362–1384

[Seg 47a] Segal, I.E.
Irreducible representations of operator algebras, *Bull. Amer. Math. Soc.* **61** (1947) 69–105

[Seg 47b] Segal, I.E.
Postulates for general quantum mechanics, *Ann. Math.* **48** (1947) 930–948

[Sew 86] Sewell, G.L.
Quantum theory of collective phenomena, Clarendon Press, Oxford 1986

[Sexl 87] Sexl, R.U. and Urbantke, H.K.
Gravitation und Kosmologie, B.I. Wissenschaftsverlag, Mannheim 1987

[Shu 79] Shupe, M.A.
A composite model of leptons and quarks, *Phys. Lett.* **86 B** (1979) 87–92

[Sil 56] Silin, W.P. and Fainberg, W.J.
Die Tamm–Dancoff Methode, *Fortschr. Phys.* **4** (1956) 233–293

[Sjö 96] Sjöstrand, T., Storrow, J.K. and Vogt, A.
Parton distributions of real and virtuell photons, *J. Phys. G: Nucl. Part. Phys.* **22** (1996) 893–901

[Sou 92] D'Souza, I.A. and Kalman, C.S.
Preons, models of leptons, quarks and gauge bosons as composite objects,
World Scientific, Singapore 1992

[Stu 70a] Stumpf, H.
Functional quantum theory of free relativistic Fermi fields, *Z. Naturforsch.* **25a** (1970) 575–586

[Stu 70b] Stumpf, H., Scheerer, K. and Märtl, H.G.
Symmetry conditions on nonlinear spinor field functionals, *Z. Naturforsch.* **25a** (1970) 1556–1561

[Stu 72] Stumpf, H.
On the functional definition and calculation of global observables in nonlinear spinorfield quantum theory, *Z. Naturforsch.* **27a** (1972) 1058–1072

[Stu 75] Stumpf, H. and Scheerer, K.
On the construction of relativistic representation spaces in functional quantum theory of the nonlinear spinorfield, *Z. Naturforsch.* **30a** (1975) 1361–1371

[Stu 82] Stumpf, H.
On some physical properties of a unified lepton–hadron field model with composite bosons, *Z. Naturforsch.* **37a** (1982) 1295–1299

[Stu 83a] Stumpf, H.
Discussion of the two-fermion sector in a unified nonlinear spinorfield model with indefinite metric I, *Z. Naturforsch.* **38a** (1983) 1064–1071

[Stu 83b] Stumpf, H.
Discussion of the two-fermion sector in a unified nonlinear spinorfield model with indefinite metric II, *Z. Naturforsch.* **38a** (1983) 1184–1188

[Stu 85a] Stumpf, H.
Effective interactions of relativistic composite particles in unified nonlinear spinorfield models I, *Z. Naturforsch.* **40a** (1985) 14–28

[Stu 85b] Stumpf, H.
Effective interactions of relativistic composite particles in unified nonlinear spinorfield models II, *Z. Naturforsch.* **40a** (1985) 183–190

[Stu 85c] Stumpf, H.
Effective interactions of relativistic composite particles in unified nonlinear spinorfield models III, *Z. Naturforsch.* **40a** (1985) 294–301

[Stu 85d] Stumpf, H.
Form factors of relativistic composite particle interactions in unified nonlinear spinorfield models, *Z. Naturforsch.* **40a** (1985) 752–774

[Stu 86a] Stumpf, H.
Yang–Mills dynamics as effective nonlinear field theory of composite vector bosons in unified spinorfield models, *Z. Naturforsch.* **41a** (1986) 683–703

[Stu 86b] Stumpf, H.
Electro–weak bosons, leptons and Han–Nambu quarks in a unified spinor–isospinor preon field model, *Z. Naturforsch.* **41a** (1986) 1399–1411

[Stu 87] Stumpf, H.
Composite gluons and effective nonabelian gluon dynamics in a unified spinor–isospinor preon field model, *Z. Naturforsch.* **42a** (1987) 213–226

[Stu 88] Stumpf, H.
Gravitation as a composite particle effect in a unified spinor–isospinor preon field model I, *Z. Naturforsch.* **43a** (1988) 345–359

[Stu 89] Stumpf, H.
Normalization and probability densites for wavefunctions of quantum fieldtheoretic energy equations, *Z. Naturforsch.* **44a** (1989) 262–268

[Stu 92] Stumpf, H. and Pfister, W.
Yang–Mills dynamics as an effective theory of composite vector bosons, *Nuovo Cim. A* **105** (1992) 677–694

[Stu 93a] Stumpf, H., Fauser, B. and Pfister, W.
Composite particle theory in quantum electrodynamics, *Z. Naturforsch.* **48a** (1993) 765–776

[Stu 93b] Stumpf, H., Borne, Th., Kaus, H.-J.
Is the gravitational force elementary? *Z. Naturforsch.* **48a** (1993) 1151–1165

[Stu 94] Stumpf, H. and Borne, Th.
Composite Particle Dynamics in Quantum Field Theory, Braunschweig/Wiesbaden, 1994

[Stu 96] Stumpf, H. and Pfister, W
Resolution of constraints and gauge equivalence in algebraic Schrödinger representation of quantum electrodynamics, *Z. Naturforsch.* **51a** (1996) 1054–1066

[Stu 97] Stumpf, H., Pfister, W.
Algebraic Schrödinger representation of quantum chromodynamics in
temporal gauge and resolution of constraints, *Z. Naturforsch.* **52a**
(1997) 220–240

[Stu 00] Stumpf, H.
Covariant regularization of nonlinear spinorfield quantum theories and
probability interpretation *Z. Naturforsch.* **55a** (2000) 415–432

[Sym 54] Symanzik, K.
Über das Schwingersche Funktional in der Feldtheorie, *Z. Naturforsch.*
9a (1954) 809–824

[Sym 73] Symanzik, K.
Infrared singularities and small-distance behavior analysis, *Comm.
Math. Phys.* **34** (1973) 7–36

[Swi 84] Swift, A.R. and Marrero, J.L.R.
Color confinement and the quantum-chromodynamic vacuum,
Phys. Rev. D **29** (1984) 1823–1838

[Thir 59] Thirring, W.
Lorentz-invariante Gravitationstheorien, *Fortschr. Physik* **7** (1959)
79–101

[Thir 61] Thirring, W.
An alternative approach to the theory of gravitation, *Ann. Phys. (USA)*
16 (1961) 96–117

[Thir 80] Thirring, W.
Lehrbuch der Mathematischen Physik Bd. 4, Springer, Wien, New York
1980

[Ton 42] Tonnelat, M.-A.
Une nouvelle forme de théorie unitaire: Étude de la particule de spin 2,
Annales de Physique **17** (1942) 158–208

[Tre 78] Treder, H.J.
Die Asymmetrie der kosmischen Zeit und Riemanns Gravitationstheorie,
Astron. Nachr. **299** (1978) 165–169

[Tsai 90] Tsai, E.C.
Gauge invariance and BRS quantum field theory (Lecture notes in
physics **361**), Springer, Heidelberg 1990, 41–51

[Völ 77] Völkel, A.M.
Fields, particles and currents (Lecture Notes in Physics **66**), Springer,
Heidelberg 1977

[Vogl 91] Vogl, U. and Weise, W.
The Nambu and Jona–Lasinio model: Its implications for hadrons and nuclei, *Prog. Part. Nucl. Phys.* **27** (1991) 195–272

[Wahl 84] Wahl, F., Duscher, R., Göbel, K. and Maichle, J. K.
Foundations of a theory concerning hydrogen centres in metals, *Z. Naturforsch.* **39a** (1984) 524–536

[Wein 65] Weinberg, S.
Photons and gravitons in perturbation theory: Derivation of Maxwell's and Einstein's equations, *Phys. Rev.* **138** (1965) B 988–B 1002

[Wein 80] Weinberg, S. and Witten, E.
Limits on massless particles, *Phys. Lett.* **96 B** (1980) 59–62

[Wein 95] Weinberg
The quantum theory of fields I, Cambridge Univ. Press 1995

[Wein 96] Weinberg
The quantum theory of fields II, Cambridge Univ. Press 1996

[Weiz 85] Weizsäcker, C.F.v.
Aufbau der Physik, Hanser Verlag, München 1985

[Wet 48] de Wet, J.S.
On the quantization of field theories derived from higher order Lagrangians, *Proc. Cambridge Phil. Soc.* **44** (1948) 546–559

[Wie 96] Wiesendanger, C.
Poincaré gauge invariance and gravitation in Minkowski space–time, *Class. Quantum Grav.* **13** (1996) 681–699

[Wigh 56] Wightman, A.S.
Quantum field theory in terms of vacuum expectation values, *Phys. Rev.* **101** (1956) 860–866

[Wign 39] Wigner, E.P.
On unitary representations of the inhomogeneous Lorentz group, *Ann. Math.* **40** (1939) 149–204

[Wild 50] Wildermuth, K.
Eine Differentialgleichungstheorie der Elementarteilchen, *Z. Naturforsch.* **5a** (1950) 373–381

[Wyss 65] Wyss, W.
Zur Unizität der Gravitationstheorie, *Helv. Phys. Acta* **38** (1965) 469–480

[Zel 86] Zel'dovich, Y.A. and Grishchuk, L.P.
Gravitation, the general theory of relativity and alternative theories,
Sov. Phys. Usp. **29** (1986) 695–707

[Zim 54] Zimmermann, W.
The relation between the Bethe–Salpeter equation and the
Tamm–Dancoff method, *Nuovo Cim. Suppl.* **11** (1954) 43–90

[Zim 66a] Zimmermann, R.L.
Equivalent field theories and elementary particle democracy, *Phys. Rev.*
141 (1966) 1554–1559

[Zim 66b] Zimmermann, R.L.
Construction of an equivalent Lagrangian for interacting fermion–boson
fields, *Phys. Rev.* **146** (1966) 955–959

[Zim 67] Zimmermann, R.L.
Equivalent field theories and bound states, *Phys. Rev.* **164** (1967)
1945–1950

Index

Fundamental Theories of Physics

Fundamental Theories of Physics

Fundamental Theories of Physics

Fundamental Theories of Physics

KLUWER ACADEMIC PUBLISHERS – DORDRECHT / BOSTON / LONDON